Springer Proceedings in Mathematics & Statistics

Volume 129

Springer Proceedings in Mathematics & Statistics

This book series features volumes composed of selected contributions from workshops and conferences in all areas of current research in mathematics and statistics, including operation research and optimization. In addition to an overall evaluation of the interest, scientific quality, and timeliness of each proposal at the hands of the publisher, individual contributions are all refereed to the high quality standards of leading journals in the field. Thus, this series provides the research community with well-edited, authoritative reports on developments in the most exciting areas of mathematical and statistical research today.

More information about this series at http://www.springer.com/series/10533

Patrícia Gonçalves · Ana Jacinta Soares
Editors

From Particle Systems to Partial Differential Equations II

Particle Systems and PDEs II, Braga, Portugal, December 2013

Springer

Editors
Patrícia Gonçalves
Centre of Mathematics
University of Minho
Braga
Portugal

Ana Jacinta Soares
Centre of Mathematics
University of Minho
Braga
Portugal

ISSN 2194-1009 ISSN 2194-1017 (electronic)
Springer Proceedings in Mathematics & Statistics
ISBN 978-3-319-16636-0 ISBN 978-3-319-16637-7 (eBook)
DOI 10.1007/978-3-319-16637-7

Library of Congress Control Number: 2014938099

Mathematics Subject Classification (2010): 60K35, 60G60, 60F17, 35R60, 35Q20, 35Q35, 82C40, 35Q53

Springer Cham Heidelberg New York Dordrecht London

Printed on acid-free paper

Springer International Publishing AG Switzerland is part of Springer Science+Business Media
(www.springer.com)

Preface

This volume presents the proceedings of the second international conference on Particle Systems and Partial Differential Equations, "PS-PDEs II", held at the Centre of Mathematics of the University of Minho, Braga, Portugal, from 12 to 13 December 2013.

The purpose of this second meeting was to bring together renowned researchers working in the fields of probability and partial differential equations, to present their recent scientific results in both areas and to discuss some of their topics of expertise. Additionally, the meeting was intended to present to a vast and varied public, including young researchers, the subject of interacting particle systems, its underlying motivation, and its relation to partial differential equations.

The present volume includes the lecture notes from the mini-courses given by Professor Cédric Bernardin (University of Nice) on "Diffusion of energy in chains of oscillators with bulk noise", and by Professor Vincent Giovangigli (École Polytechnique) on "Dissipative Reactive Fluid Models from the Kinetic Theory". It also features 13 contributed papers written by some of the participants of the conference on highly interesting topics in the fields of probability theory, partial differential equations and kinetic theory.

We believe that this volume will be of great interest to probabilists, analysts and also to those mathematicians in general whose work focuses on topics in mathematical physics, stochastic processes and differential equations, as well as to those physicists who work in statistical mechanics and kinetic theory.

We would like to take this opportunity to give our special thanks to all the participants for their active contributions to the success of this meeting.

Finally, we gratefully acknowledge the financial support provided by Fundação para a Ciência e a Tecnologia and by the Centre of Mathematics of the University of Minho, through the FCT-CMAT Strategic Project PEstC/MAT/UI0013/2011 and the research project "Non-equilibrium statistical physics", PTDC/MAT/109844/2009.

We very much hope you enjoy reading this book!

Braga, Portugal
February 2015

Patrícia Gonçalves
Ana Jacinta Soares

Contents

Part I
Mini-Courses

Diffusion of Energy in Chains of Oscillators with Conservative Noise

Cédric Bernardin

Abstract These notes are based on a mini-course given during the conference Particle systems and PDE's - II which held at the Center of Mathematics of the University of Minho in December 2013. We discuss the problem of normal and anomalous diffusion of energy in systems of coupled oscillators perturbed by a stochastic noise conserving energy.

Keywords Superdiffusion · Anomalous fluctuations · Green-Kubo formula · Non equilibrium stationary states · Heat conduction · Hydrodynamic limits · Ergodicity

The goal of statistical mechanics is to elucidate the relation between the microscopic world and the macroscopic world. *Equilibrium statistical mechanics* assume the microscopic systems studied to be in equilibrium. In this course we will be concerned with *non-equilibrium statistical mechanics* where time evolution is taken into account: our interest will not only be in the relation between the microscopic and the macroscopic scales in space but also in time.

By microscopic system we refer to molecules or atoms governed by the classical Newton's equations of motion. The question is then to understand how do these particles manage to organize themselves in such a way as to form a coherent structure on a large scale. The "structure" will be described by few variables (temperature, pressure ...) governed by autonomous equations (Euler's equations, Navier-Stokes's equation, heat equation ...). The microscopic specificities of the system will appear on this scale only through the thermodynamics (equation of state) and through the transport coefficients. Unfortunately, we are very far from understanding how to derive such macroscopic equations for physical relevant interactions.

One of the main ingredients that we need to obtain the macroscopic laws is that the particles, which evolve deterministically, have a behavior that one can consider almost as being random. The reason for this is that the dynamical system considered is expected to have a very sensitive dependence on the initial conditions and therefore

---

C. Bernardin (✉)
Université de Nice Sophia-Antipolis, Laboratoire J.A. Dieudonné,
UMR CNRS 7351, Parc Valrose 06108, Nice Cedex 02, France
e-mail: cbernard@unice.fr

P. Gonçalves and A.J. Soares (eds.), *From Particle Systems to Partial Differential Equations II*, Springer Proceedings in Mathematics & Statistics 129,
DOI 10.1007/978-3-319-16637-7_1

is chaotic. This "deterministic chaos" is a poorly understood subject for systems with many degrees of freedom and even a precise consensual formulation is missing.

A first simplification to attack these problems consists in replacing the deterministic evolution of particles ab initio by purely stochastic evolutions. Despite this simplification we notice that the derivation of the macroscopic evolution laws is far from being trivial. For example, we do not have any derivation of a system of hyperbolic conservation laws from a stochastic microscopic system after shocks. Nevertheless, since the pioneering work of Guo et al. [35] and Yau [67], important progresses have been performed in several well understood situations by the development of robust probabilistic and analytical methods (see [41, 55] for reviews).

In this course we will be mainly (but not only) interested in hybrid models for which the time evolution is governed by a combination of deterministic and stochastic dynamics. These systems have the advantage to be mathematically tractable and conserve some aspects of the underlying deterministic evolution. The stochastic noise has to be chosen in order to not destroy the main features of the Hamiltonian system that we perturb.

The central macroscopic equation of these lecture notes is the heat equation:

$$\begin{cases} \partial_t u = \partial_x (D(u)\partial_x u), & x \in \mathring{U}, \quad t > 0, \\ u(0,x) = u_0(x), & x \in U, \\ u(t,x) = b(x), & x \in \partial U, \quad t > 0. \end{cases}$$

Here $u(t,x)$ is a function of the time $t \geq 0$ and the space $x \in U \subset \mathbb{R}^d$, $d \geq 1$, starting from the initial condition u_0 and subject to boundary conditions prescribed by the function b. The advantage of the heat equation with respect to other macroscopic equations such as the Euler or Navier-Stokes equations is that the notion of solution is very well understood. The dream would be to start from a system of $N \gg 1$ particles whose interactions are prescribed by Newton's laws and to show that in the large N limit, the empirical energy converges in the diffusive time scale $t = \tau N^2$ to u (τ is the microscopic time and t the macroscopic time). In fact, this picture is expected to be valid only under suitable conditions and to fail for some low dimensional systems. In the case where the heat equation (or its variants) holds we say that the system has a normal behavior. Otherwise anomalous behavior occurs and the challenging question (even heuristically) is to know by what we shall replace the heat equation and what is the time scale over which we have to observe the system in order to see this macroscopic behavior ([19, 24, 47] for reviews).

The course is organized as follows. In this chapter we introduce the models studied. Chapter "Dissipative Reactive Fluid Models from the Kinetic Theory" is concerned with models which have a normal diffusive behavior. In chapter "Large Deviations in a Gaussian Setting: The Role of the Cameron-Martin Space" we are interested in systems producing an anomalous diffusion. An important issue not discussed here is the effect of disorder on diffusion problems. In order to deal with lecture notes of a reasonable size, many of the proofs have been suppressed or only roughly presented.

1 Chains of Oscillators

1.1 Chains of Oscillators with Bulk Noise

Chains of coupled oscillators are usual microscopic models of heat conduction in solids. Consider a finite box $\Lambda_N = \{1, \ldots, N\}^d \subset \mathbb{Z}^d$, $d \geq 1$, whose boundary $\partial \Lambda_N$ is defined as $\partial \Lambda_N = \{x \notin \Lambda_N\, ; \; \exists y \in \Lambda_N, \; |x - y| = 1\}$. Here $|\cdot|$ denotes the Euclidian norm in $\mathbb{R}^d$ and "$\cdot$" the corresponding scalar product. Let us fix a nonnegative pair interaction potential V and a pinning potential W on $\mathbb{R}$. The atoms are labeled by $x \in \Lambda_N$. The momentum of atom x is $p_x \in \mathbb{R}$ and its displacement from its equilibrium position[1] is $q_x \in \mathbb{R}$. The energy $\mathcal{E}_x$ of the atom x is the sum of the kinetic energy, the pinning energy and the interaction energy:

$$\mathcal{E}_x = \frac{|p_x|^2}{2} + W(q_x) + \frac{1}{2} \sum_{\substack{|y-x|=1,\\ y \in \Lambda_N}} V(q_x - q_y). \tag{1}$$

The Hamiltonian is given by

$$\mathcal{H}_N = \sum_{x \in \Lambda_N} \mathcal{E}_x + \partial\, \mathcal{H}_N \tag{2}$$

where $\partial\, \mathcal{H}_N$ is the part of the Hamiltonian corresponding to the boundary conditions which are imposed (Fig. 1).

We will consider the following cases:

- Periodic boundary conditions: we identify the site 1 to the site N and denote the corresponding box by $\mathbb{T}_N$, the discrete torus of length N (then $\partial \mathcal{H}_N = 0$).
- Free boundary conditions: this corresponds to the absence of boundary conditions, i.e. to $\partial \mathcal{H}_N = 0$.
- Fixed boundary conditions: introduce the positions $q_y = 0$, $y \in \partial \Lambda_N$, of some fictive walls. We add to the Hamiltonian $\mathcal{H}_N$ a boundary term $\partial \mathcal{H}_N = \partial^{\mathrm{f}}\, \mathcal{H}_N$ given by

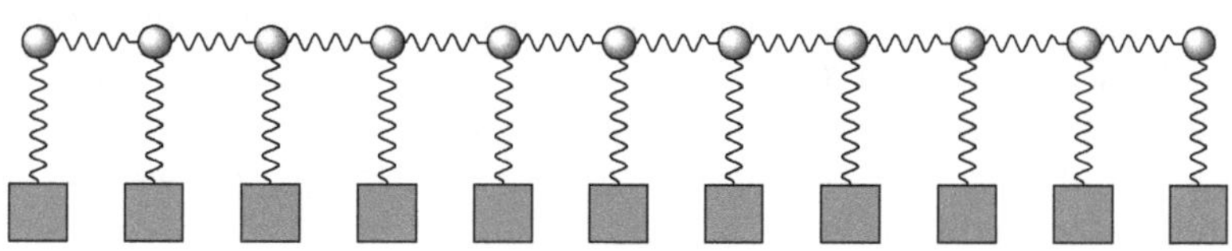

Fig. 1 A one-dimensional chain of pinned oscillators with free boundary conditions

---

[1] We restrict us to the case where $q_x \in \mathbb{R}^n$ with $n = 1$ because the relevant dimension of the system is the dimension d of the lattice. Most of the results stated in this manuscript can be generalized to the case $n \geq 1$.

$$\partial^{\mathrm{f}} \mathscr{H}_N = \sum_{\substack{|y-x|=1,\\ x\in\Lambda_N, y\in\partial\Lambda_N}} V(q_x - q_y) = \sum_{\substack{|y-x|=1,\\ x\in\Lambda_N, y\in\partial\Lambda_N}} V(q_x).$$

- Forced boundary conditions: site $\mathbf{1} = (1, \ldots, 1)$ is in contact with a wall at position $q_0 = 0$ and each site $y \in \partial\Lambda_N \setminus \{0\}$ is driven by a constant force τ_y. This results in a boundary term $\partial\mathscr{H}_N = \partial^{\tau} \mathscr{H}_N$ given by

$$\partial^{\tau} \mathscr{H}_N = \sum_{\substack{|y-x|=1,\\ x\in\Lambda_N, y\in\partial\Lambda_N}} V(q_x - q_y) - \sum_{y\in\partial\Lambda_N\setminus\{0\}} \tau_y q_y. \tag{3}$$

The equations of motion of the atoms are

$$\dot{q}_x = \partial_{p_x}\mathscr{H}_N, \quad \dot{p}_x = -\partial_{q_x}\mathscr{H}_N \tag{4}$$

and the generator $\mathscr{A}_N$ of the system is given by the Liouville operator

$$\mathscr{A}_N = \sum_{x\in\Lambda_N} \left\{\partial_{p_x}\mathscr{H}_N\,\partial_{q_x} - \partial_{q_x}\mathscr{H}_N\,\partial_{q_x}\right\}.$$

It will be also useful to consider the chain of oscillators in infinite volume, i.e. replacing Λ_N by $\mathbb{Z}^d$, $d \geq 1$, in the definitions above. The formal generator $\mathscr{A}_N$ is then denoted by $\mathscr{A}$. The dynamics can be defined for a large set of initial conditions if V and W do not behave too badly [12, 45, 50]. We define the set Ω as the subset of $\mathbb{R}^{\mathbb{Z}^d}$ given by

$$\Omega = \bigcap_{\alpha>0} \left\{\xi \in \mathbb{R}^{\mathbb{Z}^d}\,;\ \sum_{x\in\mathbb{Z}^d} e^{-\alpha|x|}|\xi_x|^2 < +\infty\right\} \tag{5}$$

and $\tilde{\Omega} = \Omega \times \Omega$. We equip Ω with its natural product topology and its Borel σ-field and $\tilde{\Omega}$ by the corresponding product topology. For $X = \Omega$ or $X = \tilde{\Omega}$, the set of Borel probability measures on X is denoted by $\mathscr{P}(X)$. A function $f : X \to \mathbb{R}$ is said to be *local* if it depends of ξ only through the coordinates $\{\xi_x\,;\ x \in \Lambda_f\}$, Λ_f being a finite box of $\mathbb{Z}$. We also introduce the sets $C_0^k(X)$, $k \geq 1$ (resp. $k = 0$), of bounded local functions on X which are differentiable up to order k with bounded partial derivatives (resp. continuous and bounded).

In the rest of the manuscript, apart from specific cases, we will assume that one of the following conditions hold:

- The potentials V and W have bounded second derivatives. Then the infinite dynamics $(\omega(t))_{t\geq 0}$ can be defined for any initial condition $\omega^0 = (\mathbf{q}^0, \mathbf{p}^0) \in \tilde{\Omega}$. Moreover $\tilde{\Omega}$ is invariant by the dynamics. This defines a semigroup $(P_t)_{t\geq 0}$ on $C_0^0(\tilde{\Omega})$ and the Chapman-Kolmogorov equations

$$(P_t f)(\omega) - f(\omega) = \int_0^t (P_s \mathscr{A} f)(\omega)\, ds \;=\; \int_0^t (\mathscr{A} P_s f)(\omega)\, ds \tag{6}$$

are valid for any $f \in C_0^1(\tilde{\Omega})$.

- The potential $W = 0$ and the interaction potential V has a second derivative uniformly bounded from above and below. It is more convenient to go over the deformation field $\eta_{(x,y)} = q_y - q_x$, $|x-y| = 1$, which by construction is constrained to have zero curl. In $d = 1$ we will denote $\eta_{(x-1,x)} = q_x - q_{x-1}$ by r_x. The dynamics (4) can be read as a dynamics for the deformation field and the momenta. Given say q_0, the scalar field $\mathbf{q} = \{q_x\}_{x\in\mathbb{Z}^d}$ can be reconstructed from η. In the sequel, when $W = 0$, we will use these coordinates without further mention. The dynamics for the coordinates $\omega = (\eta, \mathbf{p}) = (\eta_{(x,x+\mathbf{e})}, p_x)_{|\mathbf{e}|=1, x\in\mathbb{Z}}$ can be defined if the initial condition satisfies $\omega^0 \in \tilde{\Omega}$. Moreover the set $\tilde{\Omega}$ is invariant by the dynamics. This defines a semigroup $(P_t)_{t\geq 0}$ on $C_0^0(\tilde{\Omega})$ and the Chapman-Kolmogorov equations

$$(P_t f)(\omega) - f(\omega) = \int_0^t (P_s \mathscr{A} f)(\omega)\, ds \;=\; \int_0^t (\mathscr{A} P_s f)(\omega)\, ds \tag{7}$$

are valid for any $f \in C_0^1(\tilde{\Omega})$.[2]

Let us first consider the problem related to the characterization of equilibrium states. For simplicity we take the finite volume dynamics with periodic boundary conditions. Then it is easy to see that the system conserves one or two physical quantities depending on whether the chain is pinned or not. The total energy $\mathscr{H}_N$ is always conserved. If $W = 0$ the system is translation invariant and the total momentum $\sum_x p_x$ is also conserved. Notice that because of the periodic boundary conditions the sum of the deformation field $\sum_x \eta_{(x,x+e_i)}$ is automatically fixed equal to 0 for any $i = 1, \ldots, d$.

Liouville's Theorem implies that the uniform measure λ^N on the manifold Σ^N composed of the configurations with a fixed total energy (and possibly a fixed total momentum) is invariant for the dynamics. The micro canonical ensemble is defined as the probability measure λ^N. The dynamics restricted to Σ^N is not necessarily ergodic. Two examples for which one can show it is not the case are the harmonic lattice (V and W quadratic) and the Toda lattice ($d = 1$, $W = 0$, $V(r) = e^{-r} - 1 + r$) which is a completely integrable system [61]. In fact what is really needed for our purpose is not the ergodicity of the finite dynamics but of the infinite dynamics. We expect that even if the finite dynamics are never ergodic the fraction of Σ^N corresponding to non ergodic behavior decreases as N increases, and probably disappears as $N = \infty$ (apart from very peculiar cases). Therefore a good notion of ergodicity has to be stated for infinite dynamics. The definition of a conserved quantity is not straightforward in infinite volume (the total energy of the infinite chain is usually equal to $+\infty$). To give a precise definition we will use the notion of space-time invariant probability measures for the infinite dynamics defined above.

---

[2] The generator $\mathscr{A}$ has to be written in terms of the deformation field.

The infinite volume Gibbs grand canonical ensembles are such probability measures. They form a set of probability measures indexed by one (pinned chains) or $d+2$ (unpinned chains) parameters and are defined by the so-called Dobrushin-Landford-Ruelle's equations. To avoid a long discussion we just give a formal definition (see e.g. [34] for a detailed study).

- Pinned chains ($W \neq 0$): the infinite volume Gibbs grand canonical ensemble μ_β with inverse temperature $\beta > 0$ is the probability measure on $\tilde{\Omega}$ whose density with respect to the Lebesgue measure is

$$Z^{-1}(\beta) \exp\left(-\beta \sum_{x \in \mathbb{Z}^d} \mathscr{E}_x\right).$$

- Unpinned chains ($W = 0$): the infinite volume Gibbs grand canonical ensemble[3] $\mu_{\beta,\bar{p},\tau}$ with inverse temperature $\beta > 0$, average momentum $\bar{p} \in \mathbb{R}$ and tension $\tau = \beta^{-1}\lambda \in \mathbb{R}^d$ is the probability measure on $\tilde{\Omega}$ whose density with respect to the Lebesgue measure is

$$Z^{-1}(\beta, \bar{p}, \tau) \ \exp\left(-\beta \sum_{x \in \mathbb{Z}^d} \{\mathscr{E}_x - \bar{p} p_x - \sum_{i=1}^{d} \tau_i \eta_{(x,x+e_i)}\}\right). \tag{8}$$

Observe that in the one dimensional unpinned case we have simply product measures and that the tension τ is equal to the average of $V'(r_x)$.

Fix an arbitrary Gibbs grand canonical ensemble μ. A probability measure ν is said to be μ-regular if for any finite box $\Lambda \subset \mathbb{Z}^d$ whose cardinal is denoted by $|\Lambda|$, the relative entropy of $\nu|_\Lambda$ w.r.t. $\mu|_\Lambda$ is bounded above by $C|\Lambda|$ for a constant C independent of Λ. We recall that the relative entropy $H(\nu|\mu)$ of $\nu \in \mathscr{P}(X)$ with respect to $\mu \in \mathscr{P}(X)$, X being a probability space, is defined as

$$H(\nu|\mu) = \sup_\phi \left\{ \int \phi \, d\nu - \log\left(\int e^\phi \, d\mu \right) \right\}, \tag{9}$$

where the supremum is carried over all bounded measurable functions ϕ on X.

For any arbitrary Gibbs grand canonical ensembles μ and μ', μ is μ'-regular and μ' is μ-regular. Therefore ν is μ-regular is equivalent to ν is μ'-regular and we simply say that ν is regular.

A notion of ergodicity for infinite dynamics which is suitable to derive rigorously large scale limits of interacting particle systems is the following.

---

[3] They are defined with respect to the gradient fields $\eta_{(x,y)}$. It would be more coherent to call them *gradient Gibbs measures*.

Definition 1 (*Macro-Ergodicity*)[4] We say that the dynamics generated by $\mathscr{A}$ is *macro-ergodic* if and only if the only space-time invariant[5] regular measures ν for $\mathscr{A}$ are mixtures (i.e. generalized convex combinations) of Gibbs grand canonical ensembles.

If the microscopic dynamics is macro-ergodic, then, by using the relative entropy method developed in [51], we can derive the hydrodynamic equations[6] in the Euler time scale of the chain before the appearance of the shocks, at least in $d = 1$ [12]. These limits form a triplet of compressible Euler equations (for energy $\mathfrak{e}$, momentum $\mathfrak{p}$ and deformation $\mathfrak{r}$) of the form

$$\begin{cases} \partial_t \mathfrak{r} = \partial_q \mathfrak{p} \\ \partial_t \mathfrak{p} = \partial_q \tau \\ \partial_t \mathfrak{e} = \partial_q (\mathfrak{p}\tau) \end{cases} \tag{10}$$

where the pressure $\tau := \tau(\mathfrak{r}, \mathfrak{e} - \frac{\mathfrak{p}^2}{2})$ is a suitable thermodynamic function depending on the potential V. A highly challenging open question is to extend these results after the shocks. The proof can be adapted to take into account the presence of mechanical boundary conditions [20].

We do not claim that the macro-ergodicity is a necessary condition to get Euler equations for purely Hamiltonian systems. We could imagine that weaker or different conditions are sufficient but in the actual state of the art the macro-ergodicity is a clear and simple mathematical statement of what we could require from deterministic systems in order to derive Euler equations rigorously. We refer the interested reader to [21, 59] for interesting discussions about the role of ergodicity in statistical mechanics.

1.1.1 Conserving Noises

In [31], Fritz, Funaki and Lebowitz prove a weak form of macro-ergodicity for a chain of anharmonic oscillators under generic assumptions on the potentials V and W that we do not specify here (see [31]).

Theorem 1 ([31])[7] *Consider the pinned chain $W \neq 0$ generated by $\mathscr{A}$ or an unpinned chain $W = 0$ in $d = 1$. The only regular time and space invariant measures for $\mathscr{A}$ which are such that conditionally to the positions configuration $\mathbf{q} := \{q_x\,;\, x \in \mathbb{Z}^d\}$ the law of the momenta $\mathbf{p} := \{p_x\,;\, x \in \mathbb{Z}^d\}$ is exchangeable are given by mixtures of Gibbs grand canonical ensembles.*

---

[4] The name has been proposed by S. Goldstein.

[5] Observe that a probability measure ν is time invariant for the infinite dynamics if and only if $\int \mathscr{A} f\, d\nu = 0$ for any $f \in C_0^1(\tilde{\Omega})$. This is a consequence of the Chapman-Kolmogorov Eq. (7).

[6] The notion of hydrodynamic limits is detailed in Sects. 2.2.1 and 3.2.2.

[7] The proof given in [31] assumes $W \neq 0$ but it can be adapted to the unpinned one dimensional case (see [12]). It would be interesting to extend this theorem to the general unpinned case.

They also proposed to perturb the dynamics by a stochastic noise that consists in exchanging at random exponential times, independently for each pair of nearest neighbors site $x, y \in \mathbb{Z}^d$, $|x-y|=1$, the momenta p_x and p_y. The formal generator $\mathscr{L}$ of this dynamics, that we will call the *stochastic energy-momentum conserving dynamics*, is given by $\mathscr{L} = \mathscr{A} + \gamma\mathscr{S}$, $\gamma > 0$, where $\mathscr{A}$ is the Liouville operator and $\mathscr{S}$ is defined for any local function $f : \tilde{\Omega} \to \mathbb{R}$ by

$$(\mathscr{S}f)(\mathbf{q},\mathbf{p}) = \sum_{\substack{x,y\in\mathbb{Z}^d \\ |x-y|=1}} \left[f(\mathbf{q},\mathbf{p}^{x,y}) - f(\mathbf{q},\mathbf{p}) \right]. \tag{11}$$

Here the momenta configuration $\mathbf{p}^{x,y}$ is the configuration obtained from $\mathbf{p}$ by exchanging p_x with p_y. The previous discussion about existence of the dynamics on $\tilde{\Omega}$ for the deterministic case and its relation with its formal generator is also valid for this dynamics and the other dynamics defined in this section.

With some non-trivial entropy estimates we get the following result.

Theorem 2 ([31]) *Consider the pinned ($W \neq 0$) or the one-dimensional unpinned ($W \neq 0$) stochastic energy-momentum conserving dynamics. The only regular time and space invariant measures for these dynamics are given by mixtures of Gibbs grand canonical ensembles, i.e. the stochastic energy-momentum conserving dynamics is macro-ergodic.*

Consequently the stochastic energy-momentum conserving dynamics is macro-ergodic. By using the relative entropy method developed in [51], one can show it has in the Euler time scale and before the appearance of the shocks the same hydrodynamics (10) as the deterministic model. This is because the noise has some macroscopic effects only in the diffusive time scale [12].

We consider now a different stochastic perturbation. Let us define the flipping operator $\sigma_x : \mathbf{p} \in \Omega \to \mathbf{p}^x \in \Omega$ where $\mathbf{p}^x$ is the configuration such that $(\mathbf{p}^x)_z = p_z$ for $z \neq x$ and $(\mathbf{p}^x)_x = -p_x$. In [31] is also proved that the only time-space regular stationary measures for the Liouville operator $\mathscr{A}$ such that conditionally to the positions the momenta distribution is invariant by any flipping operator σ_x are mixtures of Gibbs grand canonical ensembles with zero momentum average. Then we consider the dynamics on $\tilde{\Omega}$ generated by $\mathscr{L} = \mathscr{A} + \gamma\mathscr{S}$, $\gamma > 0$, with $\mathscr{S}$ the noise defined by

$$(\mathscr{S}f)(\mathbf{q},\mathbf{p}) = \frac{1}{2}\sum_{x\in\mathbb{Z}^d} \left[f(\mathbf{q},\mathbf{p}^x) - f(\mathbf{q},\mathbf{p}) \right] \tag{12}$$

for any local function $f : \tilde{\Omega} \to \mathbb{R}$. This dynamics conserves the energy and the deformation of the lattice but destroys all the other conserved quantities. We call this system the *velocity-flip model* (sometimes the *stochastic energy conserving model*).

Theorem 3 ([31]) *Consider the pinned d-dimensional velocity-flip model or the one-dimensional unpinned velocity-flip model. The only regular time and space invariant* *measures are given by mixtures of Gibbs grand canonical ensembles. In other words the velocity-flip model is macro-ergodic.*

Since the velocity flip-model does not conserve the momentum its Gibbs invariant measures are given by (8) with $\bar{p} = 0$. In particular the average currents with respect to theses measures is zero. Therefore assuming propagation of local equilibrium in the Euler time scale we get that it has trivial hydrodynamics in this time scale: initial profile of energy does not evolve. This is only in the diffusive scale that an evolution should take place.

1.1.2 NESS of Chains of Oscillators Perturbed by an Energy Conserving Noise

The models defined in the previous sections can also be considered in a non-equilibrium stationary state (NESS) by letting them in contact with thermal baths at different temperatures and imposing various mechanical boundary conditions. Let us only give some details for the NESS of the one-dimensional velocity-flip model.

Consider a chain of N unpinned oscillators where the particle 1 (resp. N) is subject to a constant force τ_ℓ (resp. τ_r). Moreover we assume that the particle 1 (reps. N) is in contact with a Langevin thermal bath at temperature T_ℓ (resp. T_r). The generator $\mathscr{L}_N$ of the dynamics on the phase space $\Omega_N = \mathbb{R}^{N-1} \times \mathbb{R}^N$ is given by

$$\mathscr{L}_N = \mathscr{A}_N^{\tau_\ell,\tau_r} + \gamma \mathscr{S}_N + \gamma_\ell \mathscr{B}_{1,T_\ell} + \gamma_r \mathscr{B}_{N,T_r}, \quad \gamma > 0, \tag{13}$$

where $\mathscr{A}_N^{\tau_\ell,\tau_r}$ is the Liouville operator, $\mathscr{B}_{j,T}$ the generator of the Langevin bath at temperature T acting on the jth particle and $\mathscr{S}_N$ the generator of the noise. The strength of noise and thermostats are regulated by γ, γ_ℓ and γ_r respectively. The Liouville operator is defined by

$$\begin{aligned} \mathscr{A}_N^{\tau_\ell,\tau_r} &= \sum_{x=2}^{N} (p_x - p_{x-1})\, \partial_{r_x} + \sum_{x=2}^{N-1} \big(V'(r_{x+1}) - V'(r_x)\big)\, \partial_{p_x} \\ &\quad - \big(\tau_\ell - V'(r_2)\big)\, \partial_{p_1} + \big(\tau_r - V'(r_N)\big)\, \partial_{p_N}. \end{aligned} \tag{14}$$

The generators of the thermostats are given by

$$\mathscr{B}_{j,T} = T\partial_{p_j}^2 - p_j \partial_{p_j}. \tag{15}$$

The noise corresponds to independent velocity change of sign, i.e.

$$(\mathscr{S}_N f)(\mathbf{r}, \mathbf{p}) = \frac{1}{2} \sum_{x=2}^{N-1} \big(f(\mathbf{r}, \mathbf{p}^x) - f(\mathbf{r}, \mathbf{p})\big), \quad f : \Omega_N \to \mathbb{R}. \tag{16}$$

We will also consider the case where the chain has fixed boundary conditions.

Proposition 1 ([10, 11, 22]) *Consider a finite chain of pinned or unpinned oscillators with fix, free or forced boundary conditions in contact with two thermal baths at different temperatures and perturbed by one of the energy conserving noises defined above. Then, there exists a unique non-equilibrium stationary state for this dynamics which is absolutely continuous w.r.t. Lebesgue measure.*

Proof The proof of the existence of the invariant state can be obtained from the knowledge of a suitable Liapounov function. To prove the uniqueness of the invariant measure it is sufficient to prove that the dynamics is irreducible and has the strong-Feller property. Some hypoellipticity, control theory and conditioning arguments are used to achieve this goal. □

1.2 Simplified Perturbed Hamiltonian Systems

Introducing a noise into the deterministic dynamics help us to solve some ergodicity problems. Nevertheless, as we will see, several challenging problems remain open for chains of oscillators perturbed by a conservative noise. In [13] we proposed to simplify still these models and the main message addressed there is that the models introduced in [13] have qualitatively the same behaviors as the unpinned chains. For simplicity we define only the dynamics in infinite volume.

Let U and V be two potentials on $\mathbb{R}$ and consider the Hamiltonian system $(\omega(t))_{t\geq 0} = (\mathbf{r}(t), \mathbf{p}(t))_{t\geq 0}$ described by the equations of motion

$$\frac{dp_x}{dt} = V'(r_{x+1}) - V'(r_x), \qquad \frac{dr_x}{dt} = U'(p_x) - U'(p_{x-1}), \qquad x \in \mathbb{Z}, \tag{17}$$

where p_x is the momentum of particle x, q_x its position and $r_x = q_x - q_{x-1}$ the "deformation". Standard chains of oscillators are recovered for a quadratic kinetic energy $U(p) = p^2/2$. The dynamics conserves (at least) three physical quantities: the total momentum $\sum_x p_x$, the total deformation $\sum_x r_x$ and the total energy $\sum_x \mathcal{E}_x$ with $\mathcal{E}_x = V(r_x) + U(p_x)$. Consequently, every Gibbs grand canonical ensemble $\nu_{\beta,\lambda,\lambda'}$ defined by

$$d\nu_{\beta,\lambda,\lambda'}(\eta) = \prod_{x\in\mathbb{Z}} \mathcal{Z}(\beta,\lambda,\lambda')^{-1} \exp\left\{-\beta\mathcal{E}_x - \lambda p_x - \lambda' r_x\right\} dr_x\, dp_x \tag{18}$$

is invariant under the evolution. To simplify we assume that the potentials U and V are smooth potentials with second derivatives bounded by below and from above.

To overcome our ignorance about macro-ergodicity of the dynamics, as before, we add a stochastic conserving perturbation. In the general case $U \neq V$, the Hamiltonian dynamics can be perturbed by the energy-momentum conserving noise acting on the velocities (as proposed in [31]) but conserving the three physical invariants mentioned above. Then the infinite volume dynamics can be defined on the state space $\tilde{\Omega}$. Its generator $\mathcal{L}$ is given by $\mathcal{L} = \mathcal{A} + \gamma\mathcal{S}$, $\gamma > 0$, where

$$(\mathscr{A}f)(\mathbf{r},\mathbf{p}) = \sum_{x\in\mathbb{Z}} \left\{ (V'(r_{x+1}) - V'(r_x))\partial_{p_x} f + (U'(p_x) - U'(p_{x-1}))\partial_{r_x} f \right\} (\mathbf{r},\mathbf{p})$$

$$(\mathscr{S}f) = \sum_{x\in\mathbb{Z}} \left[f(\mathbf{r},\mathbf{p}^{x,x+1}) - f(\mathbf{r},\mathbf{p}) \right] \tag{19}$$

for any $f \in C_0^1(\tilde{\Omega})$.

Theorem 4 ([13]) *Assume that the potentials U and V are smooth potentials with second derivatives bounded by below and from above. The dynamics generated by $\mathscr{L} = \mathscr{A} + \gamma\mathscr{S}$ with $\gamma > 0$ and $\mathscr{A}$, $\mathscr{S}$ given by (19) is macro-ergodic. Consequently, before the appearance of the shocks, in the Euler time scale, the hydrodynamic limits are given by a triplet of compressible Euler equations.*

Our motivation being to simplify as much as possible the dynamics considered in [1, 2] without destroying the anomalous behavior of the energy diffusion, we mainly focus on the symmetric case $U = V$. Then the **p**'s and **r**'s play a symmetric role so there is no reason that momentum conservation is more important than deformation conservation. We propose thus to add a noise conserving only the energy and $\sum_x [r_x + p_x]$. It is more convenient to use the variables $\{\eta_x\,;\, x \in \mathbb{Z}\} \in \mathbb{R}^{\mathbb{Z}}$ defined by $\eta_{2x} = p_x$ and $\eta_{2x-1} = r_x$ so that (17) becomes

$$d\eta_x = \left[V'(\eta_{x+1}) - V'(\eta_{x-1}) \right] dt, \quad x \in \mathbb{Z}. \tag{20}$$

We might also interpret the dynamics for the η's as the dynamics of an interface whose height (resp. energy) at site x is η_x (resp. $V(\eta_x)$). It is then quite natural to call the quantity $\sum_x \eta_x$ the "volume".

Hence, we introduce the so-called *stochastic energy-volume conserving dynamics*, which is still described by (20) between random exponential times where two nearest neighbors heights η_x and η_{x+1} are exchanged. Observe that in the momenta-deformation picture this noise is less degenerate than the momenta exchange noise since exchange between momenta and positions is now allowed. The generator $\mathscr{L}$ of the infinite volume dynamics, well defined on the state space Ω, is given by $\mathscr{L} = \mathscr{A} + \gamma\mathscr{S}$, $\gamma > 0$, where for any $f \in C_0^1(\Omega)$,

$$(\mathscr{A}f)(\eta) = \sum_{x\in\mathbb{Z}} \left[V'(\eta_{x+1}) - V'(\eta_{x-1}) \right] (\partial_{\eta_x} f)(\eta),$$

$$(\mathscr{S}f)(\eta) = \sum_{x\in\mathbb{Z}} \left[f(\eta^{x,x+1}) - f(\eta) \right]. \tag{21}$$

The noise still conserves the total energy and the total volume but destroys the conservation of momentum and deformation. Therefore, only two quantities are conserved and the invariant Gibbs grand canonical measures of the stochastic dynamics correspond to the choice $\lambda = \lambda'$ in (18). We denote $\nu_{\beta,\lambda,\lambda}$ (resp. $\mathscr{Z}(\beta,\lambda,\lambda)$) by $\mu_{\beta,\lambda}$ (resp. $Z(\beta,\lambda)$).

2 Normal Diffusion

Normal diffusion of energy in purely deterministic homogeneous chains of oscillators is expected to hold in high dimension ($d \geq 3$) or if momentum is not conserved, i.e. in the presence of a pinning potential. The problem of anomalous diffusion will be discussed in the next chapter. In this chapter we consider the case of normal diffusion.

The first step to show such normal behavior is to prove that the transport coefficient, the thermal conductivity, is well defined. Once it has been achieved, the following non-equilibrium problems can be considered:

- Hydrodynamic limits in the diffusive time scale $t\varepsilon^{-2}$, ε being the scaling parameter: if the system has trivial hydrodynamics in the time scale $t\varepsilon^{-1}$, i.e. if momentum is not conserved, we would like to show that in the diffusive time scale, the macroscopic energy profile evolves according to a diffusion equation. If the system has non-trivial hydrodynamics given by the Euler equations in the hyperbolic scaling (i.e. if momentum is conserved), in the diffusive time scale, we would like to derive the incompressible Navier-Stokes equations. These would be obtained by starting with some initial momentum macroscopic profile of order $\mathcal{O}(\varepsilon)$ but an energy profile of order $\mathcal{O}(1)$.
- Validity of Fourier's law: we consider the NESS of the system in contact at the boundaries with thermal baths at different temperatures. Fourier's law expresses that the average of the energy current in the NESS is proportional to the gradient of the local temperature. The proportionality coefficient is called the thermal conductivity.

Assume for simplicity that $d = 1$ and that the energy is the only conserved quantity. The corresponding microscopic current, denoted by $j^e_{x,x+1}$, is defined by the local energy conservation law

$$\mathcal{L}\mathcal{E}_x = -\nabla j^e_{x-1,x} \tag{22}$$

where $\mathcal{L}$ is the generator of the infinite dynamics under investigation and ∇ is the discrete gradient defined for any $(u_x)_x \in \mathbb{R}^{\mathbb{Z}}$ by $\nabla u_x = u_{x+1} - u_x$. In the current state of the art, in all the problems mentioned above, the usual approach consists to prove that there exist functions $\varphi_x = \theta_x \varphi_0$ and $h_x = \theta_x h_0$ (actually only approximate solutions are needed) such that the following decomposition

$$j^e_{x,x+1} = \nabla \varphi_x + \mathcal{L} h_x \tag{23}$$

holds. Here θ_x denotes the shift by $x \in \mathbb{Z}^d$. Equation (23) is called a *microscopic fluctuation-dissipation equation*. Then, taking arbitrary large integer $\ell \geq 1$, by using a multi-scale analysis we replace the block averaged function $\frac{1}{2\ell+1}\sum_{|y-x|\leq \ell} \nabla \varphi_y$ by $D(\mathcal{E}_x^\ell)\nabla \mathcal{E}_x^\ell$ where the function D is identified to a diffusion coefficient which depends on the empirical energy $\mathcal{E}_x^\ell = \frac{1}{2\ell+1}\sum_{|y-x|\leq \ell} \mathcal{E}_y$ in the mesoscopic box of length $(2\ell + 1)$ centered around x. Intuitively, $\mathcal{L} h_x$ represents rapid fluctuation (integrated in time, it is a martingale) and the term $\nabla \varphi_x$ represents the dissipation. Gradient models are systems for which the current is equal to the gradient of a function ($h_x = 0$ with the previous notations).

There are at least two reasons for which the problems listed above are difficult:

- The existence of a microscopic fluctuation-dissipation equation has been given for the first time for reversible systems. It has been extended to asymmetric systems satisfying a *sector condition*. Roughly speaking this last condition means that the antisymmetric part of the generator is a bounded perturbation of the symmetric part of the generator.[8] Later this condition has been relaxed into the so-called *graded sector condition*: there exists a gradation of the space where the generator is defined and the asymmetric part is bounded by the symmetric part on each graded part (see [36, 43] and references therein). The Hamiltonian systems perturbed by a noise are non-reversible and since the noise (the symmetric part of the generator) is very degenerate, none of these conditions hold.
- The system evolves in a non compact space and one needs to show that energy cannot concentrate on a site. This technical problem turns out to be difficult since no general techniques are available. For deterministic nonlinear chains the bounds on the average energy moments are usually polynomial in the size N of the system. Typically we need bounds of order one with respect to N.

2.1 Anharmonic Chain with Velocity-Flip Noise

2.1.1 Linear Response Theory: Green-Kubo Formula

The Green-Kubo formula is one of the most important formulas of non-equilibrium statistical mechanics. In the two problems mentioned in the introduction of the chapter (hydrodynamic limits and Fourier's law) the limiting objects are defined via some macroscopic coefficients which can be expressed by a Green-Kubo formula. The latter is a formal expression and showing that it is indeed well defined is a difficult problem. It is usually introduced in the context of the linear response theory that we describe below.

Consider a one dimensional unpinned chain of N harmonic oscillators with forced boundary conditions and perturbed by the velocity-flip noise. The two external constant forces are denoted by τ_ℓ and τ_r. Furthermore on the boundary particles 1 and N, Langevin thermostats are acting at different temperature $T_\ell = \beta_\ell^{-1}$ and $T_r = \beta_r^{-1}$. The generator $\mathscr{L}_N$ of the dynamics is given by (13) and we denote the unique non-equilibrium stationary state by μ_{ss}. The expectation w.r.t. μ_{ss} is denoted by $\langle \cdot \rangle_{ss}$ (Fig. 2).

---

[8] The antisymmetric (resp. symmetric) part of the generator $\mathscr{L}$ is given by $\frac{\mathscr{L}-\mathscr{L}^*}{2}$ (resp. $\frac{\mathscr{L}+\mathscr{L}^*}{2}$) where $\mathscr{L}^*$ is the adjoint of $\mathscr{L}$ in $\mathbb{L}^2(\mu)$, μ being any Gibbs grand canonical measure. For the models considered in this course, the antisymmetric part is $\mathscr{A}$ and due to the deterministic dynamics, the symmetric part is $\mathscr{S}$ and due to the noise.

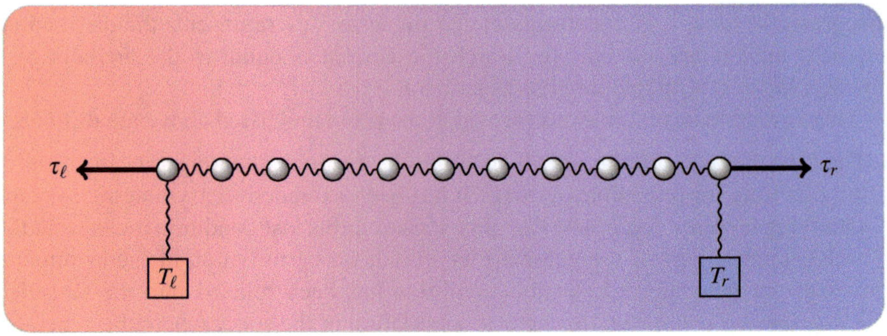

Fig. 2 The unpinned chain with boundary thermal reservoirs and forced boundary conditions

The energy[9] of atom x is defined by

$$\mathscr{E}_1 = \frac{p_1^2}{2}, \quad \mathscr{E}_x = \frac{p_x^2}{2} + V(r_x), \quad x = 2, \dots, N.$$

The local conservation of energy is expressed by the microscopic continuity equation

$$\mathscr{L}_N(\mathscr{E}_x) = -\nabla j^e_{x-1,x}, \quad x = 1, \dots, N,$$

where the energy current $j^e_{x,x+1}$ from site x to site $x+1$ is given by

$$\begin{aligned} j^e_{0,1} &= -\tau_\ell p_1 + \gamma_\ell (T_\ell - p_1^2), \\ j^e_{N,N+1} &= -\tau_r p_N - \gamma_r (T_r - p_N^2), \\ j^e_{x,x+1} &= -p_x V'(r_{x+1}),\ x = 1, \dots, N-1. \end{aligned} \tag{24}$$

The energy current $j^e_{0,1}$ (and similarly for $j^e_{N,N+1}$) is composed of two terms: the term $-\tau_\ell p_1$ corresponds to the work done on the first particle by the linear force and the term $\gamma_\ell(T_\ell - p_1^2)$ is the heat current due to the left reservoir.

Let P_s be the velocity of the center of mass of the system and J_s be the average energy current, which are defined by

$$P_s = \langle p_x \rangle_{ss} \quad \text{and} \quad J_s = \langle j^e_{x,x+1} \rangle_{ss}.$$

We have the simple relation between these two quantities

$$J_s = -\tau_\ell P_s + \gamma_\ell (T_\ell - \langle p_1^2 \rangle_{ss}), \quad J_s = -\tau_r P_s - \gamma_r (T_r - \langle p_N^2 \rangle_{ss}). \tag{25}$$

[9] The definition of the energy is slightly modified w.r.t. (1). It is more convenient since the energies are then independent random variables in the Gibbs grand canonical ensemble.

The value of P_s can be determined exactly and is independent of the nonlinearities present in the system. By writing that $\langle \mathscr{L}_N(p_x)\rangle_{ss} = 0$ for any $x = 1, \dots, N$ we get that the tension profile, defined by $\tau_x = \langle V'(r_x)\rangle_{ss}$, satisfies

$$\tau_2 - \tau_\ell = \gamma_\ell P_s, \quad \tau_r - \tau_N = \gamma_r P_s,$$
$$\tau_{x+1} - \tau_x = \gamma P_s, \quad x = 2, \dots, N-1.$$

We have then:

Lemma 1 ([11]) *The velocity P_s of the center of mass is given by*

$$P_s = \frac{\tau_r - \tau_\ell}{\gamma(N-2) + \gamma_\ell + \gamma_r} \tag{26}$$

and the tension profile is linear:

$$\tau_x = \frac{\gamma(x-2) + \gamma_\ell}{\gamma(N-2) + \gamma_\ell + \gamma_r}(\tau_r - \tau_\ell) + \tau_\ell. \tag{27}$$

Consequently

$$\lim_{n\to\infty} \tau_{[Nu]} = \tau_\ell + (\tau_r - \tau_\ell)u, \quad u \in [0,1]. \tag{28}$$

For purely deterministic chain ($\gamma = 0$), the velocity P_s is of order 1, while the tension profile is flat at the value $(\gamma_\ell + \gamma_r)^{-1}\left[\gamma_\ell \tau_r + \gamma_r \tau_\ell\right]$. The first effect of the noise is to make P_s of order N^{-1} and to give a nontrivial macroscopic tension profile.

It is expected that there exists a positive constant C independent of the size N such that $\langle \mathscr{E}_x\rangle_{ss} \leq C$ for any $x = 1, \dots, N$. Apart from the harmonic case we do not know how to prove such a bound.

We shall denote by $\tilde{f}_{ss}$ the derivative of the stationary state μ_{ss} with respect to the local Gibbs equilibrium state μ_{lg} defined by $\mu_{lg}(d\mathbf{r}, d\mathbf{p}) = g(\mathbf{r}, \mathbf{p})d\mathbf{r}d\mathbf{p}$ with

$$g(\mathbf{r}, \mathbf{p}) = \prod_{x=1}^{N} \frac{e^{-\beta_x(\mathscr{E}_x - \tau_x r_x)}}{Z(\tau_x\beta_x, \beta_x)}, \tag{29}$$

where $\beta_x = \beta_\ell + \frac{x}{N}(\beta_r - \beta_\ell)$ and $\tau_x = \tau_\ell + \frac{x}{N}(\tau_r - \tau_\ell)$. In the formula above we have introduced $r_1 = 0$ to avoid annoying notations.

The function $\tilde{f}_{ss}$ is solution, in the sense of distributions, of the equation

$$\tilde{\mathscr{L}}_N^* \tilde{f}_{ss} = 0 \tag{30}$$

where $\tilde{\mathscr{L}}_N^*$ is the adjoint of $\mathscr{L}_N$ in $\mathbb{L}^2(\mu_{lg})$. We assume that $T_r = T + \delta T$, $T_\ell = T$ and $\tau_r = \tau - \delta\tau$, $\tau_\ell = \tau$ with δT, $\delta\tau$ small. At first order in δT and $\delta\tau$, we have

$$\tilde{\mathscr{L}}_N^* = \mathscr{L}_{N,\text{eq.}}^* + \gamma_r \delta T \partial_{p_N}^2 - \delta\tau\, \partial_{p_N} - \frac{\delta T}{T^2 N} \sum_{x=1}^{N-1} \left(j_{x,x+1}^e + \tau p_x\right)$$
$$- \frac{\delta\tau}{NT} \sum_{x=1}^{N-1} p_x + o(\delta T, \delta\tau)$$

where $\mathscr{L}_{N,\text{eq.}}^* = -\mathscr{A}_N^{\tau,\tau} + \gamma \mathscr{S}_N + \gamma_\ell \mathscr{B}_{1,T} + \gamma_r \mathscr{B}_{N,T}$ is the adjoint in $\mathbb{L}^2(\mu_{\tau,T}^N)$ of

$$\mathscr{L}_{N,\text{eq.}} = \mathscr{A}_N^{\tau,\tau} + \gamma \mathscr{S}_N + \gamma_\ell \mathscr{B}_{1,T} + \gamma_r \mathscr{B}_{N,T} \tag{31}$$

and $\mu_{\tau,T}^N$ is the finite volume Gibbs grand canonical ensemble with tension τ and temperature T. We now expand $\tilde{f}_{ss}$ at the linear order in δT and $\delta\tau$:

$$\tilde{f}_{ss} = 1 + \tilde{u}\, \delta T + \tilde{v}\, \delta\tau + o(\delta T, \delta\tau) \tag{32}$$

and we get that $\tilde{u}$ and $\tilde{v}$ are solution of

$$\mathscr{L}_{N,\text{eq.}}^* \tilde{u} = \frac{1}{T^2 N} \sum_{x=1}^{N-1} \left(j_{x,x+1}^e + \tau p_x\right),$$
$$\mathscr{L}_{N,\text{eq.}}^* \tilde{v} = \frac{1}{NT} \sum_{x=1}^{N-1} p_x. \tag{33}$$

It is clear that the function h_x appearing in the microscopic fluctuation-dissipation Eq. (23) is closely related (up to a time reversal) to the functions $\tilde{u}$, $\tilde{v}$, i.e. to the *first order correction to local equilibrium*.

We can now compute the average energy current at the first order in δT and $\delta\tau$ as $N \to \infty$ but we need to introduce some notation. We recall that the generator of the infinite dynamics is given by $\mathscr{L} = \mathscr{A} + \gamma \mathscr{S}$ where, for any $f \in C_0^1(\tilde{\Omega})$,

$$(\mathscr{A} f)(\mathbf{r}, \mathbf{p}) = \sum_{x \in \mathbb{Z}} \left[(p_x - p_{x-1})\, \partial_{r_x} f + \left(V'(r_{x+1}) - V'(r_x)\right) \partial_{p_x} f \right] (\mathbf{r}, \mathbf{p}),$$
$$(\mathscr{S} f)(\mathbf{r}, \mathbf{p}) = \frac{1}{2} \sum_{x \in \mathbb{Z}} \left[f(\mathbf{r}, \mathbf{p}^x) - f(\mathbf{r}, \mathbf{p}) \right].$$

Let $\mathbb{H} := \mathbb{H}_{\tau,T}$ be the completion of the vector space of bounded local functions w.r.t. the semi-inner product $\langle\langle\cdot,\cdot\rangle\rangle$ defined for bounded local functions $f, g : \tilde{\Omega} \to \mathbb{R}$, by

$$\langle\langle f, g\rangle\rangle = \sum_{x\in\mathbb{Z}} \left\{\mu_{\tau,T}(f\theta_x g) - \mu_{\tau,T}(f)\mu_{\tau,T}(g)\right\}. \tag{34}$$

Observe that in $\mathbb{H}$ every constant $c \in \mathbb{R}$ and discrete gradient $\psi = \theta_1 f - f$ is equal to zero since for any local bounded function h we have $\langle\langle c, h\rangle\rangle = 0$ and $\langle\langle \psi, h\rangle\rangle = 0$.

Assuming they exist let $\tilde{J}_s$ and $\hat{P}_s$ be the limiting average energy current and velocity:

$$\tilde{J}_s = \lim_{N\to\infty} N\langle j^e_{0,1}\rangle_{ss}, \quad \hat{P}_s = \lim_{N\to\infty} N\langle p_0\rangle_{ss}, \tag{35}$$

and define $\hat{J}_s = \tilde{J}_s + \tau\hat{P}_s$. We expect that as N goes to infinity and, at first order in δT and $\delta\tau$,

$$\begin{pmatrix}\hat{J}_s\\ \hat{P}_s\end{pmatrix} = -\kappa(T,\tau)\begin{pmatrix}\delta T\\ \delta\tau\end{pmatrix}$$

with

$$\kappa(T,\tau) = \begin{pmatrix}\kappa^e & \kappa^{e,r}\\ \kappa^{r,e} & \kappa^r\end{pmatrix} \tag{36}$$

the *thermal conductivity* matrix. Assume for simplicity that $N = 2k$ is even. By (32) and (33), we get that

$$\begin{aligned}
N\langle p_0\rangle_{ss} = N\langle p_k\rangle_{ss} &= N\int p_k \tilde{f}_{ss}\, d\mu_{lg}\\
&= N\,\delta T\int p_k\,\tilde{u}\, d\mu_{lg} + N\,\delta\tau\int p_k\,\tilde{v}\, d\mu_{lg} + o(\delta T,\delta\tau)\\
&= -\frac{\delta T}{T^2}\int p_k\,(-\mathcal{L}^*_{N,\text{eq.}})^{-1}\Big(\sum_{x=1}^{N-1}(j^e_{x,x+1}+\tau p_x)\Big)\, d\mu_{lg}\\
&\quad -\frac{\delta\tau}{T}\int p_k\,(-\mathcal{L}^*_{N,\text{eq.}})^{-1}\Big(\sum_{x=1}^{N-1}p_x\Big)\, d\mu_{lg} + o(\delta T,\delta\tau).
\end{aligned}$$

Since $\frac{d\mu_{lg}}{d\mu^N_{\tau,T}}$ is equal to $1 + O(\delta T,\delta\tau)$, we can replace μ_{lg} by $\mu^N_{\tau,T}$ in the last terms of the previous expression. Using that $\mathcal{L}^*_{N,\text{eq.}}$ is the adjoint of $\mathcal{L}_{N,\text{eq.}}$ in $\mathbb{L}^2(\mu^N_{\tau,T})$ and denoting by $\langle\cdot,\rangle_{\tau,T}$ the scalar product in $\mathbb{L}^2(\mu^N_{\tau,T})$, we obtain that

$$\begin{aligned}
N\langle p_0\rangle_{ss} &= -\frac{\delta T}{T^2}\left\langle (-\mathscr{L}_{N,\mathrm{eq.}})^{-1}p_k\ ,\ \sum_{x=1}^{N-1}(j^e_{x,x+1}+\tau p_x)\right\rangle_{\tau,T}\\
&\quad -\frac{\delta\tau}{T}\left\langle (-\mathscr{L}_{N,\mathrm{eq.}})^{-1}p_k\ ,\ \sum_{x=1}^{N-1}p_x\right\rangle_{\tau,T} + o(\delta T,\delta\tau)\\
&= -\frac{\delta T}{T^2}\left\langle (-\mathscr{L}_{2k,\mathrm{eq.}})^{-1}p_k\ ,\ \sum_{y=-k+1}^{k-1}(j^e_{y+k,y+k+1}+\tau p_{y+k})\right\rangle_{\tau,T}\\
&\quad -\frac{\delta\tau}{T}\left\langle (-\mathscr{L}_{2k,\mathrm{eq.}})^{-1}p_k\ ,\ \sum_{y=-k+1}^{k-1}p_{y+k}\right\rangle_{\tau,T} + o(\delta T,\delta\tau)
\end{aligned}$$

In the first order terms of the previous expression we can recenter everything around k by a translation of $-k$ and we get

$$\begin{aligned}
N\langle p_0\rangle_{ss} &= -\frac{\delta T}{T^2}\left\langle (-\mathscr{L}_{\Lambda_k,\mathrm{eq.}})^{-1}p_0\ ,\ \sum_{y=-k+1}^{k-1}(j^e_{y,y+1}+\tau p_y)\right\rangle_{\tau,T}\\
&\quad -\frac{\delta\tau}{T}\left\langle (-\mathscr{L}_{\Lambda_k,\mathrm{eq.}})^{-1}p_0\ ,\ \sum_{y=-k+1}^{k-1}p_y\right\rangle_{\tau,T} + o(\delta T,\delta\tau)
\end{aligned}$$

where $\Lambda_k = \{-k+1,\ldots,k\}$ and

$$\mathscr{L}_{\Lambda_k,\mathrm{eq.}} = \mathscr{A}^{\tau,\tau}_{\Lambda_k} + \gamma\mathscr{S}_{\Lambda_k} + \gamma_\ell\mathscr{B}_{-k,T} + \gamma_r\mathscr{B}_{k,T}$$

with

$$\begin{aligned}
\mathscr{A}^{\tau,\tau}_{\Lambda_k} &= \sum_{x=-k+2}^{k}(p_x-p_{x-1})\,\partial_{r_x} + \sum_{x=-k+2}^{k-1}\big(V'(r_{x+1})-V'(r_x)\big)\,\partial_{p_x}\\
&\quad -\big(\tau - V'(r_{-k+2})\big)\,\partial_{p_{-k+1}} + \big(\tau - V'(r_k)\big)\,\partial_{p_k}
\end{aligned}$$

and

$$(\mathscr{S}_{\Lambda_k}f)(\mathbf{r},\mathbf{p}) = \frac{1}{2}\sum_{x=-k+2}^{k-1}\big(f(\mathbf{r},\mathbf{p}^x)-f(\mathbf{r},\mathbf{p})\big)\,.$$

A similar formula can be obtained for $N\langle j^e_{0,1}\rangle_{ss}$. As $k\to\infty$, the finite volume Gibbs measure converges to the infinite volume Gibbs measure. Moreover, we expect that since $k\to\infty$ the effect of the boundary operators $\mathscr{B}_{\pm k,T}$ around the site 0 disappears so that $(-\mathscr{L}_{\Lambda_k,\mathrm{eq.}})^{-1}p_0$ converges to $(-\mathscr{L})^{-1}p_0$. Therefore, in the thermodynamic limit $N\to\infty$ (*i.e.* $k\to\infty$), the transport coefficients are given by the Green-Kubo formulas

$$\kappa^{e} = T^{-2}\langle\langle j^e_{0,1} + \tau p_0 \,,\, (-\mathscr{L})^{-1}\,(j^e_{0,1} + \tau p_0)\rangle\rangle,$$
$$\kappa^{e,r} = T^{-1}\langle\langle p_0 \,,\, (-\mathscr{L})^{-1}\,(j^e_{0,1} + \tau p_0)\rangle\rangle, \tag{37}$$

and

$$\kappa^{r} = T^{-1}\langle\langle p_0 \,,\, (-\mathscr{L})^{-1}\,(p_0)\rangle\rangle,$$
$$\kappa^{r,e} = T^{-2}\langle\langle j^e_{0,1} + \tau p_0 \,,\, (-\mathscr{L})^{-1}\,(p_0)\rangle\rangle. \tag{38}$$

The argument above is formal. In fact even proving the existence of the transport coefficients defined by (37), (38) is a non-trivial task. The existence of $\hat{P}_s$ defined by the second limit in (35) can be made rigorous since we have the exact expression of P_s. From Lemma 1, we have, even for $\delta\tau$, δT that are not small,

$$\hat{P}_s = -\frac{\delta\tau}{\gamma}.$$

On the other hand we show in Theorem 5 that the quantities κ^r, $\kappa^{r,e}$, formally given by (38), can be defined in a slightly different but rigorous way, and are then equal to

$$\kappa^{r} = \gamma^{-1}, \quad \kappa^{r,e} = 0. \tag{39}$$

Thus we can rigorously establish the validity of the linear response theory for the velocity $\hat{P}_s$.

2.1.2 Existence of the Green-Kubo Formula

One of the main results of [11] is the existence of the Green-Kubo formula for the conductivity matrix. Let $\mathbb{H}^a$ (resp. $\mathbb{H}^s$) be the set of functions $f : \tilde{\Omega} \to \mathbb{R}$ antisymmetric (resp. symmetric) in $\mathbf{p}$, i.e. $f(\mathbf{r},\mathbf{p}) = -f(\mathbf{r},-\mathbf{p})$ (resp. $f(\mathbf{r},\mathbf{p}) = f(\mathbf{r},-\mathbf{p})$) for every configuration $(\mathbf{r},\mathbf{p}) \in \tilde{\Omega}$. For example, the functions $j^e_{0,1}, p_0$ and every linear combination of them are antisymmetric in $\mathbf{p}$.

Theorem 5 ([9, 11]) *Let $f, g \in \mathbb{H}^a$. The limit*

$$\sigma(f,g) = \lim_{\substack{z\to 0\\ z>0}} \langle\langle f \,,\, (z - \mathscr{L})^{-1}\, g\rangle\rangle$$

exists and $\sigma(f,g) = \sigma(g,f)$. Therefore, the conductivity matrix $\kappa(T,\tau)$ is well defined in the following sense: the limits

$$
\begin{aligned}
\kappa^e &= \lim_{\substack{z\to 0\\ z>0}} T^{-2}\langle\langle j^e_{0,1}+\tau p_0\,,\,(z-\mathscr{L})^{-1}\,(j^e_{0,1}+\tau p_0)\rangle\rangle,\\
\kappa^{e,r} &= \lim_{\substack{z\to 0\\ z>0}} T^{-1}\langle\langle p_0\,,\,(z-\mathscr{L})^{-1}\,(j^e_{0,1}+\tau p_0)\rangle\rangle,\\
\kappa^{r} &= \lim_{\substack{z\to 0\\ z>0}} T^{-1}\langle\langle p_0\,,\,(z-\mathscr{L})^{-1}\,(p_0)\rangle\rangle \;=\gamma^{-1},\\
\kappa^{r,e} &= \lim_{\substack{z\to 0\\ z>0}} T^{-2}\langle\langle j^e_{0,1}+\tau p_0\,,\,(z-\mathscr{L})^{-1}\,(p_0)\rangle\rangle
\end{aligned}
\tag{40}
$$

exist and are finite. Moreover Onsager's relation $\kappa^{e,r}=\kappa^{r,e}(=0)$ *holds.*

We have a nice thermodynamical consequence of the previous result. If δT and $\delta\tau$ are small and of the same order, the system cannot be used as a refrigerator or a boiler: at the first order, a gradient of tension does not contribute to the heat current $\hat{J}_s$. The argument above says nothing about the possibility to realize a heater or a refrigerator if $\delta\tau$ is not of the same order as δT. For the harmonic chain, we will see that it is possible to get a heater if $\delta\tau$ is of order $\sqrt{\delta T}$.

Remark 1 1. The existence of the Green-Kubo formula is also valid for a pinned or unpinned chain in any dimension.
2. Observe that with respect to the establishment of a microscopic fluctuation-dissipation Eq. (23) the computation of the Green-Kubo formula is less demanding since only the knowledge of $\sum_x h_x$ is necessary.

The proof of Theorem 5 is based on functional analysis arguments. The first main observation is that there exists a spectral gap for the operator $\mathscr{S}$ restricted to the space $\mathbb{H}^a$.

Lemma 2 *The noise operator* $\mathscr{S}$ *lets* $\mathbb{H}^a$ *and* $\mathbb{H}^s$ *invariant. For any local function* $f\in\mathbb{H}^a$ *we have that*

$$
\langle\langle f, f\rangle\rangle \;\le\; \langle\langle f, -\mathscr{S}f\rangle\rangle. \tag{41}
$$

Moreover, for any local function $f\in\mathbb{H}^a$*, there exists a local function* $h\in\mathbb{H}^a$ *such that*

$$
\mathscr{S}h=f.
$$

Proof Since the Gibbs states are Gaussian states in the p_x's it is convenient to decompose the operator $\mathscr{S}$ (which acts only on the p_x's) in the orthogonal basis of Hermite polynomials. The the lemma follows easily. □

Proof (*Theorem 5*)
We observe first that $\mathbb{H}^a$ and $\mathbb{H}^s$ are orthogonal Hilbert spaces such that $\mathbb{H}=\mathbb{H}^a\oplus\mathbb{H}^s$. It is also convenient to define the following semi-inner product

$$
\langle\langle u, w\rangle\rangle_1 \;=\; \langle\langle u, (-\mathscr{S})w\rangle\rangle.
$$

Let $\mathbb{H}^1$ be the associated Hilbert space. We also define the Hilbert space $\mathbb{H}^{-1}$ via the duality given by the $\mathbb{H}$ norm, that is

$$\|u\|_{-1}^2 = \sup_w \{ 2\langle\langle u, w\rangle\rangle - \langle\langle w, w\rangle\rangle_1 \}$$

where the supremum is taken over local bounded functions w. By Lemma 2 we have that $\mathbb{H}^a \subset \mathbb{H}^{-1}$. Thus $g \in \mathbb{H}^{-1}$.

Let w_z be the solution of the resolvent equation $(z - \mathscr{L})w_z = g$. We have to show that $\langle\langle f, w_z\rangle\rangle$ converges as z goes to 0. We decompose w_z into $w_z = w_z^- + w_z^+$, $w_z^- \in \mathbb{H}^a$ and $w_z^+ \in \mathbb{H}^s$. Since $\mathbb{H}^a$ is orthogonal to $\mathbb{H}^s$ and $f \in \mathbb{H}^a$ we have $\langle\langle f, w_z\rangle\rangle = \langle\langle f, w_z^-\rangle\rangle$. It is thus sufficient to prove that $(w_z^-)_{z>0}$ converges weakly in $\mathbb{H}$ as $z \to 0$.

Since $\mathscr{A}$ inverts the parity and $\mathscr{S}$ preserves it and $\mathbb{H}^a \oplus \mathbb{H}^s = \mathbb{H}$ and $g \in \mathbb{H}^a$, we have, for any $\mu, \nu > 0$,

$$\begin{aligned} &\nu w_\nu^+ - \mathscr{A} w_\nu^- - \gamma \mathscr{S} w_\nu^+ = 0, \\ &\mu w_\mu^- - \mathscr{A} w_\mu^+ - \gamma \mathscr{S} w_\mu^- = g. \end{aligned} \tag{42}$$

Taking the scalar product with w_μ^+ (resp. w_ν^-) on both sides of the first (resp. second) equation of (42), we get

$$\begin{aligned} \nu\langle\langle w_\mu^+, w_\nu^+\rangle\rangle - \langle\langle w_\mu^+, \mathscr{A} w_\nu^-\rangle\rangle + \gamma\langle\langle w_\mu^+, w_\nu^+\rangle\rangle_1 &= 0, \\ \mu\langle\langle w_\nu^-, w_\mu^-\rangle\rangle - \langle\langle w_\nu^-, \mathscr{A} w_\mu^+\rangle\rangle + \gamma\langle\langle w_\mu^-, w_\nu^-\rangle\rangle_1 &= \langle\langle w_\nu, g\rangle\rangle. \end{aligned} \tag{43}$$

Summing the above equations we have

$$\nu\langle\langle w_\mu^+, w_\nu^+\rangle\rangle + \mu\langle\langle w_\nu^-, w_\mu^-\rangle\rangle + \gamma\langle\langle w_\mu, w_\nu\rangle\rangle_1 = \langle\langle w_\nu, g\rangle\rangle \tag{44}$$

Putting $\mu = \nu$ we get

$$\nu\langle\langle w_\nu, w_\nu\rangle\rangle + \gamma\langle\langle w_\nu, w_\nu\rangle\rangle_1 \le \|w_\nu\|_1 \|g\|_{-1}.$$

Hence $(w_\nu)_{\nu>0}$ is uniformly bounded in $\mathbb{H}^1$ and by the spectral gap property so is $(w_\nu^-)_{\nu>0}$ in $\mathbb{H}$. Moreover, $(\nu w_\nu)_{\nu>0}$ converges strongly to 0 in $\mathbb{H}$ as $\nu \to 0$. We can then extract weakly convergent subsequences. Taking first the limit, in (44), $\nu \to 0$ and then $\mu \to 0$ along one such subsequence (converging to w_*) we have

$$\gamma\langle\langle w_*, w_*\rangle\rangle_1 = \langle\langle w_*, g\rangle\rangle.$$

Next, taking the limit along different weakly convergent subsequences (let w^* be the other limit) we have

$$\gamma\langle\langle w_*, w^*\rangle\rangle_1 = \langle\langle w^*, g\rangle\rangle$$

and, exchanging the role of the two sequences

$$2\gamma\langle\langle w_*, w^*\rangle\rangle_1 = \langle\langle w_*, g\rangle\rangle + \langle\langle w^*, g\rangle\rangle = \gamma\langle\langle w_*, w_*\rangle\rangle_1 + \gamma\langle\langle w^*, w^*\rangle\rangle_1$$

which implies $w_* = w^*$, that is all the subsequences have the same limit. Thus $(w_\nu)_{\nu>0}$ converges weakly in $\mathbb{H}^1$ as well as $(w_\nu^-)_{\nu>0}$ in $\mathbb{H}$ by Lemma 2. □

In the harmonic case, $V(r) = r^2/2$, much more is known. Indeed one easily checks that the exact microscopic fluctuation-dissipation Eq. (23) holds with

$$h_x = \frac{1}{2\gamma} r_{x+1}(p_x + p_{x+1}) - \frac{r_{x+1}^2}{4}, \quad \varphi_x = -\frac{1}{2\gamma}(r_x r_{x+1} + p_x^2). \tag{45}$$

It follows that we can compute explicitly $(z - \mathscr{L})^{-1} j_{0,1}^e$ and obtain that the value of the conductivity matrix:

$$\kappa(\tau, T) = \begin{pmatrix} \frac{1}{2\gamma} & 0 \\ 0 & \frac{1}{\gamma} \end{pmatrix}.$$

This value will be recovered by considering the hydrodynamic limits of the system (Theorem 6) and also by establishing the validity of Fourier's law (see Theorem 7).

2.1.3 Expansion of the Green-Kubo Formula in the Weak Coupling Limit

In the previous subsection we proved the existence of the Green-Kubo formula showing that the transport coefficient is well defined if some noise is added to the deterministic dynamics. We are now interested in the behavior of the Green-Kubo formula as the noise vanishes. We investigate this question in the weak coupling limit, i.e. assuming that the interaction potential is of the form εV where $\varepsilon \langle\langle 1$ is the (small) coupling parameter. For notational simplicity we consider the one dimensional infinite pinned system but the arguments given below are easily generalized to the (pinned or unpinned) $d \geq 1$-dimensional case.[10] The expansion presented in this section is formal but we will precise at the end of the section what has been rigorously proved. In order to emphasize the dependence of κ^e (denoted in the sequel by κ) in the coupling parameter ε and the noise intensity γ, we denote κ by $\kappa(\varepsilon, \gamma)$. Here we propose a formal expansion of the conductivity κ in the form

$$\kappa(\varepsilon, \gamma) = \sum_{n\geq 2} \kappa_n(\gamma)\varepsilon^n. \tag{46}$$

[10] If $W = 0$ the variables q_x have to take values in a compact manifold.

Then we study rigorously the first term of this expansion $\kappa_2(\gamma)$. It is intuitively clear that the expansion starts from ε^2 since the Green-Kubo formula is a quadratic function of the energy current and that the latter is of order ε (see (49)).

When the system is uncoupled ($\varepsilon = 0$), the dynamics is given by the generator $\mathscr{L}_0 = \mathscr{A}_0 + \gamma S$ with $\mathscr{S}$ the flip noise defined by (12) and

$$\mathscr{A}_0 = \sum_{x\in\mathbb{Z}} p_x \partial_{q_x} - W'(q_x)\partial_{p_x}.$$

When $\varepsilon > 0$, the generator of the coupled dynamics is denoted by

$$\mathscr{L}_\varepsilon = \mathscr{L}_0 + \varepsilon \mathscr{G} \tag{47}$$

where

$$\mathscr{G} = \sum_{x\in\mathbb{Z}} V'(q_x - q_{x-1})(\partial_{p_{x-1}} - \partial_{p_x}).$$

The energy of each cell, which is the sum of the internal energy and of the interaction energy, is defined by

$$\mathscr{E}_x^\varepsilon = \mathscr{E}_x + \frac{\varepsilon}{2}\left(V(q_{x+1} - q_x) + V(q_x - q_{x-1})\right), \quad \mathscr{E}_x = \frac{p_x^2}{2} + W(q_x). \tag{48}$$

Observe that $\mathscr{E}_x = \mathscr{E}_x^0$ is the energy of the isolated system x. The dynamics generated by $\mathscr{L}_0$ preserves all the individual energies $\mathscr{E}_x$. The dynamics generated by $\mathscr{L}_\varepsilon$ conserves the total energy. The corresponding energy currents $\varepsilon j_{x,x+1}$, defined by the local conservation law

$$\mathscr{L}_\varepsilon \mathscr{E}_x^\varepsilon = \varepsilon\left(j_{x-1,x} - j_{x,x+1}\right)$$

are given by

$$\varepsilon\, j_{x,x+1} = -\frac{\varepsilon}{2}\left(p_x + p_{x+1}\right) \cdot V'(q_{x+1} - q_x). \tag{49}$$

Let us denote by $\mu_{\beta,\varepsilon} = \langle\cdot\rangle_{\beta,\varepsilon}$ the canonical Gibbs measure at temperature $\beta^{-1} > 0$ defined by the Dobrushin-Lanford-Ruelle equations, which of course depends on the interaction εV. We shall assume in all the cases considered that $\mu_{\beta,\varepsilon}$ is analytical in ε for sufficiently small ε (when applied to local functions). In particular we assume that the potentials V and W are such that the Gibbs state is unique and has spatial exponential decay of correlations (this holds under great general conditions on V and W, see [33]).

In order to emphasize the dependence in ε we reintroduce some notation. For any given local functions f, g, define the semi-inner product

$$\langle\langle f, g\rangle\rangle_{\beta,\varepsilon} = \sum_{x\in\mathbb{Z}} [\langle \theta_x f, g\rangle_{\beta,\varepsilon} - \langle f\rangle_{\beta,\varepsilon}\langle g\rangle_{\beta,\varepsilon}]. \tag{50}$$

We recall that θ_x is the shift operator by x. The sum is finite in the case $\varepsilon = 0$, and converges for $\varepsilon > 0$ thanks to the exponential decay of the spatial correlations. Denote by $\mathbb{H}_\varepsilon = \mathbb{L}^2(\langle\langle \cdot, \cdot\rangle\rangle_{\beta,\varepsilon})$ the corresponding closure. We define the subspace of antisymmetric functions in the velocities

$$\mathbb{H}^a_\varepsilon = \{f \in \mathbb{H}_\varepsilon : f(\mathbf{q}, -\mathbf{p}) = -f(\mathbf{q}, \mathbf{p})\}. \tag{51}$$

Similarly we define the subspace of symmetric functions in $\mathbf{p}$ as $\mathbb{H}^s_\varepsilon$. On local functions this decomposition of a function into symmetric and antisymmetric parts is independent of ε. Let us denote by $\mathscr{P}^a_\varepsilon$ and $\mathscr{P}^s_\varepsilon$ the corresponding orthogonal projections, whose definition in fact does not depend on ε. Therefore we sometimes omit the index ε in the notation. Finally, for any function $f \in \mathbb{L}^2(\mu_{\beta,\varepsilon})$, define

$$(\Pi_\varepsilon f)(\mathscr{E}) = \mu_{\beta,\varepsilon}(f|\mathscr{E}), \quad Q_\varepsilon = \mathrm{Id} - \Pi_\varepsilon$$

where $\mathscr{E} := \{\mathscr{E}_x \, ; \, x \in \mathbb{Z}\}$. According to Theorem 5 the conductivity is defined by

$$\kappa(\varepsilon, \gamma) = \varepsilon^2 \lim_{\nu\to 0} \langle\langle j_{0,1}, (\nu - \mathscr{L}_\varepsilon)^{-1} j_{0,1}\rangle\rangle_{\beta,\varepsilon}. \tag{52}$$

It turns out that, for calculating the terms in the expansion (46), it is convenient to choose $\nu = \varepsilon^2\lambda$ in (52), for a $\lambda > 0$, and solve the resolvent equation

$$(\lambda\varepsilon^2 - \mathscr{L}_\varepsilon) u_{\lambda,\varepsilon} = \varepsilon j_{0,1} \tag{53}$$

for the unknown function $u_{\lambda,\varepsilon}$. The factor ε^2 is the natural scaling in view of the subsequent computations. We assume that a solution of (53) is in the form

$$u_{\lambda,\varepsilon} = \sum_{n\geq 0} U_{\lambda,n}\varepsilon^n = \sum_{n\geq 0} (v_{\lambda,n} + w_{\lambda,n})\varepsilon^n, \tag{54}$$

where $\Pi v_{\lambda,n} = Q w_{\lambda,n} = 0$, i.e. $w_{\lambda,n} = \Pi U_{\lambda,n}$ and $v_{\lambda,n} = Q U_{\lambda,n}$. Here $\Pi = \Pi_0$ and $Q = Q_0$ refer to the uncoupled measure $\mu_{\beta,0}$.[11] Given such an expression we can, in principle, use it in (52) to write

[11] The reason to use the orthogonal decomposition of $U_{\lambda,n} = v_{\lambda,n} + w_{\lambda,n}$ is that at some point we will have to consider, for a given function f, the solution h to the Poisson equation $\mathscr{L}_0 h = f$. The minimal requirement for the existence of h is that $\Pi f = 0$.

$$\begin{aligned}\kappa(\varepsilon,\gamma) &= \lim_{\lambda\to 0}\sum_{n\geq 0}\varepsilon^{n+1}\langle\langle j_{0,1}, v_{\lambda,n}+w_{\lambda,n}\rangle\rangle_{\beta,\varepsilon}\\ &= \sum_{n\geq 1}\lim_{\lambda\to 0}\varepsilon^{n}\langle\langle j_{0,1}, v_{\lambda,n-1}\rangle\rangle_{\beta,\varepsilon}\end{aligned}\tag{55}$$

where we have used the fact that that $\langle\langle j_{0,1}, w_{\lambda,\varepsilon}\rangle\rangle_{\beta,\varepsilon} = 0$ and we have, arbitrarily, exchanged the limit with the sum. Note that this is not yet of the type (46) since the terms in the expansion depend themselves on ε. To identify the coefficients κ_n we would need to expand in ε also the expectations. This is not obvious since the functions $v_{\lambda,n}$ are non local.

Let us consider the operator $\mathfrak{L} = \Pi\mathscr{G}\mathscr{P}^a(-\mathscr{L}_0)^{-1}\mathscr{G}\Pi$. We show below that the operator $\mathfrak{L}$ is a generator of a Markov process so that $(\lambda-\mathfrak{L})^{-1}$ is well defined for $\lambda > 0$. Pluging (54) in (53) we obtain the following hierarchy

$$\begin{aligned}&v_{\lambda,0} = 0,\\ &w_{\lambda,0} = (\lambda-\mathfrak{L})^{-1}\Pi\mathscr{G}\mathscr{P}^a(-\mathscr{L}_0)^{-1}j_{0,1},\\ &v_{\lambda,1} = (-\mathscr{L}_0)^{-1}\left[j_{0,1}+\mathscr{G}w_{\lambda,0}\right],\\ &w_{\lambda,n} = (\lambda-\mathfrak{L})^{-1}\Pi\mathscr{G}\mathscr{P}^a(-\mathscr{L}_0)^{-1}\left[-\lambda v_{\lambda,n-1}+\mathcal{Q}\mathscr{G}v_{\lambda,n}\right], \qquad n\geq 1\\ &v_{\lambda,n+1} = (-\mathscr{L}_0)^{-1}\left[-\lambda v_{\lambda,n-1}+\mathscr{G}w_{\lambda,n}+\mathcal{Q}\mathscr{G}v_{\lambda,n}\right], \qquad n\geq 1.\end{aligned}\tag{56}$$

Observe that in the previous equations the (formal) operator $(-\mathscr{L}_0)^{-1}$ is always applied to functions f such that $\Pi f = 0$ (this is the minimal requirement to have consistent equations). This is however not sufficient to make sense of the functions $v_{\lambda,n}$ and $w_{\lambda,n}$. Nevertheless, by using an argument similar to the one given in Theorem 5, we have that the local operator $\mathscr{T}_0$ on $\mathbb{H}_0^a$ defined by

$$\mathscr{T}_0 f = \lim_{\nu\to 0}\mathscr{P}^a(\nu-\mathscr{L}_0)^{-1}f, \quad f\in\mathbb{H}_0^a,$$

is well defined. Therefore, it is possible to make sense, as a distribution, of

$$\alpha_{01} = \Pi\mathscr{G}\mathscr{P}^a(-\mathscr{L}_0)^{-1}j_{0,1} := \Pi\mathscr{G}\mathscr{T}_0 j_{0,1}.\tag{57}$$

Nevertheless, the function $w_{\lambda,0}$ is still not well defined since we are not sure that $\mathscr{T}_0 j_{0,1}$ is in the domain of $\mathscr{G}$.

Even if the previous computations are formal a remarkable fact is that the operator $\mathfrak{L}$, when applied to functions of the internal energies, coincides with the Markov generator $\mathfrak{L}_{\rm GL}$ of a reversible Ginzburg-Landau dynamics on the internal energies. Let us denote by ρ_β the distribution of the internal energies $\mathscr{E} = \{\mathscr{E}_x\,;\, x\in\mathbb{Z}\}$ under the Gibbs measure $\mu_{\beta,0}$. It can be written in the form

$$d\rho_\beta(\mathscr{E}) = \prod_{x\in\mathbb{Z}} Z_\beta^{-1}\exp(-\beta\mathscr{E}_x - U(\mathscr{E}_x))d\mathscr{E}_x$$

for a suitable function U. We denote the formal sum $\sum_x U(\mathcal{E}_x)$ by $\mathcal{U} := \mathcal{U}(\mathcal{E})$. We denote also, for a given value of the internal energy $\tilde{\mathcal{E}}_x$ in the cell x, by $\nu^x_{\tilde{\mathcal{E}}_x}$ the microcanonical probability measure in the cell x. i.e. the uniform probability measure on the manifold

$$\Sigma_{\tilde{\mathcal{E}}_x} := \{(q_x, p_x) \in \Omega\, ;\, \mathcal{E}_x(q_x, p_x) = \tilde{\mathcal{E}}_x\}.$$

Then, the generator $\mathfrak{L}_{\mathrm{GL}}$ is given by

$$\mathfrak{L}_{\mathrm{GL}} = \sum_x e^{\mathcal{U}} (\partial_{\mathcal{E}_{x+1}} - \partial_{\mathcal{E}_x}) \left[e^{-\mathcal{U}} \gamma^2(\mathcal{E}_x, \mathcal{E}_{x+1})(\partial_{\mathcal{E}_{x+1}} - \partial_{\mathcal{E}_x}) \right], \tag{58}$$

where

$$\gamma^2(\tilde{\mathcal{E}}_0, \tilde{\mathcal{E}}_1) = \int_{\Sigma_{\tilde{\mathcal{E}}_0} \times \Sigma_{\tilde{\mathcal{E}}_1}} \big(j_{0,1}\; \mathcal{T}_0 j_{0,1}\big)\, d\nu^0_{\tilde{\mathcal{E}}_0} d\nu^1_{\tilde{\mathcal{E}}_1}. \tag{59}$$

The operator $\mathfrak{L}_{\mathrm{GL}}$ is well defined only if the function γ^2 has some regularity properties, that are actually proven in specific examples [25, 48]. We can show that the Dirichlet forms[12] associated to $\mathfrak{L}$ and $\mathfrak{L}_{\mathrm{GL}}$ coincide. Then in the cases where γ^2 is proven to be smooth (58) is well defined and $\mathfrak{L} = \mathfrak{L}_{\mathrm{GL}}$.

Proposition 2 ([9]) *For each local smooth functions f, g of the internal energies only we have*

$$\langle\langle g, (-\mathfrak{L})f \rangle\rangle_{\beta,0} = \langle\langle g, (-\mathfrak{L}_{\mathrm{GL}})f \rangle\rangle_{\beta,0}. \tag{60}$$

The operator $\mathfrak{L}_{\mathrm{GL}}$ is the generator of a Ginzburg-Landau dynamics which is reversible with respect to ρ_β, for any $\beta > 0$. It is conservative in the energy $\sum_x \mathcal{E}_x$ and the corresponding currents are given by $\theta_x \alpha_{0,1}$ where $\alpha_{0,1}$ has been defined in (57). The corresponding finite size dynamics appears in [26, 48] as the weak coupling limit of a finite number N (fixed) of cells weakly coupled by a potential εV in the limit $\varepsilon \to 0$ when time t is rescaled as $t\varepsilon^{-2}$. Moreover, the hydrodynamic limit of the Ginzburg-Landau dynamics is then given (in the diffusive time scale tN^2, $N \to +\infty$), by a heat equation with diffusion coefficient which coincides with κ_2 as given by (62) below [65]. This is summarized in Fig. 3.

According to the previous expansion it makes sense to define $\kappa_2(\gamma)$ by

$$\kappa_2(\gamma) = \lim_{\varepsilon \to 0} \lim_{\lambda \to 0} \left\{ \langle\langle j_{0,1}, \mathcal{T}_0 j_{0,1} \rangle\rangle_{\beta,\varepsilon} + \langle\langle j_{0,1}, \mathcal{T}_0 \mathcal{G} w_{\lambda,0} \rangle\rangle_{\beta,\varepsilon} \right\} \tag{61}$$

if the limits exist. In fact, a priori, it is not even clear that the term $\mathcal{T}_0 \mathcal{G} w_{\lambda,0}$ makes sense since $w_{\lambda,0}$ is not well defined. In [9] we argue that

$$\kappa_2(\gamma) = \left\langle \gamma_{0,1}^2 \right\rangle_\beta - \langle\langle \alpha_{0,1}, (-\mathfrak{L}_{GL})^{-1} \alpha_{0,1} \rangle\rangle_\beta. \tag{62}$$

---

12 They are well defined even if γ^2 is not regular.

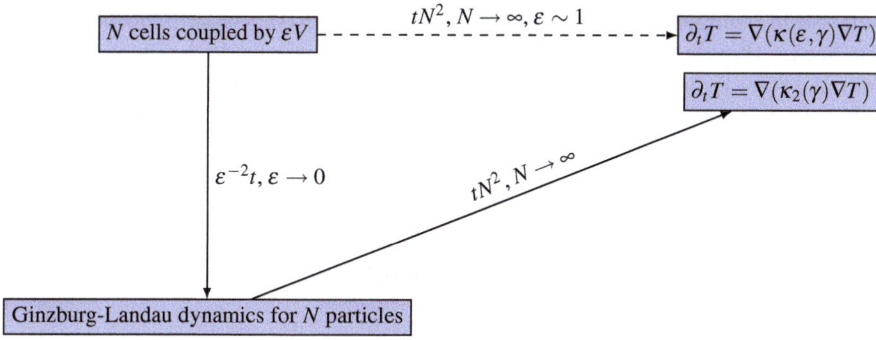

Fig. 3 The relation between the hydrodynamic limit, the weak coupling limit and the Green-Kubo expansion. The *dotted arrow* (hydrodynamic limits in the diffusive time scale) has not been proved. The weak coupling limit (*vertical arrow*) has been proved in [48] (see also [26]) and the *diagonal arrow* (hydrodynamic limits for a Ginzburg-Landau dynamics) has been obtained in [65] in some cases which however do not cover our cases. In [9] it is argued that $\kappa(\varepsilon, \gamma) \sim \varepsilon^2 \kappa_2(\gamma)$ as $\varepsilon \to 0$

Here $\langle \cdot \rangle_\beta$ and $\langle\langle \cdot \rangle\rangle_\beta$ refer to the scalar products w.r.t. ρ_β. In the special case $W = 0$,[13] we prove rigorously in [9] that we can make sense for any λ, ε of the term in the righthandside of (61) and that (62) is valid, supporting the conjecture that (62) is valid in more general situations. Observe that (62) is the Green-Kubo formula for the diffusion coefficient of the Ginzburg-Landau dynamics.

In specific examples, it is possible to study the behavior of $\kappa_2(\gamma)$ defined by (62) in the vanishing noise limit $\gamma \to 0$:

1. Harmonic chain: it is known that the conductivity of the (deterministic) harmonic chain is $\kappa(\varepsilon, 0) = \infty$. If $\gamma > 0$, $\kappa(\varepsilon, \gamma) = c\gamma^{-1}\varepsilon^{-2}$, $c > 0$ a constant, and we get thus that $\lim_{\gamma \to 0} \kappa_2(\gamma) = \infty$.
2. Disordered pinned harmonic chain: V is quadratic and the one-site potential W is site-dependent given by $W_x(q) = v_x q^2$ where $\{v_x \,;\, x \in \mathbb{Z}\}$ is a sequence of independent identically distributed positive bounded random variables.[14] It is known [8] that $\kappa(\varepsilon, 0) = 0$ so that $\kappa_2(\varepsilon, 0) = 0$. It can be proved that $\kappa_2(\gamma)$ vanishes as γ goes to 0.
3. Harmonic chain with quartic pinning potential: V is quadratic and $W(q) = q^4$. Then it can be shown that $\limsup_{\gamma \to 0} \kappa_2(\gamma) < \infty$. This upper bound does not prevent the possibility that $\lim_{\gamma \to 0} \kappa_2(\gamma) = 0$.

To prove these results we use the upper bound $\kappa_2(\gamma) \leq \left\langle \gamma_{0,1}^2 \right\rangle_\beta$. Recalling (59) we see that if we are able to compute $\mathscr{T}_0 j_{0,1}$ then we can estimate $\left\langle \gamma^2(\mathscr{E}_0, \mathscr{E}_1) \right\rangle_\beta$. It is exactly what is done in [9] for the specific cases above.

---

[13] If $W = 0$ the variables q_x have to take values in a compact manifold.

[14] Even if this model does not belong *stricto sensu* to the class of models discussed above it is easy to generalize to this case, at least formally, the previous results.

It would be highly interesting to have a rigorous derivation of the formal expansion above. Bypassing this problem, another relevant issue is to decide if genuinely $\lim_{\gamma\to 0}\kappa_2(\gamma)$ is zero or not. Some authors (see [23] and references therein) conjecture that, in some cases, the conductivity of the deterministic chain $\kappa(\varepsilon, 0)$ has a trivial weak coupling expansion ($\kappa(\varepsilon, 0) = \mathscr{O}(\varepsilon^n)$ for any $n \geq 2$). Showing that $\kappa_2(\gamma) \to 0$ as $\gamma \to 0$ would support this conjecture.

2.2 Harmonic Chain with Velocity-Flip Noise

In this section we assume that $V(r) = r^2/2$.

2.2.1 Hydrodynamic Limits

As explained in the beginning of this chapter an interesting problem consists to derive a diffusion equation for a chain of oscillators perturbed by an energy conserving noise. Consider a one dimensional unpinned chain of N harmonic oscillators with periodic boundary conditions perturbed by the velocity flip noise in the diffusive scale. In other words let $\omega(t) = (\mathbf{r}(t), \mathbf{p}(t))_{t\geq 0}$ be the process with generator $N^2\mathscr{L}_N = N^2\left[\mathscr{A}_N + \gamma\mathscr{S}_N\right]$ where $\mathscr{S}_N$ is given by (12), $\mathbb{Z}^d$ being replaced by $\mathbb{T}_N$, the discrete torus of length N, and $\mathscr{A}_N$ is the Liouville operator of a chain of unpinned harmonic oscillators with periodic boundary conditions. The system conserves two quantities: the total energy $\sum_{x\in\mathbb{T}_N}\mathscr{E}_x$, $\mathscr{E}_x = \frac{p_x^2}{2} + \frac{r_x^2}{2}$, and the total deformation of the lattice $\sum_{x\in\mathbb{T}_N} r_x$. Consequently, the Gibbs equilibrium measures $\nu_{\beta,\tau}$ are indexed by two parameters $\beta > 0$, the inverse temperature, and $\tau \in \mathbb{R}$, the pressure. They take the form

$$d\nu_{\beta,\tau}(d\mathbf{r}, d\mathbf{p}) = \prod_{x\in\mathbb{T}_N} \mathscr{Z}^{-1}(\beta,\tau)\, \exp\{-\beta(\mathscr{E}_x - \tau r_x)\}dr_x dp_x$$

where

$$\mathscr{Z}(\beta,\tau) = \frac{2\pi}{\beta}\exp(\beta\tau^2/2).$$

Observe the following thermodynamic relations

$$\int \mathscr{E}_x\, d\nu_{\beta,\tau} = \beta^{-1} + \tfrac{\tau^2}{2}, \quad \int r_x\, d\nu_{\beta,\tau} = \tau$$

or equivalently

$$\tau = \int r_x\, d\nu_{\beta,\tau}, \quad \beta = \left\{\int \mathscr{E}_x\, d\nu_{\beta,\tau} - \frac{\left(\int r_x\, d\nu_{\beta,\tau}\right)^2}{2}\right\}^{-1}.$$

Definition 2 Let $\mathbb{T} = [0, 1)$ be the continuous torus. Let $\mathfrak{e}_0 : \mathbb{T} \to \mathbb{R}$ and $\mathfrak{r}_0 : \mathbb{T} \to \mathbb{R}$ be two continuous macroscopic profiles such that $\mathfrak{e}_0 > \frac{\mathfrak{r}_0^2}{2}$. A sequence of probability measures $(\mu^N)_{N\geq 1}$ on $(\mathbb{R} \times \mathbb{R})^{\mathbb{T}_N}$ is said to be a sequence of Gibbs local equilibrium states associated to the energy profile $\mathfrak{e}_0$ and the deformation profile $\mathfrak{r}_0$ if

$$d\mu^N(d\mathbf{r}, d\mathbf{p}) = \prod_{x\in\mathbb{T}_N} \mathscr{Z}^{-1}(\beta_0(\tfrac{x}{N}), \tau_0(\tfrac{x}{N})) \exp\{-\beta_0(x/N)(\mathscr{E}_x - \tau_0(x/N)r_x)\}dr_x dp_x$$

where the functions β_0 and τ_0 are defined by

$$\tau_0 = \mathfrak{r}_0, \quad \beta_0 = \{\mathfrak{e}_0 - \tfrac{\mathfrak{r}_0^2}{2}\}^{-1}.$$

Once we have the microscopic fluctuation-dissipation equation (see (45)) and assuming the propagation of local equilibrium in the diffusive time scale it is easy to guess the hydrodynamic equations followed by the system. In [54] the following theorem is proved.

Theorem 6 ([54]) *Consider the unpinned velocity-flip model with periodic boundary conditions. Let $(\mu^N)_N$ be a sequence of Gibbs local equilibrium states*[15] *associated to a bounded energy profile $\mathfrak{e}_0$ and a deformation profile $\mathfrak{r}_0$. For every $t \geq 0$, and any test continuous functions $G, H : \mathbb{T} \to \mathbb{R}$, the random variables*

$$\Big(\frac{1}{N}\sum_{x\in\mathbb{T}_N} G(\tfrac{x}{N})r_x(tN^2), \frac{1}{N}\sum_{x\in\mathbb{T}_N} H(\tfrac{x}{N})\mathscr{E}_x(tN^2)\Big) \tag{63}$$

converge in probability as $N \to \infty$ to

$$\Big(\int_{\mathbb{T}} G(y)\mathfrak{r}(t, y)dy, \int_{\mathbb{T}} H(y)\mathfrak{e}(t, y)dy\Big)$$

where $\mathfrak{r}$ and $\mathfrak{e}$ are the (smooth) solutions to the hydrodynamical equations

$$\begin{cases} \partial_t \mathfrak{r} = \frac{1}{\gamma}\,\partial_y^2\,\mathfrak{r}, \\ \partial_t \mathfrak{e} = \frac{1}{2\gamma}\,\partial_y^2\left[\mathfrak{e} + \frac{\mathfrak{r}^2}{2}\right] \end{cases}, \quad y \in \mathbb{T}, \tag{64}$$

with initial conditions $\mathfrak{r}(0, y) = \mathfrak{r}_0(y)$, $\mathfrak{e}(0, y) = \mathfrak{e}_0(y)$.

[15] One can consider more general initial states, see [54].

The proof of this theorem is based on Yau's relative entropy method [51, 67]. The general strategy is simple. Let μ_t^N be the law of the process at time tN^2 starting from μ^N and let $\tilde{\mu}_t^N$ be a sequence of Gibbs local equilibrium state corresponding to the deformation profile $\mathfrak{r}_t(\cdot) := \mathfrak{r}(t,\cdot)$ and energy profile $\mathfrak{e}_t(\cdot) := \mathfrak{e}(t,\cdot)$ solution of (64). We expect that since $\mathfrak{e}$ and $\mathfrak{r}$ are the hydrodynamic profiles, the probability measure of the process μ_t^N is close, in some sense, to the local Gibss state $\tilde{\mu}_t^N$. Yau's relative entropy method consists to show that the entropic distance[16]

$$H_N(t) := H(\mu_t^N | \tilde{\mu}_t^N) = o(N) \tag{65}$$

between the two states is relatively small. Assuming (65), in order to prove for example the convergence of the empirical energy, we use the entropy inequality[17] which states that for any $\alpha > 0$ and test function ϕ

$$\int \phi d\mu_t^N \le \frac{H(\mu_t^N|\tilde{\mu}_t^N)}{\alpha} + \frac{1}{\alpha}\log\left(\int e^{\alpha\phi} d\tilde{\mu}_t^N\right). \tag{66}$$

We take then $\alpha = \delta N$, $\delta > 0$, and

$$\phi = \left| \frac{1}{N} \sum_{x \in \mathbb{T}_N} H(\tfrac{x}{N}) \mathcal{E}_x - \int_{\mathbb{T}} H(y) \mathfrak{e}(t,y) dy \right|.$$

Since $\tilde{\mu}_t^N$ is fully explicit and even product, by using large deviations estimates, it is possible to show that

$$\limsup_{N\to\infty} \frac{1}{\delta N} \log\left(\int e^{\delta N \phi} d\tilde{\mu}_t^N\right) = I(\delta) \tag{67}$$

where $I(\delta) \to 0$ as $\delta \to 0$. By using (65), we are done. It remains then to prove (65) and for this we rely on a Gronwall inequality for the entropy production ($C > 0$ is a constant)

$$\partial_t H_N \le C H_N(t) + o(N). \tag{68}$$

The proof of (68) is quite evolved and we refer the interested reader to [12, 54] (see also [41] for some overview on the subject). It is in this step that the macro-ergodicity of the dynamics is used in order to derive the so-called one-block estimate.

---

16 There is some abuse of language here since the relative entropy is not a distance between probability measures.

17 It is a trivial consequence of the definition (9).

For non-gradient systems, i.e. systems such that the microscopic currents of the conserved quantities are not given by discrete gradients,[18] the previous strategy has to be modified. Indeed, in order to have (65) it is necessary to replace the local equilibrium Gibbs state $\tilde{\mu}_t^N$ by a local equilibrium state with a first order correction term of the form

$$d\hat{\mu}_t^N(d\mathbf{r}, d\mathbf{p})$$
$$= Z_{t,N}^{-1} \prod_{x \in \mathbb{T}_N} \exp\left\{-\beta_t(x/N)(\mathscr{E}_x - \tau_t(x/N) r_x) + \tfrac{1}{N} F(t, x/N)(\theta_x g)(\mathbf{r}, \mathbf{p})\right\} dr_x dp_x \quad (69)$$

where $Z_{t,N}$ is a normalization constant,

$$\tau_t = \mathfrak{r}_t, \quad \beta_t = \{\mathfrak{e}_t - \tfrac{\mathfrak{r}_t^2}{2}\}^{-1}$$

and the functions F and g are judiciously chosen. The choice is guided by the fluctuation-dissipation relation (45) and done in order to obtain the first order "Taylor expansion" (71) below.

Let $\Omega^N = (\mathbb{R} \times \mathbb{R})^{\mathbb{T}_N}$ be the configurations space and denote

$$\hat{H}_N(t) := H\left(\mu_t^N | \hat{\mu}_t^N\right) = \int_{\Omega^N} f_t^N(\omega) \log \frac{f_t^N(\omega)}{\phi_t^N(\omega)} d\nu_*(\omega), \quad (70)$$

where f_t^N is the density of μ_t^N with respect to the Gibbs reference measure $\nu_* := \nu_{1,0}$. In the same way, ϕ_t^N is the density of $\hat{\mu}_t^N$ with respect to ν_* (which is fully explicit). The goal is to get (68) with H_N replaced by $\hat{H}_N$.

We begin with the following entropy production bound. Let us denote by $\mathscr{L}_N^* = -\mathscr{A}_N + \gamma \mathscr{S}_N$ the adjoint of $\mathscr{L}_N$ in $\mathbb{L}^2(\nu_*)$.

Lemma 3

$$\partial_t \hat{H}_N(t) \le \int \frac{1}{\phi_t^N}\left(N^2 \mathscr{L}_N^* \phi_t^N - \partial_t \phi_t^N\right) f_t^N d\nu_* = \int \left[\frac{1}{\phi_t^N}\left(N^2 \mathscr{L}_N^* \phi_t^N - \partial_t \phi_t^N\right)\right] d\mu_t^N .$$

Proof We have that f_t^N solves the Fokker-Plack equation $\partial_t f_t^N = N^2 \mathscr{L}_N^* f_t^N$. Assuming it is smooth to simplify, we have

$$\partial_t \hat{H}_N(t) = \int \partial_t f_t^N [1 + \log f_t^N] d\nu_* - \int \partial_t f_t^N \log \phi_t^N d\nu_* - \int \partial_t \phi_t^N \frac{f_t^N}{\phi_t^N} d\nu_*$$
$$= N^2 \int \mathscr{L}_N^* f_t^N [\log f_t^N - \log \phi_t^N] d\nu_* - \int \partial_t \phi_t^N \frac{f_t^N}{\phi_t^N} d\nu_*$$

[18] Observe that if a system is gradient then a microscopic fluctuation-dissipation Eq. (23) holds with a zero fluctuating term.

$$= N^2 \int f_t^N \mathscr{L}_N [\log \frac{f_t^N}{\phi_t^N}] d\nu_* - \int \partial_t \phi_t^N \frac{f_t^N}{\phi_t^N} d\nu_*$$

$$= N^2 \int \frac{f_t^N}{\phi_t^N} \mathscr{L}_N [\log \frac{f_t^N}{\phi_t^N}] \phi_t^N d\nu_* - \int \partial_t \phi_t^N \frac{f_t^N}{\phi_t^N} d\nu_*$$

$$\leq N^2 \int \mathscr{L}_N [\frac{f_t^N}{\phi_t^N}] \phi_t^N d\nu_* - \int \partial_t \phi_t^N \frac{f_t^N}{\phi_t^N} d\nu_*$$

$$= N^2 \int \frac{f_t^N}{\phi_t^N} \mathscr{L}_N^* \phi_t^N d\nu_* - \int \partial_t \phi_t^N \frac{f_t^N}{\phi_t^N} d\nu_*$$

where we used that for any positive function h, $h\mathscr{L}_N(\log h) \leq \mathscr{L}_N h$ (this is a consequence of Jensen's inequality). □

We define $\xi_x := (\mathscr{E}_x, r_x)$ and $\pi(t,q) := (\mathfrak{e}(t,q), \mathfrak{r}(t,q))$. If f is a vectorial function, we denote its differential by Df.

Proposition 3 ([54]) *Let* (λ, β) *be defined by* $\beta = (\mathfrak{e} - \frac{\mathfrak{r}^2}{2})^{-1}$ *and* $\lambda = -\beta\mathfrak{r}$. *The term* $(\phi_t^N)^{-1}\left(N^2 \mathscr{L}_N^* \phi_t^N - \partial_t \phi_t^N\right)$ *can be expanded as*

$$(\phi_t^N)^{-1}\left(N^2 \mathscr{L}_N^* \phi_t^N - \partial_t \phi_t^N\right)$$
$$= \sum_{k=1}^{5} \sum_{x \in \mathbb{T}_N} v_k\left(t, \frac{x}{N}\right)\left[J_x^k - H_k\left(\pi\left(t, \frac{x}{N}\right)\right) - (DH_k)\left(\pi\left(t, \frac{x}{N}\right)\right)\right.$$
$$\left. \times \left(\xi_x - \pi\left(t, \frac{x}{N}\right)\right)\right] + o(N) \qquad (71)$$

where

k	J_x^k	$H_k(\mathfrak{e}, \mathfrak{r})$	$v_k(t, q)$
1	$p_x^2 + r_x r_{x-1} + 2\gamma p_x r_{x-1}$	$\mathfrak{e} + \mathfrak{r}^2/2$	$-(2\gamma)^{-1}\partial_q^2 \beta(t,q)$
2	$r_x + \gamma p_x$	$\mathfrak{r}$	$-\gamma^{-1}\partial_q^2 \lambda(t,q)$
3	$p_x^2\,(r_x + r_{x-1})^2$	$(2\mathfrak{e} - \mathfrak{r}^2)\,(\mathfrak{e} + 3\mathfrak{r}^2/2)$	$(4\gamma)^{-1}[\partial_q \beta(t,q)]^2$
4	$p_x^2\,(r_x + r_{x-1})$	$\mathfrak{r}\,(2\mathfrak{e} - \mathfrak{r}^2)$	$\gamma^{-1}\partial_q \beta(t,q)\,\partial_q \lambda(t,q)$
5	p_x^2	$\mathfrak{e} - \mathfrak{r}^2/2$	$\gamma^{-1}[\partial_q \lambda(t,q)]^2$

Observe that $H_k(e, r)$ is equal to $\int J_x^k d\nu_{\beta,\tau}$ where β, τ are related to e, r by the thermodynamic relations. Thus, the terms appearing in the righthand side of (71) can be seen as first order "Taylor expansion". The form of the first order correction in (69) plays a crucial role in order to get such expansions.

A priori the first term on the right-hand side of (71) is of order N, but we want to take advantage of these microscopic Taylor expansions to show it is in fact of order $o(N)$.

First, we need to cut-off large energies in order to work with bounded variables only. To simplify, we assume they are bounded ab initio.

Let ℓ be some integer (dividing N). We introduce some averaging over microscopic blocks of size ℓ and we will let $\ell \to \infty$ after $N \to \infty$. We decompose $\mathbb{T}_N$ in a disjoint union of $p = N/\ell$ boxes $\Lambda_\ell(x_j)$ of length ℓ centered at $x_j, j \in \{1, \ldots, p\}$. The microscopic averaged profiles in a box of size ℓ around $y \in \mathbb{T}_N$ are defined by

$$\tilde{\xi}_\ell(y) = \frac{1}{\ell} \sum_{x \in \Lambda_\ell(y)} \xi_x.$$

Similarly we define

$$\tilde{J}_\ell^k(y) = \frac{1}{\ell} \sum_{x \in \Lambda_\ell(y)} J_x^k.$$

In (71) we rewrite the sum $\sum_{x \in \mathbb{T}_N}$ as $\sum_{j=1}^p \sum_{x \in \Lambda_\ell(x_j)}$ and, by using the smoothness of the function v_k, H_k, it is easy to replace the term

$$\frac{1}{N} \sum_{x \in \mathbb{T}_N} v_k\left(t, \frac{x}{N}\right) \left[J_x^k - H_k\left(\pi\left(t, \frac{x}{N}\right)\right) - (DH_k)\left(\pi\left(t, \frac{x}{N}\right)\right) \cdot \left(\xi_x - \pi\left(t, \frac{x}{N}\right)\right)\right]$$

by

$$\begin{aligned} &\frac{1}{p}\sum_{j=1}^{p} v_k\left(t, \frac{x_j}{N}\right) \left[\tilde{J}_\ell^k(x_j) - H_k\left(\pi\left(t, \frac{x_j}{N}\right)\right) - (DH_k)\left(\pi\left(t, \frac{x_j}{N}\right)\right)\right. \\ &\times \left.\left(\tilde{\xi}_\ell(x_j) - \pi\left(t, \frac{x_j}{N}\right)\right)\right] \end{aligned}$$

in the limit $N, \ell \to \infty$ with some error term of order $o(1)$.

Then, the strategy consists in proving the following crucial estimate, often called the *one-block estimate*: we replace the empirical average current $\tilde{J}_\ell^k(x_j)$ which is averaged over a box centered at x_j by its mean with respect to a Gibbs measure with the parameters corresponding to the microscopic averaged profiles $\tilde{\xi}_\ell(x_j)$, i.e. $H_k(\tilde{\xi}_\ell(x_j))$. This non-trivial step is achieved thanks to some compactness argument and the macro-ergodicity of the dynamics.

Consequently we have to deal with terms in the form

$$\begin{aligned} \frac{1}{p}\sum_{j=1}^{p} v_k\left(t, \frac{x_j}{N}\right) &\left[H_k\left(\tilde{\xi}_\ell(x_j)\right) - H_k\left(\pi\left(t, \frac{x_j}{N}\right)\right)\right. \\ &\left. - (DH_k)\left(\pi\left(t, \frac{x_j}{N}\right)\right) \cdot \left(\tilde{\xi}_\ell(x_j) - \pi\left(t, \frac{x_j}{N}\right)\right)\right] \end{aligned} \tag{72}$$

The final step consists then in applying the entropy inequality (66) with respect to $\hat{\mu}_t^N$ with $\phi := \phi_{\ell,N}$ given by (72) and $\alpha = \delta N$, $\delta > 0$ fixed but small. This will produce some term of order $\hat{H}_N(t)/N$ plus the term

$$\limsup_{\ell\to\infty}\limsup_{N\to\infty}\frac{1}{\delta N}\log\left(\int e^{\delta N\phi}d\hat{\mu}_t^N\right) = I(\delta).$$

By using some large deviations estimates (observe that $\hat{\mu}_t^N$ is explicit and product at first order in N) one can show that $I(\delta)$ is nonpositive for δ sufficiently small. Thus we get the desired Gronwall inequality.

There is some additional difficulty that we hid under the carpet in the sketch of the proof. Since the state space is non compact, a control of high energies is required for the initial cut-off. This is a highly non trivial problem.[19] In the harmonic case considered here this control is obtained thanks to the following remark: the set of mixtures of Gaussian probability measures[20] is preserved by the (harmonic) velocity-flip model. Since for Gaussian measures all the moments are expressed in terms of the covariance matrix, required bounds can be obtained by a suitable control of the covariance matrices appearing in the mixture.

The extension of this result in the anharmonic case is a challenging open problem (see however [52] where equilibrium fluctuations are considered for an anharmonic chain perturbed by a conservative noise acting on the momenta and positions).

2.2.2 Fourier's Law

Since in the harmonic case an exact fluctuation-dissipation equation is available Fourier's law can be obtained without too much work.[21]

Theorem 7 ([10, 11]) *Consider the one-dimensional harmonic chain in contact with two heat baths and with forced boundary conditions as in Sect. 2.1.1. Then Fourier's law holds:*

$$\tilde{J}_s := \lim_{N\to\infty} N\langle j_{0,1}^e\rangle_{ss} = \frac{1}{2\gamma}\left\{(T_\ell - T_r) + (\tau_\ell^2 - \tau_r^2)\right\} \tag{73}$$

[19] A similar problem appears in [51] where the authors derived Euler equations for a gas perturbed by some ergodic noise. There, to overcome this difficulty, the authors replace ab initio the kinetic energy by the relativistic kinetic energy.

[20] A Gibbs local equilibrium state is a Gaussian state in the harmonic case.

[21] The *a posteriori* simple but fundamental remark that an exact fluctuation-dissipation equation exists for the harmonic model (see (45)) is the real contribution of [10].

and we have

$$\hat{J}_\ell = \lim_{N\to\infty} N(\langle p_1^2\rangle_{ss} - T_\ell) = \frac{1}{2\gamma\gamma_\ell}\Big[(T_r - T_\ell) + (\tau_\ell - \tau_r)^2\Big],$$
$$\hat{J}_r = \lim_{N\to\infty} N(T_r - \langle p_N^2\rangle_{ss}) = \frac{1}{2\gamma\gamma_r}\Big[(T_r - T_\ell) - (\tau_\ell - \tau_r)^2\Big]. \qquad (74)$$

Proof We divide the proof in two steps:

- We first prove that there exists a constant C independent of N such that $|\langle j^e_{0,1}\rangle_{ss}| \leq C/N$. This is obtained by using the fluctuation-dissipation equation and the fact that $\langle j^e_{x,x+1}\rangle_{ss}$ is independent of x:

$$\begin{aligned}\langle j^e_{0,1}\rangle_{ss} &= \frac{1}{N-3}\sum_{x=2}^{N-2}\langle j^e_{x,x+1}\rangle_{ss} \qquad (75)\\ &= -\frac{1}{2\gamma}\frac{1}{N-3}\sum_{x=2}^{N-2}\Big\langle\nabla\Big[p_x^2 + r_x r_{x+1}\Big]\Big\rangle_{ss}\\ &= \frac{1}{2\gamma}\frac{1}{N-3}\Big\{(\langle p_2^2\rangle_{ss} + \langle r_2 r_3\rangle_{ss}) - (\langle p_{N-1}^2\rangle_{ss} + \langle r_{N-1}r_N\rangle_{ss})\Big\}.\end{aligned}$$

  By using simple computations, one can show that $(\langle p_2^2\rangle_{ss} + \langle r_2r_3\rangle_{ss}) - (\langle p_{N-1}^2\rangle_{ss} + \langle r_{N-1}r_N\rangle_{ss})$ is uniformly bounded in N by a positive constant.
- Now we have only to evaluate the limit of each term appearing in $(\langle p_2^2\rangle_{ss} + \langle r_2r_3\rangle_{ss}) - (\langle p_{N-1}^2\rangle_{ss} + \langle r_{N-1}r_N\rangle_{ss})$. Notice that assuming local equilibrium we easily get the result. The first step implies that $\langle j^e_{0,1}\rangle_{ss}$ and $\langle j^e_{N,N+1}\rangle_{ss}$ vanish as $N \to +\infty$. Since V_s goes to 0 by Lemma 1, one has that $\langle p_1^2\rangle_{ss}$ and $\langle p_N^2\rangle_{ss}$ converge respectively to T_ℓ and T_r. By using some "entropy production bound" one can propagate this local equilibrium information to the particles close to the boundaries and show (73). □

It follows from this Theorem that the system can be used as a heater but not as a refrigerator. Assume for example that $T_r > T_\ell$. The term $\hat{J}_\ell$ (resp. $\hat{J}_r$) is the macroscopic heat current from the left reservoir to the system (resp. from the system to the right reservoir). Whatever the values of τ_ℓ, τ_r are, $\hat{J}_\ell > 0$ and we can not realize a refrigerator. But if $(T_r - T_\ell) < (\tau_r - \tau_\ell)^2$ then $\hat{J}_r < 0$ and we realized a heater.

The proof of the validity of Fourier's law for anharmonic chains perturbed by an energy conserving noise is still open.

2.2.3 Macroscopic Fluctuation Theory for the Energy Conserving Harmonic Chain

The macroscopic fluctuation theory [17] is a general approach developed by Bertini, De Sole, Gabrielli, Jona-Lasinio and Landim to calculate the large deviation functional of the empirical profiles of the conserved quantities of Markov processes in a NESS. Its main interest is that it can be applied to a large class of boundary driven diffusive systems and does not require the explicit form of the NESS but only the knowledge of two thermodynamic macroscopic parameters of the system, the diffusion coefficient $D(\rho)$ and the mobility $\chi(\rho)$. This theory can be seen as an infinite dimensional generalization of the Freidlin-Wentzel theory [28] and is based on the large deviation principle for the hydrodynamics of the system.

In order to explain (roughly) the theory we consider for simplicity a Markovian system $\{\eta(t) := \{\eta_x(t) \in \mathbb{R}\,;\, x \in \{1, \dots N\}\}_{t\geq 0}$ with only one conserved quantity, say the density ρ, in contact with two reservoirs at each extremity. Here N is the size of the system which will be sent to infinity. We denote by μ_{ss}^N the nonequilibrium stationary state of $\{\eta(t)\}_{t\geq 0}$. For any microscopic configuration $\eta := \{\eta_x\,;\, x \in \{1, \dots, N\}\}$ let

$$\pi^N(\eta, \cdot) = \sum_{x=1}^{N-1} \eta_x \mathbf{1}_{\left[\frac{x}{N}, \frac{x+1}{N}\right)}(\cdot)$$

be the empirical density profile. In the diffusive time scale, we assume that $\pi^N(\eta(tN^2), \cdot)$ converges as N goes to infinity to $\rho_t(\cdot) := \rho(t, \cdot)$ solution of

$$\begin{cases} \partial_t \rho = \partial_y (D(\rho)\partial_y \rho), & y \in [0,1], \quad t \geq 0, \\ \rho(t, 0) = \rho_\ell, \quad \rho(t,1) = \rho_r, & t \geq 0, \\ \rho(0, \cdot) = \rho_0(\cdot) \end{cases}$$

where $\rho_0(\cdot)$ is the initial density profile, $D(\rho) > 0$ is the diffusion coefficient and ρ_ℓ, ρ_r the densities fixed by the reservoirs. As $t \to \infty$ the solution ρ_t of the hydrodynamic equation converges to a stationary profile $\bar\rho : [0,1] \to \mathbb{R}$ solution of $D(\bar\rho)\partial_y \bar\rho = J = const.$ with $\bar\rho(0) = \rho_\ell$, $\bar\rho(1) = \rho_r$. We assume that under μ_{ss}^N, the empirical density profile $\pi^N(\eta, \cdot)$ converges to $\bar\rho$. This assumption is nothing but a law of large numbers for the random variables π^N.

We are here interested in the corresponding large deviation principle. Thus, we want to estimate the probability that in the NESS μ_{ss}^N the empirical density profile π^N is close to an atypical macroscopic profile $\rho(\cdot) \neq \bar\rho$. This probability typically is of order $e^{-N\mathbb{V}(\rho)}$ where $\mathbb{V}$ is the rate function:

$$\mu_{ss}^N(\pi^N(\eta, \cdot) \approx \rho(\cdot)) \approx e^{-N\mathbb{V}(\rho)}.$$

The goal of the macroscopic fluctuation theory is to obtain information about this functional.

The condition to be fulfilled by the system to apply the theory of Bertini et al. is that it satisfies a *dynamical large deviation principle* with a rate function which takes a quadratic form[22] like (77).

Let us first explain what we mean by dynamical large deviation principle. Imagine we start the system from a Gibbs local equilibrium state corresponding to the macroscopic profile ρ_0. We want to estimate the probability that the empirical density $\pi^N(\eta(tN^2), \cdot)$ is close during the macroscopic time interval $[0, T]$, T fixed, to a smooth macroscopic profile $\gamma(t, y)$ supposed to satisfy[23] $\gamma(0, \cdot) = \rho_0$. This probability is exponentially small in N with a rate $I_{[0,T]}(\gamma \mid \rho_0)$

$$\mathbb{P}\Big[\pi^N(\eta(tN^2), y) \approx \gamma(t, y),\ (t, y) \in [0, T] \times [0, 1]\Big] \sim e^{-NI_{[0,T]}(\gamma|\rho_0)}. \tag{76}$$

The rate function is assumed to be of the form

$$I_{[0,T]}(\gamma \mid \rho_0) = \frac{1}{2}\int_0^T dt \int_0^1 dy\, \chi(\rho(t, y)) \left[(\partial_y H)(t, y)\right]^2 \tag{77}$$

where $\partial_y H$ is the extra gradient external field needed to produce the fluctuation γ, namely such that

$$\partial_t \gamma = \partial_y \left[D(\gamma)\partial_y \gamma \ - \ \chi(\gamma)\partial_y H\right]. \tag{78}$$

Thus, $I_{[0,T]}(\gamma|\rho_0)$ is the work done by the external field $\partial_y H$ to produce the fluctuation γ in the time interval $[0, T]$. The function χ appearing in (78) is the second thermodynamic parameter (with the diffusion coefficient D) mentioned in the beginning of this section. The two parameters D and χ are in fact related together by the Einstein relation so that knowing one of them and the Gibbs states of the microscopic model is sufficient to obtain the second.

To show this result the strategy is the following. We perturb the Markov process $\{\eta(t)\}_{t\geq 0}$ thanks to the function $H := H(\gamma)$, which is solution of the Poisson Eq. (78), by adding locally a small space inhomogeneous drift provided by $\partial_y H$. In doing so we obtain a new Markov process $\{\eta^H(t)\}_{t\geq 0}$ such that in the diffusive time scale $\pi^N(\eta^H(tN^2), \cdot)$ converges to $\gamma(\cdot)$. Let $\mathbb{P}^H$ (resp. $\mathbb{P}^0$) be the probability measure on the empirical density paths space induced by $\{\eta^H(tN^2)\}_{t\in[0,T]}$ (resp. $\{\eta(tN^2)\}_{t\in[0,T]}$). Then, by using hydrodynamic limits techniques similar to the ones explained in Sect. 2.2.1 we show that in the large N limit, under $\mathbb{P}^H$, the Radon-Nikodym derivative is well approximated by[24]

$$\frac{d\mathbb{P}^0}{d\mathbb{P}^H}(\pi) \approx \exp\left\{-NI_{[0,T]}(\pi|\rho_0)\right\}.$$

[22] Such property has been proved to be valid for a large class of stochastic dynamics [41, 42].

[23] This assumption avoids taking into account the cost to produce the initial profile, cost which is irrelevant for us.

[24] We use Girsanov transform to express the Radon-Nikodym derivative. A priori it is not a functional of the empirical density and we need to establish some *replacement lemma* (see [41]).

Here $\pi := \{\pi(t,y)\,;\, t \in [0,T], y \in [0,1]\}$ is any space-time density profile. Thus, since

$$\mathbb{P}^0\left[\pi^N(\eta(tN^2),\cdot) \sim \gamma(t,\cdot),\ t \in [0,T]\right] = \mathbb{E}^H\left[\frac{\mathrm{d}\mathbb{P}^0}{\mathrm{d}\mathbb{P}^H}(\pi)\,\mathbf{1}_{\{\pi(t,\cdot)\sim\gamma(t,\cdot),t\in[0,T]\}}\right]$$

we obtain (76).

The macroscopic fluctuation theory claims that the large deviations functional $\mathbb{V}(\rho)$ of the empirical density in the NESS coincides with the quasi-potential $\mathbb{W}(\rho)$ defined by

$$\mathbb{W}(\rho) = \inf_{\substack{\gamma:\gamma(-\infty)=\bar{\rho}\\ \gamma(0)=\rho}} I_{[-\infty,0]}(\gamma|\bar{\rho}).$$

Here $I_{[-\infty,0]}$ is obtained from $I_{[0,T]}$ by a shift in time by $-T$, T being sent to $+\infty$ afterwards. In words, the quasi potential determines the cost to produce a fluctuation equal to γ at $t = 0$ when the system is macroscopically in the stationary profile $\bar{\rho}$ at $t = -\infty$.

Thus, the problem is reduced to computing W. It can be shown that W solves (at least formally) the infinite-dimensional Hamilton-Jacobi equation

$$\frac{1}{2}\left\langle \partial_y\left[\frac{\delta\mathbb{W}}{\delta\rho}\right], \chi(\rho)\,\partial_y\left[\frac{\delta\mathbb{W}}{\delta\rho}\right]\right\rangle + \left\langle\frac{\delta\mathbb{W}}{\delta\rho},\ \partial_y\left[D(\rho)\partial_y\rho\right]\right\rangle = 0 \tag{79}$$

where $\langle\cdot,\cdot\rangle$ denotes the usual scalar product in $\mathbb{L}^2([0,1])$. Note that there is no uniqueness of solutions ($\mathbb{W} = 0$ is a solution) and up to now a general theory of infinite dimensional Hamilton-Jacobi equations is still missing. This implies that we have in fact to solve by hand the variational problem and the solution is only known for few systems. This is an important limitation of the macroscopic fluctuation theory. Even getting interesting qualitative properties on $\mathbb{W}$ is difficult.

The rigorous implementation of this long program has only been carried for the boundary driven Symmetric Simple Exclusion Process and extended with less rigor to a few other systems (see [15, 18, 27] for rigorous results).

Let us now try to apply this theory for the harmonic chain with velocity-flip noise. Since we have a fully explicit microscopic fluctuation-dissipation equation (even when some harmonic pinning is added) we can easily guess what is the form of the hydrodynamic equations under various boundary conditions by assuming that the propagation of local equilibrium in the diffusive time scale holds. Nevertheless, let us observe that a rigorous derivation is missing, the obstacle being a sufficiently good control of the high energies.[25] The boundary conditions we impose to the system are the following. At the left (resp. right) end we put the chain in contact with a Langevin bath at temperature T_ℓ (resp. T_r) and consider the system with fixed boundary conditions or with forced boundary conditions with the same force τ at the two boundaries. Then, for the unpinned chain, the equations (64) are still valid but

[25] This control is only available in the case of periodic boundary conditions [54].

they are supplemented with the boundary conditions [14]

$$\left[\mathfrak{e} - \frac{\mathfrak{r}^2}{2}\right](t,0) = T_\ell, \quad \left[\mathfrak{e} - \frac{\mathfrak{r}^2}{2}\right](t,1) = T_r, \tag{80}$$

since the Langevin baths fix the temperatures at the boundaries and

$$\partial_y \mathfrak{r}(t,0) = \partial_y \mathfrak{r}(t,1) = 0 \tag{81}$$

for fixed boundary conditions (the total length of the chain is constant[26]) and

$$\mathfrak{r}(t,0) = \mathfrak{r}(t,1) = \tau \tag{82}$$

for forced boundary conditions.

If the chain is pinned by the harmonic potential $W(q) = \nu q^2/2$ then only the energy is conserved and the macroscopic diffusion equation takes the form

$$\begin{cases} \partial_t \mathfrak{e} = \partial_y(\kappa \partial_y \mathfrak{e}), \\ \mathfrak{e}(0,y) = \mathfrak{e}_0(y), & y \in (0,1) \\ \mathfrak{e}(t,0) = T_\ell, \ \mathfrak{e}(t,1) = T_r, \end{cases} \tag{83}$$

where the conductivity κ is equal to [14]

$$\kappa = \frac{1/\gamma}{2 + \nu^2 + \sqrt{\nu(\nu+4)}}. \tag{84}$$

Assuming a good control of high energies, it is possible to derive the dynamical large deviations function of the empirical conserved quantities. The goal would be to compute the large deviation functional of the NESS which according to the macroscopic fluctuation theory coincides with the quasi potential. We recall that the quasi potential is defined by a variational problem and that it depends only on two thermodynamic quantities, the diffusion coefficient and the mobility (the latter are matrices if several conserved quantities are involved).

Let us first consider the pinned velocity flip model where the energy is the only conserved quantity. It turns out that the mobility is a quadratic function. Consequently, the methods exposed in Theorem 6.5 of [16] apply and the variational formula can be computed. The quasi potential $\mathbb{V}(\cdot)$ is given by [14]

$$\mathbb{V}(e) = \int_0^1 dq \left[\frac{e(q)}{F(q)} - 1 - \log\left(\frac{e(q)}{F(q)}\right) - \log\left(\frac{F'(q)}{T_r - T_\ell}\right)\right], \tag{85}$$

---

[26] Indeed, by (64), we have $\partial_t(\int_0^1 \mathfrak{r}(t,y)dy) = \gamma^{-1}\int_0^1 \partial_y^2 \mathfrak{r}(t,y)dy = \gamma^{-1}[\partial_y \mathfrak{r}(t,1) - \partial_y \mathfrak{r}(t,0)] = 0$.

where F is the unique non decreasing solution of

$$\begin{cases} F'' = \dfrac{F - e}{F^2}(F')^2, \\ F(0) = T_\ell, \ F(1) = T_r. \end{cases} \tag{86}$$

Surprisingly, the function $\mathbb{V}$ is independent of the pinning value ν and of the intensity of the noise γ. It is thus natural to conjecture that in the NESS of the unpinned velocity flip model the large deviation function of the empirical energy profile coincides with $\mathbb{V}$ but we did not succeed to prove it. Observe that at equilibrium ($T_\ell = T_r$), $F(q) = T_\ell = T_r$ and the last term in (85) disappears so that the quasi potential is local. On the other hand, if $T_\ell \neq T_r$, this is no longer the case and this reflects the presence of long-range correlations in the NESS. In particular, an approximation of the NESS by a Gibbs local equilibrium state in the form (29) would not give the correct value of the quasi potential.

For the unpinned chain we have two conserved quantities. Solving the variational problem of the quasi potential for these two conserved quantities is a very difficult open problem[27] (see [5] for a partial result for some other stochastic perturbation of the harmonic chain).

3 Anomalous Diffusion

An anomalous large conductivity is observed experimentally in carbon nanotubes and numerically in chains of oscillators without pinning, where numerical evidence shows a conductivity diverging with the size N of the system like N^α, with $\alpha < 1$ in dimension $d = 1$, and like $\log N$ in dimension $d = 2$. If some nonlinearity is present in the interaction, finite conductivity is observed numerically in all pinned case or in dimension $d \geq 3$ [24, 47]. Consequently it has been suggested that conservation of momentum is an important ingredient for the anomalous conductivity in low dimensions (see however [66]).

In chapter "Large Deviations in a Gaussian Setting: The Role of the Cameron-Martin Space" we considered chains of oscillators perturbed by a noise conserving only energy and destroying the possible momentum conservation. In the harmonic case we obtained Fourier's law and in the anharmonic case we proved existence of the Green-Kubo formula for the thermal conductivity.

In this chapter the added perturbation conserves both energy and momentum (energy and volume for the Hamiltonian systems considered in Sect. 1.2). These systems qualitatively have the same behavior as Hamiltonian chains of oscillators (without any noise), i.e. anomalous transport for unpinned chain in dimension $d = 1, 2$ and normal transport otherwise. We could even be more optimistic and hope that

[27] Here we do not have any exactly solvable model like the Symmetric Simple Exclusion Process which could give us some hints for the form of the quasi-potential.

they share with the deterministic systems common limits for the energy fluctuation fields, two point correlation functions ... This is because one expects that the microscopic details of the dynamics are irrelevant. Therefore some *universality* should hold. Recently Spohn [58], by following ideas of [64], used the nonlinear fluctuating hydrodynamics theory to classify very precisely the different expected universality classes. The nonlinear fluctuating hydrodynamics theory is based on the assumption that the microscopic dynamics evolve in the Euler time scale according to a system of conservation laws. The theory is macroscopic in the sense that all the predictions are done starting from this system of conservation laws without further references to the microscopic dynamics. Since we have seen that the presence of the energy-momentum conserving noise does not change the form of the hydrodynamic equations, the theory claims in fact that the limit of the fluctuations fields of the conserved quantities for purely deterministic chains of oscillators and for noisy energy-momentum conserving chains are exactly the same.

3.1 Harmonic Chains with Momentum Exchange Noise

Getting some information on the behavior of the energy fluctuation field in the large scale limit remains challenging. So far, satisfactory but not complete results have only been obtained in the harmonic case. The anharmonic case is much more difficult.

In [1, 2] we explicitly compute the time correlation current for a system of harmonic oscillators perturbed by an energy-momentum conserving noise[28] and we find that it behaves, for large times, like $t^{-d/2}$ in the unpinned cases, and like $t^{-d/2-1}$ when on-site harmonic potential is present.

These results are given in the Green-Kubo formalism. Their counterpart in the NESS formalism have been considered in [46] but a rigorous proof is still missing. Several variations of the Green-Kubo formula can be found in the literature: one can start with the infinite system in the canonical ensemble, as we did in Sect. 2.1.2, or with a finite system, in the canonical or micro-canonical ensembles, sending the size of the system to infinity. It is widely believed that all these definitions coincide (also in the case of infinite conductivity). As shown in [2], this is essentially true for the energy-momentum conserving harmonic chain. Here we consider the simplest possible definition avoiding to discuss the rigorous definition of the canonical ensemble in infinite volume and the problem of equivalence of ensembles.

The set-up is the following. We consider a chain perturbed by the energy-momentum conserving noise (see (11)) with periodic boundary conditions. Its Hamiltonian is given by $\mathscr{H}_N = \sum_{x \in \mathbb{T}_N^d} \mathscr{E}_x$ where the energy $\mathscr{E}_x$ of atom x is

$$\mathscr{E}_x = \frac{|p_x|^2}{2} + W(q_x) + \frac{1}{2} \sum_{|y-x|=1} V(q_x - q_y). \tag{87}$$

[28] It is straightforward to adapt the proofs given in [1, 2] to the case of the momenta exchange noise.

The system is considered at equilibrium under the Gibbs grand-canonical measure

$$d\mu_{N,T} = \frac{e^{-\mathscr{H}_N/T}}{Z_{N,T}} d\mathbf{q}d\mathbf{p}$$

where $Z_{N,T}$ is the renormalization constant.

The Green-Kubo formula for the thermal conductivity in the direction e_k, $1 \leq k \leq d$, is[29] the limiting variance of the energy current $J^{e,\gamma}_{x,x+e_k}([0,t])$ up to time t in the direction e_k in a space-time box of size $N \times t$:

$$\kappa(T) = \frac{1}{2T^2} \lim_{t\to+\infty} \lim_{N\to\infty} \mathbb{E}_{\mu_{N,T}} \left[\left(\frac{1}{\sqrt{N^d t}} \sum_{x\in\mathbb{T}_N^d} J^{e,\gamma}_{x,x+e_1}([0,t]) \right)^2 \right]. \tag{88}$$

The energy currents $\{J^{e,\gamma}_{x,x+e_k}([0,t])\,;\, k = 1,\ldots,d\}$ are defined by the energy conservation law

$$\mathscr{E}_x(t) - \mathscr{E}_x(0) = \sum_{k=1}^d \left(J^{e,\gamma}_{x-e_k,x}([0,t]) - J^{e,\gamma}_{x,x+e_k}([0,t]) \right).$$

The energy current up to time t can be written as

$$J^{e,\gamma}_{x,x+e_k}([0,t]) = \int_0^t j^{e,\gamma}_{x,x+e_k}(s)ds + M_{x,x+e_k}(t) \tag{89}$$

where $M_{x,x+e_k}(t)$ is a martingale and $j^{e,\gamma}_{x,x+e_k}$ is the instantaneous current which has the form

$$j^{e,\gamma}_{x,x+e_k} = \tilde{j}^e_{x,x+e_k} + \gamma \left[p^2_{x+e_k} - p^2_x \right], \quad \tilde{j}^e_{x,x+e_k} = -\frac{1}{2} V'(q_{x+e_k} - q_x)(p_{x+e_k} + p_x). \tag{90}$$

The term $\tilde{j}^e_{x,x+e_k}$ is the Hamiltonian contribution while the gradient term is due to the noise.

We now expand the square in (88). Notice first that since we have periodic boundary conditions the gradient term appearing in (90) does not contribute. By a time reversal argument one can show that the cross term between the martingale and the time integral of the instantaneous current vanishes. Moreover a simple computation shows that the square of the martingale term gives a contribution equal to γ (see [2] for details). Thus we obtain

---

[29] By symmetry arguments this is independent of k.

$$\kappa(T) = T^{-2} \lim_{t\to+\infty} \lim_{N\to\infty} \frac{1}{2N^d t} \mathbb{E}_{\mu_{N,T}} \left[\left(\sum_{x\in\mathbb{T}_N^d} \int_0^t \tilde{j}^e_{x,x+e_k}(s) ds \right)^2 \right] + \gamma$$

$$= T^{-2} \lim_{t\to+\infty} \lim_{N\to\infty} \sum_{x\in\mathbb{T}_N^d} \int_0^{+\infty} ds \left(1 - \frac{s}{t}\right)^+ \mathbb{E}_{\mu_{N,T}} \left[\tilde{j}^e_{0,e_k}(0) \tilde{j}^e_{x,x+e_k}(s) \right] ds + \gamma \tag{91}$$

where the last line is obtained by time and space stationarity of the Gibbs measure and u^+ denotes $\max(u, 0)$.[30] It is then clear that the divergence of the Green-Kubo formula, *i.e.* anomalous transport, is due to a slow decay of the time correlation function $C(t)$ defined by

$$C(t) = \lim_{N\to\infty} \sum_{x\in\mathbb{T}_N^d} \mathbb{E}_{\mu_{N,T}} \left[\tilde{j}^e_{0,e_k}(0) \tilde{j}^e_{x,x+e_k}(t) \right]. \tag{92}$$

Theorem 8 ([2]) *Consider the harmonic case: $V(r) = \alpha r^2$, $W(q) = \nu q^2$ where $\alpha > 0$ and $\nu \geq 0$.*
Then the limit defining $C(t)$ in (92) exists and can be computed explicitly. In particular, we have that $C(t) \sim t^{-d/2}$ if $\nu = 0$ and $C(t) \sim t^{-d/2-1}$ if $\nu > 0$.
Consequently, the limit (91) exists in $(0, +\infty]$ and is finite if and only if $d \geq 3$ or $\nu > 0$. When finite, $\kappa(T)$ is independent of T and can be computed explicitly.

Proof We compute the Laplace transform $L_N(z) = \int_0^{+\infty} e^{-zt} C_N(t) dt$, $z > 0$, of $C_N(t) = \sum_{x\in\mathbb{T}_N^d} \mathbb{E}_{\mu_{N,T}} \left[\tilde{j}_{0,e_k}(0) \tilde{j}^e_{x,x+e_k}(t) \right]$. Since we have

$$L_N(z) = N^{-1} \mu_{N,T} \left[\left(\sum_{x\in\mathbb{T}_N^d} \tilde{j}^e_{x,x+,e_k} \right) (z - \mathscr{L}_N)^{-1} \left(\sum_{x\in\mathbb{T}_N^d} \tilde{j}^e_{x,x+,e_k} \right) \right]$$

it is equivalent to solve the resolvent equation $(z - \mathscr{L}_N) h_N = \sum_{x\in\mathbb{T}_N^d} \tilde{j}^e_{x,x+e_k}$. Notice that $\mathscr{L}_N$ maps polynomial functions of degree 2 into polynomial functions of degree 2 and that $\sum_x \tilde{j}^e_{x,x+e_k}$ is a polynomial function of degree 2. Thus, the function h_N is a polynomial function of degree 2. Moreover it has to be space translation invariant since $\sum_x \tilde{j}^e_{x,x+e_k}$ is. Therefore we can look for a function h_N of the form

$$h_N = \sum_{x,y} a(y - x) p_x p_y + \sum_{x,y} b(y - x) p_x q_y + \sum_{x,y} c(y - x) q_x q_y$$

[30] Observe that replacing $(1 - \frac{s}{t})^+$ by $e^{-s/t}$ and $\lim_{N\to\infty} \sum_{x\in\mathbb{T}_N^d}$ by $\sum_{x\in\mathbb{Z}^d}$ we formally get an expression similar to the Green-Kubo formula of Theorem 5.

where a, b and c are functions from $\mathbb{T}_N^d$ into $\mathbb{R}$. We compute explicitly a, b and c and we get $a = c = 0$ while b is the solution to

$$(z + 2\nu - \gamma \Delta) b = -\alpha(\delta_{e_k} - \delta_{-e_k})$$

where Δ is the discrete Laplacian. Then we deduce $L_N(z)$, hence $C_N(t)$ by inverse Laplace transform. The limit $C(t) = \lim_{N \to +\infty} C_N(t)$ follows. □

Consequently in the unpinned harmonic cases in dimension $d = 1$ and 2, the conductivity of our model diverges as N goes to infinity. Otherwise it converges as $N \to \infty$. In the anharmonic case we obtained some upper bounds showing that the divergence cannot be worse than in the harmonic case. These upper bounds also show that the conductivity cannot be infinite if $d \geq 3$ (see [2] for details and precise statements).

3.2 A Class of Perturbed Hamiltonian Systems

In [13] is proposed a class of models for which anomalous diffusion is observed. These models have been introduced in Sect. 1.2. The goal of [13] was to show that these systems have a behavior very similar to that of the standard one-dimensional chains of oscillators conserving momentum.[31]

3.2.1 Definition of Thermodynamic Variables

Let us fix a potential V and consider the stochastic energy-volume conserving model defined by the generator $\mathscr{L} = \mathscr{A} + \gamma \mathscr{S}$, $\gamma \geq 0$, where $\mathscr{A}$ and $\mathscr{S}$ are given by (21). Recall that the Gibbs grand-canonical probability measures $\mu_{\beta,\lambda}$, $\beta > 0$, $\lambda \in \mathbb{R}$, defined on Ω by

$$d\mu_{\beta,\lambda}(\eta) = \prod_{x \in \mathbb{Z}} Z(\beta, \lambda)^{-1} \exp\{-\beta V(\eta_x) - \lambda \eta_x\} \, d\eta_x$$

form a family of invariant probability measures for the infinite dynamics. We assume that the partition function Z is well defined on $(0, +\infty) \times \mathbb{R}$. The following thermodynamic relations relate the chemical potentials β, λ to the mean volume v and the mean energy e under $\mu_{\beta,\lambda}$:

$$\begin{aligned} v(\beta, \lambda) &= \mu_{\beta,\lambda}(\eta_x) = -\partial_\lambda \Big(\log Z(\beta, \lambda) \Big), \\ e(\beta, \lambda) &= \mu_{\beta,\lambda}(V(\eta_x)) = -\partial_\beta \Big(\log Z(\beta, \lambda) \Big). \end{aligned} \tag{93}$$

[31] They could be defined in any dimension.

These relations can be inverted by a Legendre transform to express β and λ as a function of e and v. Define the thermodynamic entropy $S\,:\,(0,+\infty)\times\mathbb{R}\to[-\infty,+\infty)$ as

$$S(e,v)=\inf_{\lambda\in\mathbb{R},\beta>0}\Big\{\beta e+\lambda v+\log Z(\beta,\lambda)\Big\}.$$

Let $\mathscr{U}$ be the convex domain of $(0,+\infty)\times\mathbb{R}$ where $S(e,v)>-\infty$ and $\mathring{\mathscr{U}}$ its interior. Then, for any $(e,v):=(e(\beta,\lambda),v(\beta,\lambda))\in\mathring{\mathscr{U}}$, the parameters β,λ can be obtained as

$$\beta=(\partial_e S)(e,v),\qquad \lambda=(\partial_v S)(e,v). \tag{94}$$

We also introduce the tension $\tau(\beta,\lambda)=\mu_{\beta,\lambda}(V'(\eta_0))=-\lambda/\beta$. The microscopic energy current $j^{e,\gamma}_{x,x+1}$ and volume current $j^{v,\gamma}_{x,x+1}$ are given by

$$\begin{aligned} j^{e,\gamma}_{x,x+1}&=-V'(\eta_x)V'(\eta_{x+1})-\gamma\nabla[V(\eta_x)],\\ j^{v,\gamma}_{x,x+1}&=-[V'(\eta_x)+V'(\eta_{x+1})]-\gamma\nabla[\eta_x]. \end{aligned}\tag{95}$$

With these notations we have

$$\mu_{\beta,\lambda}(j^{e,\gamma}_{x,x+1})=-\tau^2,\qquad \mu_{\beta,\lambda}(j^{v,\gamma}_{x,x+1})=-2\tau. \tag{96}$$

In the sequel, with a slight abuse of notation, we also write τ for $\tau(\beta(e,v),\lambda(e,v))$ where $\beta(e,v)$ and $\lambda(e,v)$ are defined by relations (94).

3.2.2 Hydrodynamic Limits

Consider the finite *closed* stochastic energy-volume dynamics with periodic boundary conditions, that is the dynamics generated by $\mathscr{L}_{N,\text{per}}=\mathscr{A}_{N,\text{per}}+\gamma\mathscr{S}_{N,\text{per}}$ where

$$\Big(\mathscr{A}_{N,\text{per}}f\Big)(\eta)=\sum_{x\in\mathbb{T}_N}\Big[V'(\eta_{x+1})-V'(\eta_{x-1})\Big]\,\partial_{\eta_x}f(\eta), \tag{97}$$

and

$$\Big(\mathscr{S}_{N,\text{per}}f\Big)(\eta)=\sum_{x\in\mathbb{T}_N}\Big[f(\eta^{x,x+1})-f(\eta)\Big].$$

We choose to consider the dynamics on $\mathbb{T}_N$ rather than on $\mathbb{Z}$ to avoid (nontrivial) technicalities. We are interested in the macroscopic behavior of the two conserved quantities on a macroscopic time-scale Nt as $N\to\infty$.

Remark 2 The results of this section shall be compared to the results of Sect. 2.2.1. For the velocity-flip model, the hydrodynamic limits where trivial in the Euler time scale. It was only in the diffusive time scale that some evolution of the profiles was observed and the hydrodynamic limits were given by parabolic equations (see (64)).

Here, the evolution is not trivial in the Euler time scale and the hydrodynamic limits are given by hyperbolic equations (see below (99)).

We assume that the system is initially distributed according to a local Gibbs equilibrium state corresponding to a given energy-volume profile $X_0 : \mathbb{T} \to \mathring{\mathscr{U}}$:

$$X_0 = \begin{pmatrix} \mathfrak{e}_0 \\ \mathfrak{v}_0 \end{pmatrix},$$

in the sense that, for a given system size N, the initial state of the system is described by the following product probability measure:

$$d\mu^N_{\mathfrak{e}_0,\mathfrak{v}_0}(\eta) = \prod_{x\in\mathbb{T}_N} \frac{\exp\{-\beta_0(x/N)V(\eta_x) - \lambda_0(x/N)\eta_x\}}{Z(\beta_0(x/N), \lambda_0(x/N))}\, d\eta_x, \tag{98}$$

where $(\beta_0(x/N), \lambda_0(x/N))$ is actually a function of $(\mathfrak{e}_0(x/N), \mathfrak{v}_0(x/N))$ through relations (94).

Starting from such a state, we expect the state of the system at time Nt to be close, in a suitable sense, to a local Gibbs equilibrium measure corresponding to an energy-volume profile

$$X(t,\cdot) = \begin{pmatrix} \mathfrak{e}(t,\cdot) \\ \mathfrak{v}(t,\cdot) \end{pmatrix},$$

satisfying a suitable partial differential equation with initial condition X_0 at time $t = 0$. In view of (96), and assuming propagation of local equilibrium, it is not difficult to show that the expected partial differential equation is the following system of two conservation laws:

$$\begin{cases} \partial_t \mathfrak{e} - \partial_q \tau^2 = 0, \\ \partial_t \mathfrak{v} - 2\partial_q \tau = 0, \end{cases} \tag{99}$$

with initial conditions $\mathfrak{e}(0,\cdot) = \mathfrak{e}_0(\cdot)$, $\mathfrak{v}(0,\cdot) = \mathfrak{v}_0(\cdot)$. We write (99) more compactly as

$$\partial_t X + \partial_q \mathfrak{J}(X) = 0, \qquad X(0,\cdot) = X_0(\cdot),$$

with

$$\mathfrak{J}(X) = \begin{pmatrix} -\tau^2(\mathfrak{e},\mathfrak{v}) \\ -2\tau(\mathfrak{e},\mathfrak{v}) \end{pmatrix}. \tag{100}$$

The system of conservation laws (99) has other nontrivial conservation laws. In particular, the thermodynamic entropy S is conserved along a smooth solution of (99):

$$\partial_t S(\mathfrak{e},\mathfrak{v}) = 0. \tag{101}$$

Since the thermodynamic entropy is a strictly concave function on $\mathring{\mathscr{U}}$, the system (99) is strictly hyperbolic on $\mathring{\mathscr{U}}$ (see [53]). The two real eigenvalues of $(D\mathfrak{J})(\bar{\xi})$ are 0 and $-\left[\partial_e(\tau^2) + 2\partial_v(\tau)\right]$, corresponding respectively to the two eigenvectors

$$\begin{pmatrix} -\partial_v \tau \\ \partial_e \tau \end{pmatrix}, \quad \begin{pmatrix} \tau \\ 1 \end{pmatrix}. \tag{102}$$

It is well known that classical solutions to systems of $n \geq 1$ conservation laws in general develop shocks in finite times, even when starting from smooth initial conditions. If we consider weak solutions rather than classical solutions, then a criterion is needed to select a unique, relevant solution among the weak ones. For scalar conservation laws ($n = 1$), this criterion is furnished by the so-called entropy inequality and existence and uniqueness of solutions is fully understood. If $n \geq 2$, only partial results exist (see [53]). This motivates the fact that we restrict our analysis to smooth solutions before the appearance of shocks.

We assume that the potential V satisfies the following

Assumption 3.1 *The potential V is a smooth, non-negative function such that the partition function $Z(\beta, \lambda) = \int_{-\infty}^{\infty} \exp(-\beta V(r) - \lambda r)\, dr$ is well defined for $\beta > 0$ and $\lambda \in \mathbb{R}$ and there exists a positive constant C such that*

$$0 < V''(r) \leq C, \tag{103}$$

and

$$\limsup_{|r| \to +\infty} \frac{rV'(r)}{V(r)} \in (0, +\infty), \tag{104}$$

$$\limsup_{|r| \to +\infty} \frac{[V'(r)]^2}{V(r)} < +\infty. \tag{105}$$

Provided we can prove that the infinite volume dynamics is macro-ergodic, then we can rigorously prove (even if $\gamma = 0$), using the relative entropy method of Yau [67], that (99) is indeed the hydrodynamic limit in the smooth regime, *i.e.* for times t up to the appearance of the first shock (see for example [41, 62]). Observe that the expected hydrodynamic limits do not depend on γ. We need to assume $\gamma > 0$ to ensure the macro-ergodicity of the dynamics.

Remark 3 As argued in [62], it turns out that the conservation of thermodynamic entropy (101) is fundamental for Yau's method where, in the expansion of the time derivative of relative entropy, the cancelation of the linear terms is a consequence of the preservation of the thermodynamic entropy.

Averages with respect to the empirical energy-volume measure are defined, for continuous functions $G, H : \mathbb{T} \to \mathbb{R}$, as (similarly to (63))

$$\begin{pmatrix} \mathscr{E}_N(t,G) \\ \mathscr{V}_N(t,H) \end{pmatrix} = \begin{pmatrix} \frac{1}{N}\sum_{x\in\mathbb{T}_N} G\left(\frac{x}{N}\right) V(\eta_x(t)) \\ \frac{1}{N}\sum_{x\in\mathbb{T}_N} H\left(\frac{x}{N}\right) \eta_x(t) \end{pmatrix}.$$

We can then state the following result.

Theorem 9 ([13]) *Fix some $\gamma > 0$ and consider the dynamics on the torus $\mathbb{T}_N$ generated by $\mathscr{L}_{N,\text{per}}$ where the potential V satisfies Assumption 3.1. Assume that the system is initially distributed according to a local Gibbs state (98) with smooth energy profile $\mathfrak{e}_0$ and volume profile $\mathfrak{v}_0$. Consider a positive time t such that the solution $(\mathfrak{e},\mathfrak{v})$ to (99) belongs to $\mathring{\mathscr{U}}$ and is smooth on the time interval $[0,t]$. Then, for any continuous test functions $G, H : \mathbb{T} \to \mathbb{R}$, the following convergence in probability holds as $N \to +\infty$:*

$$\Big(\mathscr{E}_N(tN,G), \mathscr{V}_N(tN,H)\Big) \longrightarrow \left(\int_{\mathbb{T}} G(q)\mathfrak{e}(t,q)dq, \int_{\mathbb{T}} H(q)\mathfrak{v}(t,q)dq\right).$$

The derivation of the hydrodynamic limits beyond the shocks for systems of $n \geq 2$ conservation laws is very difficult and is one of the most challenging problems in the field of hydrodynamic limits. The first difficulty is of course our poor understanding of the solutions to such systems. Recently, Fritz proposed in [29] to derive hydrodynamic limits for hyperbolic systems (in the case $n = 2$) by some extension of the compensated-compactness approach [60] to stochastic microscopic models. This program has been achieved in [32] (see also the recent paper [30]), where the authors derive the classical $n = 2$ Leroux system of conservation laws. In fact, to be exact, only the convergence to the set of entropy solutions is proved, the question of uniqueness being left open. It nonetheless remains the best result available at this time. The proof is based on a strict control of entropy pairs at the microscopic level by the use of logarithmic Sobolev inequality estimates. It would be very interesting to extend these methods to systems such as the ones considered here.

3.2.3 Anomalous Diffusion

We investigate now the problem of anomalous diffusion of energy for these models.

If $V(r) = r^2$ then Theorem 8 is mutatis mutandis valid and we get the same conclusions: the time-space correlations for the current behave for large time t like $t^{-1/2}$. Thus the system is super-diffusive (see [13] for the details).

For generic anharmonic potentials, we can only provide numerical evidence of the super-diffusivity. However, it is difficult to estimate numerically the time auto-correlation functions of the currents because of their expected long-time tails, and because statistical errors are very large (in relative value) when t is large. Also, for finite systems (the only ones we can simulate on a computer), the autocorrelation is generically exponentially decreasing for anharmonic potentials, and, to obtain meaningful results, the thermodynamic limit $N \to \infty$ should be taken before the long-time limit.

A more tenable approach consists in studying a nonequilibrium system in its steady-state. We consider a finite system of length $2N + 1$ in contact with two thermostats which fix the value of the energy at the boundaries. The generator of the dynamics is given by

$$\mathscr{L}_N = \mathscr{A}_N + \gamma \mathscr{S}_N + \lambda_\ell \mathscr{B}_{-N,T_\ell} + \lambda_r \mathscr{B}_{N,T_r}, \tag{106}$$

where $\mathscr{A}_N$ and $\mathscr{S}_N$ are defined by

$$\begin{aligned}(\mathscr{A}_N f)(\eta) = &\sum_{x=-(N-1)}^{N-1} \Big(V'(\eta_{x+1}) - V'(\eta_{x-1})\Big)(\partial_{\eta_x} f)(\eta) \\ &- V'(\eta_{N-1})\,(\partial_{\eta_N} f)(\eta) + V'(\eta_{-N+1})\,(\partial_{\eta_{-N}} f)(\eta), \\ (\mathscr{S}_N f)(\eta) = &\sum_{x=-N}^{N-1} \Big[f(\eta^{x,x+1}) - f(\eta)\Big],\end{aligned}$$

and $\mathscr{B}_{x,T} = T\partial^2_{\eta_x} - V'(\eta_x)\partial_{\eta_x}$. The positive parameters λ_ℓ and λ_r are the intensities of the thermostats and T_ℓ, T_r the "temperatures" of the thermostats.

The generator $\mathscr{B}_{x,T}$ is a thermostatting mechanism. In order to fix the energy at site $-N$ (resp. N) to the value e_ℓ (resp. e_r), we have to choose $\beta_\ell = T_\ell^{-1}$ (resp. $\beta_r = T_r^{-1}$) such that $e(\beta_\ell, 0) = e_\ell$ (resp. $e(\beta_r, 0) = e_r$). We denote by $\langle\cdot\rangle_{ss}$ the unique stationary state for the dynamics generated by $\mathscr{L}_N$.

The energy currents $j^{e,\gamma}_{x,x+1}$, which are such that $\mathscr{L}_{N,\text{open}}(V(\eta_x)) = -\nabla j^{e,\gamma}_{x-1,x}$ (for $x = -N, \dots, N-1$), are given by the first line of (95) for $x = -N+1, \dots, N-1$ while

$$\begin{aligned} j^{e,\gamma}_{-N-1,-N} &= \lambda_\ell \Big[T_\ell V''(\eta_{-N}) - (V'(\eta_{-N}))^2\Big], \\ j^{e,\gamma}_{N,N+1} &= -\lambda_r \Big[T_r V''(\eta_N) - (V'(\eta_N))^2\Big]. \end{aligned}$$

Since $\langle \mathscr{L}_{N,\text{open}}(V(\eta_x))\rangle_{ss} = 0$, it follows that, for any $x = -N, \dots, N+1$, $\langle j^{e,\gamma}_{x,x+1}\rangle_{ss}$ is equal to a constant $J^\gamma_N(T_\ell, T_r)$ independent of x. In fact,

$$J^\gamma_N(T_\ell, T_r) = \big\langle \mathscr{J}^\gamma_N \big\rangle_{ss}, \qquad \mathscr{J}^\gamma_N = \frac{1}{2N} \sum_{x=-N-1}^{N} j^{e,\gamma}_{x,x+1}. \tag{107}$$

The latter equation is interesting from a numerical viewpoint since it allows to perform some spatial averaging, hence reducing the statistical error of the results. We estimate the exponent $\delta \geq 0$ such that

Table 1 Conductivity exponents

γ	$V(r) = r^2/2$	$V(r) = r^2/2 + r^4/4$	$V(r) = e^{-r} + r - 1$
0	1	0.13	1
0.01	–	0.14	0.12
0.1	0.50	0.27	0.25
1	0.50	0.43	0.33

$$\kappa(N) := NJ_N^{\gamma} \sim N^{\delta} \tag{108}$$

using numerical simulations. If $\delta = 0$, the system is a normal conductor of energy. If on the other hand $\delta > 0$, it is a superconductor.

The numerical simulations giving the value of δ are summarized in Table 1. They have been performed for the harmonic chain $V(r) = r^2/2$, the quartic potential $V(r) = r^2/2 + r^4/4$ and the exponential potential $V(r) = e^{-r} + r - 1$. In Sect. 3.4 we will motivate our interest in the exponential potential.

Exponents in the harmonic case agree with their expected values. For nonlinear potentials, except for the singular value $\delta = 1$ when $\gamma = 0$ and $V(r) = e^{-r} + r - 1$, the exponents seem to be monotonically increasing with γ. A similar behavior of the exponents is observed for Toda chains [37] with a momentum conserving noise. This strange behavior casts some doubts on the convergence of conductivity exponents δ with respect to system size N (see the comment after Theorem 3 in [6]). A detailed study, including the nonlinear fluctuating hydrodynamics predictions, is available in [57].

Note also that the value found for $\gamma = 0$ with the anharmonic FPU potential $V(r) = r^2/2 + r^4/4$ is smaller than the corresponding value for standard oscillators chains, which is around 0.33 (see [49]). We performed also numerical simulations for a "rotor" model, $V(r) = 1 - \cos(r)$, and we found $\delta \approx 0.02$, i.e. a normal conductivity. A similar picture is observed for the usual rotor[32] model which is composed of a chain of unpinned oscillators with interaction potential $V(r) = 1 - \cos(r)$. The normal behavior is conjectured to be due to the absence of long waves (that carry energy ballistically) because some rotors turning fast in between will break them [38]. See [56] and references therein for a recent study of the rotors model.

3.3 Fractional Superdiffusion for a Harmonic Chain with Bulk Noise

In this section we consider the energy-volume conserving model with quadratic potential. Fix $\lambda \in \mathbb{R}$ and $\beta > 0$, and consider the process $\{\eta(t); t \geq 0\}$ generated by (21) with $V(\eta) = \eta^2/2$ and with initial distribution $\mu_{\beta,\lambda}$. Notice that the distribution

[32] The variable r has to be interpreted as an angle and belongs to the torus $2\pi\mathbb{T}$.

of the process $\{\eta(t)+\rho; t \geq 0\}$ with initial measure $\mu_{\beta,\rho+\lambda}$ is the same for all values of $\lambda \in \mathbb{R}$. Therefore, we can assume, without loss of generality, that $\lambda = 0$. We write $\mu_\beta = \mu_{\beta,0}$ to simplify notation, and denote by $\mathbb{P}$ the law of $\{\eta(t); t \geq 0\}$ and by $\mathbb{E}$ the expectation with respect to $\mathbb{P}$. The *energy correlation function* $\{S_t(x); x \in \mathbb{Z}, t \geq 0\}$ is defined as

$$S_t(x) = \tfrac{\beta^2}{2}\,\mathbb{E}\big[\big(\eta_0(0)^2 - \tfrac{1}{\beta}\big)\big(\eta_x(t)^2 - \tfrac{1}{\beta}\big)\big] \tag{109}$$

for any $x \in \mathbb{Z}$ and any $t \geq 0$. The constant $\frac{\beta^2}{2}$ is just the inverse of the variance of $\eta_x^2 - \frac{1}{\beta}$ under μ_β. By translation invariance of the dynamics and the initial distribution μ_β, we see that

$$\tfrac{\beta^2}{2}\,\mathbb{E}\big[\big(\eta_x(0)^2 - \tfrac{1}{\beta}\big)\big(\eta_y(t)^2 - \tfrac{1}{\beta}\big)\big] = S_t(y-x) \tag{110}$$

for any $x, y \in \mathbb{Z}$.

Theorem 10 ([7]) *Let $f, g : \mathbb{R} \to \mathbb{R}$ be smooth functions of compact support. Then,*

$$\lim_{n\to\infty} \frac{1}{n} \sum_{x,y\in\mathbb{Z}} f\big(\tfrac{x}{n}\big)g\big(\tfrac{y}{n}\big)S_{tn^{3/2}}(x-y) = \iint f(x)g(y)P_t(x-y)dxdy, \tag{111}$$

where $\{P_t(x); x \in \mathbb{R}, t \geq 0\}$ is the fundamental solution[33] *of the skew fractional heat equation on $\mathbb{R}$*

$$\partial_t u = -\tfrac{1}{\sqrt{2}}\big\{(-\Delta)^{3/4} - \nabla(-\Delta)^{1/4}\big\}u. \tag{112}$$

A fundamental step in the proof of this theorem is the analysis of the correlation function $\{\mathbb{S}_t(x,y); x \neq y \in \mathbb{Z}, t \geq 0\}$ given by

$$\mathbb{S}_t(x,y) = \tfrac{\beta^2}{2}\mathbb{E}\big[\big(\eta_0(0)^2 - \tfrac{1}{\beta}\big)\eta_x(t)\eta_y(t)\big] \tag{113}$$

for any $t \geq 0$ and any $x \neq y \in \mathbb{Z}$. Notice that this definition makes perfect sense for $x = y$ and, in fact, we have $\mathbb{S}_t(x,x) = S_t(x)$. For notational convenience we define $\mathbb{S}_t(x,x)$ as equal to $S_t(x)$. However, these quantities are of different nature, since $S_t(x)$ is related to *energy fluctuations* and $\mathbb{S}_t(x,y)$ is related to *volume fluctuations* (for $x \neq y$).

Remark 4 It is not difficult to see that with a bit of technical work the techniques actually show that the distribution valued process $\{\mathscr{E}_t^n(\cdot)\,;\, t \geq 0\}$ defined for any test function f by

$$\mathscr{E}_t^n(f) = \frac{1}{\sqrt{n}}\sum_{x\in\mathbb{Z}} f\big(\tfrac{x}{n}\big)\{\eta_x(tn^{3/2})^2 - \tfrac{1}{\beta}\}$$

[33] Since the skew fractional heat equation is linear, it can be solved explicitly by Fourier transform.

converges as n goes to infinity to an infinite dimensional 3/4-fractional Ornstein-Uhlenbeck process, i.e. the centered Gaussian process with covariance prescribed by the right hand side of (111).

Remark 5 It is interesting to notice that P_t is the maximally asymmetric 3/2-Levy distribution. It has power law as $|x|^{-5/2}$ towards the diffusive peak and stretched exponential as $\exp[-|x|^3]$ towards the exterior of the sound cone ([63, Chap. 4]). As mentioned to us by H. Spohn, this reflects the expected physical property that no propagation beyond the sound cone occurs.

Remark 6 With a bit of technical work the proof of this theorem can be adapted to obtain a similar statement for a harmonic chain perturbed by the momentum exchanging noise (see [39] where such statement is proved for the Wigner function). In this case the skew fractional heat equation is replaced by the (symmetric) fractional heat equation.

Proof Denote by $\mathscr{C}_c^\infty(\mathbb{R}^d)$ the space of infinitely differentiable functions $f : \mathbb{R}^d \to \mathbb{R}$ of compact support. Then, $\|f\|_{2,n}$ denotes the weighted $\ell^2(\mathbb{Z}^d)$-norm

$$\|f\|_{2,n} = \sqrt{\frac{1}{n^d} \sum_{x\in\mathbb{Z}^d} f\left(\tfrac{x}{n}\right)^2}. \tag{114}$$

Let $g \in \mathscr{C}_c^\infty(\mathbb{R})$ be a fixed function. For each $n \in \mathbb{N}$ we define the field $\{\mathscr{S}_t^n; t \geq 0\}$ as

$$\mathscr{S}_t^n(f) = \frac{1}{n} \sum_{x,y\in\mathbb{Z}} g\left(\tfrac{x}{n}\right) f\left(\tfrac{y}{n}\right) S_{tn^{3/2}}(y-x) \tag{115}$$

for any $t \geq 0$ and any $f \in \mathscr{C}_c^\infty(\mathbb{R})$. By the Cauchy-Schwarz inequality we have the a priori bound

$$\left|\mathscr{S}_t^n(f)\right| \leq \|g\|_{2,n} \|f\|_{2,n} \tag{116}$$

for any $t \geq 0$, any $n \in \mathbb{N}$ and any $f, g \in \mathscr{C}_c^\infty(\mathbb{R})$. For a function $h \in \mathscr{C}_c^\infty(\mathbb{R}^2)$ we define $\{Q_t^n(h); t \geq 0\}$ as

$$Q_t^n(h) = \frac{1}{n^{3/2}} \sum_{x\in\mathbb{Z}} \sum_{y\neq z\in\mathbb{Z}} g\left(\tfrac{x}{n}\right) h\left(\tfrac{y}{n}, \tfrac{z}{n}\right) \mathbb{S}_{tn^{3/2}}(y-x, z-x). \tag{117}$$

Notice that $Q_t^n(h)$ depends only on the symmetric part of the function h. Therefore, we will always assume, without loss of generality, that $h(x,y) = h(y,x)$ for any

$x, y \in \mathbb{Z}$. We point out that $Q_t^n(h)$ does not depend on the values of h at the diagonal $\{x = y\}$. We have the a priori bound

$$\big|Q_t^n(h)\big| \le 2\|g\|_{2,n}\|\tilde{h}\|_{2,n}, \tag{118}$$

where $\tilde{h}$ is defined by $\tilde{h}\big(\tfrac{x}{n}, \tfrac{y}{n}\big) = h\big(\tfrac{x}{n}, \tfrac{y}{n}\big)\,\mathbf{1}_{x \neq y}$.

For a function $f \in \mathscr{C}_c^\infty(\mathbb{R})$, we define $\Delta_n f : \mathbb{R} \to \mathbb{R}$ as

$$\Delta_n f\big(\tfrac{x}{n}\big) = n^2\Big(f\big(\tfrac{x+1}{n}\big) + f\big(\tfrac{x-1}{n}\big) - 2f\big(\tfrac{x}{n}\big)\Big). \tag{119}$$

In other words, $\Delta_n f$ is a discrete approximation of the second derivative of f. We also define $\nabla_n f \otimes \delta : \frac{1}{n}\mathbb{Z}^2 \to \mathbb{R}$ as

$$\big(\nabla_n f \otimes \delta\big)\big(\tfrac{x}{n}, \tfrac{y}{n}\big) = \begin{cases} \frac{n^2}{2}\big(f\big(\tfrac{x+1}{n}\big) - f\big(\tfrac{x}{n}\big)\big); & y = x+1 \\ \frac{n^2}{2}\big(f\big(\tfrac{x}{n}\big) - f\big(\tfrac{x-1}{n}\big)\big); & y = x-1 \\ 0; & \text{otherwise.} \end{cases} \tag{120}$$

Less evident than the interpretation of $\Delta_n f$, $\nabla_n f \otimes \delta$ turns out to be a discrete approximation of the (two dimensional) distribution $f'(x) \otimes \delta(x = y)$, where $\delta(x = y)$ is the δ of Dirac at the line $x = y$. We have that

$$\tfrac{d}{dt}\mathscr{S}_t^n(f) = -2Q_t^n(\nabla_n f \otimes \delta) + \mathscr{S}_t^n\big(\tfrac{1}{\sqrt{n}}\Delta_n f\big). \tag{121}$$

In this equation we interpret the term $Q_t^n(\nabla_n f \otimes \delta)$ in the obvious way. By the a priori bound (116), the term $\mathscr{S}_t^n(\frac{1}{\sqrt{n}}\Delta_n f)$ is negligible, as $n \to \infty$. We describe now the equation satisfied by $Q_t^n(h)$. For this we need some extra definitions. For $h \in \mathscr{C}_c^\infty(\mathbb{R}^2)$ we define $\Delta_n h : \mathbb{R}^2 \to \mathbb{R}$ as

$$\Delta_n h\big(\tfrac{x}{n}, \tfrac{y}{n}\big) = n^2\Big(h\big(\tfrac{x+1}{n}, \tfrac{y}{n}\big) + h\big(\tfrac{x-1}{n}, \tfrac{y}{n}\big) + h\big(\tfrac{x}{n}, \tfrac{y+1}{n}\big) + h\big(\tfrac{x}{n}, \tfrac{y-1}{n}\big) - 4h\big(\tfrac{x}{n}, \tfrac{y}{n}\big)\Big). \tag{122}$$

In words, $\Delta_n h$ is a discrete approximation of the $2d$ Laplacian of h. We also define $\mathscr{A}_n h : \mathbb{R} \to \mathbb{R}$ as

$$\mathscr{A}_n h\big(\tfrac{x}{n}, \tfrac{y}{n}\big) = n\Big(h\big(\tfrac{x}{n}, \tfrac{y-1}{n}\big) + h\big(\tfrac{x-1}{n}, \tfrac{y}{n}\big) - h\big(\tfrac{x}{n}, \tfrac{y+1}{n}\big) - h\big(\tfrac{x+1}{n}, \tfrac{y}{n}\big)\Big). \tag{123}$$

The function $\mathscr{A}_n h$ is a discrete approximation of the directional derivative $(-2, -2) \cdot \nabla h$. Let us define $\mathscr{D}_n h : \frac{1}{n}\mathbb{Z} \to \mathbb{R}$ as

$$\mathscr{D}_n h\big(\tfrac{x}{n}\big) = n\Big(h\big(\tfrac{x}{n}, \tfrac{x+1}{n}\big) - h\big(\tfrac{x-1}{n}, \tfrac{x}{n}\big)\Big) \tag{124}$$

and $\widetilde{\mathscr{D}}_n h : \frac{1}{n}\mathbb{Z}^2 \to \mathbb{R}$ as

$$\widetilde{\mathscr{D}}_n h(\tfrac{x}{n}, \tfrac{y}{n}) = \begin{cases} n^2\big(h(\tfrac{x}{n}, \tfrac{x+1}{n}) - h(\tfrac{x}{n}, \tfrac{x}{n})\big); & y = x+1 \\ n^2\big(h(\tfrac{x-1}{n}, \tfrac{x}{n}) - h(\tfrac{x-1}{n}, \tfrac{x-1}{n})\big); & y = x-1 \\ 0; & \text{otherwise.} \end{cases} \tag{125}$$

The function $\mathscr{D}_n h$ is a discrete approximation of the directional derivative of h along the diagonal $x = y$, while $\widetilde{\mathscr{D}}_n h$ is a discrete approximation of the distribution $\partial_y h(x,x) \otimes \delta(x = y)$. Finally we can write down the equation satisfied by the field $Q_t^n(h)$:

$$\tfrac{d}{dt} Q_t^n(h) = Q_t^n\big(n^{-1/2}\Delta_n h + n^{1/2}\mathscr{A}_n h\big) - 2\mathscr{S}_t^n\big(\mathscr{D}_n h\big) + 2Q_t^n\big(n^{-1/2}\widetilde{\mathscr{D}}_n h\big). \tag{126}$$

Given $f \in \mathscr{C}_c^\infty(\mathbb{R})$, if we choose $h := h_n(f)$ such that

$$n^{-1/2}\Delta_n h + n^{1/2}\mathscr{A}_n h = 2\nabla_n f \otimes \delta$$

then summing (121) and (126) we get

$$\tfrac{d}{dt}\mathscr{S}_t^n(f) = -\tfrac{d}{dt}Q_t^n(h) + \mathscr{S}_t^n(\tfrac{1}{\sqrt{n}}\Delta_n f) - 2\mathscr{S}_t^n\big(\mathscr{D}_n h\big) + 2Q_t^n\big(n^{-1/2}\widetilde{\mathscr{D}}_n h\big).$$

We integrate in time the previous expression. By the a priori bounds, the term $\int_0^t \mathscr{S}_s^n(\frac{1}{\sqrt{n}}\Delta_n f)ds$ is small as well as $\int_0^t \frac{d}{ds}Q_s^n(h)ds = Q_t^n(h) - Q_0^n(h)$. The term

$$\int_0^t Q_s^n(n^{-1/2}\widetilde{\mathscr{D}}_n h)\, ds$$

is quite singular since it involves an approximation of a distribution but it turns out to be negligible, although this does not follow directly from the a priori bounds (see [7]). By using Fourier transform one can see that $-2\mathscr{D}_n h$ converges to $-\frac{1}{\sqrt{2}}\{(-\Delta)^{3/4} - \nabla(-\Delta)^{1/4}\}f$ and we are done. □

3.4 *Anomalous Diffusion for a Perturbed Hamiltonian System with Exponential Interactions*

We investigate here in more details the exponential case $V_{\exp}(r) = e^{-r} - 1 + r$. The deterministic system with generator (97) and with the exponential potential above is well known in the integrable systems literature.[34] It has been introduced in [40] by

[34] It seems that although the Hamiltonian structure of the Kac-van-Moerbecke system was known, the interpretation of the latter as a chain of oscillators with exponential kinetic energy and exponential interaction was not observed before [13].

Kac and van Moerbecke and was shown to be completely integrable. Consequently, using Mazur's inequality, it is easy to show that the energy transport is ballistic [13].

As we will see the situation dramatically changes when the momentum exchange noise is added: the energy transport is no more ballistic but superdiffusive. Thus the situation is similar to the harmonic case. Nevertheless we expect the time autocorrelation of the current to decay like $t^{-2/3}$. We are not able to show this but we proved in [6] lower bounds sufficient to imply superdiffusivity.

The results are stated in infinite volume: we consider the stochastic energy-volume conserving dynamics $\{\eta(t)\}_{t\geq 0}$ with potential $V := V_{\exp}$. Its generator is given by $\mathscr{L} = \mathscr{A} + \gamma\mathscr{S}$ where $\mathscr{A}$ and $\mathscr{S}$ are defined by (21). Since the exponential potential grows very fast as $r \to -\infty$, some care has to be taken to show that the infinite dynamics is well defined (see [6]). We recall that grand canonical Gibbs measures are denoted by $\mu_{\beta,\lambda}$ and take the form

$$d\mu_{\beta,\lambda}(\eta) = \prod_{x\in\mathbb{Z}} \frac{\mathrm{e}^{-\beta V(\eta_x)-\lambda\eta_x}}{Z(\beta,\lambda)} d\eta_x, \quad \beta > 0,\ \lambda + \beta < 0.$$

In this section, β and λ are fixed and we denote by e (resp. v) the average energy (resp. volume) w.r.t. $\mu_{\beta,\lambda}$ (see (93)).

The microscopic energy current $j^{e,\gamma}_{x,x+1}$ and volume current $j^{v,\gamma}_{x,x+1}$ are given by

$$j^{e,\gamma}_{x,x+1}(\eta) = -e^{-(\eta_x+\eta_{x+1})} + (e^{-\eta_x} + e^{-\eta_{x+1}}) - \gamma\nabla(V(\eta_x))$$

and

$$j^{v,\gamma}_{x,x+1}(\eta) = e^{-\eta_x} + e^{-\eta_{x+1}} - \gamma\nabla\eta_x.$$

We will use the compact notations

$$\omega_x = \begin{pmatrix} V(\eta_x) \\ \eta_x \end{pmatrix}, \quad J_{x,x+1} = \begin{pmatrix} j^{e,\gamma}_{x,x+1} \\ j^{v,\gamma}_{x,x+1} \end{pmatrix}.$$

In the hyperbolic scaling, the hydrodynamical equations for the energy profile $\mathfrak{e}$ and the volume profile $\mathfrak{v}$ take the form

$$\begin{cases} \partial_t \mathfrak{e} - \partial_q((\mathfrak{e}-\mathfrak{v})^2) = 0 \\ \partial_t \mathfrak{v} + 2\,\partial_q(\mathfrak{e}-\mathfrak{v}) = 0. \end{cases} \tag{127}$$

They can be written in the compact form $\partial_t X + \partial_q \mathfrak{J}(X) = 0$ with

$$X = \begin{pmatrix} \mathfrak{e} \\ \mathfrak{v} \end{pmatrix}, \quad \mathfrak{J}(X) = \begin{pmatrix} -(\mathfrak{e}-\mathfrak{v})^2 \\ 2(\mathfrak{e}-\mathfrak{v}) \end{pmatrix}. \tag{128}$$

The differential matrix of $\mathfrak{J}$ is given by

$$(\nabla\mathfrak{J})(X) = 2\begin{pmatrix} -(\mathfrak{e}-\mathfrak{v}) & \mathfrak{e}-\mathfrak{v} \\ 1 & -1 \end{pmatrix}.$$

For given (e, v) we denote by $(T_t^+)_{t\geq 0}$ (resp. $(T_t^-)_{t\geq 0}$) the semigroup on $S(\mathbb{R})\times S(\mathbb{R})$ generated by the linearized system

$$\partial_t \varepsilon + M^T\,\partial_q \varepsilon = 0, \quad (\text{resp. } \partial_t \varepsilon - M^T\,\partial_q \varepsilon = 0), \tag{129}$$

where

$$M := M(e, v) = [\nabla\mathfrak{J}](\omega), \quad \omega = \begin{pmatrix} e \\ v \end{pmatrix}.$$

We omit the dependence of these semigroups on (e, v) for lightness of the notations. Above $S(\mathbb{R})$ denotes the Schwartz space of smooth rapidly decreasing functions.

The first result of [6] gives a lower bound on the time-scale for which a non-trivial evolution of the energy-volume fluctuation field can be observed.

We take the infinite system at equilibrium under the Gibbs measure $\mu_{\bar{\beta},\bar{\lambda}}$ corresponding to a mean energy $\bar{e}$ and a mean volume $\bar{v}$. Our goal is to study the energy-volume fluctuation field in the time-scale $tn^{1+\alpha}$, $\alpha \geq 0$:

$$\mathcal{Y}_t^{n,\alpha}(\mathbf{G}) = \frac{1}{\sqrt{n}}\sum_{x\in\mathbb{Z}} \mathbf{G}\,(x/n)\cdot\Big(\omega_x(tn^{1+\alpha}) - \bar{\omega}\Big), \tag{130}$$

where for $q \in \mathbb{R}$, $x \in \mathbb{Z}$,

$$\mathbf{G}(q) = \begin{pmatrix} G_1(q) \\ G_2(q) \end{pmatrix}, \quad \omega_x = \begin{pmatrix} V(\eta_x) \\ \eta_x \end{pmatrix}$$

and G_1, G_2 are test functions belonging to $S(\mathbb{R})$.

We need to introduce some notation. For each $z \geq 0$, let $H_z(x) = (-1)^z e^{x^2} \frac{d^z}{dx^z} e^{-x^2}$ be the Hermite polynomial of order z and $h_z(x) = (z!\sqrt{2\pi})^{-1} H_z(x) e^{-x^2}$ the Hermite function. The set $\{h_z, z \geq 0\}$ is an orthonormal basis of $\mathbb{L}^2(\mathbb{R})$. Consider in $\mathbb{L}^2(\mathbb{R})$ the operator $K_0 = x^2 - \Delta$, Δ being the Laplacian on $\mathbb{R}$. For an integer $k \geq 0$, denote by $\mathbb{H}_k$ the Hilbert space obtained by taking the completion of $S(\mathbb{R})$ under the norm induced by the scalar product $\langle\cdot,\cdot\rangle_k$ defined by $\langle f, g\rangle_k = \langle f, K_0^k g\rangle_0$, where $\langle\cdot,\cdot\rangle_0$ denotes the inner product of $\mathbb{L}^2(\mathbb{R})$ and denote by $\mathbb{H}_{-k}$ the dual of $\mathbb{H}_k$, relatively to this inner product. Let $\langle\cdot\rangle$ represent the average with respect to the Lebesgue measure.

If E is a Polish space then $D(\mathbb{R}^+, E)$ (resp, $C(\mathbb{R}^+, E)$) denotes the space of E-valued functions, right continuous with left limits (resp. continuous), endowed with the Skorohod (resp. uniform) topology. Let $Q^{n,\alpha}$ be the probability measure on $D(\mathbb{R}^+, \mathbb{H}_{-k}\times\mathbb{H}_{-k})$ induced by the fluctuation field $\mathcal{Y}_t^{n,\alpha}$ and $\mu_{\beta,\lambda}$. Let $\mathbb{P}_{\mu_{\beta,\lambda}}$

denote the probability measure on $D(\mathbb{R}^+, \mathbb{R}^{\mathbb{Z}})$ induced by $(\eta(t))_{t\geq 0}$ and $\mu_{\beta,\lambda}$. Let $\mathbb{E}_{\mu_{\beta,\lambda}}$ denote the expectation with respect to $\mathbb{P}_{\mu_{\beta,\lambda}}$.

Theorem 11 ([6]) *Fix an integer $k > 2$. Denote by Q the probability measure on $C(\mathbb{R}^+, \mathbb{H}_{-k} \times \mathbb{H}_{-k})$ corresponding to a stationary Gaussian process with mean 0 and covariance given by*

$$\mathbb{E}_Q\,[\mathcal{Y}_t(\mathbf{H})\,\mathcal{Y}_s(\mathbf{G})] = \langle\, T_t^- \mathbf{H} \;\cdot\; \chi\; T_s^- \mathbf{G}\,\rangle$$

for every $0 \leq s \leq t$ and $\mathbf{H}, \mathbf{G}$ in $\mathbb{H}_k \times \mathbb{H}_k$. Here $\chi := \chi(\beta, \lambda)$ is the equilibrium covariance matrix of ω_0. Then, the sequence $(Q^{n,0})_{n\geq 1}$ converges weakly to the probability measure Q.

The theorem above means that in the hyperbolic scaling the fluctuations are trivial: the initial fluctuations are transported by the linearized system of (127). To see a nontrivial behavior we have to study, in the transport frame, the fluctuations at a longer time scale $tn^{1+\alpha}$, with $\alpha > 0$. Thus, we consider the fluctuation field $\widehat{\mathcal{Y}}^{n,\alpha}_{\cdot}$, $\alpha > 0$, defined, for any $\mathbf{G} \in S(\mathbb{R}) \times S(\mathbb{R})$, by

$$\widehat{\mathcal{Y}}_t^{n,\alpha}(\mathbf{G}) = \mathcal{Y}_t^{n,\alpha}\left(T^+_{tn^\alpha}\mathbf{G}\right). \tag{131}$$

Our second main theorem shows that the correct scaling exponent α is greater than $1/3$:

Theorem 12 ([6]) *Fix an integer $k > 1$ and $\alpha < 1/3$. Denote by Q the probability measure on $C(\mathbb{R}^+, \mathbb{H}_{-k} \times \mathbb{H}_{-k})$ corresponding to a stationary Gaussian process with mean 0 and covariance given by*

$$\mathbb{E}_Q\,[\mathcal{Y}_t(\mathbf{H})\,\mathcal{Y}_s(\mathbf{G})] = \langle \mathbf{H} \;\cdot\; \chi\; \mathbf{G}\rangle$$

for every $0 \leq s \leq t$ and $\mathbf{H}, \mathbf{G}$ in $\mathbb{H}_k \times \mathbb{H}_k$. Then, the sequence $(Q^{n,\alpha})_{n\geq 1}$ converges weakly to the probability measure Q.

The proofs of these theorems can be reduced to the proof of a so-called *equilibrium Boltzmann-Gibbs principle*. Let us explain what it means. Observables can be divided into two classes: non-hydrodynamical and hydrodynamical. The first ones are the non conserved quantities and they fluctuate on a much faster scale than the conserved ones. Hence, they should average out and only their projection on the hydrodynamical variables should persist in the scaling limit. For any local function $g := g(\eta)$, the projection $\mathcal{P}_{e,v}\, g$ of g on the fields of the conserved quantities is defined by

$$(\mathcal{P}_{e,v} g)(\eta) = g(\eta) - \tilde{g}(e, v) - (\nabla \tilde{g})(e, v) \cdot (\omega_0 - \omega)$$

where $\tilde{g}(e, v) = \langle g \rangle_{\mu_{\beta,\lambda}}$ and $\nabla \tilde{g}$ is the gradient of the function $\tilde{g}$. As explained above we expect that in the Euler time scale, for any test function $\mathbf{H} \in S(\mathbb{R}) \times S(\mathbb{R})$, the space-time variance

$$\lim_{n\to\infty} \mathbb{E}_{\mu_{\beta,\lambda}} \left[\left(\int_0^t \frac{1}{\sqrt{n}} \sum_{x\in\mathbb{Z}} \mathbf{H}(x/n) \cdot \left[\theta_x \mathscr{P}_{e,v}\, g(\eta(sn)) \right] ds \right)^2 \right] = 0 \qquad (132)$$

vanishes as n goes to infinity. In fact it suffices to show (132) for the function $g = J_{0,1}$.[35] Thus let us first define the *normalized* currents by

$$\hat{J}_{x,x+1} = \left[\theta_x \mathscr{P}_{\bar{e},\bar{v}} J_{0,1} \right] = \begin{pmatrix} j^{e,\gamma}_{x,x+1}(\eta) \\ j^{v,\gamma}_{x,x+1}(\eta) \end{pmatrix} - \mathfrak{J}(\omega) - (\nabla \mathfrak{J})(\omega) \begin{pmatrix} V_{\exp}(\eta_x) - e \\ \eta_x - v \end{pmatrix}. \qquad (133)$$

To estimate the space-time variance involved we use the following inequality (see [43]):

$$\mathbb{E}_{\mu_{\beta,\lambda}} \left[\left(\int_0^t f(\eta(sn^{1+\alpha}))\, ds \right)^2 \right] \le \frac{Ct}{n^{1+\alpha}} \left\langle f\, , \left(\frac{1}{tn^{1+\alpha}} - \gamma \mathscr{S} \right)^{-1} f \right\rangle_{\mu_{\beta,\lambda}} \qquad (134)$$

where $f = \dfrac{1}{\sqrt{n}} \sum_{x\in\mathbb{Z}} \mathbf{H}(x/n) \cdot \hat{J}_{x,x+1}$. Due to the very simple form of the operator $\mathscr{S}$ the RHS of (134) can be estimated and shown to vanish as n goes to infinity. Nevertheless it has to be done with some care since $\mathscr{S}$ is very degenerate so that without the term $\frac{1}{tn^{1+\alpha}}$ the RHS of (134) blows up.

Theorem 12 does not exclude the possibility of normal fluctuations, i.e. the convergence in law of the fluctuation field of the two conserved quantities to an infinite dimensional Ornstein-Uhlenbeck process in the diffusive time scale ($\alpha = 1$). To see that it is not the case we will show that the diffusion coefficient $\mathscr{D} := \mathscr{D}(e, v)$ appearing in this hypothetical limiting process would be infinite, excluding thus this possibility. Up to a constant matrix coming from a martingale term (due to the noise) and thus irrelevant for us (see [2, 13]), the matrix coefficient $\mathscr{D}$ is defined by the Green-Kubo formula

$$\mathscr{D} = \int_0^\infty \mathbb{E}_{\mu_{\beta,\lambda}} \left[\sum_{x\in\mathbb{Z}} \hat{J}_{x,x+1}(t) \left[\hat{J}_{0,1}(0) \right]^T \right] dt. \qquad (135)$$

The signature of the superdiffusive behavior of the system is seen in the divergence of $\mathscr{D}$, i.e. in a slow decay of the current-current correlation function. To study the latter we introduce its Laplace transform

$$\mathscr{F}(\gamma, z) = \int_0^\infty e^{-zt}\, \mathbb{E}_{\mu_{\beta,\lambda}} \left[\sum_{x\in\mathbb{Z}} \hat{J}_{x,x+1}(t) \left[\hat{J}_{0,1}(0) \right]^T \right] dt$$

[35] For Theorem 12 the Boltzmann-Gibbs principle has to be proved in the longer time scale $tn^{1+\alpha}$ and in the transport frame.

which is well defined for any $z > 0$. This can be rewritten as

$$\mathscr{F}(\gamma, z) = \langle\langle \hat{J}_{0,1}, (z - \mathscr{L})^{-1} \hat{J}_{0,1} \rangle\rangle_{\beta,\lambda}$$

where $\langle\langle \cdot, \cdot \rangle\rangle_{\beta,\lambda}$ is the semi-inner product defined with respect to $\mu_{\beta,\lambda}$ in the same way as in (34).

Our third theorem is the following lower bound on $\mathscr{F}(\gamma, z)$. Observe that $\mathscr{F}(\gamma, z)$ is a square matrix of size 2 whose entry (i, j) is denoted by $\mathscr{F}_{i,j}$.

Theorem 13 ([6]) *Fix $\gamma > 0$. There exists a positive constant $c := c(\gamma) > 0$ such that*

$$\mathscr{F}_{1,1}(\gamma, z) \geq c z^{-1/4}$$

and

$$\mathscr{F}_{i,j}(\gamma, z) = 0, \quad (i, j) \neq (1, 1).$$

Moreover, there exists a positive constant $C := C(\gamma)$ such that for any $z > 0$,

$$C^{-1} \mathscr{F}_{1,1}(1, z/\gamma) \leq \mathscr{F}_{1,1}(\gamma, z) \leq C \mathscr{F}_{1,1}(1, z/\gamma). \tag{136}$$

The last part of the theorem follows easily by a scaling argument and is in fact also valid for general potentials V and for generic "standard" anharmonic chains of oscillators. In [3, 13, 37], numerical simulations indicate a strange dependence w.r.t. the noise intensity $\gamma > 0$ of the exponent δ in the energy transport coefficient $\kappa(N) \sim N^{\delta}$ (N is the system size, see (108) for the definition of $\kappa(N)$): $\delta := \delta(\gamma) > 0$ is increasing with the noise intensity γ. This is very surprising since the more stochasticity in the model is introduced, the more the system is superdiffusive! The inequality (136) shows that the time decay of the current autocorrelation function is independent of γ (up to possible slowly varying functions corrections, i.e. in a Tauberian sense). It is common folklore that there should be a simple relationship between the slow long-time tail decay of the autocorrelation of the current in the Green-Kubo formula (described by some power law decay) and the divergence of the thermal conductivity of open systems in their steady states. The argument is that the autocorrelation should be integrated over times of order N. If we believe in this argument it means that the numerical simulations of [3, 13, 37] are not converged. There is however no clear mathematical result backing up this belief.

The proof of the first part of Theorem 13 is based on the three following arguments.

- The first idea consists in performing the microscopic change of variables $\xi_x = e^{-\eta_x}$, $x \in \mathbb{Z}$, that defines a new Markov process $\{\xi(t)\}_{t \geq 0} = \{\xi_x(t)\,;\, x \in \mathbb{Z}\}_{t \geq 0}$ with state space $(0, +\infty)^{\mathbb{Z}}$ and conserving $\sum_x \xi_x$ and $\sum_x \log \xi_x$. Its generator is given by $\tilde{\mathscr{L}} = \tilde{\mathscr{A}} + \gamma \tilde{\mathscr{S}}$ where for any local differentiable function f,

$$(\tilde{\mathscr{A}}f)(\xi) = \sum_{x\in\mathbb{Z}} \xi_x(\xi_{x+1} - \xi_{x-1})(\partial_{\xi_x} f)(\xi),$$

$$(\tilde{\mathscr{S}}f)(\xi) = \sum_{x\in\mathbb{Z}} \Big[f(\xi^{x,x+1}) - f(\xi)\Big].$$

The invariant measures for $(\xi(t))_{t\geq 0}$ are obtained from the Gibbs measures $\mu_{\beta,\lambda}$ by the change of variables above. They form a family $\{\nu_{\rho,\theta}\}_{\rho,\theta}$ of translation invariant product measures indexed by two parameters ρ and θ which satisfy

$$\rho = \nu_{\rho,\theta}(\xi_x), \quad \theta = \nu_{\rho,\theta}(\log \xi_x).$$

In fact the marginal of $\nu_{\rho,\theta}$ is a Gamma distribution. The parameters (ρ, θ) are in a one-to-one explicit correspondence with the parameters (e, v).
Rewriting $\hat{J}_{x,x+1}$ with these new variables we see that it is sufficient to prove a similar statement for the process $(\xi(t))_{t\geq 0}$ under the equilibrium probability measure $\nu_{\rho,\theta}$. Introducing the inner product $\langle\langle\cdot,\cdot\rangle\rangle$ defined, for any local functions f, g on $(0,+\infty)^{\mathbb{Z}}$ by

$$\langle\langle f, g\rangle\rangle = \sum_{x\in\mathbb{Z}} \{\nu_{\rho,\theta}(f\,\theta_x g) - \nu_{\rho,\theta}(f)\nu_{\rho,\theta}(g)\}$$

we can show that the proof of the first claim of Theorem 13 reduces to showing that there exists a positive constant c such that for any $z > 0$,

$$\langle\langle W_{0,1}, (z - \tilde{\mathscr{L}})^{-1} W_{0,1}\rangle\rangle \geq c z^{-1/4} \tag{137}$$

where $W_{0,1}(\xi) = (\xi_0 - \rho)(\xi_1 - \rho)$.

- The second step consists in using a variational formula to express the LHS of (137). Indeed we have

$$\langle\langle W_{0,1}, (z - \tilde{\mathscr{L}})^{-1} W_{0,1}\rangle\rangle = \sup_g \Big\{2\langle\langle W_{0,1}, g\rangle\rangle - \langle\langle g, (z - \gamma\tilde{\mathscr{S}})g\rangle\rangle$$
$$- \langle\langle \tilde{\mathscr{A}} g, (z - \gamma\tilde{\mathscr{S}})^{-1} \tilde{\mathscr{A}} g\rangle\rangle\Big\}$$

where the supremum is taken over local compactly supported smooth functions g. To get a lower bound it is sufficient to find a function g for which one can show that

$$2\langle\langle W_{0,1}, g\rangle\rangle - \langle\langle g, (z - \gamma\tilde{\mathscr{S}})g\rangle\rangle - \langle\langle \tilde{\mathscr{A}} g, (z - \gamma\tilde{\mathscr{S}})^{-1} \tilde{\mathscr{A}} g\rangle\rangle \geq c z^{-1/4}.$$

- Let $\mathbb{H}$ be the Hilbert space obtained by completion of the set of local functions w.r.t. the inner product $\langle\langle\cdot,\cdot\rangle\rangle$. Since $\nu_{\rho,\theta}$ is a product of Gamma distributions, the set of multivariate Laguerre polynomials form an orthogonal basis of $\mathbb{H}$. It is then possible to decompose $\mathbb{H}$ as an orthogonal sum $\oplus_{n\in\mathbb{N}}\mathbb{H}_n$ of subspaces $\mathbb{H}_n$ such that

$$\tilde{\mathscr{S}} : \mathbb{H}_n \to \mathbb{H}_n, \quad \tilde{\mathscr{A}} : \mathbb{H}_n \to \mathbb{H}_{n-1} \oplus \mathbb{H}_n \oplus \mathbb{H}_{n+1}.$$

The function $W_{0,1}$ belongs to $\mathbb{H}_2$. Then we restrict the variational formula to functions $g \in \mathbb{H}_2$ and we estimate the corresponding new variational problem which is still infinite dimensional but involves only functions belonging to $\mathbb{H}_1 \oplus \mathbb{H}_2 \oplus \mathbb{H}_3$. To solve this variational problem we adapt ideas developed first in the context of Asymmetric Simple Exclusion Process [4, 44] and exploited later for other models. One of the difficulties comes again from the fact that the noise is degenerate.

The extension of Theorem 13 to other interacting potentials is a challenging problem. The general strategy presented here could be carried out but the orthogonal basis (formed by Laguerre polynomials in the exponential case) is no longer explicit and only defined by some recurrence relations.

Acknowledgments This work has been supported by the Brazilian-French Network in Mathematics and the French Ministry of Education through the ANR grant EDNHS. The referees deserve thanks for careful reading and many useful comments.

References

1. Basile, G., Bernardin, C., Olla, S.: Momentum conserving model with anomalous thermal conductivity in low dimensional systems. Phys. Rev. Lett. **96**, 204303 (2006)
2. Basile, G., Bernardin, C., Olla, S.: Thermal conductivity for a momentum conservative model. Comm. Math. Phys. **287**(1), 67–98 (2009)
3. Basile, G., Delfini, L., Lepri, S., Livi, R., Olla, S., Politi, A.: Anomalous transport and relaxation in classical one-dimensional models. Eur. Phys. J. Spec. Top. **151**, 85–93 (2007). doi:10.1140/epjst/e2007-00364-7
4. Bernardin, C.: Fluctuations in the occupation time of a site in the asymmetric simple exclusion process. Ann. Probab. **32**(1B), 855–879 (2004)
5. Bernardin, C.:. Stationary nonequilibrium properties for a heat conduction model. Phys. Rev. E (3) **78**(2), 021134, 10 (2008)
6. Bernardin, C., Gonçalves, P.: Anomalous fluctuations for a perturbed Hamiltonian system with exponential interactions. Commun. Math. Phys. **235**(1), 291–332 (2014)
7. Bernardin, C., Gonçalves, P., Jara, M.: 3/4 fractional superdiffusion of energy in a system of harmonic oscillators perturbed by a conservative noise. ArXiv e-prints (February 2014)
8. Bernardin, C., Huveneers, F.: Small perturbation of a disordered harmonic chain by a noise and an anharmonic potential. Probab. Theory Relat. Fields **157**, 301–331 (2013)
9. Bernardin, C., Huveneers, F., Lebowitz, J.L., Liverani, C., Olla, S.: Green-kubo formula for weakly coupled systems with noise. Commun. Math. Phys. 1–36 (2014)
10. Bernardin, C., Olla, S.: Fourier's law for a microscopic model of heat conduction. J. Stat. Phys. **121**(3–4), 271–289 (2005)
11. Bernardin, C., Olla, S.: Transport properties of a chain of anharmonic oscillators with random flip of velocities. J. Stat. Phys. **145**(3–4), 1224–1255 (2011)

12. Bernardin, C., Olla, S.: Thermodynamics and non-equilibrium macroscopic dynamics of chains of anharmonic oscillators. In preparation (2014). https://www.ceremade.dauphine.fr/olla/springs13.pdf
13. Bernardin, C., Stoltz, G.: Anomalous diffusion for a class of systems with two conserved quantities. Nonlinearity **25**, 1099–1133 (2012)
14. Bernardin, C., Venkateshan, K., Lebowitz, J.L., Lukkarinen, J.: Harmonic systems with bulk noises. J. Stat. Phys. **146**, 800 (2012)
15. Bertini, L., De Sole, A., Gabrielli, D., Jona-Lasinio, G., Landim, C.: Large deviations for the boundary driven symmetric simple exclusion process. Math. Phys. Anal. Geom. **6**(3), 231–267 (2003)
16. Bertini, L., De Sole, A., Gabrielli, D., Jona-Lasinio, G., Landim, C.: Large deviation approach to non equilibrium processes in stochastic lattice gases. Bull. Braz. Math. Soc. (N.S.), **37**(4):611–643 (2006)
17. Bertini, L., De Sole, A., Gabrielli, D., Jona-Lasinio, G., Landim, C.: Stochastic interacting particle systems out of equilibrium. J. Stat. Mech. Theory Exp. **7**, P07014, 35 (2007)
18. Bodineau, T., Giacomin, G.: From dynamic to static large deviations in boundary driven exclusion particle systems. Stochastic Process. Appl. **110**(1), 67–81 (2004)
19. Bonetto, F., Lebowitz, J.L., Rey-Bellet, L.: Fourier's law: a challenge to theorists. In: Mathematical Physics 2000, pp. 128–150. Imperial College Press, London (2000)
20. Braxmeier-Even, N., Olla, S.: Hydrodynamic limit for a hamiltonian system with boundary conditions and conservative noise. Arch. Rational Mech. Anal. **213**, 561–585 (2014)
21. Bricmont, J., Dürr, D., Petruccione, F., Ghirardi, G., Gallavotti, G., Zanghi, N.: Chance in Physics: Foundations and Perspectives. Springer, Berlin (2001)
22. Carmona, P.: Existence and uniqueness of an invariant measure for a chain of oscillators in contact with two heat baths. Stochastic Process. Appl. **117**(8), 1076–1092 (2007)
23. De Roeck, W., Huveneers, F.: Asymptotic localization of energy in non-disordered oscillator chains. ArXiv e-prints, to appear in Comm. Pure Appl. Math (May 2013)
24. Dhar, A.: Heat transport in low-dimensional systems. Adv. Phys. **57**(5), 457–537 (2008)
25. Dhar, A., Lebowitz, J.L.: Effect of phonon-phonon interactions on localization. Phys. Rev. Lett. **100**, 134301 (2008)
26. Dolgopyat, Dmitry, Liverani, Carlangelo: Energy transfer in a fast-slow Hamiltonian system. Comm. Math. Phys. **308**(1), 201–225 (2011)
27. Farfan, J.: Static large deviations of boundary driven exclusion processes. ArXiv e-prints (August 2009)
28. Freidlin, M.I., Wentzell, A.D.: Random perturbations of dynamical systems. In: Grundlehren der Mathematischen Wissenschaften [Fundamental Principles of Mathematical Sciences], volume 260, 3rd edn. Springer, Heidelberg (2012). (Translated from the 1979 Russian original by Joseph Szücs)
29. Fritz, J.: Entropy pairs and compensated compactness for weakly asymmetric systems. In: Stochastic Analysis on Large Scale Interacting Systems. Adv. Stud. Pure Math. **39**, 143–171. Math. Soc. Japan, Tokyo (2004)
30. Fritz, J.: Microscopic theory of isothermal elastodynamics. Arch. Ration. Mech. Anal. **201**(1), 209–249 (2011)
31. Fritz, J., Funaki, T., Lebowitz, J.L.: Stationary states of random Hamiltonian systems. Probab. Theory Related Fields **99**(2), 211–236 (1994)
32. Fritz, J., Tóth, B.: Derivation of the Leroux system as the hydrodynamic limit of a two-component lattice gas. Comm. Math. Phys. **249**(1), 1–27 (2004)
33. Georgii, H.-O.: Gibbs measures and phase transitions. In: de Gruyter Studies in Mathematics, vol. 9, 2nd edn. Walter de Gruyter & Co., Berlin (2011)
34. Giacomin, G.: Anharmonic lattices, random walks and random interfaces. Recent Res. Dev. Stat. Phys. Transworld Res. Netw. **1**, 97–118 (2000)
35. Guo, M.Z., Papanicolaou, G.C., Varadhan, S.R.S.: Nonlinear diffusion limit for a system with nearest neighbor interactions. Comm. Math. Phys. **118**(1), 31–59 (1988)

36. Horváth, I., Tóth, B., Vető, B.: Relaxed sector condition. Bull. Inst. Math. Acad. Sin. (N.S.) **7**(4), 463–476 (2012)
37. Iacobucci, A., Legoll, F., Olla, S., Stoltz, G.: Thermal conductivity of the toda lattice with conservative noise. J. Stat. Phys. **140**, 336–348 (2010). doi:10.1007/s10955-010-9996-6
38. Iacobucci, A., Legoll, F., Olla, S., Stoltz, G.: Negative thermal conductivity in rotor model. Phys. Rev. E **84**, 061108 (2011)
39. Jara, M., Komorowski, T., Olla, S.: Superdiffusion of energy in a chain of harmonic oscillators with noise. ArXiv e-prints (February 2014)
40. Kac, M., van Moerbeke, P.: On an explicitly soluble system of nonlinear differential equations related to certain Toda lattices. Adv. Math. **16**, 160–169 (1975)
41. Kipnis, C., Landim, C.: Scaling limits of interacting particle systems. In: Grundlehren der Mathematischen Wissenschaften [Fundamental Principles of Mathematical Sciences], vol. 320. Springer, Berlin (1999)
42. Kipnis, C., Olla, S., Varadhan, S.R.S.: Hydrodynamics and large deviation for simple exclusion processes. Comm. Pure Appl. Math. **42**(2), 115–137 (1989)
43. Komorowski, T., Landim, C., Olla, S.: Fluctuation in Markov Processes: Time Symmetry and Martingale Approximations. To appear in Springer Grundlehern, Heidelberg (2012)
44. Landim, C., Quastel, J., Salmhofer, M., Yau, H.-T.: Superdiffusivity of asymmetric exclusion process in dimensions one and two. Comm. Math. Phys. **244**(3), 455–481 (2004)
45. Lanford III, O.E., Lebowitz, J.L., Lieb, E.H.: Time evolution of infinite anharmonic systems. J. Stat. Phys. **16**(6), 453–461 (1977)
46. Lepri, S., Mejía-Monasterio, C., Politi, A.: Nonequilibrium dynamics of a stochastic model of anomalous heat transport. J. Phys. A: Math. Theor. **43**(6), 065002 (2010)
47. Lepri, Stefano, Livi, Roberto, Politi, Antonio: Thermal conduction in classical low-dimensional lattices. Phys. Rep. **377**(1), 1–80 (2003)
48. Liverani, C., Olla, S.: Toward the Fourier law for a weakly interacting anharmonic crystal. J. Amer. Math. Soc. **25**(2), 555–583 (2012)
49. Mai, T., Dhar, A., Narayan, O.: Equilibration and universal heat conduction in fermi-pasta-ulam chains. Phys. Rev. Lett. **98**, 184301 (2007)
50. Marchioro, C., Pellegrinotti, A., Presutti, E.: Existence of time evolution for ν-dimensional statistical mechanics. Comm. Math. Phys. **40**, 175–185 (1975)
51. Olla, S., Varadhan, S.R.S., Yau, H.-T.: Hydrodynamical limit for a Hamiltonian system with weak noise. Comm. Math. Phys. **155**(3), 523–560 (1993)
52. Olla, Stefano, Sasada, Makiko: Macroscopic energy diffusion for a chain of anharmonic oscillators. Probab. Theory Related Fields **157**(3–4), 721–775 (2013)
53. Serre, D.: Systems of conservation laws. 2. Cambridge University Press, Cambridge (2000). (Geometric structures, oscillations, and initial-boundary value problems, Translated from the 1996 French original by I. N. Sneddon)
54. Simon, M.: Hydrodynamic limit for the velocity flip model. Stochastic Process. Appl. **123**, 3623–3662 (2013)
55. Spohn, H.: Large Scale Dynamics of Interacting Particles. Springer, Berlin (1991)
56. Spohn, H.: Fluctuating hydrodynamics for a chain of nonlinearly coupled rotators. ArXiv e-prints (2014)
57. Spohn, H. Stoltz, G.: Nonlinear fluctuating hydrodynamics in one dimension: the case of two conserved fields. ArXiv e-prints (2014)
58. Spohn, Herbert: Nonlinear fluctuating hydrodynamics for anharmonic chains. J. Stat. Phys. **154**(5), 1191–1227 (2014)
59. Szász, D.: Boltzmann's ergodic hypothesis, a conjecture for centuries? In: Hard ball systems and the Lorentz gas. Encyclopaedia Math. Sci. **101**, 421–448. Springer, Berlin (2000)
60. Tartar, L.: Compensated compactness and applications to partial differential equations. In: Nonlinear Analysis and Mechanics: Heriot-Watt Symposium, vol. IV. Res. Notes in Math. **39**, 136–212. Pitman, Boston (1979)
61. Toda, M.: Theory of nonlinear lattices. In: Springer Series in Solid-State Sciences, vol. 20, 2nd edn. Springer, Berlin (1989)

62. Tóth, B., Valkó, B.: Onsager relations and Eulerian hydrodynamic limit for systems with several conservation laws. J. Stat. Phys. **112**(3–4), 497–521 (2003)
63. Uchaikin, V.V., Zolotarev, V.M.: Chance and stability. Modern Probability and Statistics. VSP, Utrecht (1999). (Stable distributions and their applications, With a foreword by V. Yu. Korolev and Zolotarev)
64. Van Beijeren, H.: Exact results for anomalous transport in one dimensional hamiltonian systems. Phys. Rev. Lett. **108**, 180601 (2012)
65. Varadhan, S.R.S.: Nonlinear diffusion limit for a system with nearest neighbor interactions. II. In: Asymptotic Problems in Probability Theory: Stochastic Models and Diffusions on Fractals. Sanda, Kyoto (1990), Pitman Res. Notes Math. Ser., **283**, 75–128. Longman Sci. Tech. Harlow (1993)
66. Wang, J. Zhong, Y., Zhang, Y., Zhao, H.: Normal heat conduction in one-dimensional momentum conserving lattices with asymmetric interactions. Phys. Rev. E **85**(6), 060102 (2012)
67. Yau, H.-T.: Relative entropy and hydrodynamics of Ginzburg-Landau models. Lett. Math. Phys. **22**(1), 63–80 (1991)

Dissipative Reactive Fluid Models from the Kinetic Theory

Vincent Giovangigli

Abstract We present a kinetic framework describing mixtures of polyatomic species undergoing chemical reactions. Using a generalized Chapman-Enskog expansion, we derive the corresponding macroscopic fluid model. The hyperbolic-parabolic structure of the resulting system of partial differential equations is investigated and closely related to the underlying kinetic framework. We also discuss the Cauchy problem for smooth solutions as well as numerical algorithms for the evaluation of multicomponent transport coefficients using structural properties derived from the kinetic theory.

Keywords Kinetic theory · Multicomponent fluid · Hyperbolic-parabolic · Transport coefficients · Cauchy problem

1 Introduction

Multicomponent reactive flows undergoing chemical reactions arise in many engineering applications such as chemical reactors [1–3], Earth reentry of space vehicles [4, 5], or flames [6–8]. This is an important motivation for investigating the derivation of the corresponding fluid equations from the kinetic theory of gases as well as analyzing the mathematical structure of the resulting system of partial differential equations [9].

We first present a kinetic framework describing mixtures of polyatomic species undergoing chemical reactions. The Boltzmann equations governing the species distribution functions are presented in a semi-quantum framework [10–19] with reactive sources [20–34]. We only consider mixtures at thermodynamic equilibrium with a single temperature and thermodynamic nonequilibrium lay out of the scope of the present notes [35–46]. The kinetic entropy production is shown to be nonnegative, which yields the Boltzmann H-theorem. We present the Enskog expansion and focus

V. Giovangigli (✉)
CMAP-CNRS, Ecole Polytechnique, 91128 Palaiseau Cedex, France
e-mail: vincent.giovangigli@polytechnique.fr

P. Gonçalves and A.J. Soares (eds.), *From Particle Systems to Partial Differential Equations II*, Springer Proceedings in Mathematics & Statistics 129,
DOI 10.1007/978-3-319-16637-7_2

on regimes where the chemical characteristic times are larger than the collision times and than the times for relaxation of internal energy.

We derive the zeroth order as well as the first order corresponding macroscopic equations including conservation equations, thermodynamic properties, transport fluxes, transport coefficients and chemical production rates. The transport coefficients are defined in terms of bracket products involving solutions of integral linearized Boltzmann equations and their mathematical structure is *extracted* from the underlying kinetic framework. The macroscopic entropy conservation equation is obtained and we discuss the link between the kinetic entropy and the macroscopic fluid entropy. This kinetic study of reactive mixtures is performed in a physical framework, mathematical aspects of kinetic theory laying out of the scope of these notes [47–56]. The conservation equations, the transport fluxes as well as the source terms may also be obtained from different theories like the thermodynamic of irreversible processes or statistical thermodynamics but such theories do not yields the transport coefficients [57–64]. We further establish that upon neglecting the chemical pressure and the perturbations of the zeroth order source terms, both the Maxwellian reaction regime and the tempered reaction regime yield similar fluid models.

The transport fluxes of the resulting fluid conservation equations are written in terms of transport coefficients. The evaluation of accurate transport coefficients is therefore an important modeling and computational task [65–69]. However, evaluating the transport coefficients for gas mixtures require solving transport linear systems arising from Galerkin solution of systems of linearized Boltzmann equations. The mathematical structure of the transport linear systems is derived from the kinetic theory and the fast evaluation of accurate transport coefficients using either generalized conjugate gradient methods or stationary iterative algorithms is discussed [70–75]. The importance and/or influence of multicomponent transport for computing laminar flows is also addressed [76–85].

The fluid model derived from the kinetic theory is next embedded in a mathematical framework and recast in quasilinear form. The structural properties extracted from the kinetic underlying framework are transformed into relevant mathematical assumptions, thereby soundly founding the mathematical model. We then investigate the mathematical structure of the resulting system of partial differential equations. We discuss symmetrizability properties first using entropic variables and next using normal forms [86–104]. We explicitly evaluate the natural entropic form and next the natural normal form of the system of partial differential equations. We investigate the hyperbolic-parabolic structure using a definition of hyperbolicity from Denis Serre and further establish that for symmetric second order systems strong parabolicity is equivalent to Petrovsky parabolicity.

We then study the Cauchy problem for symmetrized systems of partial differential equations [105–118]. We present global existence theorems around constant equilibrium states as well as asymptotic stability and decay estimates [95]. The method of proof relies on the normal form of the governing equations, on entropic estimates, and on the local strict dissipativity properties of the linearized equations [89, 95, 106, 107].

The kinetic theory of reactive polyatomic gas mixtures is presented in Sects. 2 and 3. The evaluation of transport coefficients is addressed in Sect. 4. The resulting system of partial differential equations is imbedded in a mathematical framework in Sect. 5. Its hyperbolic-parabolic structure in investigated in Sect. 6 and the Cauchy problem is addressed in Sect. 7.

2 Kinetic Framework

We investigate the kinetic theory of mixtures of polyatomic gases [9–17, 19] with chemical reactions [20–34]. We present the Boltzmann equations governing the species distribution functions in a semi-quantum—or semi-classical—framework. We perform the Enskog expansion and obtain the zeroth order fluid governing equations. We only consider the situation of one temperature fluids, thermodynamic desequilibrium lying out of the scope of the present work [35–46]. Both Sects. 2 and 3 describing the physical derivation of the fluid equations are not formalized mathematically and only the macroscopic fluid equations will be investigated in a mathematical framework. Mathematical aspects concerning Boltzmann-type equations lay out of the scope of these notes and we refer the reader to [47–56].

2.1 Boltzmann Equations

We consider a reactive mixture composed of n species with internal degrees of freedom. The species equations are governed by Boltzmann equations written in a semi-quantum framework that may be obtained from Waldmann [11], Ludwig and Heil [24], Ferziger and Kaper [16], Alexeev et al. [31], Ern and Giovangigli [33], and Grunfeld [18].

The state of the mixture is described by the species distribution functions denoted by $f_i(t, \boldsymbol{x}, \boldsymbol{c}_i, \mathrm{I})$, $i \in S$, where t denotes the time, $\boldsymbol{x}$ the three-dimensional spatial coordinate, $\boldsymbol{c}_i$ the velocity of the ith species, I the index for the quantum state of the ith species, and $S = \{1, \dots, n\}$ the set of species indices. We denote by Q_i the indexing set of the quantum states of the ith species. The quantity $f_i(t, \boldsymbol{x}, \boldsymbol{c}_i, \mathrm{I})\delta\boldsymbol{x}\delta\boldsymbol{c}_i$ represents the expected number of molecules of type i in quantum state I in the volume element $\delta\boldsymbol{x}$ located at $\boldsymbol{x}$, whose velocities lie in $\delta\boldsymbol{c}_i$ about velocity $\boldsymbol{c}_i$ at time t.

We denote by $\mathfrak{m}_i$ the mass of the molecule of the ith species, $E_{i\mathrm{I}}$ the internal energy of the molecules of species i in quantum state I and $a_{i\mathrm{I}}$ the degeneracy of the Ith quantum state. The fluid macroscopic properties are directly obtained from the distribution functions. More specifically, the number of molecules of the ith species per unit volume $\mathfrak{n}_i$ is given by

$$\mathfrak{n}_i = \sum_{\mathrm{I}\in\mathrm{Q}_i} \int f_i \mathrm{d}\boldsymbol{c}_i, \tag{1}$$

and the mass dentity of the ith species is then $\rho_i = \mathfrak{m}_i \mathfrak{n}_i$. The mass averaged velocity $\boldsymbol{v}$ is obtained from

$$\rho \boldsymbol{v} = \sum_{\substack{i \in S \\ \mathrm{I} \in \mathrm{Q}_i}} \int \mathfrak{m}_i \boldsymbol{c}_i f_i \mathrm{d}\boldsymbol{c}_i, \tag{2}$$

where $\rho = \sum_{i \in S} \rho_i$ is the total mass density. The total energy per unit volume is also defined by

$$\mathcal{E} + \tfrac{1}{2}\rho|\boldsymbol{v}|^2 = \sum_{\substack{i \in S \\ \mathrm{I} \in \mathrm{Q}_i}} \int (E_{i\mathrm{I}} + \tfrac{1}{2}\mathfrak{m}_i|\boldsymbol{c}_i|^2) f_i \mathrm{d}\boldsymbol{c}_i, \tag{3}$$

where $\mathcal{E}$ is the internal energy per unit volume. The family of distribution functions will be written for convenience in the form $(f_i)_{i \in S}$. More generally, for a family of functions ξ_i, $i \in S$, where ξ_i depends on $\boldsymbol{c}_i$ and I, we will use the compact notation $\xi = (\xi_i)_{i \in S}$, the dependence on $(t, \boldsymbol{x})$ being left implicit.

The species distribution functions are solutions of generalized Boltzmann equations in the form

$$\mathcal{D}_i(f_i) = \mathcal{S}_i(f) + \mathcal{C}_i(f), \qquad i \in S, \tag{4}$$

where $\mathcal{D}_i(f_i)$ is the streaming differential operator, $\mathcal{S}_i(f)$ the scattering source term and $\mathcal{C}_i(f)$ the reactive source term. These Boltzmann equations express the conservation of particles in the phase space and may also be derived from the BBGKY-hierarchy [10–13, 15–19]. The streaming differential operator $\mathcal{D}_i(f_i)$ may be written

$$\mathcal{D}_i(f_i) = \partial_t f_i + \boldsymbol{c}_i \boldsymbol{\cdot} \boldsymbol{\nabla} f_i + \boldsymbol{b}_i \boldsymbol{\cdot} \boldsymbol{\partial}_{\boldsymbol{c}_i} f_i, \tag{5}$$

where $\boldsymbol{b}_i$ is the force per unit mass acting on the ith species.

The scattering term may be written

$$\mathcal{S}_i(f) = \sum_{j \in S} \sum_{\mathrm{I}' \in \mathrm{Q}_i} \sum_{\mathrm{J},\mathrm{J}' \in \mathrm{Q}_j} \int \Big(f_i' f_j' \frac{a_{i\mathrm{I}} a_{j\mathrm{J}}}{a_{i\mathrm{I}'} a_{j\mathrm{J}'}} - f_i f_j \Big) \mathrm{W}_{ij}^{\mathrm{IJI'J'}} \, \mathrm{d}\boldsymbol{c}_j \mathrm{d}\boldsymbol{c}_i' \mathrm{d}\boldsymbol{c}_j', \tag{6}$$

where, in a direct collision, I and J are the quantum states before collision, I′ and J′ the states after collision, $a_{i\mathrm{I}}$ the degeneracy of the Ith quantum state, and $\mathrm{W}_{ij}^{\mathrm{IJI'J'}}$ the transition probability for nonreactive collisions. We have denoted by f_i' the distribution $f_i' = f_i(t, \boldsymbol{x}, \boldsymbol{c}_i', \mathrm{I}')$ where $\boldsymbol{c}_i'$ is the velocity and I′ the quantum state after collision. The following reciprocity relations are satisfied by the transition probabilities [11]

$$a_{i\mathrm{I}} a_{j\mathrm{J}} \mathrm{W}_{ij}^{\mathrm{IJI'J'}} = a_{i\mathrm{I}'} a_{j\mathrm{J}'} \mathrm{W}_{ij}^{\mathrm{I'J'IJ}}. \tag{7}$$

These reciprocity relations are of fundamental importance in the theory since they are related to the symmetry properties of the collision operator and are also requisites for Boltzmann H-theorem. We are using transition probabilities rather than collision cross sections for convenience since the reactive collision terms are then much simpler to write [9, 24, 31]. For binary nonreactive collisions, denoting by $\sigma_{ij}^{\mathrm{IJI'J'}}$ the collision cross section, we have

$$g_{ij}\sigma_{ij}^{\mathrm{IJI'J'}}\mathrm{d}\boldsymbol{e}'_{ij} = \mathrm{W}_{ij}^{\mathrm{IJI'J'}}\mathrm{d}\boldsymbol{c}'_i\mathrm{d}\boldsymbol{c}'_j, \tag{8}$$

where g_{ij} denotes the norm of the relative velocity $\boldsymbol{c}_i - \boldsymbol{c}_j$ of the collision partners before collision and $\boldsymbol{e}'_{ij}$ the unit vector in the direction of the relative velocity $\boldsymbol{c}'_i - \boldsymbol{c}'_j$ after collision [18, 24, 31]. The conservation of mass, momentum and energy during collision is taken into account by using Dirac delta functions in the transition probabilities [18, 24, 31].

The reactive source term $\mathcal{C}_i(f)$ is due to chemical reaction between the species of the mixture. We consider an arbitrary chemical reaction mechanism including binary as well as ternary mixtures. Even though ternary collisions may be neglected in the nonreactive collision term $\mathcal{S}_i(f)$, ternary collisions play an important role in chemistry since it is often the unique chemical path in order to form some type of molecules [24, 31]. Ternary collision may also be seen as a succession of two rapid binary collisions [31]. The chemical reactions are indexed by $r \in \mathfrak{R} = \{1, \dots, n^{\mathrm{r}}\}$ where n^{r} is the number of reactions and may be written [33]

$$\sum_{i\in\mathcal{F}^r} \mathfrak{M}_i \rightleftarrows \sum_{k\in\mathcal{B}^r} \mathfrak{M}_k, \qquad r \in \mathfrak{R}, \tag{9}$$

where $\mathcal{F}^r$ and $\mathcal{B}^r$ are the indices of reactants and products with their multiplicity. We denote by ν_{ir}^{f} and ν_{ir}^{b} the stoichiometric coefficients of the ith species in the rth reaction, that is, the multiplicity of species i in $\mathcal{F}^r$ and $\mathcal{B}^r$, respectively, and by F^r and B^r the indices of the quantum energy states of the reactants and products, respectively. We also denote by $\mathcal{F}_i^r$ the subset of $\mathcal{F}^r$ where the index i has been removed once with similar notation for $\mathcal{B}_k^r$, $\mathrm{F}_{\mathrm{I}}^r$ and $\mathrm{B}_{\mathrm{K}}^r$. The reactive collision term for the ith species may then be written [9, 33]

$$\mathcal{C}_i(f) = \sum_{r\in\mathfrak{R}} \mathcal{C}_i^r(f), \tag{10}$$

where $\mathcal{C}_i^r(f)$ represents the contribution of the rth reaction. This term $\mathcal{C}_i^r(f)$ is given by

$$\begin{aligned}\mathcal{C}_i^r(f) = {} & \nu_{ir}^{\mathrm{f}} \sum_{\mathrm{F}_{\mathrm{I}}^r,\mathrm{B}^r} \int \Bigg(\prod_{k\in\mathcal{B}^r} f_k \frac{\prod_{k\in\mathcal{B}^r}\beta_{k\mathrm{K}}}{\prod_{j\in\mathcal{F}^r}\beta_{j\mathrm{J}}} - \prod_{j\in\mathcal{F}^r} f_j \Bigg) \mathcal{W}_{\mathcal{F}^r\mathcal{B}^r}^{\mathrm{F}^r\mathrm{B}^r} \prod_{j\in\mathcal{F}_i^r} \mathrm{d}\boldsymbol{c}_j \prod_{k\in\mathcal{B}^r} \mathrm{d}\boldsymbol{c}_k \\ & - \nu_{ir}^{\mathrm{b}} \sum_{\mathrm{F}^r,\mathrm{B}_{\mathrm{I}}^r} \int \Bigg(\prod_{k\in\mathcal{B}^r} f_k \frac{\prod_{k\in\mathcal{B}^r}\beta_{k\mathrm{K}}}{\prod_{j\in\mathcal{F}^r}\beta_{j\mathrm{J}}} - \prod_{j\in\mathcal{F}^r} f_j \Bigg) \mathcal{W}_{\mathcal{F}^r\mathcal{B}^r}^{\mathrm{F}^r\mathrm{B}^r} \prod_{j\in\mathcal{F}^r} \mathrm{d}\boldsymbol{c}_j \prod_{k\in\mathcal{B}_i^r} \mathrm{d}\boldsymbol{c}_k,\end{aligned} \tag{11}$$

where $\beta_{i\mathrm{I}} = \mathrm{h}_{\mathrm{P}}^3/(a_{i\mathrm{I}}\mathfrak{m}_i^3)$, h_{P} is the Planck constant, and $\mathcal{W}_{\mathcal{F}^r\mathcal{B}^r}^{\mathrm{F}^r\mathrm{B}^r}$ denotes the transition probability that a collision between the reactants $\mathcal{F}^r$ with energies F^r lead to the products $\mathcal{B}^r$ with energies B^r. The sommation over F^r in (11) represents the sum over all quantum indices J for all $j \in \mathcal{F}^r$ with similar conventions for $\mathrm{F}_{\mathrm{I}}^r$, B^r, and $\mathrm{B}_{\mathrm{I}}^r$. Finally, the following reciprocity relations hold between transition probabilities [9, 24, 31, 33]

$$\mathcal{W}_{\mathcal{F}^r\mathcal{B}^r}^{\mathrm{F}^r\mathrm{B}^r} \prod_{k\in\mathcal{B}^r} \beta_{k\mathrm{K}} = \mathcal{W}_{\mathcal{B}^r\mathcal{F}^r}^{\mathrm{B}^r\mathrm{F}^r} \prod_{j\in\mathcal{F}^r} \beta_{j\mathrm{J}}, \tag{12}$$

and generalize the relations (7) between the nonreactive transition probabilities. The reciprocity relations (12) between reactive transition probabilities are also important and imply in particular the symmetry properties of linearized chemical source at equilibrium that may be seen as Onsager relations for chemistry.

2.2 Collisional Invariants

A collisional invariant ψ of the rapid collision operator $\mathcal{S} = (\mathcal{S}_i(f))_{i\in S}$ is by definition a family $\psi = (\psi_i)_{i\in S}$ such that $\psi_i + \psi_j = \psi_i' + \psi_j'$ for any nonreactive collision between species i and j. The scalar collisional invariants of the rapid collision operator form a vector space spanned by the invariants ψ^l, $l \in \{1,\dots,n+4\}$, defined by

$$\psi^l = \begin{cases} (\delta_{li})_{i\in S}, & l \in S, \\ (\mathfrak{m}_i c_{i\nu})_{i\in S}, & l = n+\nu, \quad \nu \in \{1,2,3\}, \\ \left(\frac{1}{2}\mathfrak{m}_i \boldsymbol{c}_i\cdot\boldsymbol{c}_i + E_{i\mathrm{I}}\right)_{i\in S}, & l = n+4, \end{cases}$$

where $\mathfrak{m}_i$ is the mass of the molecule of the ith species and $c_{i\nu}$ the component of $\boldsymbol{c}_i$ in the ν spatial direction [10–13, 15, 16]. The n first invariants ψ^l, $l \in \{1,\dots,n\}$, are associated with species conservation in nonreactive collisions, the invariants ψ^{n+1}, ψ^{n+2}, and ψ^{n+3}, with momentum conservation in the three spatial directions, and ψ^{n+4} with total energy conservation [9, 11, 15, 16]. The collisional invariants of the complete collision operator $\mathcal{S} + \mathcal{C} = \big(\mathcal{S}_i(f) + \mathcal{C}_i(f)\big)_{i\in S}$ are associated with the conservation of momentum, total energy and chemical elements or atoms [33]. Denoting by $\mathfrak{a}_{il}$ the number of lth atom in the ith species, $\mathfrak{A} = \{1,\dots,n^{\mathrm{a}}\}$ the set of atom indices, and $n^{\mathrm{a}} \geqslant 1$ the number of atoms—or elements—in the mixture, the atomic collisional invariants of the complete collision operator may be written $\mathfrak{a}_l = (\mathfrak{a}_{il})_{i\in S}$ for $l \in \mathfrak{A}$.

Remark 2.1 When molecules are not spherically symmetric, there could be another summational invariant, namely angular momentum. However, we are only interested in these notes with *isotropic* distributions without micro-polarizations associated with strong magnetic fields. For such isotropic distributions the angular momentum summational invariant plays no role [16].

For two tensor families $\xi = (\xi_i)_{i\in S}$ and $\zeta = (\zeta_i)_{i\in S}$, we define the scalar product

$$\langle\!\langle \xi, \zeta \rangle\!\rangle = \sum_{\substack{i\in S\\ \mathrm{I}\in \mathrm{Q}_i}} \int \xi_i \odot \zeta_i \, \mathrm{d}\boldsymbol{c}_i,$$

where $\xi_i \odot \zeta_i$ is the full contracted product between tensor ξ_i and tensor ζ_i. We introduce families of tensors since they naturally arise in the definition of transport coefficients. The macroscopic properties naturally associated with the fluid may then be written in the compact form

$$\langle\!\langle f, \psi^l \rangle\!\rangle = \begin{cases} \mathfrak{n}_l, & l \in S, \\ \rho v_\nu, & l = n + \nu, \quad \nu \in \{1, 2, 3\}, \\ \frac{1}{2}\rho \boldsymbol{v}\cdot\boldsymbol{v} + \mathcal{E}, & l = n + 4, \end{cases}$$

where v_ν denotes the component in direction ν of the mass average velocity $\boldsymbol{v}$.

2.3 *Kinetic Entropy*

The kinetic entropy per unit volume is defined by

$$\mathcal{S}^{\mathrm{kin}} = -k_{\mathrm{B}} \sum_{\substack{i\in S\\ \mathrm{I}\in \mathrm{Q}_i}} \int f_i \big(\log(\beta_{i\mathrm{I}} f_i) - 1\big)\, \mathrm{d}\boldsymbol{c}_i, \tag{13}$$

where k_{B} is the Boltzmann constant and $\beta_{i\mathrm{I}} = \mathrm{h}_{\mathrm{P}}^3/(a_{i\mathrm{I}}\mathfrak{m}_i^3)$. Multiplying the Boltzmann equation (4) by $\log(\beta_{i\mathrm{I}} f_i)$, integrating with respect to $\mathrm{d}\boldsymbol{c}_i$, summing over the species $i \in S$ and over the quantum states $\mathrm{I} \in \mathrm{Q}_i$, we obtain a balance equation for $\mathcal{S}^{\mathrm{kin}}$ in the form

$$\partial_t \mathcal{S}^{\mathrm{kin}} + \nabla\cdot(\mathcal{S}^{\mathrm{kin}}\boldsymbol{v}) + \nabla\cdot\mathcal{J}^{\mathrm{kin}} = \mathfrak{v}^{\mathrm{kin}}, \tag{14}$$

where $\mathcal{J}^{\mathrm{kin}}$ is the entropy diffusive flux,

$$\mathcal{J}^{\mathrm{kin}} = -k_{\mathrm{B}} \sum_{\substack{i\in S\\ \mathrm{I}\in \mathrm{Q}_i}} \int (\boldsymbol{c}_i - \boldsymbol{v}) f_i \big(\log(\beta_{i\mathrm{I}} f_i) - 1\big)\, \mathrm{d}\boldsymbol{c}_i, \tag{15}$$

and $\mathfrak{v}^{\mathrm{kin}}$ the entropy source term. The entropy source term $\mathfrak{v}^{\mathrm{kin}}$ may be written $\mathfrak{v}^{\mathrm{kin}} = \mathfrak{v}^{\mathfrak{S}} + \mathfrak{v}^{\mathfrak{C}}$ with contributions from the nonreactive collision

$$\mathfrak{v}^{\mathfrak{S}} = -k_{\mathrm{B}} \sum_{\substack{i\in S\\ \mathrm{I}\in \mathrm{Q}_i}} \int \mathfrak{S}_i(f) \log(\beta_{i\mathrm{I}} f_i)\, \mathrm{d}\boldsymbol{c}_i, \tag{16}$$

and reactive collisions

$$\mathfrak{v}^{\mathcal{C}} = -k_{\mathrm{B}} \sum_{\substack{i \in S \\ \mathrm{I} \in \mathrm{Q}_i}} \int \mathcal{C}_i(f) \log(\beta_{i\mathrm{I}} f_i) \, \mathrm{d}\boldsymbol{c}_i. \tag{17}$$

After some algebra, it is obtained that

$$\mathfrak{v}^{\mathcal{S}} = \frac{k_{\mathrm{B}}}{4} \sum_{i,j \in S} \sum_{\mathrm{II'JJ'}} \int \Upsilon\Big(\frac{f_i' f_j'}{a_{i\mathrm{I}'} a_{j\mathrm{J}'}}, \frac{f_i f_j}{a_{i\mathrm{I}} a_{j\mathrm{J}}}\Big) \mathrm{W}_{ij}^{\mathrm{IJI'J'}} a_{i\mathrm{I}} a_{j\mathrm{J}} \, \mathrm{d}\boldsymbol{c}_i \mathrm{d}\boldsymbol{c}_j \mathrm{d}\boldsymbol{c}_i' \mathrm{d}\boldsymbol{c}_j', \tag{18}$$

$$\mathfrak{v}^{\mathcal{C}} = \frac{k_{\mathrm{B}}}{4} \sum_{r \in \mathfrak{R}} \sum_{\mathrm{F}^r, \mathrm{B}^r} \int \Upsilon\Big(\prod_{k \in \mathcal{B}^r} \beta_{k\mathrm{K}} f_k, \prod_{j \in \mathcal{F}^r} \beta_{j\mathrm{J}} f_j\Big) \frac{\mathcal{W}_{\mathcal{F}^r \mathcal{B}^r}^{\mathrm{F}^r \mathrm{B}^r}}{\prod_{j \in \mathcal{F}^r} \beta_{j\mathrm{J}}} \prod_{j \in \mathcal{F}^r} \mathrm{d}\boldsymbol{c}_j \prod_{k \in \mathcal{B}^r} \mathrm{d}\boldsymbol{c}_k, \tag{19}$$

where Υ denotes the function $\Upsilon(x, y) = (x - y)(\log x - \log y)$. Since this function only takes nonnegative values, we conclude that both quantities $\mathfrak{v}^{\mathcal{S}}$ and $\mathfrak{v}^{\mathcal{C}}$ are sum of nonnegative terms. All collisions, nonreactive or reactive, thus yield nonnegative entropy productions [9, 33]. The generalized Boltzmann equations are thus compatible with the Boltzmann H-theorem and will lead to a dissipative structure at the molecular level. This important property must be recovered at the macroscopic level, that is, both type of collisions should yield nonnegative macroscopic entropy production.

2.4 Enskog Expansion

An approximate solution of Boltzmann equations (4) is obtained with Enskog expansion and we assume to this aim that the chemical time scales are larger than the collision times as well as the times for relaxation of internal energy. We thus write the Boltzmann equations (4) in the form

$$\mathcal{D}_i(f_i) = \frac{1}{\epsilon} \mathcal{S}_i(f) + \epsilon^a \mathcal{C}_i(f), \qquad i \in S, \tag{20}$$

where ϵ is the formal parameter associated with Enskog expansion and a is a integer which depends upon the regime under consideration. We only consider in these notes the situation of Maxwellian reactions $a = 1$ or the situation of tempered reactions $a = 0$, whereas the kinetic equilibrium regime $a = -1$ investigated in [33] lay out of the scope of the present notes.

The species distribution functions are expanded in the form

$$f_i = f_i^0 \big(1 + \epsilon \phi_i + \mathcal{O}(\epsilon^2)\big), \qquad i \in S, \tag{21}$$

and the Chapman-Enskog method requires that f and f^0 have the same macroscopic observables

$$\langle\!\langle f^0, \psi^l \rangle\!\rangle = \langle\!\langle f, \psi^l \rangle\!\rangle, \qquad l \in \{1, \dots, n+4\}. \tag{22}$$

We establish in next section that f_i^0 is the local Maxwellian and introduce the following convenient notation $f^0 = (f_i^0)_{i\in S}$, $\mathcal{D}(\xi) = \big(\mathcal{D}_i(\xi_i)\big)_{i\in S}$, $\mathcal{S}(\xi) = \big(\mathcal{S}_i(\xi)\big)_{i\in S}$, $\mathcal{C}(\xi) = \big(\mathcal{C}_i(\xi)\big)_{i\in S}$ and $\mathcal{C}^r(\xi) = \big(\mathcal{C}_i^r(\xi)\big)_{i\in S}$, $r \in \mathfrak{R}$, where $\xi = (\xi)_{i\in S}$ is a family of functions ξ_i depending on $(\boldsymbol{c}_i, \mathrm{I})$.

2.5 Maxwellian Distributions

The zeroth order distributions $f^0 = (f_i^0)_{i\in S}$ satisfy the zeroth order equations

$$\mathcal{S}_i(f^0) = 0, \qquad i \in S. \tag{23}$$

Multiplying (23) by $\log(\beta_{i\mathrm{I}} f_i^0)$, integrating with respect to $\mathrm{d}\boldsymbol{c}_i$, summing over $i \in S$ and over $\mathrm{I} \in \mathrm{Q}_i$, it is obtained that

$$\sum_{\substack{i\in S\\ \mathrm{I}\in \mathrm{Q}_i}} \int \log(\beta_{i\mathrm{I}} f_i^0)\mathcal{S}_i(f^0)\,\mathrm{d}\boldsymbol{c}_i = 0.$$

Using the expression (18) of the source term $\mathfrak{v}^{\mathcal{S}}$ we obtain that $\big(\log(\beta_{i\mathrm{I}} f_i^0)\big)_{i\in S}$ is a collision invariant. It is then a linear combination of the invariants ψ^l, $1 \le l \le n+4$, in such a way that

$$\log(\beta_{i\mathrm{I}} f_i^0) = \alpha_i - \boldsymbol{\beta}\cdot \mathfrak{m}_i \boldsymbol{c}_i - \gamma(\tfrac{1}{2}\mathfrak{m}_i \boldsymbol{c}_i\cdot\boldsymbol{c}_i + E_{i\mathrm{I}}), \qquad i \in S,$$

where $\alpha_i \in \mathbb{R}$, $\boldsymbol{\beta} \in \mathbb{R}^3$ and $\gamma \in \mathbb{R}$, are parameters determined by the macroscopic constraints (22). After some algebra, it is obtained that

$$f_i^0 = \frac{\mathfrak{n}_i}{\beta_{i\mathrm{I}} \mathrm{Z}_i} \exp\Big(-\frac{\mathfrak{m}_i}{2k_{\mathrm{B}}T}\boldsymbol{C}_i\cdot\boldsymbol{C}_i - \frac{E_{i\mathrm{I}}}{k_{\mathrm{B}}T}\Big), \qquad i \in S, \tag{24}$$

where $\boldsymbol{C}_i = \boldsymbol{c}_i - \boldsymbol{v}$ is the relative velocity of the ith species, T the temperature and Z_i the partition function per unit volume of the ith species. This partition function Z_i of the ith species may be written

$$\mathrm{Z}_i = \mathrm{Z}_i^{\mathrm{int}} \mathrm{Z}_i^{\mathrm{tr}},$$

where $\mathrm{Z}_i^{\mathrm{int}}$ and $\mathrm{Z}_i^{\mathrm{tr}}$ denote respectively the internal partition function and the translational partition function per unit volume

$$\mathrm{z}_i^{\mathrm{int}} = \sum_{\mathrm{I}\in \mathrm{Q}_i} a_{i\mathrm{I}} \exp\Big(-\frac{E_{i\mathrm{I}}}{k_{\mathrm{B}}T}\Big), \qquad \mathrm{z}_i^{\mathrm{tr}} = \Big(\frac{2\pi \mathfrak{m}_i k_{\mathrm{B}} T}{\mathrm{h}_{\mathrm{P}}^2}\Big)^{3/2},$$

and another convenient expression of the local Maxwellian f_i^0 reads

$$f_i^0 = \Big(\frac{\mathfrak{m}_i}{2\pi k_{\mathrm{B}} T}\Big)^{3/2} \frac{a_{i\mathrm{I}} \mathfrak{n}_i}{\mathrm{z}_i^{\mathrm{int}}} \exp\Big\{-\frac{\mathfrak{m}_i}{2k_{\mathrm{B}}T} \boldsymbol{C}_i\cdot\boldsymbol{C}_i - \frac{E_{i\mathrm{I}}}{k_{\mathrm{B}}T}\Big\}. \tag{25}$$

2.6 Zeroth Order Equations

Macroscopic equations are generally obtained by taking scalar products of the Boltzmann equations by collisional invariants. At zeroth order, only the terms that are $O(\epsilon^0)$ are taken into account and it is obtained that

$$\langle\!\langle \mathcal{D}(f^0), \psi^l \rangle\!\rangle = \delta_{a0} \langle\!\langle \mathcal{C}(f^0), \psi^l \rangle\!\rangle, \tag{26}$$

for $l \in \{1, \dots, n+4\}$. After a few algebra, we deduce for $l = 1, \dots, n$, the species mass conservation equations

$$\partial_t \rho_i + \nabla\cdot(\rho_i \boldsymbol{v}) = \delta_{a0} \mathfrak{m}_i \mathfrak{w}_i^0, \qquad i \in S, \tag{27}$$

where $\rho_i = \mathfrak{m}_i \mathfrak{n}_i$ is the mass density of the ith species and $\mathfrak{w}_i^0$ the zeroth order chemical production rate

$$\mathfrak{w}_i^0 = \langle\!\langle \psi^i, \mathcal{C}(f^0) \rangle\!\rangle = \sum_{\mathrm{I}\in \mathrm{Q}_i} \int \mathcal{C}_i(f^0)\, \mathrm{d}\boldsymbol{c}_i, \qquad i \in S.$$

The momentum conservation equations are obtained for $l = n+1, n+2, n+3$, and may be written in vector form

$$\partial_t(\rho \boldsymbol{v}) + \nabla\cdot(\rho \boldsymbol{v}\otimes\boldsymbol{v} + p\boldsymbol{I}) = \rho \boldsymbol{b}, \tag{28}$$

where $\boldsymbol{I}$ is the unit tensor, p the pressure, and $\boldsymbol{b}$ the average force

$$\rho \boldsymbol{b} = \sum_{i\in S} \rho_i \boldsymbol{b}_i. \tag{29}$$

Finally, the equation obtained with $l = n+4$ express the conservation of total energy

$$\partial_t\big(\tfrac{1}{2}\rho \boldsymbol{v}\cdot\boldsymbol{v} + \mathcal{E}\big) + \nabla\cdot\Big(\big(\tfrac{1}{2}\rho \boldsymbol{v}\cdot\boldsymbol{v} + \mathcal{E} + p\big)\boldsymbol{v}\Big) = \rho \boldsymbol{b}\cdot\boldsymbol{v}. \tag{30}$$

The macroscopic zeroth order equations are thus the compressible Euler equations for a reactive mixture of polyatomic gases.

2.7 Thermal Properties

The internal energy per unit volume $\mathcal{E}$ may be evaluated by using $\mathcal{E} = \langle\langle f, \psi^{\text{int}}\rangle\rangle = \langle\langle f^0, \psi^{\text{int}}\rangle\rangle$ and the expression of the Maxwellian distributions, where the collisional invariant $\psi^{\text{int}} = (\psi_i^{\text{int}})_{i\in S}$ is defined by $\psi_i^{\text{int}} = E_{i\mathrm{I}} + \frac{1}{2}\mathfrak{m}_i|\boldsymbol{C}_i|^2$. After some algebra, it is obtained that

$$\mathcal{E} = \sum_{i\in S} \mathfrak{n}_i (\tfrac{3}{2}k_{\mathrm{B}}T + \overline{E}_i),$$

where $\overline{E}_i$ denotes the average internal energy per molecule of the ith species

$$\overline{E}_i = \frac{1}{\mathrm{Z}_i^{\text{int}}} \sum_{\mathrm{I}\in\mathrm{Q}_i} a_{i\mathrm{I}} E_{i\mathrm{I}} \exp\Big(-\frac{E_{i\mathrm{I}}}{k_{\mathrm{B}}T}\Big), \qquad i \in S.$$

Defining the energy per unit mass of the ith species by $e_i = (\frac{3}{2}k_{\mathrm{B}}T + \overline{E}_i)/\mathfrak{m}_i$ we also have $\mathcal{E} = \rho e = \sum_{i\in S} \rho_i e_i$. Similarly, the enthalpy per unit volume $\mathcal{H}$ is given by

$$\mathcal{H} = \sum_{i\in S} \mathfrak{n}_i (\tfrac{5}{2}k_{\mathrm{B}}T + \overline{E}_i), \tag{31}$$

or equivalently by $\mathcal{H} = \rho h = \sum_{i\in S} \rho_i h_i$ where $h_i = (\frac{5}{2}k_{\mathrm{B}}T + \overline{E}_i)/\mathfrak{m}_i$ is the enthalpy per unit mass of the ith species.

We define the internal specific heat $\mathfrak{c}_i^{\text{int}}$ per molecule of the ith species by $\mathfrak{c}_i^{\text{int}} = \frac{\mathrm{d}\overline{E}_i}{\mathrm{d}T}$ and from the expression of $\overline{E}_i$ we obtain that

$$\mathfrak{c}_i^{\text{int}} = k_{\mathrm{B}} \sum_{\mathrm{I}\in\mathrm{Q}_i} \frac{a_{i\mathrm{I}}}{\mathrm{Z}_i^{\text{int}}} \Big(\frac{E_{i\mathrm{I}} - \overline{E}_i}{k_{\mathrm{B}}T}\Big)^2 \exp\Big(-\frac{E_{i\mathrm{I}}}{k_{\mathrm{B}}T}\Big), \qquad i \in S.$$

The translational specific heat at constant volume $\mathfrak{c}_{\mathrm{v}}^{\text{tr}}$ and the specific heat at constant volume $\mathfrak{c}_{i\mathrm{v}}$ of the ith species are then defined by

$$\mathfrak{c}_{\mathrm{v}}^{\text{tr}} = \tfrac{3}{2}k_{\mathrm{B}}, \qquad \mathfrak{c}_{i\mathrm{v}} = \mathfrak{c}_{\mathrm{v}}^{\text{tr}} + \mathfrak{c}_i^{\text{int}}.$$

The mixture internal specific heat $\mathfrak{c}^{\text{int}}$ and heat at constant volume are also defined by

$$\mathfrak{c}^{\text{int}} = \sum_{i\in S} \frac{\mathfrak{n}_i}{\mathfrak{n}} \mathfrak{c}_i^{\text{int}}, \qquad \mathfrak{c}_{\mathrm{v}} = \mathfrak{c}_{\mathrm{v}}^{\text{tr}} + \mathfrak{c}^{\text{int}},$$

where $\mathfrak{n} = \sum_{i \in S} \mathfrak{n}_i$ denotes the total number density. On the other hand, the pressure is given by the state law

$$p = \mathfrak{n} k_B T, \tag{32}$$

and we recover the perfect gas law.

It will be necessary in the following to use a zeroth equation for the temperature T. After some algebra, the following zeroth order conservation equation is obtained for the temperature T

$$\mathfrak{n} \mathfrak{c}_v (\partial_t T + \boldsymbol{v} \cdot \nabla T) = -p \nabla \cdot \boldsymbol{v} - \delta_{a0} \sum_{i \in S} (\tfrac{3}{2} k_B T + \overline{E}_i) \mathfrak{w}_i^0. \tag{33}$$

2.8 Maxwellian Production Rates

The zeroth order source term may be written in the form [9]

$$\mathfrak{w}_i^0 = \sum_{r \in \mathfrak{R}} (\nu_{ir}^{b} - \nu_{ir}^{f}) \overline{\tau}_r, \qquad i \in S, \tag{34}$$

where $\overline{\tau}_r$ denotes the rate of progress of the rth reaction. The rate of progress $\overline{\tau}_r$ is obtained from $(\nu_{ir}^{b} - \nu_{ir}^{f}) \overline{\tau}_r = \langle\!\langle \mathcal{C}^r(f^0), \psi^i \rangle\!\rangle$ and may be written

$$\overline{\tau}_r = \mathcal{K}_r \Big(\prod_{j \in S} \Big(\frac{\mathfrak{n}_j}{\mathrm{Z}_j}\Big)^{\nu_{jr}^{f}} - \prod_{j \in S} \Big(\frac{\mathfrak{n}_j}{\mathrm{Z}_j}\Big)^{\nu_{jr}^{b}} \Big), \qquad r \in \mathfrak{R}, \tag{35}$$

where $\mathcal{K}_r$ is the rate constant of the rth reaction [9, 33]

$$\mathcal{K}_r = \sum_{\mathrm{F}^r, \mathrm{B}^r} \int \prod_{j \in \mathcal{F}^r} \exp\Big(-\frac{\mathfrak{m}_j}{2 k_B T} \boldsymbol{C}_j \cdot \boldsymbol{C}_j - \frac{E_{j\mathrm{J}}}{k_B T} \Big) \frac{\mathcal{W}_{\mathcal{F}^r \mathcal{B}^r}^{\mathrm{F}^r \mathrm{B}^r}}{\prod_{j \in \mathcal{F}^r} \beta_{j\mathrm{J}}} \prod_{\mathcal{F}^r} \mathrm{d}\boldsymbol{c}_j \prod_{\mathcal{B}^r} \mathrm{d}\boldsymbol{c}_k. \tag{36}$$

These zeroth order chemical production rates are compatible with the law of mass action and with traditional thermochemistry [9].

Denoting by $\mathfrak{a}_{il}$ the number of lth atom in the ith species, $\mathfrak{A} = \{1, \dots, n^{\mathrm{a}}\}$ the set of atom indices, and $n^{\mathrm{a}} \geqslant 1$ the number of atoms—or elements—in the mixture, the stoichiometric coefficients satisfy the atom conservation relations

$$\sum_{i \in S} \nu_{ir}^{f} \mathfrak{a}_{il} = \sum_{i \in S} \nu_{ir}^{b} \mathfrak{a}_{il}, \qquad r \in \mathfrak{R}, \qquad l \in \mathfrak{A}. \tag{37}$$

Using (34), these constraints imply that

$$\sum_{i\in S} \mathfrak{w}_i^0 \mathfrak{a}_{il} = 0, \qquad l \in \mathfrak{A}, \tag{38}$$

which states that atoms are conserved by chemical reactions. On the other hand, the species being constituted by atoms, denoting by $\widetilde{\mathfrak{m}}_l$ the mass of the lth atom, we have the relations

$$\mathfrak{m}_i = \sum_{l\in\mathfrak{A}} \widetilde{\mathfrak{m}}_l \mathfrak{a}_{il}, \qquad i \in S. \tag{39}$$

From these relations between the atom and species mass, and from the conservation of atoms, mass is also conserved during chemical reactions so that

$$\sum_{i\in S} \nu_{ir}^{\mathrm{f}} \mathfrak{m}_i = \sum_{i\in S} \nu_{ir}^{\mathrm{b}} \mathfrak{m}_i, \qquad r \in \mathfrak{R}, \tag{40}$$

as well as

$$\sum_{i\in S} \mathfrak{m}_i \mathfrak{w}_i^0 = 0, \tag{41}$$

so that there is no mass production due to chemical reactions.

3 Dissipative Regime

We obtain in this section the *first order* macroscopic equations governing polyatomic gas mixtures [10–19] with chemical reactions [20–34]. We derive in particular the transport fluxes—which yield *dissipative effects* like viscosity, diffusion or thermal conduction, and we also obtain the transport coefficients. For a single gas, the corresponding equations are the Navier-Stokes-Fourier system.

3.1 Linearized Boltzmann Equations

The linearized Boltzmann collision operator $\mathcal{I} = (\mathcal{I}_i)_{i\in S}$ is defined by

$$\mathcal{I}_i(\phi) = \sum_{j\in S} \sum_{\mathrm{I}'\in \mathrm{Q}_i} \sum_{\mathrm{J},\mathrm{J}'\in \mathrm{Q}_j} \int f_j^0 (\phi_i + \phi_j - \phi_i' - \phi_j') \mathrm{W}_{ij}^{\mathrm{IJI'J'}} \, \mathrm{d}\boldsymbol{c}_j \mathrm{d}\boldsymbol{c}_i' \mathrm{d}\boldsymbol{c}_j', \qquad i \in S.$$

An important property of this linearized collision operator is that it is isotropic so that it transforms a tensor built with $(\boldsymbol{c}_i)_{i\in S}$ into an analog tensor as in the monatomic case [11, 15, 16].

We introduce the corresponding bracket operator

$$[\![\xi,\zeta]\!] = \langle\!\langle f^0\xi, \mathcal{I}(\zeta)\rangle\!\rangle, \tag{42}$$

between two families of tensors $\xi = \big(\xi_i\big)_{i\in S}$, $\zeta = \big(\zeta_i\big)_{i\in S}$, where ξ_i and ζ_i depend on $\boldsymbol{c}_i$ and I. This bilinear operator is symmetric $[\![\xi,\zeta]\!] = [\![\zeta,\xi]\!]$, positive semi-definite $[\![\xi,\xi]\!] \geq 0$, and its nullspace is constituted by the collisional invariants, that is, $[\![\xi,\xi]\!] = 0$ if and only if ξ is a tensorial collisional invariant so that all its components are scalar collisional invariants [15, 16]. These symmetry properties are notably consequences of reciprocity relations between the transition probabilities or equivalently the collision cross sections.

The first order equations that govern the perturbed species distribution functions $\phi = (\phi_i)_{i\in S}$ are directly obtained from (20) to (21). These linearized equations are in the form

$$\mathcal{I}_i(\phi) = \Psi_i, \qquad i \in S, \tag{43}$$

where

$$\Psi_i = -\mathcal{D}_i(\log f_i^0) + \delta_{a0}\frac{\mathcal{C}_i(f^0)}{f_i^0}.$$

In addition, the relations (22) yield the scalar constraints

$$\langle\!\langle f^0\phi, \psi^l\rangle\!\rangle = 0, \qquad l \in \{1,\dots,n+4\}. \tag{44}$$

The term $\mathcal{D}_i(\log f_i^0)$ appearing in the right hand side is evaluated from the zeroth order macroscopic equations as required by the Chapman-Enskog method. After lengthy calculations, it is obtained that

$$\Psi_i = \Psi_i^{\mathcal{S}} + \delta_{a0}\Psi_i^{\mathcal{C}}, \qquad i \in S, \tag{45}$$

with

$$\Psi_i^{\mathcal{S}} = -\boldsymbol{\Psi}_i^{\eta}:\boldsymbol{\nabla}\boldsymbol{v} - \tfrac{1}{3}\Psi_i^{\kappa}\boldsymbol{\nabla}\cdot\boldsymbol{v} - \sum_{j\in S}\boldsymbol{\Psi}_i^{D_j}\cdot(\boldsymbol{\nabla} p_j - \rho_j\boldsymbol{b}_j) - \boldsymbol{\Psi}_i^{\widehat{\lambda}}\cdot\boldsymbol{\nabla}\Big(\frac{1}{k_{\mathrm{B}}T}\Big), \tag{46}$$

$$\Psi_i^{\mathcal{C}} = \frac{\mathcal{C}_i(f^0)}{f_i^0} - \frac{\mathfrak{w}_i^0}{\mathfrak{n}_i} - \frac{1}{p\mathfrak{c}_{\mathrm{v}}T}\Big(\tfrac{3}{2}k_{\mathrm{B}}T - \frac{\mathfrak{m}_i}{2}\boldsymbol{C}_i\cdot\boldsymbol{C}_i + \overline{E}_i - E_{i\mathrm{I}}\Big)\sum_{j\in S}(\tfrac{3}{2}k_{\mathrm{B}}T + \overline{E}_j)\mathfrak{w}_j^0, \tag{47}$$

where $p_i = \mathfrak{n}_i k_{\mathrm{B}}T$ is the partial pressure of the ith species and $\boldsymbol{C}_i = \boldsymbol{c}_i - \boldsymbol{v}$ the relative velocity of the ith species. In these expressions, we have denoted

$$\boldsymbol{\Psi}_i^{\eta} = \frac{\mathfrak{m}_i}{k_{\mathrm{B}}T}\Big(\boldsymbol{C}_i\otimes\boldsymbol{C}_i - \tfrac{1}{3}\boldsymbol{C}_i\cdot\boldsymbol{C}_i\boldsymbol{I}\Big), \tag{48}$$

$$\Psi_i^{\kappa} = \frac{2\mathfrak{c}^{\mathrm{int}}}{\mathfrak{c}_{\mathrm{v}}k_{\mathrm{B}}T}(\tfrac{1}{2}\mathfrak{m}_i\boldsymbol{C}_i\cdot\boldsymbol{C}_i - \tfrac{3}{2}k_{\mathrm{B}}T) + \frac{2\mathfrak{c}_{\mathrm{v}}^{\mathrm{tr}}}{\mathfrak{c}_{\mathrm{v}}k_{\mathrm{B}}T}(\overline{E}_i - E_{i\mathrm{I}}), \tag{49}$$

$$\boldsymbol{\Psi}_i^{D_j} = \frac{1}{p_i}(\delta_{ij} - \frac{\rho_i}{\rho})\boldsymbol{C}_i, \tag{50}$$

$$\boldsymbol{\Psi}_i^{\widehat{\lambda}} = (\tfrac{5}{2}k_{\mathrm{B}}T - \tfrac{1}{2}\mathfrak{m}_i\boldsymbol{C}_i\cdot\boldsymbol{C}_i + \overline{E}_i - E_{i\mathrm{I}})\boldsymbol{C}_i, \tag{51}$$

so that $\boldsymbol{\Psi}_i^{\eta}$ is a symmetric traceless tensor, Ψ_i^{κ} a scalar and $\boldsymbol{\Psi}_i^{D_j}$, $j \in S$, and $\boldsymbol{\Psi}_i^{\widehat{\lambda}}$ are vectors. In order to expand $\Psi_i^{\mathfrak{C}}$ we also write

$$\Psi_i^{\mathfrak{C}} = \sum_{r\in\mathfrak{R}} \Psi_i^{r}\overline{\tau}_r, \qquad i \in S, \tag{52}$$

where $\overline{\tau}_r$ is the zeroth order rate of the rth reaction (35) and where Ψ_i^r is given by

$$\Psi_i^{r} = \frac{1}{f_i^0\mathcal{K}_{\mathrm{r}}}\Big(\nu_{ir}^{\mathrm{b}}\sum_{\mathrm{F}^r,\mathrm{B}_{\mathrm{I}}^r}\int \mathfrak{D}_r\prod_{\mathcal{F}^r}\mathrm{d}\boldsymbol{c}_j\prod_{\mathcal{B}_i^r}\mathrm{d}\boldsymbol{c}_k - \nu_{ir}^{\mathrm{f}}\sum_{\mathrm{F}_{\mathrm{I}}^r,\mathrm{B}^r}\int \mathfrak{D}_r\prod_{\mathcal{F}_i^r}\mathrm{d}\boldsymbol{c}_j\prod_{\mathcal{B}^r}\mathrm{d}\boldsymbol{c}_k\Big) - \frac{\nu_{ir}^{\mathrm{b}} - \nu_{ir}^{\mathrm{f}}}{\mathfrak{n}_i}$$
$$-\frac{1}{p\mathfrak{c}_{\mathrm{v}}T}\Big(\sum_{j\in S}(\tfrac{3}{2}k_{\mathrm{B}}T + \overline{E}_j)(\nu_{jr}^{\mathrm{b}} - \nu_{jr}^{\mathrm{f}})\Big)\Big(\tfrac{3}{2}k_{\mathrm{B}}T - \frac{\mathfrak{m}_i}{2}\boldsymbol{C}_i\cdot\boldsymbol{C}_i + \overline{E}_i - E_{i\mathrm{I}}\Big).$$

By linearity of the operator $\mathcal{I}$, the solution $\phi = (\phi_i)_{i\in S}$ of (43) and (44) may be expanded in a similar form

$$\phi_i = \phi_i^{\mathfrak{S}} + \delta_{a0}\phi_i^{\mathfrak{C}}, \tag{53}$$

where

$$\phi_i^{\mathfrak{S}} = -\boldsymbol{\phi}_i^{\eta}:\boldsymbol{\nabla}\boldsymbol{v} - \tfrac{1}{3}\phi_i^{\kappa}\boldsymbol{\nabla}\cdot\boldsymbol{v} - \sum_{j\in S}\boldsymbol{\phi}_i^{D_j}\cdot(\boldsymbol{\nabla}p_j - \rho_j\boldsymbol{b}_j) - \boldsymbol{\phi}_i^{\widehat{\lambda}}\cdot\boldsymbol{\nabla}(\frac{1}{k_{\mathrm{B}}T}), \tag{54}$$

$$\phi_i^{\mathfrak{C}} = \sum_{r\in\mathfrak{R}}\phi_i^{r}\overline{\tau}_r. \tag{55}$$

The functions ϕ^{μ}, for $\mu \in \{\eta, \kappa, D_1, \dots, D_n, \widehat{\lambda}\} \cup \mathfrak{R}$, are now of tensor type and satisfy the integral equations

$$\mathcal{I}_i(\phi^{\mu}) = \Psi_i^{\mu}, \qquad i \in S, \tag{56}$$

with the constraints

$$\langle\!\langle f^0\phi^{\mu}, \psi^l\rangle\!\rangle = 0, \qquad l \in \{1, \dots, n+4\}. \tag{57}$$

These integral equations are generally shown to be well posed by using Fredholm alternative [17, 54, 56].

3.2 First Order Equations

The conservation equations at first order are obtained by taking the scalar product of Boltzmann equations by the collisional invariants and by keeping all terms that are $O(\epsilon^0)$ or $O(\epsilon^1)$

$$\langle\!\langle \mathcal{D}(f^0 + f^0\phi), \psi^l \rangle\!\rangle = \langle\!\langle \mathcal{C}(f^0), \psi^l \rangle\!\rangle + \delta_{a0} \langle\!\langle \partial_f \mathcal{C}(f^0) f^0 \phi, \psi^l \rangle\!\rangle, \tag{58}$$

where $l \in \{1, \dots, n+4\}$ and $\partial_f \mathcal{C}(f^0) f^0 \phi = (\partial_f \mathcal{C}_i(f^0) f^0 \phi)_{i \in S}$.

The equations for the conservation of species mass are obtained for $l = 1, \dots, n$, and are in the form

$$\partial_t \rho_i + \nabla\cdot(\rho_i \boldsymbol{v}) + \nabla\cdot(\rho_i \mathbf{v}_i) = \mathfrak{m}_i \mathfrak{w}_i, \qquad i \in S, \tag{59}$$

where $\mathbf{v}_i$ is the diffusion velocity and $\mathfrak{w}_i$ the production term for the ith species. The diffusion velocities are defined by

$$\mathbf{v}_i = \frac{1}{\mathfrak{n}_i} \sum_{\mathrm{I} \in \mathrm{Q}_i} \int \boldsymbol{C}_i f_i^0 \phi_i \, \mathrm{d}\boldsymbol{c}_i, \qquad i \in S, \tag{60}$$

and the source term by

$$\mathfrak{w}_i = \sum_{\mathrm{I} \in \mathrm{Q}_i} \int (\mathcal{C}_i(f^0) + \delta_{a0} \partial_f \mathcal{C}_i(f^0) f^0 \phi) \, \mathrm{d}\boldsymbol{c}_i, \qquad i \in S. \tag{61}$$

The mass flux $\mathcal{F}_i$ of the ith species is further defined as $\mathcal{F}_i = \rho_i \mathbf{v}_i = \mathfrak{m}_i \mathfrak{n}_i \mathbf{v}_i$ and satisfy the constraint

$$\sum_{i \in S} \mathcal{F}_i = \sum_{i \in S} \rho_i \mathbf{v}_i = 0,$$

since $f^0\phi$ is orthogonal to the vector collision invariant $(\mathfrak{m}_i \boldsymbol{C}_i)_{i \in S}$.

The momentum equations are obtained for $l = n+1, n+2, n+3$, and may be written in vector form

$$\partial_t(\rho \boldsymbol{v}) + \nabla\cdot(\rho \boldsymbol{v} \otimes \boldsymbol{v} + p \boldsymbol{I}) + \nabla\cdot\boldsymbol{\Pi} = \sum_{i \in S} \rho_i \boldsymbol{b}_i, \tag{62}$$

where the viscous tensor $\boldsymbol{\Pi}$ is defined by

$$\boldsymbol{\Pi} = \sum_{\substack{i\in S\\ \mathrm{I}\in \mathrm{Q}_i}} \int \mathfrak{m}_i \boldsymbol{C}_i \otimes \boldsymbol{C}_i f_i^0 \phi_i \, \mathrm{d}\boldsymbol{c}_i. \tag{63}$$

The energy conservation equation obtained for $l = n + 4$, is finally in the form

$$\partial_t(\tfrac{1}{2}\rho\boldsymbol{v}\cdot\boldsymbol{v}+\mathcal{E})+\boldsymbol{\nabla}\cdot\big((\tfrac{1}{2}\rho\boldsymbol{v}\cdot\boldsymbol{v}+\mathcal{E}+p)\boldsymbol{v}\big)+\boldsymbol{\nabla}\cdot(\boldsymbol{\mathcal{Q}}+\boldsymbol{\Pi}\cdot\boldsymbol{v}) = \sum_{i\in S} \rho_i \boldsymbol{b}_i\cdot(\boldsymbol{v}+\mathbf{v}_i), \tag{64}$$

where $\boldsymbol{\mathcal{Q}}$ is the heat flux

$$\boldsymbol{\mathcal{Q}} = \sum_{\substack{i\in S\\ \mathrm{I}\in \mathrm{Q}_i}} \int \big(\tfrac{1}{2}\mathfrak{m}_i \boldsymbol{C}_i\cdot\boldsymbol{C}_i + E_{i\mathrm{I}}\big)\boldsymbol{C}_i f_i^0 \phi_i \, \mathrm{d}\boldsymbol{c}_i. \tag{65}$$

Once the transport fluxes $\mathbf{v}_i$, $i \in S$, $\boldsymbol{\Pi}$, and $\boldsymbol{\mathcal{Q}}$ are expressed in terms of macroscopic quantities and their gradients, these are the conservation equation governing multicomponent reactive flows. Many simplifications are then possible depending on the particular application under concern but we are only interested here in the general equations.

3.3 *Transport Fluxes and Coefficients*

Using the definition (60) of the species diffusion velocities $\mathbf{v}_i$ and the expression (50) of $\boldsymbol{\Psi}_i^{D_j}$ one may establish after some algebra that [9, 19]

$$\mathbf{v}_i = k_{\mathrm{B}} T \langle\!\langle \boldsymbol{\Psi}^{D_i}, f^0\phi \rangle\!\rangle, \qquad i \in S. \tag{66}$$

Substituting the expansion (53)–(55) in Eq. (66) and using the isotropy of the linearized collision operator, only the terms ϕ^{D_j}, $j \in S$ and $\phi^{\widehat{\lambda}}$ yield nonzero contributions—in agreement with the Curie principle—and we obtain that

$$\mathbf{v}_i = -\sum_{j\in S} D_{ij} \widehat{\boldsymbol{d}}_j - \theta_i \boldsymbol{\nabla} \log T, \qquad i \in S, \tag{67}$$

where

$$\widehat{\boldsymbol{d}}_j = \frac{1}{p}(\boldsymbol{\nabla} p_j - \rho_j \boldsymbol{b}_j), \qquad j \in S, \tag{68}$$

is the unconstrained diffusion driving force of the jth species and where the transport coefficients are defined by

$$D_{ij} = \tfrac{1}{3} p k_B T \langle\langle \boldsymbol{\Psi}^{D_i}, \phi^{D_j} \rangle\rangle = \tfrac{1}{3} p k_B T [\![\phi^{D_i}, \phi^{D_j}]\!], \qquad i, j \in S, \tag{69}$$

$$\theta_i = -\tfrac{1}{3} \langle\langle \boldsymbol{\Psi}^{D_i}, \phi^{\widehat{\lambda}} \rangle\rangle = -\tfrac{1}{3} [\![\phi^{D_i}, \phi^{\widehat{\lambda}}]\!], \qquad i \in S. \tag{70}$$

The coefficients D_{ij}, $i, j \in S$, are termed the multicomponent diffusion coefficients, the coefficients θ_i, $i \in S$, the thermal diffusion or Soret coefficients, and mass diffusion due to temperature gradients is termed the Soret effect. A fundamental property of the multicomponent diffusion coefficients is that they are symmetric since

$$D_{ij} = \tfrac{1}{3} p k_B T [\![\phi^{D_i}, \phi^{D_j}]\!] = \tfrac{1}{3} p k_B T [\![\phi^{D_j}, \phi^{D_i}]\!] = D_{ji},$$

from the symmetry of the bracket operator. We only consider in these notes such symmetric diffusion coefficients—more interesting both theoretically and numerically—that have been obtained by many authors [9, 11–13, 15, 16, 19]. The symmetric diffusion coefficients have been introduced by Waldmann [11] and used in particular by Chapman and Cowling [15] and Ferziger and Kaper [16]. Following Hirschfelder, Curtiss, and Bird, various authors have considered nonsymmetric coefficients [10] hereby destroying the natural symmetries associated with kinetic processes [13]. But after the remarks of Van de Ree [13], symmetric coefficients have also been used by Curtiss [14]. We also introduce the flux diffusion coefficients

$$C_{ij} = \rho_i D_{ij}, \qquad i, j \in S, \tag{71}$$

that are such that

$$\mathcal{F}_i = -\sum_{j \in S} C_{ij} \widehat{\boldsymbol{d}}_j - \rho_i \theta_i \nabla \log T, \qquad i \in S. \tag{72}$$

We denote by y the mass fraction vector $\mathsf{y} = (\mathsf{y}_1, \dots, \mathsf{y}_n)^t$ and by $\langle \xi, \zeta \rangle = \sum_{i \in S} \xi_i \zeta_i$ the Euclidean product between two vectors $\xi = (\xi_1, \dots, \xi_n)^t$ and $\zeta = (\zeta_1, \dots, \zeta_n)^t$. Regrouping the diffusion coefficient in a matrix $D = (D_{ij})_{i,j \in S}$ and the thermal diffusion coefficients in a vector $\theta = (\theta_1, \dots, \theta_n)^t$, we have established that $D = D^t$ and it is shown in the next section that D is positive semi-definite with nullspace $\mathbb{R}\mathsf{y}$ and that $\langle \mathsf{y}, \theta \rangle = 0$. Taking into account these mass conservation constraints $D\mathsf{y} = 0$ and $\langle \mathsf{y}, \theta \rangle = 0$, instead of using $\widehat{\boldsymbol{d}}_i$, $i \in S$, it is possible to equivalently use the constrained diffusion driving forces

$$\boldsymbol{d}_i = \widehat{\boldsymbol{d}}_i - \frac{\rho_i}{\rho} \sum_{k \in S} \widehat{\boldsymbol{d}}_k = \nabla \mathsf{x}_i + (\mathsf{x}_i - \mathsf{y}_i) \nabla \log p + \mathsf{y}_i (\boldsymbol{b}_i - \boldsymbol{b}), \tag{73}$$

which sum up to zero [11, 15, 16]. From the constraints $D\mathsf{y} = 0$ and $\langle \mathsf{y}, \theta \rangle = 0$ it is also directly obtained that $\sum_{i \in S} \mathcal{F}_i = \sum_{i \in S} \rho_i \mathbf{v}_i = 0$ independently of the diffusion driving forces and the temperature gradients.

With the definition (65) of the heat flux $\mathcal{Q}$ and the expression (51) of $\boldsymbol{\Psi}_i^{\widehat{\lambda}}$ one may establish that [9, 19]

$$\mathcal{Q} = -\langle\langle \boldsymbol{\Psi}^{\widehat{\lambda}}, f^0 \phi \rangle\rangle + \sum_{i\in S} (\tfrac{5}{2} k_B T + \overline{E}_i) \mathfrak{n}_i \mathbf{v}_i. \tag{74}$$

Substituting the expansion (53)–(55) in (74) we obtain that

$$\mathcal{Q} = -\widehat{\lambda} \nabla T - p \sum_{i\in S} \theta_i \widehat{\boldsymbol{d}}_i + \sum_{i\in S} (\tfrac{5}{2} k_B T + \overline{E}_i) \mathfrak{n}_i \mathbf{v}_i, \tag{75}$$

where the transport coefficients are defined by

$$\widehat{\lambda} = \frac{1}{3 k_B T^2} \langle\langle \boldsymbol{\Psi}^{\widehat{\lambda}}, \phi^{\widehat{\lambda}} \rangle\rangle = \frac{1}{3 k_B T^2} [\![\phi^{\widehat{\lambda}}, \phi^{\widehat{\lambda}}]\!], \tag{76}$$

$$\theta_i = -\tfrac{1}{3} \langle\langle \boldsymbol{\Psi}^{\widehat{\lambda}}, \phi^{D_i} \rangle\rangle = -\tfrac{1}{3} [\![\phi^{\widehat{\lambda}}, \phi^{D_i}]\!], \qquad i \in S. \tag{77}$$

The coefficient $\widehat{\lambda}$ is termed the partial thermal conductivity and since

$$\langle\langle \boldsymbol{\Psi}^{D_i}, \phi^{\widehat{\lambda}} \rangle\rangle = [\![\phi^{D_i}, \phi^{\widehat{\lambda}}]\!] = [\![\phi^{\widehat{\lambda}}, \phi^{D_i}]\!] = \langle\langle \boldsymbol{\Psi}^{\widehat{\lambda}}, \phi^{D_i} \rangle\rangle,$$

the coefficient θ_i is the same in the diffusion velocities and in the heat flux. The diffusion of heat due to concentration gradients in termed the Dufour effect and is reciprocal of the Soret effect.

There are many alternative expressions for the diffusion velocities and the heat flux in a multicomponent mixture [10, 11, 15, 16]. We present here the relations involving the thermal diffusion ratios $\chi = (\chi_i)_{i\in S}$ and the thermal conductivity λ which are interesting both from a mathematical and a computational point of view. The thermal diffusion ratios $\chi = (\chi_i)_{i\in S}$ have been introduced by Waldmann [11] and are defined by

$$\begin{cases} D\chi = \theta, \\ \langle \chi, \mathbb{I} \rangle = 0, \end{cases} \tag{78}$$

where $\mathbb{I}$ is the vector with n components unity $\mathbb{I} = (1)_{i\in S}$, whereas the thermal conductivity is defined by

$$\lambda = \widehat{\lambda} - \frac{p}{T} \langle \theta, \chi \rangle. \tag{79}$$

The coefficients λ and χ may also be defined through solutions of integral equations as for the other coefficients [19]. More specifically, letting

$$\boldsymbol{\Psi}^{\lambda} = \boldsymbol{\Psi}^{\widehat{\lambda}} + p k_B T \sum_{i\in S} \chi_i \boldsymbol{\Psi}^{D_i}, \tag{80}$$

$$\phi^{\lambda} = \phi^{\widehat{\lambda}} + pk_{\mathrm{B}}T \sum_{i\in S} \chi_i \phi^{D_i}, \tag{81}$$

we have $\mathcal{I}_i(\phi^{\lambda}) = \boldsymbol{\Psi}^{\lambda}$ with the constraints $\langle\!\langle f^0\phi^{\lambda}, \psi^l \rangle\!\rangle = 0$ for $l \in \{1, \dots, n+4\}$, and

$$\lambda = \frac{1}{3k_{\mathrm{B}}T^2} \langle\!\langle \boldsymbol{\Psi}^{\lambda}, \phi^{\lambda} \rangle\!\rangle = \frac{1}{3k_{\mathrm{B}}T^2} [\![\phi^{\lambda}, \phi^{\lambda}]\!], \tag{82}$$

$$\chi_i = \frac{1}{3pk_{\mathrm{B}}T} [\![\mathfrak{V}_i, \phi^{\lambda}]\!], \qquad i \in S, \tag{83}$$

where $\mathfrak{V}_k = (\mathfrak{m}_i \boldsymbol{C}_i \delta_{ki})_{i\in S}$. From these definitions, and after a little algebra, the following alternative expressions for the diffusion velocities and the heat flux are obtained

$$\mathbf{v}_i = -\sum_{j\in S} D_{ij}(\widehat{\boldsymbol{d}}_j + \chi_j \nabla \log T), \qquad i \in S, \tag{84}$$

$$\boldsymbol{\mathcal{Q}} = -\lambda \nabla T + p \sum_{j\in S} \chi_j \mathbf{v}_j + \sum_{i\in S} (\tfrac{5}{2}k_{\mathrm{B}}T + \overline{E}_i)\mathfrak{n}_i \mathbf{v}_i. \tag{85}$$

These alternative formulations are interesting from a computational point of view since it is faster to directly evaluate λ and χ rather that to evaluate $\widehat{\lambda}$ and θ [19]. When the Soret and Dufour effects are neglected, that is when $\chi = 0$ and $\theta = 0$, the partial thermal conductivity $\widehat{\lambda}$ and the traditional thermal conductivity λ then coincide.

Finally, with the relations (48) and (49) and the expression (63) one may establish the following expression for the viscous tensor $\boldsymbol{\Pi}$

$$\boldsymbol{\Pi} = k_{\mathrm{B}}T \langle\!\langle \boldsymbol{\Psi}^{\eta}, f^0\phi \rangle\!\rangle + \tfrac{1}{3}k_{\mathrm{B}}T \langle\!\langle \Psi^{\kappa}, f^0\phi \rangle\!\rangle\, \boldsymbol{I}. \tag{86}$$

Keeping in mind the isotropy of the linearized collision operator, the term $\langle\!\langle \Psi^{\kappa}, f^0\phi \rangle\!\rangle$ is evaluated in the form

$$\tfrac{1}{3}k_{\mathrm{B}}T \langle\!\langle \Psi^{\kappa}, f^0\phi \rangle\!\rangle = -\tfrac{1}{9}k_{\mathrm{B}}T \langle\!\langle \Psi^{\kappa}, f^0\phi^{\kappa} \rangle\!\rangle \nabla\cdot \boldsymbol{v} + \delta_{a0}\tfrac{1}{3}k_{\mathrm{B}}T \sum_{r\in\mathfrak{R}} \langle\!\langle \Psi^{\kappa}, f^0\phi^{r} \rangle\!\rangle \overline{\tau}_r.$$

Defining the volume viscosity by κ by

$$\kappa = \tfrac{1}{9}k_{\mathrm{B}}T \langle\!\langle \Psi^{\kappa}, f^0\phi^{\kappa} \rangle\!\rangle = \tfrac{1}{9}k_{\mathrm{B}}T [\![\phi^{\kappa}, \phi^{\kappa}]\!], \tag{87}$$

the reactive pressure by

$$p^{\mathrm{reac}} = \delta_{a0}\tfrac{1}{3}k_{\mathrm{B}}T \sum_{r\in\mathfrak{R}} [\![\phi^{\kappa}, \phi^{r}]\!] \overline{\tau}_r, \tag{88}$$

and the shear viscosity by

$$\eta = \tfrac{1}{10} k_{\mathrm{B}} T [\![\phi^{\eta}, \phi^{\eta}]\!], \tag{89}$$

it is obtained that

$$\boldsymbol{\Pi} = -\kappa \boldsymbol{\nabla}\cdot\boldsymbol{v}\, \boldsymbol{I} + p^{\mathrm{reac}}\, \boldsymbol{I} - \eta\Big(\boldsymbol{\nabla}\boldsymbol{v} + (\boldsymbol{\nabla}\boldsymbol{v})^t - \tfrac{2}{3}(\boldsymbol{\nabla}\cdot\boldsymbol{v})\, \boldsymbol{I}\Big),$$

where $\boldsymbol{\nabla}\boldsymbol{v} + (\boldsymbol{\nabla}\boldsymbol{v})^t - \frac{2}{3}(\boldsymbol{\nabla}\cdot\boldsymbol{v})\, \boldsymbol{I}$ is the deviatoric part of the rate of strain tensor.

3.4 Properties of Transport Coefficients

The mathematical properties of the transport coefficients may be *extracted* from the linearized Boltzmann equations [19, 72, 75]. These coefficients satisfy symmetry properties, mass conservation constraints, as well as positivity properties associated with the bracket operator or equivalently with entropy production.

From the definition (69) of the multicomponent transport coefficients D_{ij} and the symmetry of the bracket operator it has already been obtained that $D_{ij} = D_{ji}$ and that the thermal diffusion coefficients θ_i, $i \in S$, in the diffusion velocities are identical with the coefficients relating the heat flux to the diffusion driving forces. These properties, which may be interpreted as Onsager type relations, are direct consequences of the symmetry of the bracket operator $[\![\ ,\ ]\!]$, and hence of the reciprocity relations for transition probabilities or collision cross sections.

Concerning the diffusion matrix D, we also have for any $x \in \mathbb{R}^n$

$$\langle Dx, x\rangle = \tfrac{1}{3} p k_{\mathrm{B}} T \Big[\!\!\Big[\sum_{i\in S} x_i \phi^{D_i}, \sum_{i\in S} x_i \phi^{D_i} \Big]\!\!\Big].$$

The matrix D is thus symmetric positive semi-definite since the bracket operator is positive semi-definite. Moreover, the family of right hand sides $\boldsymbol{\Psi}^{D_1}, \dots, \boldsymbol{\Psi}^{D_n}$ of the integral equations defining $\phi^{D_1}, \dots, \phi^{D_n}$ is easily shown to be of rank $n-1$ with the constraint $\sum_{i\in S} \mathsf{y}_i \boldsymbol{\Psi}^{D_i} = 0$ [19]. By linearity, keeping in mind that $\phi^{D_1}, \dots, \phi^{D_n}$ are orthogonal to the nullspace of the linearized collision operator, the family $\phi^{D_1}, \dots, \phi^{D_n}$ is also of rank $n-1$ and satisfy

$$\sum_{i\in S} \mathsf{y}_i \phi^{D_i} = 0.$$

Then $Dx = 0$ if and only if $\langle Dx, x\rangle = 0$ which is equivalent to the property that $\sum_{i\in S} x_i \phi^{D_i} = 0$ (since by construction all ϕ^{D_i} are orthogonal to the collisional invariants), and we obtain that $Dx = 0$ if and only if x is proportional to the mass fraction vector y and $N(D) = \mathbb{R}\mathsf{y}$. We also obtain from the definition of thermal diffusion coefficients θ_i, $i \in S$, that $\sum_{i\in S} \mathsf{y}_i\theta_i = -\frac{1}{3}[\![\sum_{i\in S} \mathsf{y}_i\phi^{D_i}, \phi^{\widehat{\lambda}}]\!] = 0$ so that $\langle\theta, \mathsf{y}\rangle = 0$.

Similarly, defining the matrix

$$A = \begin{pmatrix} D & \theta \\ \widehat{\theta}^t & \frac{T}{p}\widehat{\lambda} \end{pmatrix},$$

we observe that for any $x' = (x, x_0) \in \mathbb{R}^{n+1}$ with $x \in \mathbb{R}^n$, $x_0 \in \mathbb{R}$, we have

$$\langle Ax', x'\rangle = \tfrac{1}{3} pk_{\mathrm{B}}T \Big[\!\!\Big[\sum_{i\in S} x_i\phi^{D_i} - \frac{x_0}{pk_{\mathrm{B}}T}\phi^{\widehat{\lambda}}, \sum_{i\in S} x_i\phi^{D_i} - \frac{x_0}{pk_{\mathrm{B}}T}\phi^{\widehat{\lambda}} \Big]\!\!\Big],$$

so that A is positive semi-definite with nullspace spanned by $(\mathsf{y}, 0)^t$. Evaluating then $\langle Ax', x'\rangle$ for $x' = (0, \dots, 0, 1)^t$ and $x' = (-\chi, 1)^t$ it is obtained that $\widehat{\lambda} > 0$ and $\lambda > 0$. The positivity of $\widehat{\lambda}$ and λ may also directly be deduced from $\widehat{\lambda} = [\![\phi^{\widehat{\lambda}}, \phi^{\widehat{\lambda}}]\!]/3k_{\mathrm{B}}T^2$ and $\lambda = [\![\phi^{\lambda}, \phi^{\lambda}]\!]/3k_{\mathrm{B}}T^2$ since neither ϕ^{λ} nor $\phi^{\widehat{\lambda}}$ are collision invariants because neither $\boldsymbol{\Psi}^{\lambda}$ nor $\boldsymbol{\Psi}^{\widehat{\lambda}}$ are zero.

Finally, we deduce from $\eta = [\![\phi^{\eta}, \phi^{\eta}]\!]/10k_{\mathrm{B}}T$ that η is positive and from $\kappa = [\![\phi^{\kappa}, \phi^{\kappa}]\!]/9k_{\mathrm{B}}T$ that κ is nonnegative, and that κ is positive unless $\Psi^{\kappa} = 0$ when there are only monatomic species.

Remark 3.1 The evaluation of the transport coefficients will be discussed in Sect. 4 and the structural properties of the transport coefficients may also be obtained from the transport linear systems [19].

3.5 Perturbed Production Terms

The first order chemical production rates for the ith species (61) may be written in the form

$$\mathfrak{w}_i = \mathfrak{w}_i^0 + \mathfrak{w}_i^1, \qquad i \in S,$$

where $\mathfrak{w}_i^0$ is the zeroth order rate $\mathfrak{w}_i^0 = \sum_{r\in\mathfrak{R}}(\nu_{ir}^{\mathrm{b}} - \nu_{ir}^{\mathrm{f}})\overline{\tau}_r$ already discussed in Sect. 2.8 and $\mathfrak{w}_i^1$ is the perturbed rate in the dissipative or Navier-Stokes regime

$$\mathfrak{w}_i^1 = \delta_{a0} \sum_{\mathrm{I}} \int \partial_f \mathfrak{C}_i(f^0) f^0\phi \, \mathrm{d}\boldsymbol{c}_i.$$

The structure of the perturbed chemical source terms $\mathfrak{w}_i^1$ has been investigated in [9]. The perturbed source term is a quadratic expression of the zeroth order rates $\overline{\tau}_r$, $r \in \mathfrak{R}$, plus a linear combination of the same quantities multiplied by the divergence of the velocity field $\boldsymbol{\nabla\cdot v}$. A few estimates have been made of these perturbed terms in the monoatomic case by Prigogine and Mathieu [20], Prigogine and Xhrouet [21],

Present [23], Takanayaki [22], Shizghal and Karplus [26–28] and they are generally believed to be small. In the polyatomic case, however, these perturbed terms may be significant behind shocks, especially with strong thermodynamic nonequilibrium [84]. Nevertheless, we will assume in the following that they are negligible so that

$$\mathfrak{w}_i^1 = 0, \qquad i \in S, \tag{90}$$

and furthermore that

$$p^{\mathrm{reac}} = 0. \tag{91}$$

Note that this is automatic when $a = 1$ in the Boltzmann equations. In this situation, neglecting both the perturbed source terms $\mathfrak{w}_i^1$ as well as the chemical pressure p^{reac}, we have

$$\mathfrak{w}_i = \mathfrak{w}_i^0, \qquad i \in S, \tag{92}$$

$$\boldsymbol{\Pi} = -\kappa \boldsymbol{\nabla}\cdot\boldsymbol{v}\,\boldsymbol{I} - \eta\Big(\boldsymbol{\nabla}\boldsymbol{v} + (\boldsymbol{\nabla}\boldsymbol{v})^t - \tfrac{2}{3}(\boldsymbol{\nabla}\cdot\boldsymbol{v})\,\boldsymbol{I}\Big), \tag{93}$$

and the equations obtained in both regime $a = 0$ and $a = 1$ coincide. These equations will be investigated mathematically in Sects. 5, 6, and 7.

3.6 Thermodynamics

In the framework of Enskog expansion, one may expand the kinetic entropy (13) up to second order in order to obtain that $\mathcal{S}^{\mathrm{kin}} = \mathcal{S} + \mathcal{O}(\epsilon^2)$ and the zeroth order term $\mathcal{S}$ is the fluid entropy of the mixture

$$\mathcal{S} = -k_{\mathrm{B}} \sum_{\substack{i\in S\\ \mathrm{I}\in \mathrm{Q}_i}} \int f_i^0\big(\log(\beta_{i\mathrm{I}} f_i^0) - 1\big)\,\mathrm{d}\boldsymbol{c}_i = \sum_{i\in S} \rho_i \Big(\frac{5}{2}\frac{k_{\mathrm{B}}}{\mathfrak{m}_i} + \frac{\overline{E}_i}{T\mathfrak{m}_i} - \frac{k_{\mathrm{B}}}{\mathfrak{m}_i}\log\Big(\frac{\mathfrak{n}_i}{\mathrm{Z}_i}\Big)\Big). \tag{94}$$

The second order terms $\mathcal{O}(\epsilon^2)$ have been investigated in [123, 124] but lay out of the scope of the present work. Note that such a thermodynamics obtained in the framework of the kinetic theory of gases is valid out of static equilibrium and has, therefore, a wider range of validity than classical thermodynamics introduced for stationary homogeneous equilibrium states.

Defining the entropy per unit mass of the ith species by

$$s_i = \frac{5}{2}\frac{k_{\mathrm{B}}}{\mathfrak{m}_i} + \frac{\overline{E}_i}{T\mathfrak{m}_i} - \frac{k_{\mathrm{B}}}{\mathfrak{m}_i}\log\Big(\frac{\mathfrak{n}_i}{\mathrm{Z}_i}\Big), \tag{95}$$

we may also express the fluid entropy in the form $\mathcal{S} = \sum_{i\in S}\rho_i s_i$. We may similarly define the Gibbs function per unit mass of the ith species

$$g_i = \frac{k_B T}{\mathfrak{m}_i} \log\Big(\frac{\mathfrak{n}_i}{\mathrm{Z}_i}\Big), \tag{96}$$

and the Gibbs function per unit volume of the mixture $\mathcal{G}$ is then given by $\mathcal{G} = \sum_{i\in S} \rho_i g_i$. We also introduce the reduced chemical potential of the ith species

$$\mu_i = \frac{\mathfrak{m}_i g_i}{k_B T} = \log\Big(\frac{\mathfrak{n}_i}{\mathrm{Z}_i}\Big), \tag{97}$$

and the reaction rates of progress may then be rewritten in the convenient form

$$\overline{\tau}_r = \mathcal{K}_r\big(\exp\langle\mu, \nu_r^{\mathrm{f}}\rangle - \exp\langle\mu, \nu_r^{\mathrm{b}}\rangle\big), \qquad r \in \mathfrak{R}, \tag{98}$$

where $\nu_i^{\mathrm{f}} = (\nu_{1i}^{\mathrm{f}}, \ldots, \nu_{ni}^{\mathrm{f}})^t$, $\nu_i^{\mathrm{b}} = (\nu_{1i}^{\mathrm{b}}, \ldots, \nu_{ni}^{\mathrm{b}})^t$, and $\mu = (\mu_1, \ldots, \mu_n)^t$.

The balance equation for the macroscopic fluid entropy $\mathcal{S}$ is obtained in the form

$$\partial_t \mathcal{S} + \nabla\cdot(\mathcal{S}\boldsymbol{v}) + \nabla\cdot\Big(\frac{\mathcal{Q}}{T} - \sum_{i\in S} \frac{g_i}{T}\mathcal{F}_i\Big) = \mathfrak{v}, \tag{99}$$

where $\mathcal{F}_i = \rho_i \mathbf{v}_i$, and is $\mathfrak{v}$ the entropy production term given by

$$\mathfrak{v} = -\sum_{i\in S} \frac{g_i \mathfrak{m}_i \mathfrak{w}_i^0}{T} - \frac{\boldsymbol{\Pi}:\nabla\boldsymbol{v}}{T} - \big(\mathcal{Q} - \sum_{i\in S} \rho_i h_i \mathbf{v}_i\big)\cdot\frac{\nabla T}{T^2} - \sum_{i\in S} \frac{p}{T}\mathbf{v}_i\cdot\widehat{\boldsymbol{d}}_i. \tag{100}$$

Using (34) and (98) the entropy production due to chemical reactions reads

$$-\sum_{i\in S} \frac{g_i \mathfrak{m}_i \mathfrak{w}_i^0}{T} = -\sum_{i\in S}\sum_{r\in\mathfrak{R}} (\nu_{ir}^{\mathrm{b}} - \nu_{ir}^{\mathrm{f}})\overline{\tau}_r \frac{g_i \mathfrak{m}_i}{T} = -k_B \sum_{i\in S}\sum_{r\in\mathfrak{R}} \mu_i (\nu_{ir}^{\mathrm{b}} - \nu_{ir}^{\mathrm{f}})\overline{\tau}_r$$

so that

$$-\sum_{i\in S} \frac{g_i \mathfrak{m}_i \mathfrak{w}_i^0}{T} = \sum_{r\in\mathfrak{R}} k_B \mathcal{K}_r\big(\langle\mu, \nu_r^{\mathrm{f}}\rangle - \langle\mu, \nu_r^{\mathrm{b}}\rangle\big)\big(\exp\langle\mu, \nu_r^{\mathrm{f}}\rangle - \exp\langle\mu, \nu_r^{\mathrm{b}}\rangle\big).$$

Similarly, we have

$$-\frac{\boldsymbol{\Pi}:\nabla\boldsymbol{v}}{T} = \frac{1}{T}\Big(\kappa + \eta\frac{2(3-d)}{3d}\Big)(\nabla\cdot\boldsymbol{v})^2 + \frac{\eta}{2T}\Big|\nabla\boldsymbol{v} + \nabla\boldsymbol{v}^t - \frac{2}{d}\nabla\cdot\boldsymbol{v}\Big|^2,$$

and

$$-\big(\mathcal{Q} - \sum_{i\in S} \rho_i h_i \mathbf{v}_i\big)\cdot\frac{\nabla T}{T^2} - \sum_{i\in S} \frac{p}{T}\mathbf{v}_i\cdot\widehat{\boldsymbol{d}}_i = \frac{\lambda}{T^2}|\nabla T|^2$$

$$+\frac{p}{T}\sum_{k,l\in S} D_{kl}\big(\widehat{\boldsymbol{d}}_k + \chi_k \nabla \log T\big)\cdot\big(\widehat{\boldsymbol{d}}_l + \chi_l \nabla \log T\big).$$

The structure of the chemical source term and of the transport coefficients then guarantee that the two first terms as well as the sum of the two last terms in the expression (100) of entropy production $\mathfrak{v}$ are nonnegative [9]. The entropy production due to macroscopic gradients may also be written in the form $k_B[\![\phi^S, \phi^S]\!]$ [9]. We recover here the important property that macroscopic gradients as well as chemical production independently lead to nonnegative entropy production at the fluid level as at the molecular level.

3.7 From Molecules to Moles

It is traditional to write macroscopic fluid equations in terms of mass or mole densities rather that number densities and we summarize here the corresponding new notation. Denoting by $\mathcal{N}$ the Avogadro number, we define the molar production rate of the kth species by

$$\omega_k = \mathfrak{w}_k^0/\mathcal{N}, \qquad k \in S,$$

and the molar mass by

$$m_k = \mathcal{N}\mathfrak{m}_k. \qquad k \in S,$$

The number of mole per unit volume is then defined by

$$n_k = \mathfrak{n}_k/\mathcal{N}, \qquad k \in S,$$

and we have the traditional relations $\rho_k = m_k n_k$. We also define the molar rate of progress $\tau_r = \overline{\tau}_r/\mathcal{N}$ in such a way that $\omega_k = \sum_{r\in\mathfrak{R}} \nu_{kr}\tau_r$ and we also define

$$\mathcal{K}_r^{\mathrm{s}} = \mathfrak{K}_{\mathrm{r}}/\mathcal{N}, \qquad r \in \mathfrak{R}.$$

The specific heats per unit mole are also given by

$$c_i^{\mathrm{int}} = \mathfrak{c}_i^{\mathrm{int}}\mathcal{N}, \qquad c_{\mathrm{V}i} = \mathfrak{c}_{\mathrm{V}i}\mathcal{N}, \qquad i \in S,$$

as well as

$$c_{\mathrm{V}}^{\mathrm{tr}} = \mathfrak{c}_{\mathrm{V}}^{\mathrm{tr}}\mathcal{N}, \qquad c^{\mathrm{int}} = \mathfrak{c}^{\mathrm{int}}\mathcal{N}, \qquad c_{\mathrm{V}} = \mathfrak{c}_{\mathrm{V}}\mathcal{N},$$

and we also have $R = k_B\mathcal{N}$.

4 Evaluation of Transport Coefficients

The transport fluxes appearing in the conservation equations governing multicomponent flows are expressed in terms of transport coefficients. The fast and accurate evaluation of these coefficients is therefore an important modeling and computational task [65–69].

The transport coefficients are expressed in terms bracket bilinear products involving solutions of systems of integral equations under constraints. These systems of integral equations are generally solved with Galerkin variational approximation procedure and the structure of the resulting transport linear systems may be deduced from the kinetic theory. These linear systems are typically semi-definite systems under constraints. Generalized conjugate gradient algorithms as well as stationary methods are then shown to be convergent using the structural properties obtained from the kinetic theory [70–75].

4.1 Transport Linear Systems

The Chapman-Enskog method requires solving the systems of integral linearized Boltzmann equations with constraints governing the perturbed distribution functions $\phi^\mu = (\phi_i^\mu)_{i\in S}$. These integral equations have been shown to be in the generic form

$$\begin{cases} \mathcal{I}(\phi^\mu) = \Psi^\mu, \\ \langle\!\langle f^{(0)}\phi^\mu, \psi^l \rangle\!\rangle = 0, \qquad 1 \le l \le n+4, \end{cases} \tag{101}$$

where $\mu \in \{\kappa, \eta, \widehat{\lambda}\} \cup \{D_1, \dots, D_n\}$ and the various right hand sides have been evaluated in Sect. 3.3. These systems of integral equations are of matrix type for $\mu = \eta$, of vector type for $\mu = \widehat{\lambda}$ or $\mu \in \{D_1, \dots, D_n\}$, and of scalar type for $\mu = \kappa$. The corresponding transport coefficients are then typically obtained through bracket products in the form $\mu = [\![\phi^\mu, \phi^\mu]\!] = \langle\!\langle f^{(0)}\Psi^\mu, \phi^\mu \rangle\!\rangle$ as detailed in Sect. 3.3.

A Galerkin variational approximation procedure is generally used to solve the system of integral Eq. (101). A variational approximation space is first selected

$$\Xi^\mu = \mathrm{span}\{\, \xi^{rk},\ rk \in \mathcal{B}^\mu \,\}, \tag{102}$$

where ξ^{rk}, $rk \in \mathcal{B}^\mu$, are basis functions of the same tensorial type than ϕ^μ and Ψ^μ. The set $\mathcal{B}^\mu$ is the basis indexing set and $\mathcal{B}^\mu \subset \mathcal{F} \times S$ where $\mathcal{F}$ is the indexing set of function type and S the species indexing set, that is, when $rk \in \mathcal{B}^\mu$ then $r \in \mathcal{F}$ and $k \in S$. We denote by v the dimension of Ξ^μ, that is $v = \dim(\Xi^\mu) = \mathrm{Card}(\mathcal{B}^\mu)$. The unknown ϕ^μ is then expanded in the form

$$\phi^\mu = \sum_{rk\in\mathcal{B}^\mu} \alpha_k^{r\mu} \xi^{rk}, \tag{103}$$

and the variational equations read $\langle\!\langle f^{(0)}\xi^{rk}, \mathcal{I}(\phi^\mu)\rangle\!\rangle = \langle\!\langle f^{(0)}\xi^{rk}, \Psi^\mu\rangle\!\rangle$ for $rk \in \mathcal{B}^\mu$. Letting $G_{kl}^{rs} = \langle\!\langle f^{(0)}\xi^{rk}, \mathcal{I}(\xi^{sl})\rangle\!\rangle = [\![\xi^{rk}, \xi^{sl}]\!]$, and $\beta_k^{r\mu} = \langle\!\langle f^{(0)}\xi^{rk}, \Psi^\mu\rangle\!\rangle$ we have obtained the linear system

$$\sum_{sl\in\mathcal{B}^\mu} G_{kl}^{rs}\alpha_l^{s\mu} = \beta_k^{r\mu}, \qquad rk \in \mathcal{B}^\mu. \tag{104}$$

The linear constraints $\langle\!\langle f^{(0)}\phi^\mu, \psi^l\rangle\!\rangle = 0$ are also rewritten as

$$\sum_{rk\in\mathcal{B}^\mu} \mathcal{G}_k^{rl\nu}\alpha_k^{r\mu} = 0, \qquad 1 \le l \le n+4, \quad 1 \le \nu \le a_\mu, \tag{105}$$

where $\mathcal{G}_k^{rl\nu} = \langle\!\langle f^{(0)}\xi^{rk}, \mathcal{T}_\nu\psi^l\rangle\!\rangle$ and $\mathcal{T}_\nu$ denotes the canonical basis for tensor of type Ψ^μ and ϕ^μ.

Defining now the constrained space by

$$\mathcal{C} = \Big(\mathrm{span}\{\ \mathcal{G}^{l\nu};\quad 1 \le l \le n+4,\quad 1 \le \nu \le a_\mu\ \}\Big)^\perp, \tag{106}$$

where $\mathcal{G}^{l\nu} = (\mathcal{G}_k^{rl\nu})_{rk\in\mathcal{B}^\mu}$, the vectors $\alpha^\mu = (\alpha_k^{r\mu})_{rk\in\mathcal{B}^\mu}$ and $\beta^\mu = (\beta_k^{r\mu})_{rk\in\mathcal{B}^\mu}$, and the matrix G by $G = \big(G_{kl}^{rs}\big)_{rk,sl\in\mathcal{B}^\mu}$, the transport linear system is in the form

$$\begin{cases} G\alpha^\mu = \beta^\mu, \\ \alpha^\mu \in \mathcal{C}, \end{cases} \tag{107}$$

and the bracket $\mu = [\![\phi^\mu, \phi^\mu]\!] = \langle\!\langle f^{(0)}\Psi^\mu, \phi^\mu\rangle\!\rangle$ is typically obtained with a scalar product

$$\mu = \sum_{rk\in\mathcal{B}^\mu} \alpha_k^{r\mu}\beta_k^{r\mu} = \langle\alpha, \beta\rangle. \tag{108}$$

We note then that the matrix G of the transport linear system is symmetric since

$$G_{kl}^{rs} = [\![\xi^{rk}, \xi^{sl}]\!] = [\![\xi^{sl}, \xi^{rk}]\!] = G_{lk}^{sr},$$

and positive semi-definite from $\langle Gx, x\rangle = [\![\xi, \xi]\!]$ where $\xi = \sum_{rk\in\mathcal{B}^\mu} x_k^r\xi^{rk}$. Moreover, the nullspace of G is directly associated with the collisional invariants of the same tensorial type that Ψ^μ and ϕ^μ that are in the variational space Ξ^μ. One may further establish that $\beta \in R(G)$ is in the range of G using that Ψ^μ is orthogonal to the collisional invariants [19]. Throughout these notes, for any matrix A, we denote by $N(A)$ its nullspace and $R(A)$ its range. Symmetric transport linear systems have been considered by many authors [11, 12, 15, 16]. Nevertheless, following Hirschfelder, Curtiss, and Bird, various authors have considered nonsymmetric transport linear systems [10] hereby again destroying the natural symmetries associated with kinetic processes. The explicit calculation of symmetric transport linear systems for mixtures of *polyatomic* species has first been performed in [19].

Denoting by $\mathfrak{I}^\mu$ these collisional invariants of the same tensorial type than Ψ^μ and ϕ^μ, we assume that the following perpendicularity property holds

$$\mathfrak{I}^\mu = \mathfrak{I}^\mu \cap \Xi^\mu \;\oplus\; \mathfrak{I}^\mu \cap (\Xi^\mu)^\perp, \tag{109}$$

where the orthogonal $(\Xi^\mu)^\perp$ of the variational approximation space Ξ^μ is taken with respect to the scalar product $\langle\!\langle f^0\xi, \zeta \rangle\!\rangle$. In this situation, one may establish [19] that that the well posedness condition holds

$$N(G) \oplus \mathcal{C} = \mathbb{R}^v. \tag{110}$$

We further introduce the sparse transport matrix [19]

$$db(G)_{kl}^{rs} = G_{kl}^{rs}\, \delta_{kl}, \qquad rk, sl \in \mathcal{B}^\mu, \tag{111}$$

and one may establish that when

$$\xi_i^{rk} = 0 \qquad i \neq k, \tag{112}$$

that is, when the basis functions are orthogonal to the constant collisional invariants ψ^i, $i \in S$, then for $x = (x_k^r)_{rk\in\mathcal{B}^\mu}$ we have

$$\begin{aligned}
&\big\langle (2db(G) - G)x, x \big\rangle = \\
&\quad \tfrac{1}{4} \sum_{i\in S} \sum_{\substack{\mathrm{I}\mathrm{I}'\in \mathrm{Q}_i \\ \tilde{\mathrm{I}},\tilde{\mathrm{I}}'\in \mathrm{Q}_i i}} \int |\xi_i + \tilde{\xi}_i - \xi_i' - \tilde{\xi}'_i|^2 f_i^{(0)} \tilde{f}_i^{(0)} \mathrm{W}_{ij}^{\mathrm{I}\tilde{\mathrm{I}}\mathrm{I}'\tilde{\mathrm{I}}'} \mathrm{d}\boldsymbol{c}_i \mathrm{d}\tilde{\boldsymbol{c}}_i \mathrm{d}\boldsymbol{c}_i' \mathrm{d}\tilde{\boldsymbol{c}}_i{}' \\
&\quad + \tfrac{1}{4} \sum_{\substack{i,j\in S \\ i\neq j}} \sum_{\substack{\mathrm{I}\mathrm{I}'\in \mathrm{Q}_i \\ \mathrm{J},\mathrm{J}'\in \mathrm{Q}_j}} \int |\xi_i - \xi_j - \xi_i' + \xi_j'|^2 f_i^{(0)} f_j^{(0)} \mathrm{W}_{ij}^{\mathrm{I}\mathrm{J}\mathrm{I}'\mathrm{J}'} \mathrm{d}\boldsymbol{c}_i \mathrm{d}\boldsymbol{c}_j \mathrm{d}\boldsymbol{c}_i' \mathrm{d}\boldsymbol{c}_j,
\end{aligned}$$

where $\xi = \sum_{rk\in\mathcal{B}^\mu} x_k^r \xi^{rk}$ and the superscript $\tilde{}$ is used to distinguish the collision partners when $i = j$. In comparison, it is interesting to note that

$$\begin{aligned}
&\big\langle Gx, x \big\rangle = \\
&\quad \tfrac{1}{4} \sum_{i\in S} \sum_{\substack{\mathrm{I}\mathrm{I}'\in \mathrm{Q}_i \\ \tilde{\mathrm{I}},\tilde{\mathrm{I}}'\in \mathrm{Q}_i i}} \int |\xi_i + \tilde{\xi}_i - \xi_i' - \tilde{\xi}'_i|^2 f_i^{(0)} \tilde{f}_i^{(0)} \mathrm{W}_{ij}^{\mathrm{I}\tilde{\mathrm{I}}\mathrm{I}'\tilde{\mathrm{I}}'} \mathrm{d}\boldsymbol{c}_i \mathrm{d}\tilde{\boldsymbol{c}}_i \mathrm{d}\boldsymbol{c}_i' \mathrm{d}\tilde{\boldsymbol{c}}_i' \\
&\quad + \tfrac{1}{4} \sum_{\substack{i,j\in S \\ i\neq j}} \sum_{\substack{\mathrm{I}\mathrm{I}'\in \mathrm{Q}_i \\ \mathrm{J},\mathrm{J}'\in \mathrm{Q}_j}} \int |\xi_i + \xi_j - \xi_i' - \xi_j'|^2 f_i^{(0)} f_j^{(0)} \mathrm{W}_{ij}^{\mathrm{I}\mathrm{J}\mathrm{I}'\mathrm{J}'} \mathrm{d}\boldsymbol{c}_i \mathrm{d}\boldsymbol{c}_j \mathrm{d}\boldsymbol{c}_i' \mathrm{d}\boldsymbol{c}_j
\end{aligned}$$

in such a way that

$$N(G) = \{\, x;\ (\xi_i, \xi_j) \in \Im_{ij} \quad i, j \in S\ i \neq j\},$$

where $\xi = \sum_{rk\in\mathcal{B}^\mu} x_k^r \xi^{rk}$ and $\Im_{ij}$ denotes the collision invariants for the species pair (i, j), whereas

$$N\big(2db(G) - G\big) = \{\, x;\ (\xi_i, -\xi_j) \in \Im_{ij} \quad i, j \in S\ i \neq j\}.$$

Using these properties,and since the only linear subspaces of collisional invariants that may lay in the variational approximation spaces $\varXi^\mu$ are at most one-dimensional, being either proportional to $(\mathfrak{m}_i \boldsymbol{C}_i)_{i\in S}$ in the vector case or $(\frac{1}{2}\mathfrak{m}_i|\boldsymbol{C}_i|^2 + E_{i\mathrm{I}})_{i\in S}$ in the scalar case, the invariant ψ^i, $i \in S$ being excluded from (112), it is established that when there are at least $n \geq 3$ species, then $2db(G) - G$ is positive definite [19]. In this situation, the matrix $db(G)$, which is easily invertible, is also positive definite [19].

We have thus established that when the perpendicularity property (109) holds, and when the variational space $\varXi^\mu$ is orthogonal to constants (112), then G is symmetric positive semi-definite, the well posedness condition $N(G) \oplus \mathcal{C} = \mathbb{R}^v$ holds, $\beta \in R(G)$, and $2db(G) - G$ and $db(G)$ are positive definite when $n \geq 3$. In the special cases $n = 1$ or $n = 2$ the corresponding nullspaces are explicitely evaluated. Various variational approximation spaces may also be used as reduced spaces [19] or spaces for a direct evaluation of the thermal conductivity and the thermal diffusion ratios [71].

4.2 Transport Algorithms

The transport linear systems have been obtained in their natural symmetric form for most useful transport coefficients [11, 12, 15, 16, 19]. These linear system associated with any coefficient μ then take on either a regular form or a singular form [9, 19]. Only the later singular form is discussed here since the regular case is easier to treat. The singular form can be written in the form

$$\begin{cases} G\alpha = \beta, \\ \langle \mathcal{G}, \alpha \rangle = 0, \end{cases} \tag{113}$$

where $G \in \mathbb{R}^{v,v}$, $\alpha, \beta, \mathcal{G} \in \mathbb{R}^v$, v is the dimension of the variational space and the coefficient is obtained with a scalar product $\mu = \langle \alpha, \beta' \rangle$ [16, 19]. In other words, in practice, the constrained subspace is found to be one dimensional so that $\mathcal{C} = \mathcal{G}^\perp$. The matrix G is symmetric positive semi-definite, its nullspace is one dimensional $N(G) = \mathbb{R}\mathcal{N}$, $\beta \in R(G)$, and the well posedness condition $N(G) \oplus \mathcal{G}^\perp = \mathbb{R}^v$ holds [19]. The sparse transport matrix $db(G)$ is a submatrix [19] composed of diagonals of blocks of G, and $2db(G) - G$ and $db(G)$ are symmetric positive definite for $n \geq 3$. All these properties have been extracted from the properties of the Boltzmann linearized collision operator and that of the variational approximation spaces [19].

The solution of the transport linear system can then be obtained either from the symmetric positive definite system $(G + \mathcal{G} \otimes \mathcal{G})\alpha = \beta$ or from iterative techniques. Among iterative techniques, we may use generalized conjugate gradients algorithms *that are possible because the natural symmetries of transport processes have been taken into account*. A very good preconditioner is then the sparse transport matrix $db(G)$.

Stationary techniques are also feasible and are associated with a splitting $G = \mathcal{M} - \mathcal{Z}$ where $\mathcal{M} = db(G)$. These methods yields that

$$\alpha = \sum_{0 \leq j < \infty} (\mathcal{P}\mathcal{T})^j \mathcal{P}\mathcal{M}^{-1}\mathcal{P}^t \beta, \tag{114}$$

where $\mathcal{T} = \mathcal{M}^{-1}\mathcal{Z}$ and $\mathcal{P} = I - \mathcal{N} \otimes \mathcal{G} / \langle \mathcal{N}, \mathcal{G} \rangle$. It can then be shown that the spectral radius of the product $\mathcal{P}\mathcal{T}$ is strictly lower than unity since the matrix $\mathcal{M} + \mathcal{Z} = 2db(G) - G$ is positive definite *but this is a consequence of Boltzmann linearized equations*. These stationary and generalized conjugate gradients methods have been found to be efficient for mixture of neutral gases [7, 70–74].

The situation of ionized mixtures is more complex since the convergence rate of stationary iterative techniques deteriorate as the ionization level increases. On the contrary, the convergence properties of generalized conjugate gradient algorithms do not depend on the ionization level. New algorithms have thus been introduced with *more singular* versions of the transport linear systems. These algorithms have led to fast convergence rates for all ionization levels and magnetic field intensities [75].

The assumptions for transport coefficients when some mass fractions are vanishing are more complex and lay out of the scope of these notes [9, 19]. Zero mass fractions lead to artificial singularities in the transport linear systems which are eliminated by considering rescaled systems [19]. Provided the diffusion matrix is replaced by the flux diffusion matrix $C_{kl} = \rho \mathsf{y}_k D_{kl}$, $k, l \in S$, all transport coefficients are smooth functions of the mass fractions and admit finite limits when some mass fractions become arbitrarily small. The iterative algorithms obtained for positive mass fractions can also be rewritten in terms of rescaled systems that are still defined for nonnegative mass fractions [19].

4.3 Stefan-Maxwell Equations

As a typical illustration of transport linear systems, we discuss in this section the multicomponent diffusion matrix $D = (D_{ij})_{i,j \in S}$. We assume that a state of the mixture is given with $T > 0$, $p > 0$, and $\mathsf{y} > 0$, that is, $\mathsf{y}_k > 0$ for $k \in S$. We assume that the mass fraction sum up to unity $\langle \mathsf{y}, \mathbb{I} \rangle = 1$ and the mole fractions are denoted by $\mathsf{x}_1, \dots, \mathsf{x}_n$. The mole fractions may be evaluated from $\mathsf{x}_i = m\mathsf{y}_i/m_i$ where m is the mean molar weight given by $\langle \mathsf{y}, \mathbb{I} \rangle / m = \sum_{i \in S} \mathsf{y}_i/m_i$.

The usual diffusion matrix D, obtained with linear systems of size $v = n$, satisfy $R(D) \subset \mathsf{y}^{\perp}$ and

$$\Delta D = I_n - \mathsf{y} \otimes \mathbb{I}, \tag{115}$$

where I_n is the identity of size n and Δ the Stefan-Maxwell matrix [16, 19]. Diffusion coefficients associated with larger transport linear systems—required for plasmas—lay out of the scope of these notes [19, 75]. The matrix Δ reads

$$\Delta_{kk} = \sum_{l \neq k} \frac{\mathsf{x}_k \mathsf{x}_l}{\mathcal{D}_{kl}^{\mathrm{bin}}}, \qquad k \in S, \tag{116}$$

$$\Delta_{kl} = -\frac{\mathsf{x}_k \mathsf{x}_l}{\mathcal{D}_{kl}^{\mathrm{bin}}}, \qquad k, l \in S, \quad k \neq l, \tag{117}$$

where $\mathcal{D}_{kl}^{\mathrm{bin}}(T, p)$ is the binary diffusion coefficient of the species pair (k, l) depending on pressure and temperature. The structure of the matrix Δ is investigated in the following Lemma [9, 70]

Lemma 4.1 *Assume that the molar masses m_k, $k \in S$, are positive constants, that the coefficients $\mathcal{D}_{kl}^{\mathrm{bin}}$, $k, l \in S$, $k \neq l$, are positive and symmetric, and that $\mathsf{y} > 0$. Then Δ is symmetric positive semidefinite, $N(\Delta) = \mathbb{R}\,\mathbb{I}$, Δ is irreducible and is a singular M-matrix.*

We will need the following lemma about generalized inverses with prescribed range and nullspace that may be found in [9, 70].

Proposition 4.2 *Let $G \in \mathbb{R}^{v,v}$ be a matrix, and let $\mathfrak{p}$ and $\mathfrak{q}$ be two subspaces of $\mathbb{R}^{v}$ such that $N(G) \oplus \mathfrak{p} = \mathbb{R}^{v}$ and $R(G) \oplus \mathfrak{q} = \mathbb{R}^{v}$. Then, there exists a unique matrix Z such that $GZG = G$, $ZGZ = Z$, $N(Z) = \mathfrak{q}$, and $R(Z) = \mathfrak{p}$. The matrix Z—termed the generalized inverse of G with prescribed range $\mathfrak{p}$ and nullspace $\mathfrak{q}$—satisfies $GZ = P_{R(G),\mathfrak{q}}$ and $ZG = P_{\mathfrak{p},N(G)}$, where $P_{\mathfrak{a},\mathfrak{b}}$ is defined for linear spaces $\mathfrak{a}$ and $\mathfrak{b}$, such that $\mathfrak{a} \oplus \mathfrak{b} = \mathbb{R}^{v}$, and denotes the projector onto $\mathfrak{a}$ along $\mathfrak{b}$.*

If in addition G is symmetric positive semi-definite and $\mathfrak{p} = \mathfrak{q}^{\perp}$ then Z is symmetric positive semi-definite.

The diffusion matrix may then be defined as a generalized inverse of Δ with prescribed range and nullspace [9, 19, 70].

Proposition 4.3 *Keeping the assumptions of Lemma 4.1, there exists a unique matrix D such that $\Delta D = I_n - \mathsf{y} \otimes \mathbb{I}$ and $R(D) \subset \mathsf{y}^{\perp}$. This matrix D is the generalized inverse of Δ with prescribed range $\mathsf{y}^{\perp}$ and nullspace $\mathbb{R}\mathsf{y}$. The matrix D is symmetric positive semidefinite, $N(D) = \mathbb{R}\mathsf{y}$, D is irreductible, and for any $a > 0$ we have $D = (\Delta + a\mathsf{y} \otimes \mathsf{y})^{-1} - (1/a)\mathbb{I} \otimes \mathbb{I}$.*

The transport linear systems associated with the flux matrix C are of a similar nature [19, 70, 74] and are well posed for $\mathsf{y} \geq 0, \mathsf{y} \neq 0$. With the mathematical properties of the matrix C it is then possible to establish the following diffusion inequality involving the entropy production quadratic form $\langle Dx, x\rangle$ on the hyperplane of zero sum gradients [9, 119].

Proposition 4.4 *Let T be a fixed temperature and $\mathcal{Y} = \mathrm{diag}(\mathsf{y}_1, \ldots, \mathsf{y}_n)$. There exists a positive constant δ such that*

$$\forall \mathsf{y} > 0 \text{ with } \langle \mathsf{y}, \mathbb{I}\rangle = 1, \quad \forall x \in \mathbb{I}^{\perp} \quad \delta\langle \mathcal{Y}^{-1}x, x\rangle \leq \langle \rho D x, x\rangle. \tag{118}$$

In other words, the natural entropic production norm associated with diffusion processes involve expressions in the form $\sum_{i\in S} |\nabla \mathsf{y}_i|^2/\mathsf{y}_i$. Such norms have first been used for an existence theorem of traveling waves with complex chemistry and detailed transport [119]. This lemma also implies that the nonzero eigenvalues of $\mathcal{Y}^{1/2}\rho D\mathcal{Y}^{1/2}$ are bounded away form zero since $\mathcal{Y}^{1/2}\rho D\mathcal{Y}^{1/2}x = \lambda x$ with $\lambda \neq 0$ implies that $x \in R(\mathcal{Y}^{1/2}D) = (\mathsf{y}^{1/2})^{\perp}$ and $\sum_{i\in S} \mathsf{y}^{1/2}x_i = 0$ and from (118) that $\delta \leq \lambda$.

One can also derive from (115) to (117) after some algebra that for any $k \in S$

$$\boldsymbol{d}_k + \chi_k \boldsymbol{\nabla} \log T = \sum_{l\neq k} \frac{\mathsf{x}_k\mathsf{x}_l}{\mathcal{D}_{kl}^{\mathrm{bin}}}\mathbf{v}_l - \sum_{l\neq k} \frac{\mathsf{x}_k\mathsf{x}_l}{\mathcal{D}_{kl}^{\mathrm{bin}}}\mathbf{v}_k. \tag{119}$$

These equations are usually termed Stefan–Maxwell equations and must be completed by the constraint $\sum_{k\in S} \mathsf{y}_k\mathbf{v}_k = 0$ associated with mass conservation to define uniquely the diffusion velocities. An elementary derivation of these equations has been given by Williams [66].

During a multicomponent flow computation, when an explicit time algorithm is used, it is sufficient to solely evaluate the diffusion velocities $\mathbf{v}_k$, $k \in S$, by solving the Stefan-Maxwell equations for each spatial direction, say by using a projected conjugate gradient method [74]. However, when an implicit time marching technique is used, evaluating the diffusion matrix D is generally required. Accurate approximation of the diffusion matrix D may be obtained by considering $\mathcal{M} = \mathrm{diag}(\mathsf{x}_1/D_1^*, \ldots, \mathsf{x}_n/D_n^*)$ where

$$D_k^* = (1 - \mathsf{y}_k) \Big/ \sum_{l\neq k} \mathsf{x}_l/\mathcal{D}_{kl}^{\mathrm{bin}}, \tag{120}$$

and $\Delta = \mathcal{M} - \mathcal{Z}$, $\mathcal{T} = \mathcal{M}^{-1}\mathcal{Z}$, and $\mathcal{P} = I_n - \mathbb{I}\otimes\mathsf{y}$. The spectral radius of $\mathcal{PT}$ is then strictly lower than unity and we have the *convergent* series expansion [70]

$$D = \sum_{j=0}^{\infty} (\mathcal{PT})^j \mathcal{PM}^{-1}\mathcal{P}^t.$$

One may then introduce the approximate diffusion matrices

$$D^{[i]} = \sum_{0 \le j \le i} (\mathcal{P}\mathcal{T})^j \mathcal{P}\mathcal{M}^{-1}\mathcal{P}^t,$$

that are symmetric, satisfy the mass constraint $D^{[i]}\mathsf{y} = 0$ and yields a positive entropy production. The first approximation $D^{[0]} = \mathcal{P}\mathcal{M}^{-1}\mathcal{P}^t$ corresponds to the Curtiss–Hirschfelder approximation $\mathbf{v}_k = -D_k^* \boldsymbol{d}_k / \mathsf{x}_k + \mathbf{v}_{\text{cor}}$ with a mass corrector $\mathbf{v}_{\text{cor}}$ ensuring the constraint $\sum_{k \in S} \mathsf{y}_k \mathbf{v}_k = 0$ and arising here from the projector $\mathcal{P}$ [19, 70].

4.4 Impact of Multicomponent Transport

Recent numerical investigations have brought further support for the importance of accurate transport property in various multicomponent reactive flows [76–85]. Thermal diffusion effects have been shown to be important in the study of vortex-flame interaction, catalytic effects near walls, interfacial phenomena, gaseous or spray diffusion flames [83], chemical vapor deposition reactors [3] and reentry [85]. The impact of multicomponent diffusion has also been shown to be important in multidimensional hydrogen/air and methane/air Bunsen flames [7], in freely propagating flames—especially with oxygen as pure oxydizer—as well as in direct numerical simulation of turbulent flames.

Theoretical calculations and experimental measurements have also shown that the ratio κ/η is not small for polyatomic gases [9, 80–82, 84]. Volume viscosity also arises in dense gases and in liquids, and its absence in dilute monatomic gases is an exception rather than a rule. Despite its potential importance, volume or bulk viscosity has seldom been included in computational models of multidimensional reactive flows. For small Mach number flows, however, the whole term $\boldsymbol{\nabla}\cdot\big(\kappa(\boldsymbol{\nabla}\cdot\boldsymbol{v})\boldsymbol{I}\big)$ has a weak influence because of its *structure*, even though both the ratio κ/η and the dilatation $\boldsymbol{\nabla}\cdot\boldsymbol{v}$ may not be small [9]. However, it has been shown that volume viscosity has an important impact during a shock/hydrogen bubble interaction [82] and its influence on shock heated and expanding flows in investigated in [81, 84].

5 Mathematical Framework

We first summarize in this section the system of partial differential equations modeling reactive fluids derived from the kinetic theory of gases in the previous sections. We also introduce a mathematical framework, notably the assumptions on the transport coefficients, and recast the system in quasilinear form. The precise form of the thermodynamic functions obtained in the previous sections is not specifically required and they are simply defined here in terms of internal specific heats and formation constants. Furthermore the mathematical properties of the transport coefficients are extracted from the kinetic theory. Note that we only investigate here the equations governing ideal mixture of perfect gases derived from the kinetic theory and we refer to [109, 111, 112, 114, 115] for other models. We have often rewritten some of the relations deduced from the kinetic theory in the previous sections in order to facilitate an independent lecture of the following sections.

5.1 Conservation Equations

The equations for conservation of species mass, momentum and energy may be recast in the form [9]

$$\partial_t \rho_k + \boldsymbol{\nabla}\cdot(\rho_k \boldsymbol{v}) + \boldsymbol{\nabla}\cdot\boldsymbol{\mathcal{F}}_k = m_k \omega_k, \qquad k \in S, \tag{121}$$

$$\partial_t(\rho \boldsymbol{v}) + \boldsymbol{\nabla}\cdot(\rho \boldsymbol{v}\otimes\boldsymbol{v} + p\boldsymbol{I}) + \boldsymbol{\nabla}\cdot\boldsymbol{\Pi} = \sum_{i\in S} \rho_i \boldsymbol{b}_i, \tag{122}$$

$$\partial_t(\mathcal{E} + \tfrac{1}{2}\rho \boldsymbol{v}\cdot\boldsymbol{v}) + \boldsymbol{\nabla}\cdot\big((\mathcal{E} + \tfrac{1}{2}\rho\boldsymbol{v}\cdot\boldsymbol{v} + p)\boldsymbol{v}\big) + \boldsymbol{\nabla}\cdot(\boldsymbol{Q} + \boldsymbol{\Pi}\cdot\boldsymbol{v}) = \sum_{i\in S} \rho_i \boldsymbol{b}_i\cdot(\boldsymbol{v} + \mathbf{v}_i), \tag{123}$$

where ∂_t denotes the time derivative, $\boldsymbol{\nabla}$ the space derivative operator, ρ_k the mass density of the kth species, $\boldsymbol{v}$ the mass average flow velocity, $\boldsymbol{\mathcal{F}}_k$ the diffusion flux of the kth species, m_k the molar mass of the kth species, ω_k the molar production rate of the kth species, $S = \{1, \dots, n\}$ the set of species indices, $n \geqslant 1$ the number of species, $\rho = \sum_{k\in S} \rho_k$ the total mass density, p the pressure, $\boldsymbol{\Pi}$ the viscous tensor, $\boldsymbol{b}_i$ the force per unit mass acting on the ith species, $\mathcal{E}$ the internal energy per unit volume and $\boldsymbol{Q}$ the heat flux. These equations have to be completed by the relations expressing the thermodynamic properties like p and $\mathcal{E}$, the chemical production rates ω_k, $k \in S$, and the transport fluxes $\boldsymbol{\Pi}$, $\boldsymbol{\mathcal{F}}_k$, $k \in S$, and $\boldsymbol{Q}$.

Assuming that the force acting on the chemical species are species independent $\boldsymbol{b}_i = \boldsymbol{b}, i \in S$, as gravity for instance, then the energy production term $\sum_{i\in S} \rho_i \boldsymbol{b}_i\cdot(\boldsymbol{v}+\mathbf{v}_i)$ simplifies into $\rho\boldsymbol{b}\cdot\boldsymbol{v}$. In the following, we will assume that there is no force acting on the chemical species so that

$$\boldsymbol{b}_i = \boldsymbol{b} = 0, \qquad i \in S. \tag{124}$$

Such zeroth order force terms $\rho\boldsymbol{b}$ and $\rho\boldsymbol{b}\cdot\boldsymbol{v}$ do not significantly influence the mathematical structure of the resulting set of partial differential equations. The spatial dimension is denoted by d and the components of $\boldsymbol{v}$ and $\boldsymbol{\nabla}$ are written $\boldsymbol{v} = (v_1, \dots, v_d)^t$ and $\boldsymbol{\nabla} = (\partial_1, \dots, \partial_d)^t$ where v_i denotes the velocity in the ith spatial direction, ∂_i the derivation in the ith spatial direction and bold symbols are used for vector or tensor quantities in the physical space $\mathbb{R}^d$.

5.2 Thermodynamics

We will use for convenience the state variable $T, \rho_1, \ldots, \rho_n$ where T is the absolute temperature and also denote $\varrho = (\rho_1, \ldots, \rho_n)^t$. Other state variables could be used as well and may lead to slightly different mathematical formalisms [9]. The internal energy per unit volume $\mathcal{E}$ and the pressure p can be written in terms of the state variables $T, \rho_1, \ldots, \rho_n$ as

$$\mathcal{E}(T, \rho_1, \ldots, \rho_n) = \sum_{k \in S} \rho_k e_k(T), \qquad p(T, \rho_1, \ldots, \rho_n) = \sum_{k \in S} RT \frac{\rho_k}{m_k},$$

where e_k is the internal energy per unit mass of the kth species and R the gas constant. The internal energy e_k of the kth species is given by

$$e_k(T) = e_k^{\mathrm{st}} + \int_{T^{\mathrm{st}}}^{T} c_{\mathrm{v}k}(\tau)\, d\tau, \qquad k \in S,$$

where e_k^{st} is the standard formation energy of the kth species at the standard temperature T^{st} and $c_{\mathrm{v}k}$ the constant volume specific heat of the kth species. We also define the formation energy at zero temperature by letting $e_k^0 = e_k(0) = e_k^{\mathrm{st}} - \int_0^{T^{\mathrm{st}}} c_{\mathrm{v}k}(\tau)\, d\tau$. The (physical) entropy per unit volume $\mathcal{S}$ can be written in the form

$$\mathcal{S}(T, \rho_1, \ldots, \rho_n) = \sum_{k \in S} \rho_k s_k(T, \rho_k),$$

where s_k is the entropy per unit mass of the kth species. This quantity is in the form

$$s_k(T, \rho_k) = s_k^{\mathrm{st}} + \int_{T^{\mathrm{st}}}^{T} \frac{c_{\mathrm{v}k}(T')}{T'}\, dT' - \frac{R}{m_k} \log\Big(\frac{\rho_k}{\gamma^{\mathrm{st}} m_k}\Big), \quad k \in S,$$

where s_k^{st} is the formation entropy of the kth species at the standard temperature T^{st} and standard pressure $p^{\mathrm{st}} = p^{\mathrm{atm}}$ and $\gamma^{\mathrm{st}} = p^{\mathrm{st}}/RT^{\mathrm{st}}$ is the standard concentration. Similarly, one can introduce the mixture enthalpy $\mathcal{H} = \sum_{k \in S} \rho_k h_k(T)$ with

$$h_k(T) = e_k(T) + RT/m_k, \qquad k \in S,$$

the mixture Gibbs function per unit volume $\mathcal{G} = \sum_{k \in S} \rho_k g_k(T, \rho_k)$, with

$$g_k(T, \rho_k) = h_k(T) - T s_k(T, \rho_k), \qquad k \in S,$$

as well as the reduced chemical potential

$$\mu_k(T, \rho_k) = \frac{m_k g_k}{RT}, \qquad k \in S.$$

Finally, the species mass fractions y_k, $k \in S$, partial pressures p_k, $k \in S$, and mole fractions x_k, $k \in S$, are defined by

$$\mathsf{y}_k = \frac{\rho_k}{\rho}, \qquad p_k = \frac{\rho_k RT}{m_k}, \qquad \mathsf{x}_k = \frac{p_k}{p}, \qquad k \in S.$$

The mole fractions may also be evaluated from $\mathsf{x}_i = m\mathsf{y}_i/m_i$ where m is the mean molar weight given by $\langle \mathsf{y}, \mathbb{I} \rangle / m = \sum_{i \in S} \mathsf{y}_i/m_i$.

5.3 Chemical Sources

We consider a system of $n^{\mathrm{r}} \geqslant 1$ elementary reactions for $n \geqslant 1$ species which can be written formally

$$\sum_{k \in S} \nu_{ki}^{\mathrm{f}} \, \mathfrak{M}_k \rightleftarrows \sum_{k \in S} \nu_{ki}^{\mathrm{b}} \, \mathfrak{M}_k, \qquad i \in \mathfrak{R},$$

where $\mathfrak{M}_k$ is the chemical symbol of the kth species, ν_{ki}^{f} and ν_{ki}^{b} the forward and backward stoichiometric coefficients of the kth species in the ith reaction, $\mathfrak{R} = \{1, \dots, n^{\mathrm{r}}\}$ the set of reaction indices, and $\nu_{ki} = \nu_{ki}^{\mathrm{b}} - \nu_{ki}^{\mathrm{f}}$ the overall stoichiometric coefficients. The species of the mixture are assumed to be constituted by atoms, and we denote by $\mathfrak{a}_{il}$ the number of lth atom in the ith species, $\mathfrak{A} = \{1, \dots, n^{\mathrm{a}}\}$ the set of atom indices, and $n^{\mathrm{a}} \geqslant 1$ the number of atoms—or elements—in the mixture. It is convenient to introduce at this point some vector notation by letting

$$\omega = \begin{pmatrix} \omega_1 \\ \vdots \\ \omega_n \end{pmatrix}, \qquad \nu_i = \begin{pmatrix} \nu_{1i} \\ \vdots \\ \nu_{ni} \end{pmatrix}, \qquad \nu_i^{\mathrm{f}} = \begin{pmatrix} \nu_{1i}^{\mathrm{f}} \\ \vdots \\ \nu_{ni}^{\mathrm{f}} \end{pmatrix}, \qquad \nu_i^{\mathrm{b}} = \begin{pmatrix} \nu_{1i}^{\mathrm{b}} \\ \vdots \\ \nu_{ni}^{\mathrm{b}} \end{pmatrix},$$

$$\mu = \begin{pmatrix} \mu_1 \\ \vdots \\ \mu_n \end{pmatrix}, \qquad \mathfrak{a}_l = \begin{pmatrix} \mathfrak{a}_{1l} \\ \vdots \\ \mathfrak{a}_{nl} \end{pmatrix}, \qquad \mathfrak{m} = \begin{pmatrix} m_1 \\ \vdots \\ m_n \end{pmatrix}, \qquad \varrho = \begin{pmatrix} \rho_1 \\ \vdots \\ \rho_n \end{pmatrix}.$$

We will denote by $\mathcal{R}$ the vector space spanned by the reaction vectors $\mathcal{R} = \mathrm{Span}\{ \nu_i, \ i \in \mathfrak{R} \}$ and by $\mathcal{A}$ the vector space spanned by the atom vectors $\mathcal{A} = \mathrm{Span}\{ \mathfrak{a}_l, \ l \in \mathfrak{A} \}$. The molar production rates that we consider are the Maxwellian production rates obtained from the kinetic theory [9, 19] when the chemical characteristic times are larger than the mean free times of the molecules and the characteristic times of internal energy relaxation. These rates ω_k, $k \in S$, are compatible with the law of mass action and are in the form

$$\omega_k = \sum_{i \in \mathfrak{R}} (\nu_{ki}^{\mathrm{b}} - \nu_{ki}^{\mathrm{f}})\tau_i, \qquad k \in S, \tag{125}$$

where τ_i is the rate of progress of the ith reaction given by

$$\tau_i = \mathcal{K}_i^{\mathrm{s}}\big(\exp\langle \mu, \nu_i^{\mathrm{f}} \rangle - \exp\langle \mu, \nu_i^{\mathrm{b}} \rangle\big), \tag{126}$$

as written by Marcelin [57], Gorban [63] and Keizer [64] and also deduced from the kinetic theory (98). These rates of progress are compatible with the law of mass action and may be rewritten

$$\tau_i = \mathcal{K}_i^{\mathrm{f}} \prod_{l \in S} \Big(\frac{\rho_l}{m_l}\Big)^{\nu_{li}^{\mathrm{f}}} - \mathcal{K}_i^{\mathrm{b}} \prod_{l \in S} \Big(\frac{\rho_l}{m_l}\Big)^{\nu_{li}^{\mathrm{b}}}, \tag{127}$$

where $\mathcal{K}_i^{\mathrm{f}}$ and $\mathcal{K}_i^{\mathrm{b}}$ are the forward and backward rate constants of the ith reaction, respectively. The reaction constants $\mathcal{K}_i^{\mathrm{f}}$, $\mathcal{K}_i^{\mathrm{b}}$ and $\mathcal{K}_i^{\mathrm{s}}$ are functions of temperature and are Maxwellian averaged values of molecular chemical transition probabilities [19]. In particular, forward and backward chemical transition probabilities are always proportional—as in any Boltzmann equation—and this implies the reciprocity relations [9, 19]

$$\mathcal{K}_i^{\mathrm{e}}(T) = \frac{\mathcal{K}_i^{\mathrm{f}}(T)}{\mathcal{K}_i^{\mathrm{b}}(T)}, \qquad \log \mathcal{K}_i^{\mathrm{e}}(T) = -\langle \nu_i, \mu^{\mathrm{u}} \rangle, \qquad i \in \mathfrak{R}, \tag{128}$$

where $\mathcal{K}_i^{\mathrm{e}}(T)$ is the equilibrium constant of the ith reaction, $\mu^{\mathrm{u}} = (\mu_1^{\mathrm{u}}, \dots, \mu_n^{\mathrm{u}})^t$ and $\mu_k^{\mathrm{u}}(T) = \mu_k(T, m_k)$, $k \in S$. These reciprocity relations are closely associated with the reciprocity relations between reactive transition probabilities as well as with symmetric representation of the rate of progress and may be seen as Onsager relations for chemistry. These reactions constants are also related by $\log \mathcal{K}_i^{\mathrm{s}} = \log \mathcal{K}_i^{\mathrm{f}} - \langle M\nu_i^{\mathrm{f}}, \mu^{\mathrm{u}} \rangle = \log \mathcal{K}_i^{\mathrm{b}} - \langle M\nu_i^{\mathrm{b}}, \mu^{\mathrm{u}} \rangle$. On the other hand, in practice, the forward reaction constants $\mathcal{K}_i^{\mathrm{f}}$, $i \in \mathfrak{R}$, are often approximated with Arrhenius law

$$\mathcal{K}_i^{\mathrm{f}} = \mathfrak{A}_i T^{\mathfrak{b}_i} \exp\big(-\mathfrak{E}_i / RT\big), \qquad i \in \mathfrak{R},$$

where $\mathfrak{A}_i$ is the preexponential factor, $\mathfrak{b}_i$ the temperature exponent and $\mathfrak{E}_i$ the activation energy of the ith reaction. Note incidentally that the thermodynamics of irreversible processes only yields rates of progress that are linear in terms of affinities $\langle \nu_i, \mu \rangle$ instead of exponential as in (126).

5.4 Transport Fluxes

The transport fluxes $\boldsymbol{\Pi}, \mathcal{F}_k, k \in S$, and $\boldsymbol{Q}$ due to macroscopic variable gradients can be written in the form [9, 11, 12, 15, 16, 19]

$$\boldsymbol{\Pi} = -\kappa \nabla \cdot \boldsymbol{v} \boldsymbol{I} - \eta \big(\nabla \boldsymbol{v} + \nabla \boldsymbol{v}^t - \tfrac{2}{3} (\nabla \cdot \boldsymbol{v}) \boldsymbol{I} \big), \tag{129}$$

$$\mathcal{F}_k = -\sum_{l \in S} C_{kl} (\widehat{\boldsymbol{d}}_l + \mathrm{x}_l \widetilde{\chi}_l \boldsymbol{\partial}_x \log T), \qquad k \in S, \tag{130}$$

$$\boldsymbol{Q} = -\lambda \nabla T + \sum_{k \in S} (RT \frac{\widetilde{\chi}_k}{m_k} + h_k) \mathcal{F}_k, \tag{131}$$

where κ denotes the volume viscosity, η the shear viscosity, $\boldsymbol{I}$ the three dimensional identity tensor, C_{kl}, $k, l \in S$, the multicomponent flux diffusion coefficients, $\widehat{\boldsymbol{d}}_k$, $k \in S$, the unconstrained species diffusion driving forces, ^{t} the transposition operator, $\widetilde{\chi}_k$, $k \in S$, the rescaled thermal diffusion ratios and λ the thermal conductivity. The diffusion driving forces are defined by

$$\widehat{\boldsymbol{d}}_k = \frac{\nabla p_k}{p}, \qquad k \in S,$$

keeping in mind that the force term acting on the species are assumed to be zero $\boldsymbol{b}_k = 0$, $k \in S$. When the mass fractions are nonzero, it is also possible to define the species diffusion velocities $\mathbf{v}_k$, $k \in S$, by

$$\mathbf{v}_k = \frac{\mathcal{F}_k}{\rho_k} = -\sum_{l \in S} D_{kl} (\widehat{\boldsymbol{d}}_l + \mathrm{x}_l \widetilde{\chi}_l \nabla \log T).$$

where $D_{kl} = C_{kl}/\rho_k$, $k, l \in S$.

The transport coefficients have important properties inherited from the underlying kinetic framework [9, 12, 19]. They satisfy symmetry properties, mass conservation constraints, as well as positivity properties as derived in Sect. 3.4 and detailed in the next section. The multicomponent transport coefficients κ, η, λ, $C = (C_{kl})_{k,l \in S}$, $D = (D_{kl})_{k,l \in S}$, or $\widetilde{\chi} = (\widetilde{\chi}_k)_{k \in S}$, are also smooth functions of the state variables. Note that the matrices C and D are generally irreducible and the governing equations have thus a complex structure [9].

5.5 Mathematical Assumptions

The assumptions on the thermodynamic properties and the transport coefficients have been extracted from the kinetic theory of gases. There are recast in the following form where $\varkappa$ denotes a regularity class of transport coefficients and thermodynamic functions [9]. Throughout these notes, for any matrix A, we denote by $N(A)$ its nullspace and $R(A)$ its range.

(H$_1$) *The molar masses m_k, $k \in S$, and the perfect gas constant R are positive constants. The formation energies $e_k^{\rm st}$, $k \in S$, and entropies $s_k^{\rm st}$, $k \in S$, are real constants. The specific heats $c_{\rm v}{}_k$, $k \in S$, are $C^{\varkappa-1}$ functions of $T \in [0, \infty)$. There exist positive constants $\underline{c}_{\rm v}$ and $\overline{c}_{\rm v}$ such that $0 < \underline{c}_{\rm v} \leqslant c_{{\rm v}k}(T) \leqslant \overline{c}_{\rm v}$ for $T \geq 0$ and $k \in S$.*

(H$_2$) *The stoichiometric coefficients $\nu_{ki}^{\rm f}$ and $\nu_{ki}^{\rm b}$, $k \in S$, $i \in \mathfrak{R}$, the atomic coefficients $\mathfrak{a}_{kl}$, $k \in S$, $l \in \mathfrak{A}$, are nonnegative integers. The atom vectors $\mathfrak{a}_l$, $l \in \mathfrak{A}$, and the reaction vectors $\nu_i = \nu_i^{\rm b} - \nu_i^{\rm f}$, $i \in \mathfrak{R}$, satisfy the atom conservation constraints $\langle \nu_i, \mathfrak{a}_l \rangle = 0$, $i \in \mathfrak{R}$, $l \in \mathfrak{A}$. The atom masses $\widetilde{m}_l$, $l \in \mathfrak{A}$, are positive constants and the vector of species molar masses $\mathfrak{m}$ is given by $\mathfrak{m} = \sum_{l \in \mathfrak{A}} \widetilde{m}_l\, \mathfrak{a}_l$.*

(H$_3$) *The symmetric reaction constants $\mathcal{K}_i^{\rm s}$ are $C^{\varkappa}$ positive functions of $T > 0$ for $i \in \mathfrak{R}$.*

(H$_4$) *The flux diffusion matrix $C = (C_{kl})_{k,l \in S}$, the rescaled thermal diffusion ratios vector $\widetilde{\chi} = (\widetilde{\chi}_1, \ldots, \widetilde{\chi}_n)^t$, the volume viscosity κ, the shear viscosity η, and the thermal conductivity λ are $C^{\varkappa}$ functions of $(T, \rho_1, \ldots, \rho_n)$ for $T > 0$ and $\rho_i > 0$, $i \in S$. These coefficients satisfy the mass conservation constraints $N(C) = \mathbb{R}\mathsf{y}$, $R(C) = \mathbb{I}^{\perp}$, and $\widetilde{\chi} \in \mathsf{x}^{\perp}$.*

(H$_5$) *The thermal conductivity λ and the shear viscosity η are positive. The volume viscosity κ is nonnegative. The diffusion matrix $D = (1/\rho)\mathcal{Y}^{-1}C$ is symmetric positive semi-definite and its nullspace is $N(D) = \mathbb{R}\mathsf{y}$ where $\mathcal{Y} = \mathrm{diag}(\mathsf{y}_1, \ldots, \mathsf{y}_n)$.*

Remark 5.1 All coefficients C, λ, η, $\widetilde{\chi}$ and κ have smooth extensions to the domain $\rho_i \geq 0$, $i \in S$, and $\rho > 0$. This is also the case for the non diagonal coefficients D_{ij} for $i \neq j$ whereas the coefficient $\rho_i D_{ii}$ has a finite positive limit when $\rho_i \to 0$ [19, 70].

Remark 5.2 We generally have $\mathcal{R} \subset \mathcal{A}^{\perp}$ but chemical reaction mechanisms are usually sufficiently rich so that reaction vectors ν_i, $i \in \mathfrak{R}$, are spanning the maximum space and $\mathcal{R} = \mathcal{A}^{\perp}$.

5.6 Entropy Production

From Gibbs' relation $T\, \mathbb{D}\mathcal{S} = \mathbb{D}\mathcal{E} - \sum_{k \in S} g_k\, \mathbb{D}\rho_k$, where $\mathbb{D}$ denotes the total derivative, the conservation equations, and the properties of transport coefficients and chemical production rate, one may derive the following balance equation for $\rho s = \mathcal{S}$

$$\begin{aligned}\partial_t(\rho s) + \nabla\cdot(\rho \boldsymbol{v} s) + \nabla\cdot\Big(\frac{\boldsymbol{Q}}{T} - \sum_{k\in S}\frac{g_k}{T}\boldsymbol{\mathcal{F}}_k\Big) &= \frac{3d\kappa + 2\eta(3-d)}{3dT}(\nabla\cdot\boldsymbol{v})^2\\
&+ \frac{\eta}{2T}\big(\nabla\boldsymbol{v} + \nabla\boldsymbol{v}^t - \tfrac{2}{d}(\nabla\cdot\boldsymbol{v})\,\boldsymbol{I}\big):\big(\nabla\boldsymbol{v} + \nabla\boldsymbol{v}^t - \tfrac{2}{d}(\nabla\cdot\boldsymbol{v})\,\boldsymbol{I}\big)\\
&+ \frac{\lambda}{T^2}\nabla T\cdot\nabla T + \frac{p}{T}\sum_{k,l\in S} D_{kl}\big(\widehat{\boldsymbol{d}}_k + \chi_k\nabla\log T\big)\cdot\big(\widehat{\boldsymbol{d}}_l + \chi_l\nabla\log T\big)\\
&+ \sum_{i\in\mathfrak{R}} R\mathcal{K}_i^{\mathrm{s}}\big(\langle\mu,\nu_i^{\mathrm{f}}\rangle - \langle\mu,\nu_i^{\mathrm{b}}\rangle\big)\,\big(\exp\langle\mu,\nu_i^{\mathrm{f}}\rangle - \exp\langle\mu,\nu_i^{\mathrm{b}}\rangle\big). \qquad (132)\end{aligned}$$

The viscous tensor has been rewritten for convenience in the form

$$\boldsymbol{\Pi} = -\big(\kappa + \frac{2\eta(3-d)}{3d}\big)\nabla\cdot\boldsymbol{v}\,\boldsymbol{I} - \eta\Big(\nabla\boldsymbol{v} + \nabla\boldsymbol{v}^t - \tfrac{2}{d}(\nabla\cdot\boldsymbol{v})\,\boldsymbol{I}\Big),$$

keeping in mind that $1 \le d \le 3$. Entropy production (132) therefore appears as a sum of nonnegative terms and the last term represents the entropy production due to chemical reactions $-R\langle\mu,\omega\rangle$. From this expression of $\langle\mu,\omega\rangle$ we also deduce the following result concerning chemical equilibrium.

Proposition 5.3 *Assume that the Properties* (H_1–H_3) *hold. Then for any* $(T, \rho_1,\dots,\rho_n) \in (0,\infty)^{1+n}$ *the following statements are equivalent :*

(i) The entropy production due to chemistry vanishes $-R\langle\mu,\omega\rangle = 0$.

(ii) The reaction rates of progress vanish $\tau_j = 0, \quad j \in \mathfrak{R}$.

(iii) The species production rates vanish $\omega_k = 0, \quad k \in S$.

(iv) The vector $\mu = (\mu_1,\dots,\mu_n)^t$ *belongs to* $\mathcal{R}^\perp$ *where*

$$\mathcal{R} = \mathrm{span}\{\,\nu_i,\ i \in \mathfrak{R}\,\}.$$

Proof From the expression of entropy production due to chemical reactions

$$-R\langle\mu,\omega\rangle = \sum_{i\in\mathfrak{R}} R\mathcal{K}_i^{\mathrm{s}}\big(\langle\mu,\nu_i^{\mathrm{f}}\rangle - \langle\mu,\nu_i^{\mathrm{b}}\rangle\big)\,\big(\exp\langle\mu,\nu_i^{\mathrm{f}}\rangle - \exp\langle\mu,\nu_i^{\mathrm{b}}\rangle\big),$$

and (H_3) we obtain that $\langle\mu,\omega\rangle = 0$ implies $\langle\mu,\nu_j\rangle = 0$, $j \in \mathfrak{R}$, and so $\tau_j = 0$, $j \in \mathfrak{R}$, and we have established that (i) implies (ii). The fact that (ii) implies (iii) is a consequence of $\omega = \sum_{j\in\mathfrak{R}}\tau_j\nu_j$. We also deduce from the expression of entropy production $-R\langle\mu,\omega\rangle$ that (iii) implies (i) so that the three statements (i), (ii), and (iii) are equivalent. Finally, it is easily established that (iv) is equivalent to $\langle\mu,\nu_j\rangle = 0$, $j \in \mathfrak{R}$, so that (ii) and (iv) are also equivalent. □

Definition 5.4 A point $T^\star > 0$, $\varrho^\star \in (0,\infty)^n$, which satisfies the equivalent properties of Proposition 5.3 is termed an equilibrium point.

We are only interested here in positive equilibrium states with $\rho_i > 0, i \in S$, which are in the interior of the composition space. Spurious points with zero mass fractions where the source terms ω_k, $k \in n$, also vanish—termed 'boundary equilibrium points'—are of a different nature [9]. Properly structured chemical kinetic mechanisms automatically exclude such spurious points unless some element is missing in the mixture [9].

5.7 Vector Notation

The equations governing multicomponent flows (121)–(123) can be rewritten in the compact vector form

$$\partial_t \mathsf{u} + \sum_{i \in C} \partial_i \mathsf{F}_i + \sum_{i \in C} \partial_i \mathsf{F}_i^{\mathrm{diss}} = \Omega, \tag{133}$$

where u is the conservative variable, ∂_i the spatial derivative operator in the ith spatial direction, $C = \{1, \dots, d\}$ the indexing set of spatial directions, $d \in \{1, 2, 3\}$ the spatial dimension, F_i the convective flux in the ith direction, $\mathsf{F}_i^{\mathrm{diss}}$ the dissipative flux in the ith direction, and Ω the source term. Letting $\mathsf{n} = n+d+1$, the conservative variable $\mathsf{u} \in \mathbb{R}^{\mathsf{n}}$ is found to be

$$\mathsf{u} = \big(\rho_1, \dots, \rho_n, \rho\boldsymbol{v}, \mathcal{E} + \tfrac{1}{2}\rho\boldsymbol{v}\cdot\boldsymbol{v}\big)^t, \tag{134}$$

and the natural variable $\mathsf{z} \in \mathbb{R}^{\mathsf{n}}$ is defined by

$$\mathsf{z} = \big(\rho_1, \dots, \rho_n, \boldsymbol{v}, T\big)^t. \tag{135}$$

For convenience, the velocity components of vectors in $\mathbb{R}^{\mathsf{n}} = \mathbb{R} \times \mathbb{R}^d \times \mathbb{R}^2$ are generally written as vectors of $\mathbb{R}^d$ and bold symbols are used for vector or tensor quantities in the physical space $\mathbb{R}^d$. The map $\mathsf{z} \to \mathsf{u}$ is a $C^{\varkappa}$ diffeomorphism from the open set

$$\mathcal{O}_{\mathsf{z}} = (0, \infty)^n \times \mathbb{R}^d \times (0, \infty),$$

onto a convex open set $\mathcal{O}_{\mathsf{u}}$ of $\mathbb{R}^{\mathsf{n}}$ [9, 95].

Proposition 5.5 *The map $\mathsf{z} \longmapsto \mathsf{u}$ is a $C^{\varkappa}$ diffeomorphism from the open set $\mathcal{O}_{\mathsf{z}}$ onto an open set $\mathcal{O}_{\mathsf{u}}$. The open set $\mathcal{O}_{\mathsf{u}}$ is convex and given by*

$$\mathcal{O}_{\mathsf{u}} = \{\, u \in \mathbb{R}^{\mathsf{n}};\, u_i > 0,\ 1 \le i \le n,\ \ u_{\mathsf{n}} - \phi(u_1, \dots, u_{n+d}) > 0 \,\},$$

where

$$\phi(u_1, \dots, u_{n+d}) = \frac{1}{2} \frac{u_{n+1}^2 + \cdots + u_{n+d}^2}{\sum_{i \in S} u_i} + \sum_{i \in S} u_i e_i^0,$$

and e_i^0 is the energy of formation of the ith species at zero temperature.

Proof The map $\mathsf{z} \rightarrow \mathsf{u}$ is easily shown to be one to one from $\mathcal{O}_\mathsf{z}$ onto $\mathcal{O}_\mathsf{u}$ using the positivity of the constant volume specific heats. The jacobian matrix $\partial_\mathsf{z}\mathsf{u}$, has a triangular structure and is invertible so that we may use the inverse function theorem. □

The convective and diffusive fluxes F_i and $\mathsf{F}_i^{\mathrm{diss}}$ in the ith direction are defined by

$$\mathsf{F}_i = \big(\rho_1 v_i, \dots, \rho_n v_i,\ \rho v_i \boldsymbol{v} + p\boldsymbol{e}_i,\ (\mathcal{E} + \tfrac{1}{2}\rho \boldsymbol{v}\cdot\boldsymbol{v} + p)v_i\big)^t,$$
$$\mathsf{F}_i^{\mathrm{diss}} = \big(\mathcal{F}_{1i}, \dots, \mathcal{F}_{ni},\ \boldsymbol{\Pi}_i,\ \mathcal{Q}_i + \boldsymbol{\Pi}_i\cdot\boldsymbol{v}\big)^t,$$

where $\boldsymbol{e}_i$ denotes the ith basis vector in the physical space $\mathbb{R}^d$, v_i the velocity in the ith direction, $\mathcal{F}_{ki}$ the diffusion flux of the kth species in the ith direction, $\mathcal{Q}_i$ the heat flux in the ith direction, $\boldsymbol{\Pi} = (\Pi_{ij})_{i,j\in C}$ the viscous tensor, and $\boldsymbol{\Pi}_i$ the vector $\boldsymbol{\Pi}_i = (\Pi_{1i}, \dots, \Pi_{di})^t$, so that $\boldsymbol{v} = (v_1, \dots, v_d)^t$, $\boldsymbol{\mathcal{F}}_k = (\mathcal{F}_{k1}, \dots, \mathcal{F}_{kd})^t$, and $\boldsymbol{Q} = (\mathcal{Q}_1, \dots, \mathcal{Q}_d)^t$.

The dissipative fluxes may further be expressed in the form $\mathsf{F}_i^{\mathrm{diss}} = -\sum_{j\in C} \widehat{\mathsf{B}}_{ij}(\mathsf{u})\,\partial_j\mathsf{z}$ since all transport fluxes are linear expressions in terms of the gradients of the natural variable z. Since $\mathsf{z} \rightarrow \mathsf{u}$ is a smooth diffeomorphism, defining $\mathsf{B}_{ij}(\mathsf{u}) = \widehat{\mathsf{B}}_{ij}(\mathsf{u})\partial_\mathsf{u}\mathsf{z}$, for $i, j \in C$, we obtain that

$$\mathsf{F}_i^{\mathrm{diss}} = -\sum_{j\in C} \mathsf{B}_{ij}(\mathsf{u})\partial_j \mathsf{u}, \qquad i \in C,$$

where the dissipation matrix B_{ij} relates the dissipative flux in the ith direction $\mathsf{F}_i^{\mathrm{diss}}$ to the gradient of the conservative variable in the jth direction $\partial_j\mathsf{u}$. Further denoting by $\mathsf{A}_i = \partial_\mathsf{u} F_i$, $i \in C$, the convective flux Jacobian matrices, and Ω the source term

$$\Omega = \big(m_1\omega_1, \dots, m_n\omega_n, \mathbf{0}, 0\big)^t,$$

we obtain the quasilinear system

$$\partial_t \mathsf{u} + \sum_{i\in C} \mathsf{A}_i(\mathsf{u})\partial_i \mathsf{u} = \sum_{i,j\in C} \partial_i\Big(\mathsf{B}_{ij}(\mathsf{u})\partial_j \mathsf{u}\Big) + \Omega(\mathsf{u}), \tag{136}$$

and all the system coefficients $\mathsf{A}_i(\mathsf{u})$, $i \in C$, $\mathsf{B}_{ij}(\mathsf{u})$, $i, j \in C$, and $\Omega(\mathsf{u})$, are smooth functions of u on the open convex set $\mathcal{O}_\mathsf{u}$. The mathematical structure of such quasilinear systems is discussed in the next section in an abstract framework.

6 Hyperbolic-Parabolic Structure

We discuss symmetrization with entropic variables and normal variables for abstract systems in quasilinear form. We next investigate the hyperbolic-parabolic structure of the resulting systems of partial differential equations. We further explicitly evaluate the natural entropic symmetrized form as well as the natural normal form for the system of partial differential equations modeling multicomponent reactive fluids.

6.1 Entropic Variables

We consider an abstract second order quasilinear system of conservation laws in the general form

$$\partial_t \mathsf{u} + \sum_{i\in C} \mathsf{A}_i(\mathsf{u})\partial_i \mathsf{u} - \sum_{i,j\in C} \partial_i \big(\mathsf{B}_{ij}(\mathsf{u})\partial_j \mathsf{u}\big) - \Omega(\mathsf{u}) = 0, \tag{137}$$

where $\mathsf{u} \in \mathcal{O}_\mathsf{u}$, $\mathcal{O}_\mathsf{u}$ is an open convex set of $\mathbb{R}^\mathsf{n}$, and $\mathsf{n} \geq 1$. The system coefficients are assumed such that $\mathsf{A}_i = \partial_\mathsf{u}\mathsf{F}_i$ and the fluxes F_i, $i \in C$, the dissipation matrices B_{ij}, $i, j \in C$, and the source term Ω, are assumed to be $C^\varkappa$ over $\mathcal{O}_\mathsf{u}$ where $\varkappa \geq 3$.

A mathematical entropy for the system of partial differential Eq. (137) must be compatible with the convective terms, the dissipative terms as well as the source term and we use the definition presented in [101, 103] simplified to the situation where the set $\mathcal{O}_\mathsf{u}$ is convex. In the following definition, properties (E_1), (E_2) concerning the convective terms have been adapted from [86, 88], properties (E_3), (E_4) associated with the dissipative terms have been adapted from [89, 92, 106, 107] and properties (E_5)–(E_7) concerning the source terms have been adapted from [93, 98] and we denote by Σ^{d-1} the sphere in d dimension.

Definition 6.1 Consider a $C^\varkappa$ function $\mathsf{u} \to \sigma(\mathsf{u})$ defined over the open convex domain $\mathcal{O}_\mathsf{u}$. The function σ is said to be an entropy function for the system (137) if the following properties hold.

- (E_1) The Hessian matrix $\partial^2_\mathsf{u}\sigma(\mathsf{u}) = \partial_\mathsf{u}(\partial_\mathsf{u}\sigma)^t(\mathsf{u})$ is positive definite over $\mathcal{O}_\mathsf{u}$.
- (E_2) There exist $C^\varkappa$ functions $\mathsf{u} \to \mathsf{q}_i(\mathsf{u})$ such that $\partial_\mathsf{u}\sigma(\mathsf{u})\,\mathsf{A}_i(\mathsf{u}) = \partial_\mathsf{u}\mathsf{q}_i(\mathsf{u})$ for $\mathsf{u} \in \mathcal{O}_\mathsf{u}$ and $i \in C$.
- (E_3) We have $\big(\mathsf{B}_{ij}(\mathsf{u})\,\big(\partial^2_\mathsf{u}\sigma(\mathsf{u})\big)^{-1}\big)^t = \mathsf{B}_{ji}(\mathsf{u})\big(\partial^2_\mathsf{u}\sigma(\mathsf{u})\big)^{-1}$ for $\mathsf{u} \in \mathcal{O}_\mathsf{u}$ and $i, j \in C$.
- (E_4) The matrix $\sum_{i,j\in C} \mathsf{B}_{ij}(\mathsf{u})\big(\partial^2_\mathsf{u}\sigma(\mathsf{u})\big)^{-1}\,\xi_i\xi_j$ is positive semi-definite for $\mathsf{u} \in \mathcal{O}_\mathsf{u}$ and $\boldsymbol{\xi} \in \Sigma^{d-1}$.
- (E_5) There exists a fixed vector space $\mathcal{E} \subset \mathbb{R}^\mathsf{n}$ such that $\Omega(\mathsf{u}) \in \mathcal{E}^\perp$ for $\mathsf{u} \in \mathcal{O}_\mathsf{u}$ and $\Omega(\mathsf{u}) = 0$ if and only if $\big(\partial_\mathsf{u}\sigma(\mathsf{u})\big)^t \in \mathcal{E}$ and if and only if $\partial_\mathsf{u}\sigma(\mathsf{u})\ \Omega(\mathsf{u}) = 0$.
- (E_6) If $\Omega(\mathsf{u}) = 0$, then the matrix $\partial_\mathsf{u}\Omega(\mathsf{u})\,\big(\partial^2_\mathsf{u}\sigma(\mathsf{u})\big)^{-1}$ is symmetric and its nullspace is given by $N\big(\partial_\mathsf{u}\Omega(\mathsf{u})\,(\partial^2_\mathsf{u}\sigma(\mathsf{u}))^{-1}\big) = \mathcal{E}$.
- (E_7) We have $\partial_\mathsf{u}\sigma(\mathsf{u})\ \Omega(\mathsf{u}) \leq 0$ for $\mathsf{u} \in \mathcal{O}_\mathsf{u}$.

Existence of an entropy is closely associated with symmetrization properties [86, 88, 89, 92, 93, 98, 101–104, 106, 107]. We do not encounter here the difficulty associated with nonideal fluids where only *local* symmetrization are feasible and where $\mathcal{O}_\mathsf{u}$ may not be convex [101]. Note also that more general source terms with no symmetry properties at equilibrium have been considered by Chen et al. [93] and Yong [110].

Definition 6.2 Consider a $C^{\varkappa-1}$ diffeomorphism $\mathsf{u} \to \mathsf{v}$ from $\mathcal{O}_{\mathsf{u}}$ onto an open domain $\mathcal{O}_{\mathsf{v}}$ and the system in the v variable

$$\widetilde{\mathsf{A}}_0(\mathsf{v})\partial_t \mathsf{v} + \sum_{i \in C} \widetilde{\mathsf{A}}_i(\mathsf{v})\partial_i \mathsf{v} - \sum_{i,j \in C} \partial_i \big(\widetilde{\mathsf{B}}_{ij}(\mathsf{v})\partial_j \mathsf{v}\big) - \widetilde{\Omega}(\mathsf{v}) = 0, \tag{138}$$

where $\widetilde{\mathsf{A}}_0 = \partial_{\mathsf{v}}\mathsf{u}$, $\widetilde{\mathsf{A}}_i = \mathsf{A}_i\partial_{\mathsf{v}}\mathsf{u} = \partial_{\mathsf{v}}\mathsf{F}_i$, $\widetilde{\mathsf{B}}_{ij} = \mathsf{B}_{ij}\partial_{\mathsf{v}}\mathsf{u}$, and $\widetilde{\Omega} = \Omega$, have at least regularity $\varkappa - 2$. The system is said of the symmetric form if properties (S_1–S_7) hold.

(S_1) The matrix $\widetilde{\mathsf{A}}_0(\mathsf{v})$ is symmetric positive definite for $\mathsf{v} \in \mathcal{O}_{\mathsf{v}}$.
(S_2) The matrices $\widetilde{\mathsf{A}}_i(\mathsf{v})$, $i \in C$, are symmetric for $\mathsf{v} \in \mathcal{O}_{\mathsf{v}}$.
(S_3) We have $\widetilde{\mathsf{B}}_{ij}^t(\mathsf{v}) = \widetilde{\mathsf{B}}_{ji}(\mathsf{v})$ for $i, j \in C$ and $\mathsf{v} \in \mathcal{O}_{\mathsf{v}}$.
(S_4) The matrix $\widetilde{\mathsf{B}}(\mathsf{v}, \boldsymbol{\xi}) = \sum_{i,j \in C} \widetilde{\mathsf{B}}_{ij}(\mathsf{v})\xi_i\xi_j$ is positive semi-definite for $\mathsf{v} \in \mathcal{O}_{\mathsf{v}}$ and $\boldsymbol{\xi} \in \Sigma^{d-1}$.
(S_5) There exists a fixed vector space $\mathcal{E} \subset \mathbb{R}^{\mathsf{n}}$ such that $\widetilde{\Omega}(\mathsf{v}) \in \mathcal{E}^{\perp}$ for $\mathsf{v} \in \mathcal{O}_{\mathsf{v}}$ and $\widetilde{\Omega}(\mathsf{v}) = 0$ if and only if $\mathsf{v} \in \mathcal{E}$ and if and only if $\langle \mathsf{v}, \widetilde{\Omega}(\mathsf{v})\rangle = 0$.
(S_6) If $\widetilde{\Omega}(\mathsf{v}) = 0$, then $\partial_{\mathsf{v}}\widetilde{\Omega}(\mathsf{v})$ is symmetric and $N\big(\partial_{\mathsf{v}}\widetilde{\Omega}(\mathsf{v})\big) = \mathcal{E}$.
(S_7) We have $\langle \mathsf{v}, \widetilde{\Omega}(\mathsf{v})\rangle \leq 0$ for $\mathsf{v} \in \mathcal{O}_{\mathsf{v}}$.

The manifold $\mathcal{E}$ is naturally termed the equilibrium manifold since $\widetilde{\Omega}(\mathsf{v}) = 0$ when $\mathsf{v} \in \mathcal{E}$. The equivalence between symmetrization (S_1–S_7) and entropy (E_1–E_7) for hyperbolic-parabolic systems of conservation laws is obtained with $\mathsf{v} = (\partial_{\mathsf{u}}\sigma)^t$ [101].

Theorem 6.3 *Assume that the system* (137) *admits a* $C^{\varkappa}$ *entropy function* σ *defined over an open convex domain* $\mathcal{O}_{\mathsf{u}}$. *Then the system can be symmetrized with the entropic variable* $\mathsf{v} = (\partial_{\mathsf{u}}\sigma)^t$. *Conversely, assume that the system can be symmetrized with the* $C^{\varkappa-1}$ *diffeomorphism* $\mathsf{u} \to \mathsf{v}$. *Then there exists a* $C^{\varkappa}$ *entropy over the open convex set* $\mathcal{O}_{\mathsf{u}}$ *such that* $\mathsf{v} = (\partial_{\mathsf{u}}\sigma)^t$.

Sketch of the proof. The equivalence of (S_1–S_2) and (E_1–E_2) is classical and is essentially obtained with Poincaré lemma. Then (S_3–S_7) and (E_3–E_7) become identical statements with $\mathsf{v} = (\partial_{\mathsf{u}}\sigma)^t$. □

6.2 Normal Variables

In order to split the variables between hyperbolic and parabolic variables, we further have to put the system into a normal form, that is,in the form of a symmetric hyperbolic–parabolic composite system [89, 92, 95].

Definition 6.4 Consider a symmetrized system as in Definition 6.2 and let $\mathsf{v} \to \mathsf{w}$ be a $C^{\varkappa-1}$ diffeomorphism from the open set $\mathcal{O}_{\mathsf{v}}$ onto an open set $\mathcal{O}_{\mathsf{w}}$. Letting $\mathsf{v} = \mathsf{v}(\mathsf{w})$ in the symmetrized system (138) and multiplying on the left side by $(\partial_{\mathsf{w}}\mathsf{v})^t$ we obtain a new system in the variable w

$$\overline{\mathsf{A}}_0(\mathsf{w})\partial_t\mathsf{w} + \sum_{i\in C}\overline{\mathsf{A}}_i(\mathsf{w})\partial_i\mathsf{w} - \sum_{i,j\in C}\partial_i\big(\overline{\mathsf{B}}_{ij}(\mathsf{w})\partial_j\mathsf{w}\big) - \overline{\Omega}(\mathsf{w}) = \overline{\mathsf{b}}(\mathsf{w},\partial_{\mathsf{x}}\mathsf{w}), \quad (139)$$

where $\overline{\mathsf{A}}_0 = (\partial_{\mathsf{w}}\mathsf{v})^t\ \widetilde{\mathsf{A}}_0\ (\partial_{\mathsf{w}}\mathsf{v})$, $\overline{\mathsf{B}}_{ij} = (\partial_{\mathsf{w}}\mathsf{v})^t\ \widetilde{\mathsf{B}}_{ij}\ (\partial_{\mathsf{w}}\mathsf{v})$, $\overline{\mathsf{A}}_i = (\partial_{\mathsf{w}}\mathsf{v})^t\ \widetilde{\mathsf{A}}_i\ (\partial_{\mathsf{w}}\mathsf{v})$, $\overline{\Omega} = (\partial_{\mathsf{w}}\mathsf{v})^t\widetilde{\Omega}$, have at least regularity $\varkappa-2$ and where $\overline{\mathsf{b}}$ is quadratic in the gradients $\overline{\mathsf{b}} = -\sum_{i,j\in C}\partial_i(\partial_{\mathsf{w}}\mathsf{v})^t\ \widetilde{\mathsf{B}}_{ij}\ (\partial_{\mathsf{w}}\mathsf{v})\partial_j\mathsf{w}$. This system satisfies in particular properties $(\overline{\mathsf{S}}_1)$–$(\overline{\mathsf{S}}_4)$, that is, properties $(\mathsf{S}_1$–$\mathsf{S}_4)$ rewritten in terms of overbar matrices. This system (139) is said to be of the normal form if there exists a partition of $\{1,\dots,\mathsf{n}\}$ into $\mathrm{I} = \{1,\dots,\mathsf{n}_{\mathrm{I}}\}$ and $\mathrm{II} = \{\mathsf{n}_{\mathrm{I}}+1,\dots,\mathsf{n}_{\mathrm{I}}+\mathsf{n}_{\mathrm{II}}\}$ with $\mathsf{n} = \mathsf{n}_{\mathrm{I}}+\mathsf{n}_{\mathrm{II}}$ such that the following properties hold.

(N_1) The matrices $\overline{\mathsf{A}}_0$ and $\overline{\mathsf{B}}_{ij}$ have the block structure

$$\overline{\mathsf{A}}_0 = \begin{bmatrix} \overline{\mathsf{A}}_0^{\mathrm{I},\mathrm{I}} & 0_{\mathsf{n}_{\mathrm{I}},\mathsf{n}_{\mathrm{II}}} \\ 0_{\mathsf{n}_{\mathrm{II}},\mathsf{n}_{\mathrm{I}}} & \overline{\mathsf{A}}_0^{\mathrm{II},\mathrm{II}} \end{bmatrix}, \qquad \overline{\mathsf{B}}_{ij} = \begin{bmatrix} 0_{\mathsf{n}_{\mathrm{I}},\mathsf{n}_{\mathrm{I}}} & 0_{\mathsf{n}_{\mathrm{I}},\mathsf{n}_{\mathrm{II}}} \\ 0_{\mathsf{n}_{\mathrm{II}},\mathsf{n}_{\mathrm{I}}} & \overline{\mathsf{B}}_{ij}^{\mathrm{II},\mathrm{II}} \end{bmatrix}.$$

(N_2) The matrix $\overline{\mathsf{B}}^{\mathrm{II},\mathrm{II}}(\mathsf{w},\boldsymbol{\xi}) = \sum_{i,j\in C}\overline{\mathsf{B}}_{ij}^{\mathrm{II},\mathrm{II}}(\mathsf{w})\xi_i\xi_j$ is positive definite for $\mathsf{w} \in \mathcal{O}_{\mathsf{w}}$ and $\boldsymbol{\xi} \in \Sigma^{d-1}$.

(N_3) We have $\overline{\mathsf{b}}(\mathsf{w},\partial_{\mathsf{x}}\mathsf{w}) = \big(\overline{\mathsf{b}}_{\mathrm{I}}(\mathsf{w},\partial_{\mathsf{x}}\mathsf{w}_{\mathrm{II}}), \overline{\mathsf{b}}_{\mathrm{II}}(\mathsf{w},\partial_{\mathsf{x}}\mathsf{w})\big)^t$.

We have used here the vector and matrix block structure induced by the partitioning of $\mathbb{R}^{\mathsf{n}}$ into $\mathbb{R}^{\mathsf{n}} = \mathbb{R}^{\mathsf{n}_{\mathrm{I}}} \times \mathbb{R}^{\mathsf{n}_{\mathrm{II}}}$ so that we have $\mathsf{w} = (\mathsf{w}_{\mathrm{I}}, \mathsf{w}_{\mathrm{II}})^t$ for instance and denoted by $0_{i,j}$ the zero matrix with i lines and j columns.

The quadratic residual may also be written in the more elegant form

$$\overline{\mathsf{b}} = \sum_{i,j\in C}\overline{\mathrm{M}}_{ij}(\mathsf{w})\,\partial_i\mathsf{w}\,\partial_j\mathsf{w}, \quad (140)$$

where $\overline{\mathrm{M}}_{ij}(\mathsf{w})$ are third order tensors that are functions of $\mathsf{w} \in \mathcal{O}_{\mathsf{w}}$. From the regularity assumptions of the original system (137), the coefficients of both symmetrized systems (138) and (139) have at least regularity $\varkappa - 2$ and the coefficients $\overline{\mathrm{M}}_{ij}$, $i, j \in C$, of $\overline{\mathsf{b}}$ have at least regularity $\varkappa - 3$. A sufficient condition for system (138) to be recast into a normal form is that the nullspace naturally associated with dissipation matrices $\widetilde{\mathsf{B}}$ is a fixed subspace of $\mathbb{R}^{\mathsf{n}}$. This is Condition (N) introduced by Kawashima and Shizuta [92] which has been strengthened in [95].

(N) *The nullspace $N(\widetilde{\mathsf{B}})$ of the matrix $\widetilde{\mathsf{B}}(\mathsf{v},\boldsymbol{\xi}) = \sum_{i,j\in C}\widetilde{\mathsf{B}}_{ij}(\mathsf{v})\xi_i\xi_j$ does not depend on $\mathsf{v} \in \mathcal{O}_{\mathsf{v}}$ and $\boldsymbol{\xi} \in \Sigma^{d-1}$ and $\widetilde{\mathsf{B}}_{ij}(\mathsf{v})N(\widetilde{\mathsf{B}}) = 0$, for $i, j \in C$.*

Letting $\mathsf{n}_{\mathrm{I}} = \dim\big(N(\widetilde{\mathsf{B}})\big)$ and $\mathsf{n}_{\mathrm{II}} = \mathsf{n} - \mathsf{n}_{\mathrm{I}}$ we denote by P an arbitrary constant nonsingular matrix of dimension n such that its first n_{I} columns span the nullspace $N(\widetilde{\mathsf{B}})$. In order to characterize more easily normal forms for symmetric systems of conservation laws satisfying (N) we may introduce the auxiliary variables [9, 95] $\mathsf{u}' = \mathsf{P}^t\mathsf{u}$ and $\mathsf{v}' = \mathsf{P}^{-1}\mathsf{v}$. The dissipation matrices corresponding to these auxiliary variables have nonzero coefficients only in the lower right block of size $\mathsf{n}_{\mathrm{II}} = \mathsf{n} - \mathsf{n}_{\mathrm{I}}$. Normal symmetric forms are then equivalently—and more easily—obtained from the v' symmetric equation [9, 95].

Theorem 6.5 *Consider a system of conservation laws (138) that is symmetric in the sense of Definition 6.2 and assume that the nullspace invariance property* (N) *is satisfied. Denoting by* $\mathsf{u}' = \mathsf{P}^t\mathsf{u}$ *and* $\mathsf{v}' = \mathsf{P}^{-1}\mathsf{v}$*, the auxiliary variable, any normal form of the system (138) is given by a change of variable in the form* $\mathsf{w} = \big(\mathcal{F}_{\mathrm{I}}(\mathsf{u}'_{\mathrm{I}}), \mathcal{F}_{\mathrm{II}}(\mathsf{v}'_{\mathrm{II}})\big)^t$ *where* $\mathcal{F}_{\mathrm{I}}$ *and* $\mathcal{F}_{\mathrm{II}}$ *are two diffeomorphisms of* $\mathbb{R}^{\mathsf{n}_{\mathrm{I}}}$ *and* $\mathbb{R}^{\mathsf{n}_{\mathrm{II}}}$*, respectively, and we have*

$$\overline{\mathsf{b}} = \Big(0, \overline{\mathsf{b}}_{\mathrm{II}}(\mathsf{w}, \partial_x \mathsf{w}_{\mathrm{II}})\Big)^t = \Big(0, \sum_{i,j\in C} \overline{\mathrm{M}}_{ij}^{\mathrm{II},\mathrm{II},\mathrm{II}}(\mathsf{w})\, \partial_i \mathsf{w}_{\mathrm{II}}\, \partial_j \mathsf{w}_{\mathrm{II}}\Big)^t, \tag{141}$$

where $\overline{\mathrm{M}}_{ij}^{\mathrm{II},\mathrm{II},\mathrm{II}}(\mathsf{w})$ *are third order tensors depending on* w *with regularity at least* $\varkappa - 3$*. Finally, when* $\mathcal{F}_{\mathrm{II}}$ *is linear, the quadratic residual* $\overline{\mathsf{b}}$ *is zero.*

The main interest of normal forms is that the resulting subsystem of partial differential equations governing the variable w_{I} is symmetric hyperbolic [89, 94] whereas the subsystem governing w_{II} is symmetric strongly parabolic [89, 92] as discussed in the next section.

Remark 6.6 It is also possible to investigate situations where the general structure of the symmetrized source term $\widetilde{\Omega}$ is transfered to the source term $\overline{\Omega}$ of the normal variable [104].

6.3 Hyperbolicity and Parabolicity

Consider a first-order abstract system of partial differential equations written in the form

$$\overline{\mathsf{A}}_0(\mathsf{w})\partial_t \mathsf{w} + \sum_{i\in C} \overline{\mathsf{A}}_i(\mathsf{w})\partial_i \mathsf{w} = \overline{\Omega}(\mathsf{w}), \tag{142}$$

where $\overline{\mathsf{A}}_0$, $\overline{\mathsf{A}}_i$, $i \in C$, and $\overline{\Omega}$ are smooth functions of w over an open set $\mathcal{O}_{\mathsf{w}}$, $C = \{1, \ldots, d\}$ the set of direction indices of $\mathbb{R}^d$, and where $\overline{\mathsf{A}}_0$ is assumed to be invertible. The following definition of hyperbolicity can be found in the book of Denis Serre [94].

Definition 6.7 The system (142) is said to be hyperbolic at a given point w if

$$\sup_{\xi \in \mathbb{R}^d} \left\| \exp\left(-\mathrm{i}\left(\overline{\mathsf{A}}_0(\mathsf{w})\right)^{-1} \overline{\mathsf{A}}(\mathsf{w}, \boldsymbol{\xi})\right) \right\| < \infty, \tag{143}$$

where for $\boldsymbol{\xi} \in \mathbb{R}^d$ we have defined $\overline{\mathsf{A}}(\mathsf{w}, \boldsymbol{\xi}) = \sum_{i \in C} \overline{\mathsf{A}}_i(\mathsf{w})\xi_i$.

When the system (142) is hyperbolic, it is easily established that the matrix $\left(\overline{\mathsf{A}}_0(\mathsf{w})\right)^{-1} \overline{\mathsf{A}}(\mathsf{w}, \xi)$ is diagonalizable with real eigenvalues so that it is hyperbolic in the classical sense [94, 96]. We also have the following sufficient condition in terms of eigenvalues and eigenvector matrices established by Serre [94].

Proposition 6.8 *Assume that $\left(\overline{\mathsf{A}}_0(\mathsf{w})\right)^{-1} \overline{\mathsf{A}}(\mathsf{w}, \boldsymbol{\xi})$ is diagonalizable at w for any $\boldsymbol{\xi} \in \mathbb{R}^d$ with real eigenvalues. Let $P(\boldsymbol{\xi})$ denote a matrix of eigenvectors and assume that*

$$\sup_{\xi \in \mathbb{R}^d} \|P(\boldsymbol{\xi})\| \; \|P(\boldsymbol{\xi})^{-1}\| < \infty. \tag{144}$$

Then the system (142) *is also hyperbolic at* w.

A fundamental property of Definition 6.7 is its invariance under a change of variable [94]. Moreover, when a first-order system is symmetrizable, it is hyperbolic [94, 96].

Definition 6.9 The system (142) is said to be symmetric at a given point w when $\overline{\mathsf{A}}_0(\mathsf{w})$ is positive definite and the matrices $\overline{\mathsf{A}}_i(\mathsf{w})$, $i \in C$, are symmetric.

Proposition 6.10 *A symmetric system of partial differential Eq.* (142) *is hyperbolic.*

Since the existence of an entropy function also implies symmetrizability, it automatically implies hyperbolicity in the sense of Definition 6.7 as well as in the classical sense [94, 96].

We now discuss parabolicity and consider a second-order system in the form

$$\overline{\mathsf{A}}_0(\mathsf{w})\partial_t \mathsf{w} = \sum_{i,j \in C} \overline{\mathsf{B}}_{ij}(\mathsf{w})\partial_i \partial_j \mathsf{w} + \overline{\Omega}(\mathsf{w}, \partial_{\mathsf{x}} \mathsf{w}), \tag{145}$$

where $\overline{\mathsf{A}}_0$, $\overline{\mathsf{B}}_{ij}$, $i, j \in C$, are smooth functions of w over an open set $\mathcal{O}_{\mathsf{w}}$, $\overline{\Omega}$ is a smooth function over $\mathcal{O}_{\mathsf{w}} \times \mathbb{R}^{d\mathsf{n}}$, and where $\overline{\mathsf{A}}_0$ is invertible. We will generally consider second-order symmetric systems in the following sense.

Definition 6.11 The system (145) is said to be symmetric at a given point w when $\overline{\mathsf{A}}_0(\mathsf{w})$ is symmetric positive definite and $\left(\overline{\mathsf{B}}_{ij}(\mathsf{w})\right)^t = \overline{\mathsf{B}}_{ji}(\mathsf{w})$, for $i, j \in C$.

We then have the following definition for strongly parabolic systems of second-order partial differential equations involving the Legendre-Hadamard condition. Note that this definition should only concern the parabolic subsystem in the variable w_{II} but we have suppressed all II indices to simplify the presentation.

Definition 6.12 Assume that the system (145) is symmetric at a given point w. This system is said to be strongly parabolic at w if there exists a positive constant $\delta > 0$ such that for any $\boldsymbol{\xi} = (\xi_1, \dots, \xi_d)^t$ and $w = (w_1, \dots, w_{\mathsf{n}})^t$ we have

$$\sum_{\substack{1 \le i,j \le d \\ 1 \le k,l \le \mathsf{n}}} \big(\overline{\mathsf{B}}_{ij}(\mathsf{w})\big)_{kl} \xi_i \xi_j w_k w_l \ge \delta |\boldsymbol{\xi}|^2 |w|^2. \tag{146}$$

Remark 6.13 It is often the case that systems of partial differential equations of physical origin satisfy a stronger property—the Legendre condition—than (146) at a given point w and are indeed such that

$$\sum_{\substack{1 \le i,j \le d \\ 1 \le k,l \le \mathsf{n}}} \big(\overline{\mathsf{B}}_{ij}(\mathsf{w})\big)_{kl} \zeta_{ik} \zeta_{jl} \ge \delta |\zeta|^2, \tag{147}$$

for any $\zeta \in \mathbb{R}^{d\mathsf{n}}$. The condition (146) then simply corresponds to the situation where ζ is constrained to be a tensor product $\zeta = \xi \otimes w$, so that $\zeta_{ik} = \xi_i w_k$, for $i \in \{1, \dots, d\}$ and $k \in \{1, \dots, \mathsf{n}\}$.

The definition of strong parabolicity is only given here for symmetric systems and will be applied to the symmetrized forms like (138) or (139), thereby naturally involving entropy Hessians. Indeed, the definition of strong parabolicity *in the usual sense*, which neither require symmetry properties nor entropy hessians, only has a meaning for *particular forms* of systems of partial differential equation under consideration which *need to be specified* as shown by the following counter example. Consider the system $\partial_t \mathsf{w} - \mathcal{D} \Delta \mathsf{w} = 0$ where $\mathsf{w} = (\mathsf{w}_1, \mathsf{w}_2)^t$, $\Delta \mathsf{w} = (\Delta \mathsf{w}_1, \Delta \mathsf{w}_2)^t$, and $\mathcal{D} = \operatorname{diag}(\delta_1, \delta_2)$, with $\delta_1 > 0$, $\delta_2 > 0$, and $\delta_1 \neq \delta_2$. In other words, consider two uncoupled heat equations which of course form a symmetric strongly parabolic system. Introduce next the modified variable $\mathsf{w}'^{*} = (\mathsf{w}_1 + \mathsf{w}_2, \delta \mathsf{w}_2)^t$ where $\delta > 0$ is a positive parameter. We then have $\partial_t \mathsf{w}'^{*} - \mathcal{D}' \Delta \mathsf{w}'^{*} = 0$ with

$$\mathcal{D}' = \begin{bmatrix} \delta_1 & \frac{\delta_2 - \delta_1}{\delta} \\ 0 & \delta_2 \end{bmatrix},$$

so that if $0 < \delta < |\delta_2 - \delta_1|/(2\sqrt{\delta_1 \delta_2})$ the quadratic form associated with $\mathcal{D}'$ is not positive definite and the system in the w'^{*} variable is not strongly parabolic *in the usual sense* even though it is obtained from a trivially strongly parabolic system.

Definition 6.14 Denoting $\overline{\mathsf{B}}(\mathsf{w}, \boldsymbol{\xi}) = \sum_{i,j \in C} \xi_i \xi_j \overline{\mathsf{B}}_{ij}(\mathsf{w})$, a system (145) is said to be parabolic in the sense of Petrovsky at a given point w if there exists a positive constant δ such that for any $\boldsymbol{\xi} \in \mathbb{R}^d$, the eigenvalues λ of $\big(\overline{\mathsf{A}}_0(\mathsf{w})\big)^{-1} \overline{\mathsf{B}}(\mathsf{w}, \boldsymbol{\xi})$, are such that

$$\Re(\lambda) \ge \delta |\boldsymbol{\xi}|^2. \tag{148}$$

Various other generalized definitions of parabolicity are discussed in the book of Ladyženskaja et al. [87], in particular that of Douglis and Nirenberg, Shirota, and Eĭdel'man, but these definitions coincide with that of Petrovsky for second-order systems [87] and also coincide with the notion of normal ellipticity. For symmetric systems, we now have the following equivalence result [101].

Proposition 6.15 *Consider a second-order system in the form*

$$\overline{A}_0(w)\partial_t w = \sum_{i,j\in C} \overline{B}_{ij}(w)\partial_i\partial_j w + \overline{\Omega}(w, \partial_x w),$$

and assume that the system is symmetric. Then the system is strongly parabolic at w *if and only if it is Petrovsky parabolic at* w.

Sketch of the proof. The eigenvalues of $\big(\overline{\mathsf{A}}_0(\mathsf{w})\big)^{-1}\overline{\mathsf{B}}(\mathsf{w}, \boldsymbol{\xi})$ are essentially Rayleigh quotients with respect to the scalar product $\langle\langle x, y\rangle\rangle = \langle \overline{\mathsf{A}}_0 x, y\rangle$. □

From a practical point of view, for systems of partial differential equations derived from physics, thanks to the existence of a mathematical entropy, we can use symmetrized systems of partial differential equations and then rely of the proper definition of strongly parabolic systems 6.12. In addition, strongly parabolicity is then invariant by a change of variable for symmetric systems, after multiplication of the left by the transpose of the jacobian matrix. Considering for instance the previous system $\partial_t \mathsf{w} - \mathcal{D}\Delta \mathsf{w} = 0$ where $\mathsf{w} = (\mathsf{w}_1, \mathsf{w}_2)^t$, $\Delta \mathsf{w} = (\Delta \mathsf{w}_1, \Delta \mathsf{w}_2)^t$, and $\mathcal{D} = \mathrm{diag}(\delta_1, \delta_2)$, with $\delta_1 > 0$, $\delta_2 > 0$, and letting $\mathsf{w}'^* = (\mathsf{w}_1 + \mathsf{w}_2, \delta \mathsf{w}_2)^t$ where $\delta > 0$ is a positive parameter and

$$\mathsf{P} = \begin{bmatrix} 1 & 1 \\ 0 & \delta \end{bmatrix}, \qquad \mathsf{Q} = \mathsf{P}^{-1},$$

we then have

$$\mathsf{Q}^t\mathsf{Q}\partial_t \mathsf{w}'^* - \mathsf{Q}^t\mathcal{D}\mathsf{Q}\Delta \mathsf{w}'^* = 0,$$

which remains symmetric strongly parabolic.

6.4 Natural Entropic form for Multicomponent Flows

We evaluate in this section the natural entropic symmetrized form for the system of partial differential equations modeling multicomponent reative fluids. We use the mathematical entropy $\sigma = -\mathcal{S}/R$ where the $1/R$ factor is introduced for convenience. For this particular system of partial differential equations we have $\mathsf{n} = n + d + 1$, the velocity components of all quantities in $\mathbb{R}^{n+d+1}$ are denoted as vectors of $\mathbb{R}^d$ and the corresponding partitioning is also used for matrices.

Theorem 6.16 *Assume that* (H_1–H_5) *hold. Then the function* $\sigma = -\mathcal{S}/R$ *is a mathematical entropy for the system (136) governing multicomponent fluids and the corresponding entropic variable is*

$$\mathsf{v} = (\partial_{\mathsf{u}}\sigma)^t = \frac{1}{RT}\Big(g_1 - \tfrac{1}{2}|\boldsymbol{v}|^2, \dots, g_n - \tfrac{1}{2}|\boldsymbol{v}|^2, \boldsymbol{v}, -1\Big)^t. \tag{149}$$

The map $\mathsf{u} \to \mathsf{v}$ *is a* $C^{\varkappa-1}$ *diffeomorphism from* $\mathcal{O}_{\mathsf{u}}$ *onto* $\mathcal{O}_{\mathsf{v}}$. *The system written in term of the entropic variable* v *is*

$$\widetilde{A}_0(\mathsf{v})\partial_t \mathsf{v} + \sum_{i\in C} \widetilde{A}_i(\mathsf{v})\partial_i \mathsf{v} = \sum_{i,j\in C} \partial_i\big(\widetilde{B}_{ij}(\mathsf{v})\partial_j \mathsf{v}\big) + \widetilde{\Omega}(\mathsf{v}), \tag{150}$$

with $\widetilde{A}_0 = \partial_{\mathsf{v}}\mathsf{u}$, $\widetilde{A}_i = A_i\partial_{\mathsf{v}}\mathsf{u}$, $\widetilde{B}_{ij} = B_{ij}\partial_{\mathsf{v}}\mathsf{u}$, *and* $\widetilde{\Omega} = \Omega$, *and is of the symmetric form. The matrix* $\widetilde{A}_0$ *is given by*

$$\widetilde{A}_0 = \begin{bmatrix} \Lambda & & Sym \\ \boldsymbol{v}\otimes\Lambda\mathbb{I} & \langle\Lambda\mathbb{I}, \mathbb{I}\rangle\boldsymbol{v}\otimes\boldsymbol{v} + \rho RT\boldsymbol{I} & \\ \Lambda e^{\mathrm{tl}} & \langle\Lambda e^{\mathrm{tl}}, \mathbb{I}\rangle\boldsymbol{v}^t + \rho RT\boldsymbol{v}^t & \Upsilon \end{bmatrix}, \tag{151}$$

where Λ *is the diagonal matrix of size n given by*

$$\Lambda = \mathrm{diag}(m_1\rho_1, \dots, m_n\rho_n),$$

e^{tl} *is the vector of size n given by*

$$e^{\mathrm{tl}} = (e_1^{\mathrm{tl}}, \dots, e_n^{\mathrm{tl}})^t,$$

and $\Upsilon = \langle\Lambda e^{\mathrm{tl}}, e^{\mathrm{tl}}\rangle + \rho RT|\boldsymbol{v}|^2 + RT^2 c_{\mathrm{v}}$. *Since* $\widetilde{A}_0$ *is symmetric, we only give its left lower triangular part and write "Sym" in the upper triangular part. Denoting by* $\boldsymbol{\xi} = (\xi_1, \dots, \xi_d)^t$ *an arbitrary vector of* $\mathbb{R}^d$ *and letting* $\widetilde{A} = \sum_{i\in C}\xi_i\widetilde{A}_i$, *we have*

$$\widetilde{A} = \boldsymbol{v}\cdot\boldsymbol{\xi}\,\widetilde{A}_0 + RT\begin{bmatrix} 0 & & Sym \\ \boldsymbol{\xi}\otimes\varrho & \rho(\boldsymbol{\xi}\otimes\boldsymbol{v} + \boldsymbol{v}\otimes\boldsymbol{\xi}) & \\ \boldsymbol{v}\cdot\boldsymbol{\xi}\,\varrho^t & \boldsymbol{v}\cdot\boldsymbol{\xi}\,\rho\boldsymbol{v}^t + \rho h^{\mathrm{tl}}\boldsymbol{\xi}^t & 2\rho h^{\mathrm{tl}}\boldsymbol{v}\cdot\boldsymbol{\xi} \end{bmatrix}. \tag{152}$$

Moreover, we have the decomposition

$$\widetilde{B}_{ij} = \widetilde{B}^{D\lambda}\delta_{ij} + \kappa RT\,\widetilde{B}_{ij}^{\kappa} + \eta RT\,\widetilde{B}_{ij}^{\eta}, \tag{153}$$

with

$$\widetilde{B}^{D\lambda} = \frac{RT}{p}\begin{bmatrix} \mathcal{D} & & Sym \\ 0_{d,n} & 0_{d,d} & \\ (\mathcal{D}\mathit{h})^t & 0_{1,d} & \lambda pT + \langle\mathcal{D}\mathit{h}, \mathit{h}\rangle \end{bmatrix}, \tag{154}$$

where $\mathcal{D} = (\rho_k \rho_l D_{kl})_{k,l \in S}$ *is the matrix of size n with components* $\rho_k \rho_l D_{kl}$ *and* $\hbar$ *is the vector of size n with components* $\hbar_i = h_i + \frac{RT}{m_i}\widetilde{\chi}_i$. *Moreover, denoting by* $\boldsymbol{\xi} = (\xi_1, \ldots, \xi_d)^t$ *and* $\boldsymbol{\zeta} = (\zeta_1, \ldots, \zeta_d)^t$ *arbitrary vectors of* $\mathbb{R}^d$, *the matrices* $\widetilde{\mathsf{B}}^{\kappa}_{ij}$ *and* $\widetilde{\mathsf{B}}^{\eta}_{ij}$, $i, j \in C$, *are given by*

$$\sum_{i,j\in C} \xi_i \zeta_j \widetilde{\mathsf{B}}^{\kappa}_{ij} = \begin{bmatrix} 0_{n,n} & 0_{n,d} & 0_{n,1} \\ 0_{d,n} & \boldsymbol{\xi}\otimes\boldsymbol{\zeta} & \boldsymbol{v}\cdot\boldsymbol{\zeta}\,\boldsymbol{\xi} \\ 0_{1,n} & \boldsymbol{v}\cdot\boldsymbol{\xi}\,\boldsymbol{\zeta}^t & \boldsymbol{v}\cdot\boldsymbol{\xi}\,\boldsymbol{v}\cdot\boldsymbol{\zeta} \end{bmatrix}, \tag{155}$$

$$\sum_{i,j\in C} \xi_i \zeta_j \widetilde{\mathsf{B}}^{\eta}_{ij} = \begin{bmatrix} 0_{n,n} & 0_{n,d} & 0_{n,1} \\ 0_{d,n} & \boldsymbol{\xi}\cdot\boldsymbol{\zeta}\,\boldsymbol{I} + \boldsymbol{\zeta}\otimes\boldsymbol{\xi} - \frac{2}{3}\boldsymbol{\xi}\otimes\boldsymbol{\zeta} & \boldsymbol{\xi}\cdot\boldsymbol{\zeta}\,\boldsymbol{v} + \boldsymbol{v}\cdot\boldsymbol{\xi}\,\boldsymbol{\zeta} - \frac{2}{3}\boldsymbol{v}\cdot\boldsymbol{\zeta}\,\boldsymbol{\xi} \\ 0_{1,n} & \boldsymbol{\xi}\cdot\boldsymbol{\zeta}\,\boldsymbol{v}^t + \boldsymbol{v}\cdot\boldsymbol{\zeta}\,\boldsymbol{\xi}^t - \frac{2}{3}\boldsymbol{v}\cdot\boldsymbol{\xi}\,\boldsymbol{\zeta}^t & \boldsymbol{\xi}\cdot\boldsymbol{\zeta}\,\boldsymbol{v}\cdot\boldsymbol{v} + \frac{1}{3}\boldsymbol{v}\cdot\boldsymbol{\xi}\,\boldsymbol{v}\cdot\boldsymbol{\zeta} \end{bmatrix}. \tag{156}$$

Finally, the equilibrium manifold is given by

$$\mathcal{E} = (M\mathcal{R})^{\perp} \times \mathbb{R}^d \times \mathbb{R}, \tag{157}$$

where $\mathcal{R} = span\{\nu_i,\ i \in \mathfrak{R}\} \subset \mathbb{R}^n$ *is spanned by the reaction vectors and* $M = \mathrm{diag}(m_1, \ldots, m_n)$. *Moreover, at an equilibrium point* $\mathsf{v}^\star$ *according to Definition 5.3 where* $\widetilde{\Omega}(\mathsf{v}^\star) = 0$, *the linearized source term* $\widetilde{\mathsf{L}}(\mathsf{v}^\star) = -\partial_{\mathsf{v}}\widetilde{\Omega}(\mathsf{v}^\star)$ *is in the form*

$$\widetilde{\mathsf{L}}(\mathsf{v}^\star) = \sum_{i\in\mathfrak{R}} \Lambda_i \mathfrak{p}_i \otimes \mathfrak{p}_i, \tag{158}$$

where $\mathfrak{p}_i = (m_1\nu_{1i}, \ldots, m_n\nu_{ni}, \mathbf{0}, 0)^t$ *and* $\Lambda_i = \mathcal{K}^{\mathrm{f}}_i(T^\star) \prod_{k\in S}(\rho^\star_k/m_k)^{\nu^{\mathrm{f}}_{ki}}$.

Sketch of the proof. The proof is lengthy but present no serious difficulty and we refer the reader to [9, 95, 101]. □

Remark 6.17 For ideal fluids, the symmetrizing change of variable $\mathsf{u} \to \mathsf{v}$ is one to one and is thus a global change of variable [9, 95]. On the contrary, for nonideal fluid, even though the entropy σ is globally defined, a typical situation is that of distinct points $\mathsf{u}^\sharp$ and $\mathsf{u}^\flat$ such that $\mathsf{v}^\sharp = \mathsf{v}^\flat$. Indeed, we see from (149) that the equality $\mathsf{v}^\sharp = \mathsf{v}^\flat$ corresponds to the chemical equilibrium between the two stable phases $\mathsf{u}^\sharp$ and $\mathsf{u}^\flat$ with identical pressure, temperature and Gibbs functions, that may notably be observed for nonideal fluids.

For mixtures of ideal gases, there is also a uniqueness theorem for mathematical entropies that are independent of transport coefficients [103]. This result strengthen the representation theorem of normal variable as well as the importance of the natural entropic symmetrized form.

Theorem 6.18 *Let* $\widetilde{\sigma}$ *be a* $C^{\varkappa}$ *function defined on the open set* $\mathcal{O}_{\mathsf{u}}$ *satisfying* (E_2), (E_3) *and such that* $\partial^2_{\mathsf{u}}\widetilde{\sigma}$ *is invertible. Assuming that* $\widetilde{\sigma}$ *is independent of the mass and heat diffusion parameters, then* $\widetilde{\sigma}$ *is in the form*

$$\widetilde{\sigma} = \alpha_{\mathcal{S}} \mathcal{S} + \sum_{i \in S} \alpha_i \rho_i + \alpha_{\boldsymbol{v}} \rho \boldsymbol{v} + \alpha_{\mathcal{E}} (\mathcal{E} + \tfrac{1}{2} \rho |\boldsymbol{v}|^2), \tag{159}$$

where $\alpha_{\mathcal{S}}$ α_i, $i \in S$, $\alpha_{\boldsymbol{v}}$ and $\alpha_{\mathcal{E}}$ are constants.

This shows in particular that mathematical entropies independent of transport parameters—a somewhat natural condition—are indeed unique up to an affine transform, once the trivial factors proportional to conserved quantities have been eliminated and the corresponding entropic variables are then proportional.

6.5 Natural Normal form for Multicomponent Flows

The symmetric system (150) may be rewritten into a normal form, that is, in the form of a symmetric hyperbolic-parabolic composite system, where hyperbolic and parabolic variables are split [89, 92, 95, 100, 105–108, 113]. We first establish that the nullspace invariance property holds for multicomponent flows.

Lemma 6.19 *The nullspace of the matrix*

$$\widetilde{\mathsf{B}}(\mathsf{v}, \boldsymbol{\xi}) = \sum_{i,j \in C} \widetilde{\mathsf{B}}_{ij}(\mathsf{v}) \xi_i \xi_j,$$

is independent of $\mathsf{v} \in \mathcal{O}_{\mathsf{v}}$ and $\boldsymbol{\xi} \in \Sigma^{d-1}$ and given by

$$N(\widetilde{\mathsf{B}}) = \mathbb{R}(\mathbb{I}, \mathbf{0}, 0)^t,$$

and we have $\widetilde{\mathsf{B}}_{ij}(\mathsf{v}) N(\widetilde{\mathsf{B}}) = 0$, $i, j \in C$, for $\mathsf{v} \in \mathcal{O}_{\mathsf{v}}$.

Proof Letting $\mathsf{x} = (\mathsf{x}_1, \dots, \mathsf{x}_n, \mathsf{x}_{\boldsymbol{v}}, \mathsf{x}_T)^t$, with $\mathsf{x}_{\varrho} = (\mathsf{x}_1, \dots, \mathsf{x}_n)^t$, $\mathsf{x}_{\boldsymbol{v}} = (\mathsf{x}_{n+1}, \dots, \mathsf{x}_{n+d})^t$, it is obtained after some algebra with $\xi \in \Sigma^{d-1}$ that

$$\langle \widetilde{\mathsf{B}}\mathsf{x}, \mathsf{x} \rangle = \kappa RT \big(\boldsymbol{\xi} \cdot (\mathsf{x}_{\boldsymbol{v}} + \boldsymbol{v} \mathsf{x}_T)\big)^2 + \eta RT \Big(\tfrac{1}{3} \big(\boldsymbol{\xi} \cdot (\mathsf{x}_{\boldsymbol{v}} + \boldsymbol{v} \mathsf{x}_T)\big)^2 + \big|\mathsf{x}_{\boldsymbol{v}} + \boldsymbol{v} \mathsf{x}_T\big|^2 \Big)$$

$$+ \frac{RT}{p} \sum_{i,j \in S} \mathcal{D}_{ij} (\mathsf{x}_i + \hbar_i \mathsf{x}_T)(\mathsf{x}_j + \hbar_j \mathsf{x}_T) + \lambda R T^2 \mathsf{x}_T^2.$$

Assuming that $\langle \widetilde{\mathsf{B}}\mathsf{x}, \mathsf{x} \rangle = 0$ we thus obtain that successively that $\mathsf{x}_T = 0$ and $\mathsf{x}_{\boldsymbol{v}} = \mathbf{0}$ and next since $N(\mathcal{D}) = \mathbb{R}\mathbb{I}$, we deduce that $(\mathsf{x}_1, \dots, \mathsf{x}_n)^t \in \mathbb{R}\mathbb{I}$. We have thus established that $N(\widetilde{\mathsf{B}})$ is spanned by $(1, \dots, 1, \mathbf{0}, 0)^t$ and it is easily checked that $\widetilde{\mathsf{B}}_{ij}(\mathsf{v}) N(\widetilde{\mathsf{B}}) = 0$, $i, j \in C$, for $\mathsf{v} \in \mathcal{O}_{\mathsf{v}}$. □

Since $N(\widetilde{\mathsf{B}})$ is spanned by $(\mathbb{I}, \mathbf{0}, 0)^t$, we define the matrix P by

$$\mathsf{P} = \begin{bmatrix} 1 & 0_{1,n-1} & 0_{1,d} & 0 \\ 1_{n-1,1} & I_{n-1} & 0_{n-1,d} & 0_{n-1,1} \\ 0_{d,1} & 0_{d,n-1} & \boldsymbol{I} & 0_{d,1} \\ 0 & 0_{1,n-1} & 0_{1,d} & 1 \end{bmatrix}, \tag{160}$$

and we may introduce the auxiliary variable $\mathsf{u}' = \mathsf{P}^t\mathsf{u}$ and the corresponding entropic variable $\mathsf{v}' = \mathsf{P}^{-1}\mathsf{v}$ given by

$$\mathsf{u}' = \Big(\rho, \rho_2, \ldots, \rho_n, \rho\boldsymbol{v}, \mathcal{E} + \tfrac{1}{2}\rho|\boldsymbol{v}|^2\Big)^t$$

and

$$\mathsf{v}' = \frac{1}{RT}\Big(g_1 - \tfrac{1}{2}|\boldsymbol{v}|^2, g_2 - g_1, \ldots g_n - g_1, \boldsymbol{v}, -1\Big)^t.$$

From the representation theorem 6.5, we deduce that all normal forms of the system (138) are obtained with variables w in the form

$$\mathsf{w} = \left(\mathcal{F}_{\mathrm{I}}(\rho), \mathcal{F}_{\mathrm{II}}\Big(\frac{g_2-g_1}{RT}, \ldots, \frac{g_n-g_1}{RT}, \frac{\boldsymbol{v}}{T}, \frac{-1}{RT}\Big)\right)^t, \tag{161}$$

where $\mathcal{F}_{\mathrm{I}}$ and $\mathcal{F}_{\mathrm{II}}$ are diffeomorphism of $\mathbb{R}$ and $\mathbb{R}^{n+d}$. In the following theorem, we evaluate the normal form corresponding to the natural normal variable.

Theorem 6.20 *Assume that* $(\mathsf{H}_1\text{–}\mathsf{H}_5)$ *hold and consider the normal variable*

$$\mathsf{w} = \Big(\rho, \frac{g_2-g_1}{RT}, \ldots, \frac{g_n-g_1}{RT}, \frac{\boldsymbol{v}}{RT}, \frac{-1}{RT}\Big)^t, \tag{162}$$

and the diffeomorphism $\mathsf{v} \longmapsto \mathsf{w}$ *from* $\mathcal{O}_{\mathsf{v}}$ *onto the open set* $\mathcal{O}_{\mathsf{w}} = (0, \infty) \times \mathbb{R}^{n-1} \times \mathbb{R}^d \times (-\infty, 0)$. *Then the system of partial differential equations in normal form may be written*

$$\overline{\mathsf{A}}_0(\mathsf{w})\partial_t \mathsf{w} + \sum_{i\in C} \overline{\mathsf{A}}_i(\mathsf{w})\partial_i \mathsf{w} = \sum_{i,j\in C} \partial_i\big(\overline{\mathsf{B}}_{ij}(\mathsf{w})\partial_j \mathsf{w}\big) + \overline{\Omega}(\mathsf{w}), \tag{163}$$

where the matrix $\overline{\mathsf{A}}_0$ *is given by*

$$\overline{\mathsf{A}}_0 = \begin{bmatrix} \overline{\mathsf{A}}_0^{\mathrm{I,I}} & Sym \\ 0_{n+d,1} & \overline{\mathsf{A}}_0^{\mathrm{II,II}} \end{bmatrix},$$

with

$$\overline{A}_0^{\mathrm{I,I}} = \frac{1}{\langle \Lambda \mathbb{I}, \mathbb{I} \rangle} \qquad \overline{A}_0^{\mathrm{II,II}} = \frac{1}{\langle \Lambda \mathbb{I}, \mathbb{I} \rangle} \begin{bmatrix} \overline{\mathcal{A}}^{\mathrm{II,II}} & & Sym \\ 0_{d,n-1} & \langle \Lambda \mathbb{I}, \mathbb{I} \rangle \rho RT \boldsymbol{I} & \\ a^t & \langle \Lambda \mathbb{I}, \mathbb{I} \rangle \rho RT \boldsymbol{v}^t & \overline{\Upsilon} \end{bmatrix}.$$

The matrix $\overline{\mathcal{A}}^{\mathrm{II,II}}$ *is the square matrix of dimension* $n-1$ *with coefficients*

$$\overline{\mathcal{A}}_{kl}^{\mathrm{II,II}} = \langle \Lambda \mathbb{I}, \mathbb{I} \rangle \Lambda_{kl} - (\Lambda \mathbb{I})_k (\Lambda \mathbb{I})_l, \quad 2 \le k, l \le n,$$

a *is the vector of dimension* $n-1$ *with coefficients*

$$a_l = \langle \Lambda \mathbb{I}, \mathbb{I} \rangle (\Lambda e^{\mathrm{tl}})_l - (\Lambda \mathbb{I})_l \langle \Lambda e^{\mathrm{tl}}, \mathbb{I} \rangle, \qquad 2 \le l \le n,$$

and $\overline{\Upsilon}$ *is given by*

$$\overline{\Upsilon} \langle \Lambda \mathbb{I}, \mathbb{I} \rangle \langle \Lambda e^{\mathrm{tl}}, e^{\mathrm{tl}} \rangle - \langle \Lambda e^{\mathrm{tl}}, \mathbb{I} \rangle^2 + \langle \Lambda \mathbb{I}, \mathbb{I} \rangle \rho RT (c_{\mathrm{v}} T + |\boldsymbol{v}|^2),$$

keeping in mind that $\mathbb{I} = (1, \dots, 1)^t$, $e^{\mathrm{tl}} = (e_1^{\mathrm{tl}}, \dots, e_n^{\mathrm{tl}})^t$, *and that* $\Lambda = \mathrm{diag}(m_1\rho_1, \dots, m_n\rho_n)$. *Denoting by* $\boldsymbol{\xi} = (\xi_1, \dots, \xi_d)^t$ *an arbitrary vector of* $\mathbb{R}^d$, *the matrices* $\overline{A}_i$, $i \in C$, *are given by*

$$\sum_{i \in C} \xi_i \overline{A}_i = \overline{A}_0 \boldsymbol{v}\cdot\boldsymbol{\xi} + \frac{\rho RT}{\langle \Lambda \mathbb{I}, \mathbb{I} \rangle} \begin{bmatrix} 0 & & & Sym \\ 0_{n-1,1} & 0_{n-1,n-1} & & \\ \boldsymbol{\xi} & \boldsymbol{\xi} \otimes y & 0_{d,d} & \\ \boldsymbol{v}\cdot\boldsymbol{\xi} & \boldsymbol{v}\cdot\boldsymbol{\xi} y^t & \gamma \boldsymbol{\xi}^t & 2\gamma \boldsymbol{v}\cdot\boldsymbol{\xi} \end{bmatrix},$$

where y *is the vector of dimension* $n-1$ *with components*

$$y_l = \langle \Lambda \mathbb{I}, \mathbb{I} \rangle y_l - (\Lambda \mathbb{I})_l, \qquad 2 \le l \le n,$$

and

$$\gamma = \langle \Lambda \mathbb{I}, \mathbb{I} \rangle h^{\mathrm{tl}} - \langle \Lambda e^{\mathrm{tl}}, \mathbb{I} \rangle.$$

The matrices $\overline{B}_{ij}$ *have the structure*

$$\overline{B}_{ij} = \delta_{ij} \overline{B}^{D\lambda} + RT\kappa \overline{B}_{ij}^{\kappa} + RT\eta \overline{B}_{ij}^{\eta},$$

where

$$\overline{B}_{ij}^{\kappa} = \widetilde{B}_{ij}^{\kappa}, \qquad \overline{B}_{ij}^{\eta} = \widetilde{B}_{ij}^{\eta}, \qquad i, j \in C,$$

whereas $\overline{B}^{D\lambda}$ *has its first line and first column composed of zeros and its lower right block* $\overline{B}^{D\lambda\,\mathrm{II,II}}$ *equal to* $\widetilde{B}^{D\lambda\,\mathrm{II,II}}$ *so that*

$$\overline{\mathsf{B}}^{D\lambda} = \frac{RT}{p}\begin{bmatrix} 0 & & & Sym \\ 0 & \mathcal{D}_{n-1,n-1} & & \\ 0 & 0_{d,n} & 0_{d,d} & \\ 0 & (\mathcal{D}\hbar)^t_{n-1} & 0_{1,d} & \lambda p T + \langle \mathcal{D}\hbar, \hbar\rangle \end{bmatrix},$$

where $\mathcal{D}_{n-1,n-1}$ is the matrix of size $n-1$ with coefficients $(\rho_k\rho_l D_{kl})_{2\le k,l\le n}$ and $(\mathcal{D}\hbar)_{n-1}$ are the $n-1$ last components of $\mathcal{D}\hbar$. Finally, the source term for the normal form is given by

$$\overline{\Omega} = \Big(0, m_2\omega_2, \dots, m_n\omega_n, \mathbf{0}, 0\Big)^t.$$

Proof The proof is lengthy and tedious but presents no serious difficulties. □

7 The Cauchy Problem

The equations governing multicomponent reactive flows have been derived from the kinetic theory of gases and rewritten in normal form. These equations have local smooth solutions [108], global smooth solutions around constant equilibrium states [95] and also traveling wave solutions [119]. We discuss in this section existence theorem around equilibrium states and asymptotic stability. The smooth dependence on a parameter has also been investigated in [97]. We first discuss local strict dissipativity in an abstract setting and then discuss the situation of multicomponent flows and the Cauchy problem [89, 92, 94, 95, 97–102].

7.1 Local Dissipativity

We present in this section the dissipativity properties around equilibrium states that are needed in order to establish global existence and asymptotic stability [89, 95, 106]. We consider a symmetrizable quasilinear system, assume that the nullspace invariance property (N) holds, and that the system has been rewritten in normal form.

We assume that there exists an equilibrium state $\mathsf{u}^\star \in \mathcal{O}_\mathsf{u}$ with $\Omega(\mathsf{u}^\star) = 0$ and we denote by $\mathsf{v}^\star$ and $\mathsf{w}^\star$ the corresponding constant state in the v and w variables respectively. If we linearize system (139) around the constant stationary state $\mathsf{w}^\star$, we obtain a linear system in the variable $\delta\mathsf{w} = \mathsf{w} - \mathsf{w}^\star$ in the form

$$\overline{\mathsf{A}}_0(\mathsf{w}^\star)\partial_t\delta\mathsf{w} + \sum_{i\in C}\overline{\mathsf{A}}_i(\mathsf{w}^\star)\partial_i\delta\mathsf{w} - \sum_{i,j\in C}\overline{\mathsf{B}}_{ij}(\mathsf{w}^\star)\partial_i\partial_j\delta\mathsf{w} + \overline{\mathsf{L}}(\mathsf{w}^\star)\delta\mathsf{w} = 0, \quad (164)$$

where $\overline{\mathsf{L}}(\mathsf{w}^\star)$ is defined by $\overline{\mathsf{L}}(\mathsf{w}^\star) = -\partial_{\mathsf{w}}\overline{\Omega}(\mathsf{w}^\star)$. Investigating smooth global solutions around constant equilibrium states require such linearized normal forms to be strictly dissipative [89]. When the source term of the normal form is in quasilinear form, the matrix $\overline{\mathsf{L}}$ is positive semi-definite. However, since $\mathsf{w}^\star$ is an equilibrium point, this is a general property deduced form (S_7) [101].

Lemma 7.1 *Consider a system of conservation law with an entropy and assume that the system is written in normal form* (139). *Assuming that* $\mathsf{w}^\star$ *is such that* $\overline{\Omega}(\mathsf{w}^\star) = 0$, *then, letting* $\overline{\mathsf{L}} = -\partial_{\mathsf{w}}\overline{\Omega}$, *the matrix* $\overline{\mathsf{L}}(\mathsf{w}^\star)$ *is symmetric positive semi-definite.*

By Fourier transform, the spectral problem associated with the linear system of partial differential equations (164) reads

$$\gamma\overline{\mathsf{A}}_0(\mathsf{w}^\star)\phi + \Big(\mathrm{i}\zeta\overline{\mathsf{A}}(\mathsf{w}^\star,\boldsymbol{\xi}) + \zeta^2\overline{\mathsf{B}}(\mathsf{w}^\star,\boldsymbol{\xi}) + \overline{\mathsf{L}}(\mathsf{w}^\star)\Big)\phi = 0, \tag{165}$$

where $\zeta \in \mathbb{R}$, $\mathrm{i}^2 = -1$, $\boldsymbol{\xi} \in \Sigma^{d-1}$, $\phi \in \mathbb{C}^{\mathsf{n}}$, $\overline{\mathsf{A}}(\mathsf{w}^\star,\boldsymbol{\xi}) = \sum_{i\in C}\overline{\mathsf{A}}_i(\mathsf{w}^\star)\xi_i$ and

$$\overline{\mathsf{B}}(\mathsf{w}^\star,\boldsymbol{\xi}) = \sum_{i,j\in C}\overline{\mathsf{B}}_{ij}(\mathsf{w}^\star)\xi_i\xi_j, \qquad \overline{\mathsf{L}}(\mathsf{w}^\star) = -\partial_{\mathsf{w}}\overline{\Omega}(\mathsf{w}^\star).$$

We denote by $\mathcal{S}(\zeta, w)$ the set of complex numbers γ such that there exists $\phi \in \mathbb{C}^{\mathsf{n}}$, $\phi \neq 0$, satisfying (165). The following equivalent forms of the 'Kawashima condition' have been established by Shizuta and Kawashima [106] for $(\mathsf{K}_1$–$\mathsf{K}_4)$ and Beauchard and Zuazua [117] for (K_5).

Theorem 7.2 *Assume that the matrix* $\overline{\mathsf{A}}_0(\mathsf{w}^\star)$ *is symmetric positive definite, that the matrices* $\overline{\mathsf{A}}_i(\mathsf{w}^\star)$, $i \in C$, *are symmetric, that the reciprocity relations* $\overline{\mathsf{B}}_{ij}(\mathsf{w}^\star)^t = \overline{\mathsf{B}}_{ji}(\mathsf{w}^\star)$, $i, j \in C$ *hold, that the matrix* $\overline{\mathsf{B}}(\mathsf{w}^\star,\boldsymbol{\xi}) = \sum_{i,j\in C}\overline{\mathsf{B}}_{ij}(\mathsf{w}^\star)\xi_i\xi_j$ *is positive semi-definite for* $\boldsymbol{\xi} \in \Sigma^{d-1}$, *that* $\overline{\mathsf{L}}(\mathsf{w}^\star)$ *is symmetric positive semi-definite, and denote* $\overline{\mathsf{A}}(\mathsf{w}^\star,\boldsymbol{\xi}) = \sum_{i\in C}\overline{\mathsf{A}}_i(\mathsf{w}^\star)\xi_i$. *The system of partial differential equations is said to be strictly dissipative at* $\mathsf{w}^\star$ *when any of the following equivalent properties holds.*

(K_1) *There exists a* C^∞ *map* $K : \Sigma^{d-1} \to \mathbb{R}^{d,d}$ *such that for any* $\boldsymbol{\xi} \in \Sigma^{d-1}$ *the product* $K(\boldsymbol{\xi})\overline{\mathsf{A}}_0(\mathsf{w}^\star)$ *is skew-symmetric,* $K(-\boldsymbol{\xi}) = -K(\boldsymbol{\xi})$, *and* $K(\boldsymbol{\xi})$ $\overline{\mathsf{A}}(\mathsf{w}^\star,\boldsymbol{\xi}) + \overline{\mathsf{B}}(\mathsf{w}^\star,\boldsymbol{\xi}) + \overline{\mathsf{L}}(\mathsf{w}^\star)$ *is positive definite.*

(K_2) *For any* $\zeta \in \mathbb{R}$, $\zeta \neq 0$, *and any* $\boldsymbol{\xi} \in \Sigma^{d-1}$, *the eigenvalues* $\gamma \in \mathcal{S}(\zeta,\boldsymbol{\xi})$ *have a negative real part* $\Re(\gamma) < 0$.

(K_3) *Let* $\phi \in \mathbb{R}^n\backslash\{0\}$ *such that* $\overline{\mathsf{B}}(\mathsf{w}^\star,\boldsymbol{\xi})\phi = 0$ *and* $\overline{\mathsf{L}}(\mathsf{w}^\star)\phi = 0$ *for some* $\boldsymbol{\xi} \in \Sigma^{d-1}$. *Then we have* $\zeta\overline{\mathsf{A}}_0(\mathsf{w}^\star)\phi + \overline{\mathsf{A}}(\mathsf{w}^\star,\boldsymbol{\xi})\phi \neq 0$ *for any* $\zeta \in \mathbb{R}$.

(K_4) *There exists* $\delta > 0$ *such that for any* $\zeta \in \mathbb{R}$, $\boldsymbol{\xi} \in \Sigma^{d-1}$, *the eigenvalue* $\gamma \in \mathcal{S}(\zeta,\boldsymbol{\xi})$ *have their real part majorized by* $\Re(\gamma) \leq -\delta\frac{|\zeta|^2}{1+|\zeta|^2}$.

(K_5) *Letting* $\widehat{A}^\star = \big(\overline{A}_0(w^\star)\big)^{-1}\overline{A}(w^\star, \xi)$ *and*

$$\widehat{B}^\star = \big(\overline{A}_0(w^\star)\big)^{-1}\big(\overline{B}(w^\star, \xi) + \overline{L}(w^\star)\big),$$

the Kalman condition $\operatorname{rank}\big[\, \widehat{B}^\star, \widehat{A}^\star \widehat{B}^\star, \ldots, (\widehat{A}^\star)^{n-1} \widehat{B}^\star \,\big] = n$ *holds.*

A physical interpretation of the 'Kawashima condition' (K_1–K_5) is that all waves associated with the hyperbolic operator $\overline{A}_0(w^\star)\partial_t + \sum_{i \in C} \overline{A}_i(w^\star)\partial_i$ lead to dissipation, i.e., entropy production, since there are not in the nullspace of $\overline{B}$, as shown by (K_3). Only the symmetric part of the product $K(\xi)\,\overline{A}(w^\star, \xi)$ plays a role in (K_1). The traditional Kalman condition involving the $n^2 \times n$ matrix with first block $\widehat{B}^\star$, second block $\widehat{B}^\star\widehat{A}^\star$, and kth block $\widehat{B}^\star\,(\widehat{A}^\star)^{k-1}$ has been rewritten in the form (K_5) with a $n \times n^2$ matrix thanks to the symmetry of $\overline{A}_0(w^\star)$, $\overline{B}(w^\star, \xi)$ and $\overline{L}(w^\star)$.

Remark 7.3 It is not known in general if the matrix $K(\xi)$ may be written $\sum_{j \in C} K_j \xi_j$ although it is generally possible to obtain compensating matrices in this form in practical applications.

7.2 Existence of Solutions

Local in time solutions [108] may first be obtained by using a normal from as well as a general existence theorem from Volpert and Hujaev [105]. The corresponding existence result in [108] has been presented for a more general fluid with vibrational desequilibrium but also directly applies to the system of partial differential equations presented in the previous sections.

On the other hand, global solutions around equilibrium states may also be obtained using the local strict dissipative properties for multicomponent flows. The existence of chemical equilibrium points is first a consequence of the structural properties of thermochemistry and is traditionally obtained by minimizing a thermodynamic function [9].

Proposition 7.4 *For* $T^\star > 0$ *and* $\varrho^c \in (0, \infty)^n$ *there exists a unique equilibrium point* $u^\star$ *associated with* $z^\star = \big(\rho_1^\star, \ldots, \rho_n^\star, \mathbf{0}, T^\star\big)^t$ *such that* $\varrho^\star - \varrho^c \in \mathcal{MR}$.

The system of partial differential equations governing multicomponent reactive flows written in normal form is then strictly dissipative [89, 92, 95, 97, 106].

Proposition 7.5 *Consider an equilibrium state* $w^\star$ *as obtained in Proposition 7.4. Then the linearized normal form at* $w^\star$ *is strictly dissipative.*

Proof Using for convenience the characterization (K_3), we consider $\xi \in \Sigma^{d-1}$ and assume that $\phi \neq 0$ is such that $\overline{B}(w^\star, \xi)\phi = \overline{L}(w^\star)\phi = 0$. We first establish that $\phi = \alpha(1, 0, \ldots, 0)^t$ for some $\alpha \neq 0$.

From the expression of $\overline{\mathsf{B}}$ in Theorem 6.20, letting

$$\phi = (\phi_1, \ldots, \phi_n, \boldsymbol{\phi_v}, \phi_T)^t,$$

we obtain

$$\langle \overline{\mathsf{B}}\phi, \phi\rangle = \kappa RT\big(\boldsymbol{\xi}\cdot(\boldsymbol{\phi_v}+\boldsymbol{v}\phi_T)\big)^2 + \eta RT\Big(\tfrac{1}{3}\big(\boldsymbol{\xi}\cdot(\boldsymbol{\phi_v}+\boldsymbol{v}\phi_T)\big)^2 + \big|\boldsymbol{\phi_v}+\boldsymbol{v}\phi_T\big|^2\Big)$$

$$+\frac{RT}{p}\sum_{2\le i,j\le n} \mathcal{D}_{ij}(\phi_i + \mathfrak{h}_i\phi_T)(\phi_j + \mathfrak{h}_j\phi_T) + \lambda RT^2\phi_T^2.$$

This is a sum of nonnegative terms and if it is zero, we successively deduce that $\phi_T = 0$ and then that $\boldsymbol{\phi_v} = 0$ and finally that $\sum_{2\le i,j\le n}\mathcal{D}_{ij}\phi_i\phi_j = 0$. Since $N(D) = \mathbb{R}\mathsf{y}$ we have $N(\mathcal{D}) = \mathbb{R}\mathbb{I}$ so that $\mathcal{D}$ is positive definite for vectors in the form $(0, \phi_2, \ldots, \phi_n)^t$ and $\phi_2 = \ldots = \phi_n = 0$. We have established that ϕ is proportional to $(1, 0, \ldots, 0)^t$ and such vectors are also in the nullspace of $\overline{\mathsf{L}}$. Indeed, using (158) and $\overline{\mathsf{L}}(\mathsf{w}^\star) = \big(\partial_{\mathsf{w}}\mathsf{v}(\mathsf{w}^\star)\big)^t\,\widetilde{\mathsf{L}}(\mathsf{v}^\star)\,\partial_{\mathsf{w}}\mathsf{v}(\mathsf{w}^\star)$, a direct calculation yields

$$\overline{\mathsf{L}}(\mathsf{w}^\star) = \sum_{i\in\mathfrak{R}} \Lambda_i \mathfrak{p}_i'\otimes\mathfrak{p}_i',$$

where $\mathfrak{p}_i' = (0, m_2\nu_{2i}, \ldots, m_n\nu_{ni}, \boldsymbol{0}, 0)^t$ so that $(1, 0, \ldots, 0)^t$ is in the nullspace of $\overline{\mathsf{L}}(\mathsf{w}^\star)$.

We must now establish that for any $\zeta \in \mathbb{R}$ we have $\zeta\overline{\mathsf{A}}_0(\mathsf{w}^\star)\phi + \overline{\mathsf{A}}(\mathsf{w}^\star, \boldsymbol{\xi})\,\phi \neq 0$ where $\overline{\mathsf{A}}(\mathsf{w}^\star, \boldsymbol{\xi}) = \sum_{i\in C}\overline{\mathsf{A}}_i(\mathsf{w}^\star)\xi_i$. However, a direct calculation keeping the notation of Theorem 6.16 yields that

$$\frac{\langle \Lambda\mathbb{I}, \mathbb{I}\rangle}{\alpha}\Big(\zeta\overline{\mathsf{A}}_0(\mathsf{w}^\star)\phi + \overline{\mathsf{A}}(\mathsf{w}^\star, \boldsymbol{\xi})\phi\Big) = (\zeta + \boldsymbol{v}\cdot\boldsymbol{\xi})\begin{bmatrix}\mathsf{e}_1\\ \boldsymbol{0}\\ 0\end{bmatrix} + \rho RT\begin{bmatrix}0_{1,n}\\ \boldsymbol{\xi}\\ \boldsymbol{v}\cdot\boldsymbol{\xi}\end{bmatrix},$$

where $\mathsf{e}_1 \in \mathbb{R}^n$ is the first basis vector $\mathsf{e}_1 = (1, 0, \ldots, 0)^t$ of $\mathbb{R}^n$. As a consequence, the vector $\zeta\overline{\mathsf{A}}_0(\mathsf{w}^\star)\phi + \overline{\mathsf{A}}(\mathsf{w}^\star, \boldsymbol{\xi})\phi$ may have its first component zero with $\zeta = -\boldsymbol{v}\cdot\boldsymbol{\xi}$ but its velocity components never vanish since $\boldsymbol{\xi} \in \Sigma^{d-1}$ and (K_3) holds. □

The local strict dissipativity properties now imply global existence and asymptotic stability of equilibrium states [89, 95, 97]. The existence proof mainly consists in establishing a priori estimates for linearized equations, then a local existence theorem, next a priori estimates independent of the time interval using strict dissipativity and finally using the local existence repeatedly [89, 95, 97, 106, 107, 118]. We present such a theorem for the equations governing multicomponent reactive flows in their natural normal form.

Theorem 7.6 *Let $d \geqslant 1$, $l \geqslant [d/2] + 2$, and consider the equations governing multicomponent reactive flows in normal form $\mathsf{w} = (\mathsf{w}_{\mathrm{I}}, \mathsf{w}_{\mathrm{II}})^t$ with*

$$\mathsf{w}_{\mathrm{I}} = \rho, \qquad \mathsf{w}_{\mathrm{II}} = \Big(\frac{g_2 - g_1}{RT}, \ldots, \frac{g_n - g_1}{RT}, \frac{\boldsymbol{v}}{RT}, \frac{-1}{RT}\Big)^t. \tag{166}$$

Let $\mathsf{w}^\star$ be a chemical equilibrium state as in Proposition 7.4. If the initial condition $\mathsf{w}^0(x)$ is such that $\|\mathsf{w}^0 - \mathsf{w}^\star\|_{H^l}$ is small enough, then the Cauchy problem with $\mathsf{w}(0, x) = \mathsf{w}^0(x)$ has a global solution with

$$\mathsf{w}_{\mathrm{I}} - \mathsf{w}_{\mathrm{I}}^\star \in C^0\big([0, \infty); H^l\big) \cap C^1\big([0, \infty); H^{l-1}\big), \tag{167}$$

$$\mathsf{w}_{\mathrm{II}} - \mathsf{w}_{\mathrm{II}}^\star \in C^0\big([0, \infty); H^l\big) \cap C^1\big([0, \infty); H^{l-2}\big). \tag{168}$$

Moreover we have the estimates

$$\|\mathsf{w}(t) - \mathsf{w}^\star\|_{H^l}^2 + \int_0^t \big(\|\nabla \mathsf{w}_{\mathrm{I}}(\tau)\|_{H^{l-1}}^2 + \|\nabla \mathsf{w}_{\mathrm{II}}(\tau)\|_{H^l}^2\big)\, d\tau \leqslant C\|\mathsf{w}^0 - \mathsf{w}^\star\|_{H^l}^2,$$

and $\sup_{\mathbb{R}^d} |\mathsf{w}(t) - \mathsf{w}^\star|$ *goes to zero as $t \to \infty$.*

From Theorem 7.6 the equilibrium point $\mathsf{w}^\star$ is asymptotically stable and with stronger assumptions it is also possible to obtain decay estimates [95].

Theorem 7.7 *Let $d \geqslant 1$, $l \geqslant [d/2] + 3$ and assume that the initial condition $\mathsf{w}^0(x)$ is such that $\mathsf{w}^0 - \mathsf{w}^\star \in H^l(\mathbb{R}^d) \cap L^p(\mathbb{R}^d)$, where $p = 1$, if $d = 1$, and $p \in [1, 2)$, if $d \geqslant 2$. Then if $\|\mathsf{w}^0 - \mathsf{w}^\star\|_{H^l} + \|\mathsf{w}^0 - \mathsf{w}^\star\|_{L^p}$ is small enough, the global solution satisfy the dacay estimate*

$$\|\mathsf{w}(t) - \mathsf{w}^\star\|_{H^{l-2}} \leqslant \beta(1 + t)^{-\gamma}\big(\|\mathsf{w}^0 - \mathsf{w}^\star\|_{H^{l-2}} + \|\mathsf{w}^0 - \mathsf{w}^\star\|_{L^p}\big),$$

for $t \in [0, \infty)$, where β is a positive constant and $\gamma = d \times (1/2p - 1/4)$.

Such theorems for hyperbolic-parabolic systems may be used for various other fluid models. They have been notably used for ambipolar plasmas where Poisson Equation is replaced by the zero current limit [97], partial equilibrium flows where some group of chemical reactions are assumed infinitely fast [116], as well as for Saint-Venant equations modeling thin viscous layers over fluid substrates [120].

8 Conclusion and Future Directions

We have presented in these notes the kinetic theory of reactive gas mixtures. We have extracted from this molecular setting the corresponding fluid system of partial differential equations as well as the natural assumptions for the system coefficients. We have established that these fluid equations have a hyperbolic-parabolic structure in such a way that they have local in time solutions [108] as well as global solutions around equilibrium states [9, 95, 101]. These equations also have traveling wave solutions in the low Mach number limit [119].

The models developed in the previous sections may further be generalized to describe gas mixtures in full vibrational nonequilibrium when each vibrational quantum level is treated as a separate "chemical species" allowing detailed state-to-state relaxation models [108]. When the vibrational quantum levels are partially at equilibrium between them but not at equilibrium with the translational/rotational states—allowing the definition of a vibrational temperature—a different structure is obtained. Models at thermodynamic nonequilibrium with two temperatures have also been investigated and the apparition of a volume viscosity term has been justified mathematically [104, 118].

The mathematical analysis of chemical equilibrium flows has been extended to the situation of partial chemical equilibrium [116]. However, the mathematical structure of numerous simplified chemistry methods is still obscure from a mathematical point of view at variance with partial equilibrium.

Various extensions could also consider initial-boundary value problems [121] with the possibility of inflow or outflow conditions, heat losses, surface reactions with complex heterogeneous chemistry, species surface diffusion or heat surface conduction. Various numerical analysis theoretical results could also be extended to the case of mixtures like convergence results of Petrov-Galerkin 'Streamline–Diffusion' finite element techniques [90, 122].

The notion of higher order entropy may also be generalized to the situation of multicomponent flows [123–126] as well as the singular limit of small Mach number flow [127, 128]. Multiphase flows with sprays governed by Boltzmann type equations [6], or derived multifluid sectional models for droplets [129], may also be investigated mathematically.

References

Applications

1. Rosner, D.E.: Transport Processes in Chemically Reacting Flow Systems. Butterworths, Boston (1986)
2. Roenigk, K.F., Jensen, K.F.: Low pressure CVD of silicon nitride. J. Electrochem. Soc. **134**, 1777–1785 (1987)
3. Ern, A., Giovangigli, V., Smooke, M.: Numerical study of a three-dimensional chemical vapor deposition reactor with detailed chemistry. J. Comp. Phys. **126**, 21–39 (1996)
4. Anderson Jr, J.D.: Hypersonics and High Temperature Gas Dynamics. McGraw-Hill Book Company, New York (1989)
5. Laboudigue, B., Giovangigli, V., Candel, S.: Numerical solution of a free-boundary problem in hypersonic flow theory: nonequilibrium viscous shock layers. J. Comp. Phys. **102**, 297–309 (1992)
6. Williams, F.A.: Combustion Theory, 2nd edn. The Benjamin/Cummings Pub. Co., Inc., Menlo Park (1985)
7. Ern, A., Giovangigli, V.: Thermal diffusion effects in hydrogen/air and methane/air flames. Comb. Theor. Mod. **2**, 349–372 (1998)
8. Poinsot, T., Veynante, D.: Theoretical and Numerical Combustion. Edwards, Philadelphia (2005)
9. Giovangigli, V.: Multicomponent Flow Modeling. Birkhäuser, Boston (1999)

Kinetic Theory

10. Hirschfelder, J.O., Curtiss, C.F., Bird, R.B.: Molecular Theory of Gases and Liquids. Wiley, New York (1954)
11. Waldmann, L.: Transporterscheinungen in Gasen von mittlerem Druck. Handbuch der Physik **12**, 295–514 (1958)
12. Waldmann, L., Trübenbacher, E.: Formale Kinetische Theorie von Gasgemischen aus Anregbaren Molekülen. Zeitschr. Naturforschg. **17**(a), 363–376 (1962)
13. Van de Ree, J.: On the definition of the diffusion coefficients in reacting gases. Physica **36**, 118–126 (1967)
14. Curtiss, C.F.: Symmetric gaseous diffusion coefficients. J. Chem. Phys. **49**, 2917–2919 (1968)
15. Chapman, S., Cowling, T.G.: The Mathematical Theory of Non-Uniform Gases. Cambridge University Press, Cambridge (1970)
16. Ferziger, J.H., Kaper, H.G.: Mathematical Theory of Transport Processes in Gases. North Holland Pub. Co., Amsterdam (1972)
17. Cercignani, C.: The Boltzmann Equation and Its Applications, Applied Mathematical Sciences, vol. 67. Springer, Berlin (1988)
18. Grunfeld, C.: On a class of Kinetic equations for reacting gas mixtures with multiple collisions. C. R. Acad. Sci. Paris **316**(Série I), 953–958 (1993)
19. Ern, A., Giovangigli, V.: Multicomponent Transport Algorithms, Lecture Notes in Physics, New Series "Monographs", m 24 (1994)

Reactive Kinetic Theory

20. Prigogine, I., Xhrouet, E.: On the perturbation of Maxwell distribution function by chemical reactions in gases. Physica **15**, 913–932 (1949)
21. Prigogine, I., Mathieu, M.: Sur la Perturbation de la Distribution de Maxwell par des Réactions Chimiques en Phase Gazeuse. Physica **16**, 51–64 (1950)
22. Takayanagi, K.: On the theory of chemically reacting gas. Prog. Theor. Phys. **VI**, 486–497 (1951)
23. Present, R.D.: On the velocity distribution in a chemically reacting gas. J. Chem. Phys. **31**, 747–750 (1959)
24. Ludwig, G., Heil, M.: Boundary Layer Theory with Dissociation and Lonization, vol. VI, pp. 39–118. Academic Press, New York (1960) (In Advances in Applied Mechanics)
25. Ross, J., Mazur, P.: Some deductions from a formal statistical mechanical theory of chemical kinetics. J. Chem. Phys. **35**, 19–28 (1961)
26. Shizgal, B., Karplus, M.: Nonequilibrium contributions to the rate of reaction. I. Perturbation of the velocity distribution function. J. Chem. Phys. **52**, 4262–4278 (1970)
27. Shizgal, B., Karplus, M.: Nonequilibrium contributions to the rate of reaction. II. Isolated multicomponent systems. J. Chem. Phys. **54**, 4345–4356 (1971)
28. Shizgal, B., Karplus, M.: Nonequilibrium contributions to the rate of reaction. III. Isothermal multicomponent systems. J. Chem. Phys. **54**, 4357–4362 (1971)
29. Vallander, S., Nagnibeda, E., Rydalevskaya, M.: Some Questions of the Theory of the Chemically Reacting Gas Mixture. Leningrad University Press, Leningrad (1977) (English translation U.S. Air Force FASTC-ID (RS) TO-0608-93)
30. Alexeev, B.V.: Mathematical Kinetics of Reacting Gases. Nauka, Moskow (1982)
31. Alexeev, B.V., Chikhaoui, A., Grushin, I.T.: Application of the generalized Chapman-Enskog method to the transport-coefficient calculation in a reacting gas mixture. Phys. Rev. E **49**, 2809–2825 (1994)
32. Chikhaoui, A., Dudon, J.P., Kustova, E.V., Nagnibeda, E.A.: Transport properties in reacting mixture of polyatomic gases. Physica A **247**, 526–552 (1997)

33. Ern, A., Giovangigli, V.: The Kinetic equilibrium regime. Physica-A **260**, 49–72 (1998)
34. Monaco, R., Pandolfi Bianchi, M., Soares, A.J.: BGK-type models in strong reaction and kinetic chemical equilibrium regimes. J. Phys. A **38**, 10413 (2005)

Thermodynamic Nonequilibrium

35. Kogan, M.: Rarefied Gas Dynamics. Nauka, Moskow (1967) (English translation Plenum Press, New York (1969))
36. Brun, R.: Transport et Relaxation dans les Ecoulements Gazeux. Masson, Paris (1986)
37. Zdhanov, V., Alievskiy, M.: Relaxation and Transport i Molecular Gases. Nauka, Moskow (1989)
38. Bruno, D., Capitelli, M., Longo, S.: Direct simulation of non-equilibrium kinetics under shock conditions in nitrogen. Chem. Phys. Lett. **360**, 31–37 (2002)
39. Zhdanov, V.M.: Transport Processes in Multicomponent Plasmas. Taylor and Francis, London (2002)
40. Nagnibeda, E.A., Kustova, E.A.: Kinetic Theory of Transport and Relaxation Processes in Non-Equilibrium Reacting Flows. Saint-Petersburg University Press, Russian (2003)
41. Colonna, G., Armenise, I., Bruno, D., Capitelli, M.: Reduction of state-to-state kinetics to macroscopic models in hypersonic flows. J. Thermophys. Heat Transf. **20**, 477–486 (2006)
42. Capitelli, M., Armenise, I., Bruno, D., Cacciatore, M., Celiberto, R., Colonna, G., de Pascale, O., Diomede, P., Esposito, F., Gorse, C., Hassouni, K., Laricchiuta, A., Longo, S., Pagano, D., Pietanza, D., Rutigliano, M.: Non-equilibrium plasma kinetics: a state-to-state approach. Plasma Sourc. Sci. Tech. **16**, S30–S44 (2007)
43. Nagnibeda, E., Kustova, E.: Non-equilibrium Reacting Gas Flows. Springer, Berlin (2009)
44. Bruno, D., Giovangigli, V.: Relaxation of internal temperature and volume viscosity. Phys. Fluids **23**, 093104 (2011)
45. Bruno, D., Giovangigli, V.: Erratum: Relaxation of internal temperature and volume viscosity. Phys. Fluids **25**, 039902 (2013)
46. Bruno, D., Esposito, F., Giovangigli, V.: Relaxation of rotational-vibrational energy and volume viscosity in H-H_2 mixtures. J. Chem. Phys. **138**, 084302 (2013)

Mathematical Kinetic Theory

47. DiPerna, R.J., Lions, P.L.: On the global existence for boltzmann equations: global existence and weak stability. Ann. Math. **130**, 321–366 (1989)
48. Cercignani, C., Illner, R., Pulvirenti, M.: The Mathematical Theory of Dilute Gases, Applied Mathematical Sciences, vol. 106. Springer, Berlin (1994)
49. Grunfeld, C., Georgescu, E.: On a class of Kinetic equations for reacting gas mixtures. Mat. Fiz. Analiz. Geom. **2**, 408–435 (1995)
50. Lions, P.-L., Masmoudi, N.: From Boltzmann equation to the Navier-Stokes and Euler equations I. Arch. Ration. Mech. Anal. **158**, 173–193 (2001)
51. Golse, F., Levermore, C.D.: The Stokes-Fourier and acoustic limits for the Boltzmann equation. Commun. Pure Appl. Math. **55**, 336–393 (2002)
52. Saint-Raymond, L.: Convergence of solutions to the Boltzmann equation in the incompressible Euler limit. Arch. Ration. Mech. Anal. **166**, 47–80 (2003)
53. Golse, F., Saint-Raymond, L.: The Navier-Stokes limit of the Boltzmann equation for bounded collision kernels. Invent. Math. **155**, 81–161 (2004)

54. Golse, F., Saint-Raymond, L.: Hydrodynamic limits for the Boltzmann equation. Riv. Mat. Univ. Parma **7**, 1–144 (2005)
55. Saint-Raymond, L.: Hydrodynamic limits of the Boltzmann equation. In: Lecture Notes in Mathematics 1971. Springer, Berlin-Heidelberg (2009)
56. Brull, S., Pavan, V., Schneider, J.: Derivation of a BGK model for mixtures. Eur. J. Mech. B/Fluids **33**, 74–86 (2012)

Macroscopic Theories

57. Marcelin, M.R.: Sur la Mécanique des Phénomènes Irréversibles. C. R. Acad. Sci. Paris 1052–1055 (1910)
58. Meixner, J.: Zur Thermodynamik der irreversiblen Prozesse in Gasen mit chemisch reagierenden, dissoziierenden und anregbaren Komponenten. Ann. Phys. **43**, 244–270 (1943)
59. Prigogine, I.: Etude thermodynamique des phénomènes irréversibles. Dunod, Paris (1947)
60. Irving, J.H., Kirkwood, J.G.: The statistical mechanics of transport processes. IV. The equations of hydrodynamics. J. Chem. Phys. **18**, 817–829 (1950)
61. Bearman, R.J., Kirkwood, J.G.: The statistical mechanics of transport processes. XI. Equations of transport in multicomponent systems. J. Chem. Phys. **28**, 136–145 (1958)
62. de Groot, S.R., Mazur, P.: Non-equilibrium thermodynamics. Dover Publications.Inc, New York (1984)
63. Gorban, A.N., Bykov, V.I., Yablonski, G.S.: Essays on Chemical Relaxation (in Russian). Nauka, Novosibirsk (1986)
64. Keizer, J.: Statistical Thermodynamics of Nonequilibrium Processes. Springer, New York (1987)

Multicomponent Transport Algorithms

65. Hirschfelder, J.O., Curtiss, C.F.: Theory of propagation of flames. Part I: general equations. Proc. Comb. Inst. **3**,121–127 (1949)
66. Williams, F.A.: Elementary derivation of the multicomponent diffusion equation. Am. J. Phys. **26**, 467–469 (1958)
67. Dixon-Lewis, G.: Flame structure and flame reaction Kinetics, II. Transport phenomena in multicomponent systems. Proc. Roy. Soc. A **307**, 111–135 (1968)
68. Oran, E.S., Boris, J.P.: Detailed modeling of combustion systems. Prog. Energy Combust. Sci. **7**, 1–72 (1981)
69. Giovangigli, V.: Mass conservation and singular multicomponent diffusion algorithms. IMPACT Comput. Sci. Eng. **2**, 73–97 (1990)
70. Giovangigli, V.: Convergent Iterative methods for multicomponent diffusion. IMPACT Comput. Sci. Eng. **3**, 244–276 (1991)
71. Ern, A., Giovangigli, V.: Thermal conduction and thermal diffusion in dilute polyatomic gas mixtures. Physica-A **214**, 526–546 (1995)
72. Ern, A., Giovangigli, V.: The structure of transport linear systems in dilute isotropic gas mixtures. Phys. Rev. E **53**, 485–492 (1996)
73. Ern, A., Giovangigli, V.: Optimized transport algorithms for flame codes. Comb. Sci. Tech. **118**, 387–395 (1996)
74. Ern, A., Giovangigli, V.: Projected iterative algorithms with application to multicomponent transport. Linear Algebra Appl. **250**, 289–315 (1997)
75. Giovangigli, V.: Multicomponent transport algorithms for partially ionized plasmas. J. Comp. Phys. **229**, 4117–4142 (2010)

Impact of Multicomponent Transport

76. Hancock, R.D., Schauer, F.R., Lucht, R.P., Katta, V.R., Hsu, K.Y.: Thermal diffusion effects and vortex-flame interactions in hydrogen jet diffusion flames. Proc. Combust. Inst. **26**, 1087–1093 (1996)
77. Ern, A., Giovangigli, V.: Impact of multicomponent transport on planar and counterflow hydrogen/air and methane/air flames. Comb. Sci. Tech. **149**, 157–181 (1999)
78. Bendahklia, R., Giovangigli, V., Rosner, D.: Soret effects in laminar counterflow spray diffusion flames. Comb. Theory Mod. **6**, 1–17 (2002)
79. Hilbert, R., Tap, F., El-Rabii, H., Thevenin, D.: Impact of detailed chemistry and transport models on turbulent combustion simulations. Prog. Ener. Comb. Sci. **30**, 61–117 (2004)
80. Galkin, V.S., Rusakov, S.V.: On the theory of bulk viscosity and relaxation pressure. Appl. Math. Mech. **69**, 943–954 (2005)
81. Elizarova, T., Kholkov, A., Montero, S.: Numerical simulaition of shock waves structure in nitrogen. Phys. Fluids **19**, 068102 (2007)
82. Billet, G., Giovangigli, V., de Gassowski, G.: Impact of volume viscosity on a shock/hydrogen bubble interaction. Comb. Theor. Mod. **12**, 221–248 (2008)
83. Dworkin, S., Smooke, M.D., Giovangigli, V.: The impact of detailed multicoponent transport and thermal diffusion effects on soot formation in ethylene/air flames. Proc. Comb. Inst. **32**, 1165–1172 (2009)
84. Kustova, E.: On the role of bulk viscosity and relaxation pressure in nonequilibrium flows. In: Abe (ed.) 26th International Symposiuù on Rarefied Gas Dynamic, AIP Conference Proceedings 1024, pp. 807–812 (2009)
85. Armenise, I., Kustova, E.: On different contributions to the heat flux and diffusion in non-equilibrium flows. Chem. Phys. **428**, 90–104 (2014)

Hyperbolic-Parabolic Systems

86. Godunov, S.: An interesting class of quasilinear systems. Sov. Math. Dokl **2**, 947–949 (1961)
87. Ladyženskaja, O.A., Solonikov, V.A., Ural'ceva, N.N.: Linear and Quasilinear Equations of Parabolic Type. Translations of Mathematical Monographs, 23, American Mathematical Society, Providence, Rhode Island (1968)
88. Friedrichs, K.O., Lax, P.D.: Systems of conservation laws with a convex extension. Proc. Nat. Acad. Sci. USA **68**, 1686–1688 (1971)
89. Kawashima, S.: Systems of Hyperbolic-Parabolic Composite type, with Application to the Equations of Magnetohydrodynamics. In: Doctoral Thesis, Kyoto University (1984)
90. Hughes, T.J.R., Franca, L.P., Mallet, M.: A new finite element formulation for computational fluid dynamics: I. Symmetric forms of the compressible Euler and Navier-Stokes equations and the second law of thermodynamics. Comp. Meth. Appl. Mech. Eng. **54**, 223–234 (1986)
91. Ruggeri, T.: Thermodynamics and Symmetric Hyperbolic Systems, pp. 167–183. Rend. Sem. Mat. Univ. Torino, Hyperbolic Equations (1987)
92. Kawashima, S., Shizuta, Y.: On the normal form of the symmetric hyperbolic-parabolic systems associated with the conservation laws. Tôhoku Math. J. **40**, 449–464 (1988)
93. Chen, G.Q., Levermore, C.D., Liu, T.P.: Hyperbolic conservation laws with stiff relaxation terms and entropy. Commun. Pure Appl. Math. **47**, 787–830 (1994)
94. Serre, D.: Systèmes de Lois de Conservation I and II. Diderot Editeur. Art et Science, Paris (1996)
95. Giovangigli, V., Massot, M.: Asymptotic stability of equilibrium states for multicomponent reactive flows. Math. Mod. Meth. Appl. Sci. **8**, 251–297 (1998)
96. Dafermos, C.: Hyperbolic Conservation Laws in Continuum Physics. Springer, Heidelberg (2000)

97. Giovangigli, V., Graille, B.: Asymptotic stability of equilibrium states for ambipolar plasmas. Math. Mod. Meth. Appl. Sci. **14**, 1361–1399 (2004)
98. Kawashima, S., Yong, W.-A.: Dissipative structure and entropy for hyperbolic systems of conservation laws. Arch. Ration. Mech. Anal. **174**, 345–364 (2004)
99. Kawashima, S., Yong, W.A.: Decay estimates for hyperbolic balance laws. J. Anal. Appl. **28**, 1–33 (2009)
100. Serre, D.: The structure of dissipative viscous system of conservation laws. Physica D **239**, 1381–1386 (2010)
101. Giovangigli, V., Matuszewski, L.: Mathematical modeling of supercritical multicomponent reactive fluids. Math. Mod. Meth. Appl. Sci. **23**, 2193–2251 (2013)
102. Dafermos, C.M.: Hyperbolic systems of balance laws with weak dissipation II. J. Hyp. Diff. Equat. **10**, 173–179 (2013)
103. Giovangigli, V., Matuszewski, L.: Structure of entropies in dissipative multicomponent fluids. Kin. Relat. Mod. **6**, 373–406 (2013)
104. Giovangigli, V., Yong, W.A.: Volume viscosity and fast internal energy relaxation: symmetrization and Chapman-Enskog expansion. Kin. Relat. Mod. (in press)

Cauchy Problem

105. Vol'Pert, A.I., Hudjaev, S.I.: On the Cauchy problem for composite systems of nonlinear differential equations. Math. USSR Sbornik. **16**, 517–544 (1972)
106. Shizuta, Y., Kawashima, S.: Systems of equations of hyperbolic-parabolic type with applications to the discrete Boltzmann equation. Hokkaido Math. J. **14**, 249–275 (1985)
107. Kawashima, S.: Large-time behavior of solutions to hyperbolic-parabolic systems of conservations laws and applications. Proc. Roy. Soc. Edinb. **106A**, 169–1944 (1987)
108. Giovangigli, V., Massot, M.: The local Cauchy problem for multicomponent reactive flows in full vibrational nonequilibrium. Math. Meth. Appl. Sci. **21**, 1415–1439 (1998)
109. Feireisl, E.: Dynamics of Viscous Compressible Fluids. Oxford University Press, Oxford (2004)
110. Yong, W.-A.: Entropy and global existence for hyperbolic balance laws. Arch. Ration. Mech. Anal. **172**, 247–266 (2004)
111. Feireisl, E., Novotný, A.: The low mach number limit for the full Navier-Stokes-Fourier system. Arch. Ration. Mech. Anal. **186**, 77–107 (2007)
112. Feireisl, E., Petzeltoá, H., Trivisa, K.: Multicomponent reactive flows: global-in-time existence for large data. Commun. Pure Appl. Anal. **7**, 1017–1047 (2008)
113. Yong, W.A.: An interesting class of partial differential equations. J. Math. Phys. **49**, 033503 (2008)
114. Kwong, Y.S., Trivisa, K.: Stability and large-time behavior for multicomponent reactive flows. Nonlinearity **22**, 2443–2471 (2009)
115. Hopf, D.: Asymptotic behavior of solutions to a model for the flow of a reacting fluid. Arch. Ration. Mech. Anal. **196**, 951–979 (2010)
116. Giovangigli, V., Massot, M.: Entropic structure of multicomponent reactive flows with partial equilibrium reduced chemistry. Math. Meth. Appl. Sci. **27**, 739–768 (2004)
117. Beauchard, K., Zuazua, E.: Large time asymptotics for partially dissipative hyperbolic systems. Arch. Ration. Mech. Anal. **199**, 177–227 (2011)
118. Giovangigli, V., Yong, W.A.: Volume Viscosity and Fast Internal Energy Relaxation: Error Estimates (submitted for publication)

Miscelaneous

119. Giovangigli, V.: Plane flames with multicomponent transport and complex chemistry. Math. Mod. Meth. Appl. Sci. **9**, 337–378 (1999)
120. Giovangigli, V., Tran, B.: Mathematical analysis of a Saint-Venant model with variable temperature. Math. Mod. Meth. Appl. Sci. **20**, 1–47 (2010)
121. Matsumura, A., Nishida, T.: The initial value problem for the equations of motion of a viscous and heat-conductive fluids. J. Math. Kyoto Univ. **200**, 67–104 (1980)
122. Johnson, C., Szepessy, A., Hansbo, P.: On the convergence of shock-capturing streamline diffusion finite element methods for hyperbolic conservation laws. Math. Comput. **54**, 107–129 (1990)
123. Giovangigli, V.: Higher order entropies. Arch. Ration. Mech. Anal. **187**, 221–285 (2008)
124. Giovangigli, V.: Higher order entropies for compressible fluid models. Math. Mod. Meth. Appl. Sci. **19**, 67–125 (2009)
125. Evans, L.C.: A survey of entropy methods for partial differential equations. Bull. AMS **41**, 409–438 (2004)
126. Bresch, D., Desjardins, B.: On the existence of global weak solutions to the Navier-Stokes equations for viscous compressible and heat conducting fluids. J. Math. Pure Appl. **87**, 57–90 (2007)
127. Alazard, T.: Low mach number limit of the full Navier-Stokes equations. Arch. Ration. Mech. Anal. **180**, 1–73 (2006)
128. Feireisl, E., Novotnỳ, A.: Singular Limits in Thermodynamics of Viscous Fluids. Birkhäuser, Basel (2009)
129. Laurent, F., Massot, M.: Multi-fluid modeling of laminar poly-dispersed spray flames: origin, assumptions and comparison of the sectional and sampling methods. Comb. Theor. Mod. **5**, 537–572 (2001)

Part II
Short Papers

Large Deviations in a Gaussian Setting: The Role of the Cameron-Martin Space

Diogo Bessam

Abstract We discuss the use of the Cameron-Martin space of a Gaussian measure in the description of the large deviations rates in a Gaussian setting. Three classical examples of large deviations rates in this setting are surveyed and related. We further illustrate the main ideas by directly verifying the Freidlin-Wentzell upper and lower estimates for εX, as $\varepsilon \to 0$, where X is a $\mathbb{R}^d$ Gaussian vector and by presenting explicitly the Cameron-Martin space associated with X.

Keywords Large deviations rates · Gaussian measures · Cameron-Martin space · Reproducing kernel Hilbert space

1 Introduction

This article is a short review of the relation between the Cameron-Martin spaces and the large deviations principle associated some Gaussian families. The intended readers are graduate students of probability with some knowledge of functional analysis or researchers in probability and analysis unfamiliar with the mentioned relation. Generally speaking, the specific Gaussian setting here is a family of probability measures $\mathbb{P}_\varepsilon(\cdot) = \mathbb{P}(\varepsilon X \in \cdot)$, as $\varepsilon \to 0$, where X is some Gaussian random element or stochastic process. In fact, for the class of examples we have in mind, Gaussian measures induce and are induced by Gaussian processes, due to the function space where the stochastic process has its paths, the space of square integrable functions $L_2(0, 1)$: see [4] that discusses this correspondence for a number of function spaces usually used in the analysis of stochastic processes, including $L_2(0, T)$, for $T \in (0, \infty)$. In the most general setting of Gaussian measures in Banach spaces, the definition of Cameron-Martin space is somewhat technical involving some standard notions of probability in a functional analytic setting. On the other hand, in finite dimensions,

D. Bessam (✉)
Departamento de Matemática/PUC-Rio, Rua Marquês de São Vicente 225,
Edifício Cardeal Leme, sala 862, Gávea - Rio de Janeiro CEP 22451-900, Brazil
e-mail: bessam@mat.puc-rio.br

P. Gonçalves and A.J. Soares (eds.), *From Particle Systems to Partial Differential Equations II*, Springer Proceedings in Mathematics & Statistics 129,
DOI 10.1007/978-3-319-16637-7_3

we believe the knowledge of multivariate calculus, some real analysis and probability is enough to convey most of the concepts and ideas of the proofs without the need of going into the technicalities related either to stochastic processes or measures in abstract spaces. Furthermore, in this simpler case we can readily identify what is the Cameron-Martin space, say $\mathscr{H}_\gamma$, associated with the Gaussian distribution, say γ—it is the support of the distribution. In particular, if the distribution is non-degenerate, then $\mathscr{H}_\gamma$ is all $\mathbb{R}^d$. This situation contrasts strikingly with the infinite dimensional case, where $\gamma(\mathscr{H}_\gamma) = 0$, as soon as $\dim \mathscr{H}_\gamma = \infty$.

With the above objective in mind, we discuss the underlying structure of the large deviations rates for certain Gaussian families and some characterizations of this structure. The referred structure is the Cameron-Martin space: given a Gaussian measure γ on the Borel sets of a separable Banach space E, the Cameron-Martin space is a Hilbert space continuously embedded in E which characterizes the measure γ. A property that identifies this space is that it consists of the directions $h \in E$ under which the translated measures $\gamma(\cdot + h)$ are equivalent to γ—this property is an essential fact behind the proof of the large deviations principle using a change of measure method. Right away, a notable example of this structure is that of the triplet $(\mathscr{H}_\gamma, E, \gamma_W)$ where: γ_W is the Wiener measure (law of the Wiener process) on $E = C_0([0, 1])$, the space of continuous paths on $[0, 1]$ starting at 0, and the Cameron-Martin space is $\mathscr{H}_\gamma = H_0^1([0, 1]) = \{f : [0, 1] \to \mathbb{R}, f(\cdot) = \int_0^{\cdot} g(s)ds, g \in L_2(0, 1)\}$ (see [7] and for a more detailed discussion of the properties of this triplet see [16, 20]). Various aspects of Gaussian measures, the Cameron-Martin space and connections of these objects with Gaussian processes can be found, e.g., in [3, 6, 17] and references therein.

The characterization of the large deviations rates in a Gaussian setting through the Cameron-Martin space is a classical result (see, e.g., [2, Chap. 2] or [9, 18]). Here we survey three classical examples appearing in the literature and the relation between one another. We further illustrate the main ideas of the method of proof of each large deviations principle in those examples by directly verifying the Freidlin-Wentzell upper and lower estimates for εX, as $\varepsilon \to 0$, where X is a $\mathbb{R}^d$ Gaussian vector.

In the specific Gaussian setting discussed in this survey, the problem of proving and representing large deviations rates is satisfactorily solved once we know how to correctly represent the Cameron-Martin space (this is not a problem of large deviation type anymore): for Gaussian measures with non-degenerate covariance operator we may look for the series representation (and we may need to solve Fredholm equations); for Gaussian measures induced by Gaussian processes on $C([0, T])$, $T < \infty$, we look at ways to characterize the reproducing kernel Hilbert spaces associated with the covariance function $q(s, t)$. The problem of representing the large deviations rates is solved in abstract, but, in each case, explicitation may be non trivial.

The remaining is organized as follows: in Sect. 2 we discuss Varadhan's large deviations principle and an equivalent formulation (the Freidlin-Wentzell estimates), Gaussian measures and the Cameron-Martin space, how to relate three large deviations rates appearing in the literature, furthermore we discuss the issue of describing the Cameron-Martin space through series representations and the relation to reproducing kernel Hilbert spaces; in Sect. 3 we discuss the Freidlin-Wentzell estimates, prove these to hold for the laws of the family εX, as $\varepsilon \to 0$, where X is a $\mathbb{R}^d$ Gaussian vector, and we also explicitly identify the associated Cameron-Martin space and prove it satisfies the change of measure property mentioned above.

2 Three Examples

We discuss the large deviations principle, Shilder's theorem and two other generalizations.

2.1 Large Deviations Principle

Let E be a complete, separable, metric space and $(\mathbb{P}_\varepsilon)_{\varepsilon\geq 0}$ a family of probability measures on the Borel subsets $\mathscr{B}(E)$ of E. Also, consider a lower semi-continuous function $I : E \to [0, +\infty]$ and $r(\varepsilon)$ be a positive real valued function converging to $+\infty$, as $\varepsilon \to 0$.

Definition 1 We say $(\mathbb{P}_\varepsilon)_{\varepsilon\geq 0}$ satisfies a large deviations principle (LDP) on E with rate function I (and scaling or speed $r(\varepsilon)$) if

(i) (upper bound) for each closed set $C \subset E$,

$$\limsup_{\varepsilon\to 0} \frac{1}{r(\varepsilon)} \log \mathbb{P}_\varepsilon(C) \leq -\inf_{x\in C} I(x),$$

(ii) (lower bound) for each open set $U \subset E$,

$$\liminf_{\varepsilon\to 0} \frac{1}{r(\varepsilon)} \log \mathbb{P}_\varepsilon(U) \geq -\inf_{x\in U} I(x).$$

We say I is a good rate function on E if

(0) for each $c < \infty$, $\{x \in E;\ I(x) \leq c\}$ is a compact set of E.

This formulation can be traced back to the works of M. Donsker and S.R.S. Varadhan in [12, 13, 21].

The upper and lower bounds introduced express an asymptotic exponential decay of the probability of such events, as $\varepsilon \to 0$. Here $I(x)$ should be seen as expressing a local (in a small ball around x) asymptotic decay rate, while $\inf_A I$ is the global decay rate for A. Let us make these ideas a little more definite: suppose the set S is such that $\inf_{S^\circ} I = \inf_{\overline{S}} I$, where S° is the interior of S and $\overline{S}$ is the closure of S, then

$$\lim_{\varepsilon\to 0} \frac{1}{r(\varepsilon)} \log \mathbb{P}_\varepsilon(S) = -\inf_{x\in S} I(x). \tag{1}$$

In an informal notation,

$$P_\varepsilon(S) \cong e^{-r(\varepsilon)\inf_{x\in S} I(x)}, \text{ as } \varepsilon \to 0. \tag{2}$$

In heuristic and incomplete terms, to interpret (2), we may think of $I(x)$ as the cost of moving, say some point, from its typical position $\overline{x}$ to a position at x. As $r(\varepsilon) \to \infty$ the event S becomes unlikely, unless its typical value is contained in S. In fact, the event S is unlikely with an exponentially decaying rate $\inf_{x\in S} I(x)$. On the other hand, this decaying rate is the optimal one, that is, the one which costs the less. This explanation is, of course, a simplification: there could be more than one typical position.

The following quote from [8] may be appropriate here:

Any large deviation occurs in the least unlikely of all unlikely ways.

It can be seen that, since I is lower semi continuous, the $\inf_S I$ is attained on every non-empty compact S (or every non-empty closed set if I is a good rate).

For convenience, let us introduce yet another formulation of the same principle that can be traced back to the works of M. Freidlin and A. Wentzell in the context of the study of small random perturbations of dynamical systems (see [15]). It is an equivalent formulation that will be used in Sect. 3.

Definition 2 We say that $I(x)$ is a (normalized) action function for $(\mathbb{P}_\varepsilon)_{\varepsilon\geq 0}$ if

(I) (upper bound) for any $\delta > 0$, any $\gamma > 0$ and any $s > 0$ there exists an $\varepsilon_0 > 0$ such that

$$\mathbb{P}_\varepsilon\{y;\ \rho(y, K(s)) \geq \delta)\} \leq \exp\{-r(\varepsilon)(s-\gamma)\}$$

for all $\varepsilon \leq \varepsilon_0$;

(II) (lower bound) for any $\delta > 0$, any $\gamma > 0$ and any $x \in X$ there exists an $\varepsilon_0 > 0$ such that

$$\mathbb{P}_\varepsilon\{y;\ \rho(x, y) < \delta\} \geq \exp\{-r(\varepsilon)(I(x)+\gamma)\}$$

for all $\varepsilon \leq \varepsilon_0$;

(0) for every $s < \infty$, $K(s) := \{x;\ I(x) \leq s\}$ is compact set of E.

Here $\rho(y, K(s))$ denotes the distance of y to the set $K(s)$, that is $\rho(y, K(s)) = \inf_{x\in K(s)} \rho(y, x)$. The normalized action function in this formulation is the rate function in previous Definition 1. Moreover, the lower bound conditions are equivalent (II) $\Leftrightarrow$ (ii), which elucidates the local nature of the conditions. The first upper bound condition is stronger, (i) $\Rightarrow$ (I). Conversely, if we have a good rate function, (I) $\Rightarrow$ (i) (see [15, Theorem 3.3]).

2.2 Gaussian Measures and the Cameron-Martin Space

Let $(E, \|\cdot\|)$ be a real separable Banach space. Denote its topological dual by E^* and the norm by $\|\varphi\|_{E^*} = \sup_{x\in E} |\varphi(x)|$. We will define a Gaussian measure on the Borel sets of E and then discuss its Cameron-Martin space.

2.2.1 Gaussian Measures

A Gaussian measure on a finite dimensional space is characterized by knowing its projections onto the one dimensional subspaces and this property is suited for generalization to infinite dimensional spaces.

Definition 3 A Borel measure γ on E is said a Gaussian measure if $\gamma \circ \varphi^{-1}$ is a Gaussian probability measure on $\mathbb{R}$, for every $\varphi \in E^*$. We say γ is centered if $\gamma \circ \varphi^{-1}$ has zero mean for any $\varphi \in E^*$.

We will only consider real vector spaces and centered Gaussian measures.

Remark 1 If γ, ν are two Borel probability measures on E, then $\gamma = \nu$ if, and only if, $\gamma \circ \varphi^{-1} = \nu \circ \varphi^{-1}$, for all $\varphi \in E^*$ (see [17, Proposition 3.6]).

2.2.2 Cameron-Martin Space

To each Gaussian measure γ on E, we can associate a Hilbert space $\mathscr{H}_\gamma$ continuously embedded in E which identifies γ—the Cameron-Martin space. This space has a characteristic geometric property: it consists of the directions of E along which we are allowed to translate γ and still obtain an equivalent measure, i.e., a measure with exactly the same null sets. The construction of $\mathscr{H}_\gamma$ is summarized next and for details we direct the reader to [17, Sect. 3.2] which we follow closely. We will revisit this construction for the simpler case of $E = \mathbb{R}^d$ in Sect. 3.

Definition 4 Let $C_\gamma : E^* \times E^* \to \mathbb{R}$ by

$$C_\gamma(\varphi, \psi) = \int_E \varphi(x)\psi(x)\gamma(dx).$$

We say C_γ is the covariance of γ. The Cameron-Martin space $\mathscr{H}_\gamma$ of γ is the completion of the subspace of E defined by

$$\mathring{\mathscr{H}}_\gamma = \{x \in E;\ \exists\, x^* \in E^*,\ C_\gamma(x^*, \psi) = \psi(x)\ \forall \psi \in E^*\}$$

under the norm $|x|_\gamma^2 = C_\gamma(x^*, x^*)$.

We make the following comments.

(a) C_γ takes values in $\mathbb{R}$, since

$$|C_\gamma(\varphi, \psi)| \leq \|\varphi\|_{E^*}\|\psi\|_{E^*}\int_E \|x\|^2\gamma(dx) < \infty.$$

Observe that $\int_E \|x\|^2\gamma(dx) < \infty$. In fact, Fernique's theorem implies more strongly that there exists $\lambda > 0$ such that $\int_E e^{\lambda\|x\|^2}\gamma(dx) < \infty$.

(b) $\mathscr{H}_\gamma$ is a Hilbert space under the inner product $(x, y)_\gamma = C_\gamma(x^*, y^*)$, and, even though, the correspondence $x \mapsto x^*$ may not be one to one, both the inner product and the norm are well defined. Note also, that this inner product is just the $L_2(E, \gamma)$ inner product.

(c) It can be shown that $\mathscr{H}_\gamma$ is identified with a subspace of E, even though, it is obtained by completion. Furthermore $\|\cdot\| \leq C|\cdot|_\gamma$, for some constant C.

(d) It can be shown that if $\dim \mathscr{H}_\gamma = \infty$, then $\gamma(\mathscr{H}_\gamma) = 0$. This shows that in infinite dimensions the space $\mathscr{H}_\gamma$ is, in a measure sense, negligible.

The geometric property alluded to at the beginning of this section is

Theorem 1 (Cameron-Martin) *Let $h \in E$, then*

$$\gamma(\cdot - h) \text{ is equivalent to } \gamma \text{ if and only if } h \in \mathscr{H}_\gamma.$$

If $h \in \mathscr{H}_\gamma$, the Radon-Nikodym derivative is given by

$$\frac{d\gamma(\cdot - h)}{d\gamma}(\varphi) = e^{(\varphi,h)_\gamma - \frac{1}{2}|h|_\gamma^2}.$$

If $h \notin \mathscr{H}_\gamma$, then $\gamma(\cdot - h)$ and γ are mutually singular.

For a proof, except the statement about singularity, see [17, Proposition 3.41, p. 19]. For a complete proof (even in a more general setting than Banach spaces) see [6, Corollary 2.4.3, Theorem 2.4.5].

2.2.3 The Case of Hilbert Spaces

Now let $E = H$ be a separable Hilbert space with inner product $(\cdot, \cdot)_H$. This yields some specializations due to the identification $H^* \equiv H$ (Riesz representation).

Theorem 2

1. If γ is a centered Gaussian measure, then there exists a positive, symmetric, trace class, bounded operator Q such that:

$$C_\gamma(u, v) = (Qu, v)_H, \; u, \; v \in H. \tag{3}$$

Here Q is called the covariance operator of γ. Furthermore, the trace is given by

$$tr(Q) = \int_H |x|_H^2 \gamma(dx).$$

2. *Reciprocally, given Q bounded, positive, symmetric, trace class operator in H, there is a centered Gaussian measure γ such that (3) holds.*

See a proof in [9, Sect. 3.3.2]. This result allows us to represent properties of a Gaussian measure in terms of its covariance operator Q. Note in particular that $\mathscr{H}_\gamma$ is the completion of

$$\mathring{\mathscr{H}}_\gamma = \{u \in H;\ \exists u^* \in H,\ (Qu^*, v)_H = (u, v)_H,\ \forall v \in H\}, \text{ i.e., the range of } Q,$$

under the inner product

$$(Qu, Qv)_\gamma = (Qu, v)_H. \tag{4}$$

We denote this range by QH and we denote the completion of QH under $(\cdot,\cdot)_\gamma$ by $QH_{(\cdot,\cdot)_\gamma}$. Equivalently,

Proposition 1 *Let Q be the covariance operator of Gaussian measure γ on a separable Hilbert space H, then*

1. $\mathscr{H}_\gamma = QH_{(\cdot,\cdot)_\gamma}$,
2. $(u, v)_\gamma = (u, Q^{-1}v)_H,\ u,\ v \in QH$.

Note that we are not assuming Q is injective: by $Q^{-1}v$ we mean any element v^* such that $Qv^* = v$. It turns out that $(\cdot,\cdot)_\gamma$ is a well defined inner product (as stated in the more general case of Banach spaces E, in (b)). In fact, to see it is well defined: let $Qv^* = v = Q\tilde{v}$, $Qu^* = u$ then by symmetry of Q

$$(u, v^*)_H = (Qu^*, v^*)_H = (u^*, Qv^*)_H = (u^*, Q\tilde{v})_H = (u, \tilde{v})_H.$$

Similar arguments show that $(\cdot,\cdot)_\gamma$ is definite positive symmetric bilinear form. Alternatively, we can use the notion of generalized inverse as proposed in the sense of [15, p. 93].

An alternative formulation is

Proposition 2 *Let Q be the covariance operator of Gaussian measure γ on a separable Hilbert space H, then*

1. $\mathscr{H}_\gamma = Q^{1/2}H_{((\cdot,\cdot))_\gamma}$,
2. $\|u\|_\gamma = |Q^{-1/2}u|_H,\ u \in Q^{1/2}H$.

Proposition 2 follows from Proposition 1 and the symmetry of Q: we are stating that the Cameron-Martin space $\mathscr{H}_\gamma$ associated with the Gaussian measure γ can be represented as $Q^{1/2}H_{((\cdot,\cdot))_\gamma}$—the completion of $Q^{1/2}H$ under the inner product

$$((u, v))_\gamma = (Q^{-1/2}u, Q^{-1/2}v)_H, \quad u, v \in Q^{1/2}H. \tag{5}$$

In fact, $(\cdot,\cdot)_\gamma$ and $|\cdot|_\gamma$ are the restrictions to QH of $((\cdot,\cdot))_\gamma$ and $\|\cdot\|_\gamma$, respectively.

When Q is injective, it turns out that $Q^{1/2}H$ is dense in H and this fact can be used to show $\mathcal{H}_\gamma = Q^{1/2}H$.

When $H = \mathbb{R}^d$, $\mathcal{H}_\gamma = Q^{1/2}H$ for any Q (see Sect. 3).

2.3 Shilder's Theorem and Generalizations

Next we discuss three large deviations (good) rates, all with speed $r(\varepsilon) = \varepsilon^{-2}$.

Example 1 (*Shilder*) Given a standard Wiener process $(W(t); t \in [0, 1])$, the distribution laws $\mathbb{P}_\varepsilon = \mathcal{L}(\varepsilon W)$ satisfy a LDP on $L_2(0, 1)$, as $\varepsilon \to 0$, with rate function

$$I(\varphi) = \begin{cases} \frac{1}{2}\int_0^1 |\dot{\varphi}(t)|^2 dt, \ \varphi \in H_0^1([0, 1]) \\ \infty, \text{otherwise}, \end{cases}$$

where $H_0^1([0, 1]) = \{\varphi;\ \varphi(t) = \int_0^t g(s)ds, g \in L_2(0, 1)\}$. See [15, Chap. 1, Sect. 2] or [14, Theorem 5.2.3]. See also [19].

Example 2 More generally, consider a real valued, square integrable, zero-mean Gaussian process $X = (X(t); t \in [0, 1])$. The laws $\mathcal{L}(\varepsilon X)$, $\varepsilon > 0$, satisfy a LDP on $(L_2([0, 1]), |\cdot|_2)$, as $\varepsilon \to 0$, with rate function

$$I(\varphi) = \begin{cases} \frac{1}{2}|Q^{-1/2}\varphi(t)|_2^2, \ \varphi \in Q^{1/2}L_2([0, 1]) \\ \infty, \ \text{otherwise}. \end{cases}$$

Here $q(s, t) = \mathbb{E}X(s)X(t)$, Q is the integral operator $f(\cdot) \mapsto \int_S f(s)q(s, \cdot)ds$ and $Q^{1/2}$ is its square root. In the case when Q is not injective, the $Q^{-1/2}$ is a generalized inverse. See [15, Chap. 3, Sect. 4].

Example 3 Lastly, even more generally, given a Gaussian measure γ on a separable Banach space E, the family $\gamma(\frac{\cdot}{\varepsilon})$, $\varepsilon \geq 0$, satisfies a LDP on E, as $\varepsilon \to 0$, with rate function

$$I(x) = \begin{cases} \frac{1}{2}|x|_\gamma^2, x \in \mathcal{H}_\gamma \\ \infty, \text{otherwise}, \end{cases}$$

where $(\mathcal{H}_\gamma, (\cdot,\cdot)_\gamma)$ is a certain Hilbert space continuously embedded in E. See [9, Theorem 12.7], [18], [11, Chap. III, Sect. 3.4].

From inspection, we naturally arrive at the following

Claim 1 *The three examples above have a correspondence in the following sense:*

1. *Let* $X = (X(t); t \in [0, 1])$ *be a zero-mean Gaussian stochastic process such that* $\mathbb{E} \int_0^1 X^2(t)dt < \infty$ *and let* $\gamma = \mathscr{L}(X)$ *on* $E = L_2(0, 1)$. *Then the rate function for the laws* $\gamma(\frac{\cdot}{\varepsilon}) = \mathscr{L}(\varepsilon X)$ *obtained in Example 2 is the rate function obtained from Example 3.*
2. *In particular, if* γ *is the Wiener measure on* $L_2(0, 1)$, *then the rate function for the laws* $\gamma(\frac{\cdot}{\varepsilon}) = \mathscr{L}(\varepsilon W)$ *obtained in Example 2 is the rate function obtained in Example 1.*

The second statement in display can be verified by looking at the series representations of both spaces (see, e.g., [6, Example 2.3.15]). We will say more about this type of representation in Sect. 2.3.1 below.

The first statement in display follows from Proposition 2. More specifically, it follows from verifying that, in the given conditions, the covariance Q_X of $X = (X(t) : t \in [0, 1])$ as a stochastic process coincides with the covariance Q of X as a $L_2(0, 1)$-valued random element as follows: denote $(\cdot, \cdot)_2$ the $L_2(0, 1)$ inner product,

$$
\begin{aligned}
(Q_X f, g)_2 &:= \int_0^1 \left(\int_0^1 q(s, t) f(s) ds \right) g(t) dt \\
&= \mathbb{E} \int_0^1 f(s) X(s) ds \int_0^1 g(t) X(t) dt \\
&= \int_{L_2(0,1)} (f, x)_2 (g, x)_2 \gamma(dx) \\
&= (Qf, g)_2
\end{aligned}
$$

for all $f, g \in L_2(0, 1)$. Hence, $Q_X = Q$ and Examples 2 and 3 coincide.

The next two sections survey two descriptions of the Cameron-Martin space and point directions on how to represent it.

2.3.1 The Cameron-Martin Space via Series Representation

Suppose a Gaussian process X is given in the conditions of the second example above. In particular the associated covariance operator Q is a positive, symmetric, trace class operator and the covariance function $q(s, t)$ is a positive definite kernel (meaning that the matrices $(q(s_i, s_j))_{i,j}$ are positive semi-definite, symmetric).

If the covariance operator Q is injective the following series representation is available

$$
H_\gamma = \{h = \sum_k h_k e_k \in H;\ \sum_k \frac{|h_k|^2}{\lambda_k} < \infty\},\quad |h|_\gamma^2 = \sum_k \frac{|h_k|^2}{\lambda_k},\quad h \in H_\gamma,
$$

where (e_k) is a complete orthonormal system of eigenvectors of Q, with corresponding eigenvalues $\lambda_k > 0$. In turn, the diagonalization of Q amounts to solving equations of the type

$$Qf = \lambda f \Leftrightarrow (Q - \lambda I)f = 0, f \neq 0$$

with Q positive, self-adoint and trace class, or equivalently

$$\int_0^T q(s,t)f(s)ds = \lambda f(t), \quad f \neq 0, \tag{6}$$

with $q(s,t)$ positive definite kernel such that $\int_S q(s,s)ds < \infty$. Equations of type (6) are said Fredholm integral equations of the second kind. Direct methods such as differentiation and algebraic manipulation of both sides of the equation are of use in the case of the Wiener process ([6, Example 2.3.15]). For more general cases like the Ornstein-Uhlenbeck process, those methods are already less useful.

In [5] a partial result in this direction was shown: a differential equation for the eigenvalues and the eigenvectors of the associated covariance operator of the Ornstein-Uhlenbeck.

Proposition 3 *Let $a \neq 0$, $c > 0$, $0 < T < \infty$ and let $X = (X(t); t \in [0,T])$ solve*

$$dX(t) = -aX(t)dt + \sqrt{c}dW(t), \ t \in [0,T], \quad X(0) = 0.$$

Let Q_X be the covariance associated with X. If $Q_X f = \lambda f, f \neq 0$, then

$$f'' + (\frac{c}{\lambda} - a^2)f = 0,$$

$$f(0) = 0, \quad f'(T) = -af(T).$$

2.3.2 The Cameron-Martin Space via RKHS

When we are dealing with the Gaussian law of a certain stochastic process over a set S, we may use the notion of reproducing kernel Hilbert space (RKHS) to describe the Cameron-Martin space. Indeed, associated with the covariance function $q(s,t)$ there exists a unique RKHS $\mathcal{H}_q$ of functions on S. The space $\mathcal{H}_q$ is obtained canonically through the Moore-Aronszajn construction (see [1]): it is the completion of the finite linear combinations of $\{q(t,\cdot);\ t \in S\}$ under the inner product $(\cdot,\cdot)_q$ determined by requiring the reproducing property $(q(t,\cdot), q(s,\cdot))_q = q(t,s)$. It turns out that the RKHS is very often isometrically isomorphic to the Cameron-Martin space, so that in the literature various authors use the terms fairly indistinguishably. Furthermore, the very term *Cameron-Martin space* often refers to the specific space associated with the Wiener measure, first considered in the article by Cameron and Martin [7]. To avoid confusion, we will adopt the terminology that: a RKHS comes associated with a positive definite kernel and the Cameron-Martin space comes associated with a Gaussian law. As just mentioned, these notions are equivalent under mild assumptions. Sufficient conditions for equivalence have been established in [22, Theorem

2.1]: if the process has paths in a separable Banach subspace of the bounded functions equipped with the uniform norm, then $\mathscr{H}_\gamma = \mathscr{H}_q$.

3 A Finite Dimensional Tour

We now consider the case when γ is the distribution law of a standard Gaussian vector $X \in E = \mathbb{R}^d$: first, we present the associated Cameron-Martin space—which in this case is the support of γ—and prove that Theorem 1 holds; next, we provide a proof of the LDP for the laws of εX, as $\varepsilon \to 0$.

Let $X = (X_1, \dots, X_d)$ be a zero mean Gaussian vector on a probability space $(\Omega, \mathscr{F}, \mathbb{P})$ with values in the Euclidean space $(\mathbb{R}^d, |\cdot|_{\mathbb{R}^d})$. Also, denote by γ the distribution law of X on $\mathbb{R}^d$ and denote by Q its covariance matrix:

$$Q = \operatorname{cov}(X) = (\mathbb{E}X_iX_j)_{i,j=1}^d.$$

The covariance matrix Q is symmetric and positive ($x^tQx \geq 0$), so there exists a positive symmetric square root matrix $Q^{1/2}$. It is straightforward to verify that

$$X = Q^{1/2}\xi,$$

where $\xi \in R^d$ is a standard Gaussian, hence X has measure support $Q^{1/2}\mathbb{R}^d$.

3.1 The Cameron-Martin Space in Finite Dimensions

We now find that the Cameron-Martin space associated with the Gaussian distribution γ is $Q^{1/2}\mathbb{R}^d$. Firstly, we observe the following facts: $Q^{1/2}\mathbb{R}^d = Q\mathbb{R}^d$ and the previously defined inner products $(\cdot, \cdot)_\gamma$ in (4) and $((\cdot, \cdot))_\gamma$ in (5) coincide. Indeed, $Q\mathbb{R}^d$ is a subspace of $Q^{1/2}\mathbb{R}^d$ and we have the linear isometry

$$Q^{1/2} : \left(Q^{1/2}\mathbb{R}^d, |\cdot|_{\mathbb{R}^d}\right) \to \left(Q\mathbb{R}^d, |\cdot|_\gamma\right), \quad |Q^{1/2}Q^{1/2}u|_\gamma = |Q^{1/2}u|_{\mathbb{R}^d}.$$

Proposition 4 *The space $\mathscr{H}_\gamma$ is the set*

$$Q^{1/2}\mathbb{R}^d \quad \textit{with inner product} \quad (u, v)_\gamma := (Q^{-1/2}u, Q^{-1/2}v)_{\mathbb{R}^d}.$$

Proof From Proposition 2 and the facts stated above it is sufficient to show that

$$Q^{1/2}\mathbb{R}^d = Q^{1/2}\mathbb{R}^d_{(\cdot,\cdot)_\gamma}.$$

Observe first that we have the continuous inclusion

$$(Q^{1/2}\mathbb{R}^d, (\cdot,\cdot)_\gamma) \hookrightarrow (\mathbb{R}^d, |\cdot|_{\mathbb{R}^d}),$$

i.e., on $Q^{1/2}\mathbb{R}^d$ we have $|\cdot|_{\mathbb{R}^d} \leq C|\cdot|_\gamma$, for some constant C: in fact for all $u \in \mathbb{R}^d$,

$$|Q^{1/2}u|_{\mathbb{R}^d} \leq C|Q^{1/2}u|_\gamma,$$

where C is the operator norm of the matrix $Q^{1/2}$. Now, take a Cauchy sequence $u_n \in Q^{1/2}\mathbb{R}^d$ for the norm $|\cdot|_\gamma$. By the continuous inclusion, it is also a Cauchy sequence for the Euclidean norm $|\cdot|_{\mathbb{R}^d}$. Since $Q^{1/2}\mathbb{R}^d$ is closed for $|\cdot|_{\mathbb{R}^d}$, u_n converges in $Q^{1/2}\mathbb{R}^d$.

We now verify that $\mathscr{H}_\gamma$ indeed characterizes the vectors h for which X and $X+h$ have equivalent distributions, that is, γ and $\gamma(\cdot - h)$ have exactly the same zero probability events, i.e., we sketch the proof of Theorem 1, when $E = \mathbb{R}^d$ and γ is the distribution law of X.

We think it is instructive to pay attention to the case where X is non degenerate first.

3.1.1 The Non-degenerate Case

If Q is non-degenerate, that is, the kernel $\mathrm{N}(Q)$ is $\{0\}$ (or $x^tQx > 0,\ x \neq 0$), then $\mathscr{H}_\gamma = \mathbb{R}^d$ and $(\cdot,\cdot)_\gamma$ is just the inner product determined by the matrix Q^{-1} on R^d. This inner product may be written in a number of equivalent ways:

$$(x,y)_\gamma = (x, Q^{-1}y)_{\mathbb{R}^d} = (Q^{-1/2}x, Q^{-1/2}y)_{\mathbb{R}^d} = \sum_{i,j=1}^{d} x_i y_j q^{i,j}, \quad Q^{-1} = (q^{i,j})_{i,j=1}^d.$$

We visualize the geometry induced by $(\cdot,\cdot)_\gamma$ by identifying the unit ball under this metric: it is the set $\{x \in \mathbb{R}^d;\ x^tQ^{-1}x \leq 1\}$, an ellipsoidal level curve of the quadratic form $x \mapsto x^tQ^{-1}x$ (see [6, Chap. 1]).

Furthermore, the distribution γ has a density given by $\frac{d\gamma}{d\lambda}(x) = \exp(-\frac{1}{2}x^tQ^{-1}x)$ (here λ is the Lebesgue measure of $\mathbb{R}^d$). In terms of the objects just defined

$$\frac{d\gamma}{d\lambda}(x) = \exp(-\frac{1}{2}|x|_\gamma^2).$$

For any $h \in \mathbb{R}^d$, $X + h$ is a Gaussian vector and its law ν is related to the law γ: $\nu(A) = \gamma(A - h)$, that is, $\nu = \gamma(\cdot - h)$. The density of $\gamma(\cdot - h)$ is readily seen to be

$$\frac{d\gamma(\cdot - h)}{d\lambda}(x) = \exp(-\frac{1}{2}|x - h|_\gamma^2).$$

Also, $\gamma(\cdot - h)$ and γ (and also λ) are equivalent measures and the Radon-Nykodim derivative is

$$\begin{aligned}
\frac{d\gamma(\cdot - h)}{d\gamma}(x) &= \frac{d\gamma(\cdot - h)}{d\lambda}(x)\frac{d\lambda}{d\gamma}(x) \\
&= \exp\left(-\frac{1}{2}|x - h|_\gamma^2 + \frac{1}{2}|x|_\gamma^2\right) \\
&= \exp\left(-\frac{1}{2}|x|_\gamma^2 + x^t Qh - \frac{1}{2}|h|_\gamma^2 + \frac{1}{2}|x|_\gamma^2\right) \\
&= \exp((x, h)_\gamma - \frac{1}{2}|h|_\gamma^2).
\end{aligned}$$

3.1.2 The Degenerate Case

If Q does not have inverse (we say Q is degenerate), some structure is lost and the argument is not so direct. Contrary to the non-degenerate case, $\mathscr{H}_\gamma$ and $\mathbb{R}^d$ can now be very different from one another (see Example 4 and Fig. 1 at the end). Let r be the dimension of $\mathscr{H}_\gamma$. Here γ is supported on $\mathscr{H}_\gamma$ hence it is singular to λ and there is no density. Also, the shifted measure $\gamma(\cdot - h)$ has support on $\mathscr{H}_\gamma + h$.

If $h \notin \mathscr{H}_\gamma$, the supports are disjoint and γ and $\gamma(\cdot - h)$ are singular.

If $h \in \mathscr{H}_\gamma$, then $\mathscr{H}_\gamma - h = \mathscr{H}_\gamma$ and we will use the following result, which essentially says that γ can be seen as a standard Gaussian multivariate distribution on $\mathbb{R}^r \equiv \mathscr{H}_\gamma$, with the inner product $(\cdot, \cdot)_\gamma$.

Theorem 3 *For all bounded measurable $f : \mathbb{R}^d \to \mathbb{R}$*

$$\int_{\mathbb{R}^d} f(x) d\gamma(x) = \frac{1}{(2\pi)^{r/2}} \int_{\mathscr{H}_\gamma} f(y) \exp\left(-\frac{1}{2}|y|_\gamma^2\right) dy.$$

See [10, Theorem 4.3].

Continuing, we conclude, by a change of variables, that for all bounded measurable f

$$\begin{aligned}
\int_{\mathbb{R}^d} f(x) d\gamma(x - h) &= \int_{R^d} f(x + h) d\gamma(x) \\
&= \frac{1}{(2\pi)^{r/2}} \int_{\mathscr{H}_\gamma} f(x + h) \exp\left(-\frac{1}{2}|x|_\gamma^2\right) dx
\end{aligned}$$

$$= = \frac{1}{(2\pi)^{r/2}} \int_{\mathscr{H}_\gamma - h} f(x) \exp\left(-\frac{1}{2}|x-h|_\gamma^2\right) dx$$

$$= \frac{1}{(2\pi)^{r/2}} \int_{\mathscr{H}_\gamma} f(x) \exp\left(-\frac{1}{2}|x|^2 + (x,h)_\gamma - \frac{1}{2}|h|_\gamma^2\right) dx$$

$$= \int_{\mathscr{H}_\gamma} f(x) \exp\left((x,h)_\gamma - \frac{1}{2}|h|_\gamma^2\right) d\gamma(x).$$

Conclusion, $\gamma(\cdot - h)$ and γ are equivalent if and only if $h \in \mathscr{H}_\gamma$.
If $h \in \mathscr{H}_\gamma$ the Radon-Nykodim derivative is

$$\frac{d\gamma(\cdot - h)}{d\gamma}(x) = \exp((x,h)_\gamma - \frac{1}{2}|h|_\gamma^2).$$

If $h \notin \mathscr{H}_\gamma^d$, then $\gamma(\cdot - h)$ and γ are singular.

Summarizing, we showed Theorem 1, when $E = \mathbb{R}^d$ and γ is the distribution law of X.

3.2 *The LDP in Finite Dimensions*

There are various ways to show that the family εX, $\varepsilon \geq 0$, satisfies a LDP as $\varepsilon \to 0$. Here we illustrate on a simple finite dimensional setting a method that has been generalized for more general Gaussian elements. We state right away the result

Theorem 4 (Little Shilder) *The family of Gaussian vectors εX, $\varepsilon > 0$ satisfies a LDP on $\mathbb{R}^d$ with rate function*

$$I(x) = \begin{cases} \frac{1}{2}|Q^{-1/2}x|_{\mathbb{R}^d}^2, & x \in Q^{1/2}\mathbb{R}^d \\ \infty, & \text{otherwise.} \end{cases} \tag{7}$$

From what was shown in the beginning of this section, the rate function from Theorem 4 can be written as

$$I(x) = \begin{cases} \frac{1}{2}|x|_\gamma^2, & x \in \mathscr{H}_\gamma \\ \infty, & \text{otherwise.} \end{cases} \tag{8}$$

Proof (*of Theorem 4*) We will show that (8) is the (normalized) action function of εX (see Definition 2). For the lower bound, let $\delta > 0$, $h \in \mathscr{H}_\gamma$, then

$$
\begin{aligned}
\mathbb{P}\left(|\varepsilon X - h|_{\mathbb{R}^d} < \delta\right) &= \mathbb{P}\left(|X - \frac{h}{\varepsilon}|_{\mathbb{R}^d} < \frac{\delta}{\varepsilon}\right) \\
&= \gamma\left(B\left(0, \frac{\delta}{\varepsilon}\right) - \left(-\frac{h}{\varepsilon}\right)\right) \\
&= \int_{B\left(0,\frac{\delta}{\varepsilon}\right)} \exp\left(-\varepsilon^{-1}(x,h)_\gamma - \varepsilon^{-2} I(h)\right) \gamma(dx) \\
&= \exp\left(-\varepsilon^{-2} I(h)\right) \int_{B\left(0,\frac{\delta}{\varepsilon}\right)} \exp\left(-\varepsilon^{-1}(x,h)_\gamma\right) \gamma(dx),
\end{aligned}
$$

where in the passage to the integral we used Theorem 1. Now, by Chebyshev's inequality

$$
\gamma\left(B\left(0, \frac{\delta}{\varepsilon}\right)\right) = \mathbb{P}\left(|X|_{\mathbb{R}^d} < \frac{\delta}{\varepsilon}\right) \geq 1 - \frac{\varepsilon^2}{\delta^2}\mathbb{E}|X|^2_{\mathbb{R}^d} \geq 3/4, \tag{9}
$$

for sufficiently small ε. Since $X = Q^{1/2}\xi$, with ξ being a d-dimensional standard Gaussian, it is straightforward to verify that

$$
\mathbb{E}|(X,h)_\gamma|^2 = |h|^2_\gamma = 2I(h)
$$

so that

$$
\begin{aligned}
\mathbb{P}\left(-\varepsilon^{-1}(X,h)_\gamma \leq C\right) &= \mathbb{P}\left(|\varepsilon^{-1}(X,h)_\gamma| \geq C\right) \\
&\leq \frac{\varepsilon^{-2}\mathbb{E}|(X,h)^2_\gamma|}{C^2} = 1/4
\end{aligned}
$$

for $C = 2\sqrt{2}\varepsilon^{-1}\sqrt{I(h)}$. Therefore,

$$
\mathbb{P}\left(\exp(-\varepsilon^{-1}(X,h)_\gamma) \geq \exp(2\sqrt{2}\varepsilon^{-1}\sqrt{I(h)})\right) \geq 3/4. \tag{10}
$$

From estimates (9) and (10) it follows that

$$
\int_{B\left(0,\frac{\delta}{\varepsilon}\right)} \exp\left(-\varepsilon^{-1}(x,h)_\gamma\right) \gamma(dx) > 1/2 \exp(-2\sqrt{2}\varepsilon^{-1}\sqrt{I(h)}),
$$

consequently

$$
\mathbb{P}\left(|\varepsilon X - h|_{\mathbb{R}^d} < \delta\right) > 1/2 \exp(-\varepsilon^{-2} I(h) - 2\sqrt{2}\varepsilon^{-1}\sqrt{I(h)}),
$$

which implies the wanted lower bound.

As for the upper bound estimate, let $\Phi(s) = \{x \in \mathbb{R}^d;\ I(x) = \frac{1}{2}|x|_\gamma^2 \leq s\}$ and denote $\rho(x, F) = \inf\{|x - y|_{\mathbb{R}^d}; y \in F\}$, for any F closed and $x \in \mathbb{R}^d$

$$
\begin{aligned}
\mathbb{P}(\rho(\varepsilon X, \Phi(s)) \geq \delta) \leq \mathbb{P}(\varepsilon X \notin \Phi(s)) &= \mathbb{P}(\varepsilon Q^{1/2}\xi \notin \Phi(s)) \\
&= \mathbb{P}(|\varepsilon Q^{1/2}\xi|_\gamma^2 > 2s) \\
&= \mathbb{P}(\varepsilon^2 \sum_{i=1}^{d} \xi_i^2 > 2s).
\end{aligned}
$$

Since $Y = \sum_{i=1}^d \xi_i^2$, Y is a chi-squared random variable with d degrees of freedom with moment generating function $\mathbb{E}e^{tY} = (1-2t)^{-d/2}$, $t < 1/2$. By Chernoff-Markov inequality, for all $\lambda > 0$

$$
\begin{aligned}
P(\varepsilon^2 Y > 2s) &\leq e^{-\lambda 2s} E e^{\lambda \varepsilon^2 Y} \\
&= e^{-\lambda 2s}(1 - 2\lambda\varepsilon^2)^{-d/2} \\
&\leq \inf_{\lambda > 0}\{e^{-\lambda 2s}(1 - 2\lambda\varepsilon^2)^{-d/2}\},
\end{aligned}
$$

for sufficiently small $\varepsilon > 0$. With some calculus we obtain the optimal $\lambda = \frac{1}{2\varepsilon^2} - \frac{d}{4s}$. Substituting we obtain the bound

$$
\mathbb{P}(\rho(\varepsilon X, \Phi(s)) > \delta) \leq C\varepsilon^{-d} e^{-s/\varepsilon^2},
$$

where $C = e^{d/2}(\frac{d}{2s})^{-d/2}$. This suffices to show the wanted upper bound, for sufficiently small $\varepsilon < 0$.

Remark 2 As seen the argument has two main ideas: (a) for the lower bound, a convenient change of measure, based on the Cameron-Martin theorem; (b) for the upper bound, Chernoff-Markov type of estimates. With convenient modifications, the previous result and method of proof can be generalized to various settings of a Gaussian type and in fact the previous proof is an adaptation to Gaussian vectors of the result appearing in [15, Theorem 4.2].

Example 4 Let $X \in \mathbb{R}^2$ be a Gaussian such that $\mathbb{E}X_1X_1 = 1$, and $\mathbb{E}X_iX_j = 0, i, j \neq 1$. Then $Q \equiv Q^{1/2} \equiv P_1$ is the projection onto the first coordinate and $r = \dim \mathcal{H}_\gamma = 1$. Translation along $(2, 0)$ originates an equivalent Gaussian measure but not translation along $(2, -1)$. See also Fig. 1. Moreover, the rate function associated with the LDP for εX, as $\varepsilon \to 0$, is

$$
I(x, y) = \begin{cases} x^2/2, & y = 0 \\ \infty, & \text{otherwise.} \end{cases}
$$

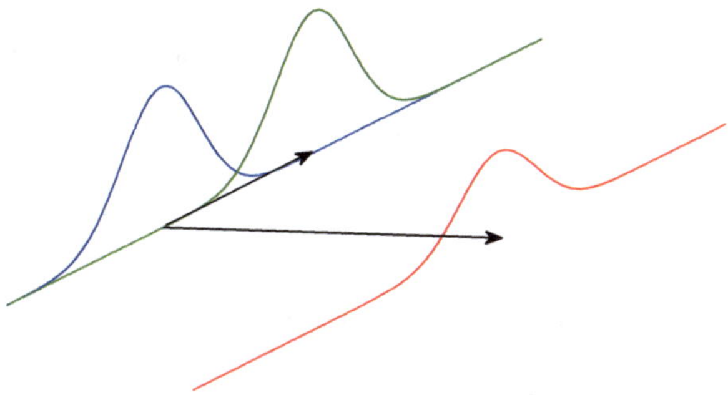

Fig. 1 Translated Gaussian measures give rise to equivalent measures if and only if the translation is done along the support of the measure

Acknowledgments The author was partially supported by FCT–Portugal fellowship SFRH/ BD/43 286/2008 and CAPES/IMPA. The author is grateful to Sergey Lototsky and Remigijus Mikulevicius for useful discussions on the topics presented in this article. The author would also like to thank the referees for their comments and improvements.

References

1. Aronszajn, N.: Theory of reproducing kernels. Trans. Amer. Math. Soc. **68**, 337–404 (1950)
2. Azencott, R.: Grandes déviations et applications. In: Lecture Notes in Mathematics, vol. 774, pp. 1–176. Springer, Berlin (1980)
3. Baxendale, P.: Gaussian measures on function spaces. Amer. J. Math. **98**(4), 891–952 (1976)
4. Balram, R.S., Cambanis, S.: Gaussian processes and gaussian measures. Ann. Math. Stat. **43**, 1944–1952 (1972)
5. Bessam, D.: Large deviations rates in a Gaussian setting and related topics, pp. 126. ProQuest LLC, Ann Arbor, MI (2014)
6. Bogachev, V.I.: Gaussian measures. In: Mathematical Surveys and Monographs, vol. 62. American Mathematical Society, Providence, RI (1998)
7. Cameron, R.H., Martin, W.T.: Transformations of Wiener integrals under translations. Ann. Math. **2**(45), 386–396 (1944)
8. den Hollander, F.: Large deviations. In: Fields Institute Monographs, vol. 14. American Mathematical Society, Providence, RI (2000)
9. Da Prato, G., Zabczyk, J.: Stochastic equations in infinite dimensions. In: Encyclopedia of Mathematics and its Applications, vol. 44. Cambridge University Press, Cambridge (1992)
10. Driver., B.: Heat kernel l_2-spaces. In: Lecture Notes (2010)
11. Deuschel, J., Stroock, D.W.: Large deviations. In: Pure and Applied Mathematics, vol. 137. Academic Press Inc., Boston (1989)
12. Donsker, M.D., Varadhan, S.R.S.: Asymptotic evaluation of certain Markov process expectations for large time. I. II. Comm. Pure Appl. Math. **28**, 1–47, 279–301 (1975)
13. Donsker, M.D., Varadhan, S.R.S.: Asymptotic evaluation of certain Markov process expectations for large time. III. Commun. Pure Appl. Math. **29**(4), 389–461 (1976)
14. Dembo, A., Zeitouni, O.: Large deviations techniques and applications, 2nd edn. In: Applications of Mathematics (New York), vol. 38. Springer, New York (1998)
15. Freidlin, M.I., Wentzell, A.D.: Random perturbations of dynamical systems, 2nd edn. In: Grundlehren der Mathematischen Wissenschaften, vol. 260. Springer, New York (1998)
16. Gross, L.: Abstract Wiener spaces. In: Proceedings of Fifth Berkeley Symposium on Mathematical Statistics and Probability (Berkeley, Calif., 1965/66), vol. II. Contributions to Probability Theory, Part 1, pp. 31–42. University of California Press, Berkeley, California (1967)

17. Hairer, M.: An introduction to stochastic PDEs. ArXiv e-prints July 2009
18. Kallianpur, G., Oodaira, H.: Freĭdlin-Wentzell type estimates for abstract Wiener spaces. Sankhyā Ser. A **40**(2), 116–137 (1978)
19. Schilder, M.: Some asymptotic formulas for Wiener integrals. Trans. Amer. Math. Soc. **125**, 63–85 (1966)
20. Stroock, D.W.: Abstract Wiener space, revisited. Commun. Stoch. Anal. **2**(1), 145–151 (2008)
21. Varadhan, S.R.S.: Asymptotic probabilities and differential equations. Commun. Pure Appl. Math. **19**, 261–286 (1966)
22. van der Vaart, A.W., van Zanten, J.H.: Reproducing kernel Hilbert spaces of Gaussian priors. In: Pushing the limits of contemporary statistics: contributions in honor of Jayanta K. Ghosh, vol. 3 of Inst. Math. Stat. Collect., pp. 200–222. Inst. Math. Statist., Beachwood, OH (2008)

Kinetic Theory of Simple Reacting Spheres: An Application to Coloring Processes

Filipe Carvalho, Jacek Polewczak and Ana Jacinta Soares

Abstract We consider a simplified version of the kinetic model of simple reacting spheres (SRS) for a quaternary reactive mixture of hard-spheres in the dilute-gas limit. The model mimics a coloring process occurring with probability α_R, described by the reversible chemical law $A_1 + A_2 \rightleftharpoons A_3 + A_4$. We provide the linearized collisional operators of our model and investigate some of their mathematical properties. In particular we obtain an explicit and symmetric representation of the elastic and reactive kernels and use this to prove the compactness of the linearized collisional operator in $(L^2(\mathbb{R}^3))^4$.

Keywords Chemically reacting mixtures · Linearized Boltzmann equation · Simple reacting spheres · Compact operators

1 Introduction

In these proceedings we continue the work developed by the authors in the context of the simple reacting spheres (SRS) model in the kinetic theory of chemically reacting gases (see [1, 2]). The SRS kinetic theory has been initially proposed by Marron [3] and then developed by Xystris, Dahler and Qin in a series of papers devoted to the kinetic modeling, non-equilibrium processes, transport properties, and extension to

F. Carvalho
Centre of Mathematics, University of Minho, Braga, Portugal
e-mail: filipecarvalho@esce.ipvc.pt

F. Carvalho
Polytechnic Institute of Viana do Castelo, Viana do Castelo, Portugal

J. Polewczak
Department of Mathematics, California State University, Northridge, USA
e-mail: jacek.polewczak@csun.edu

A.J. Soares (✉)
Centre of Mathematics, University of Minho, Braga, Portugal
e-mail: ajsoares@math.uminho.pt

P. Gonçalves and A.J. Soares (eds.), *From Particle Systems to Partial Differential Equations II*, Springer Proceedings in Mathematics & Statistics 129,
DOI 10.1007/978-3-319-16637-7_4

dense-gas limit (see [4–6]). Some advances related to the mathematical analysis of the SRS model have been done by Polewczak in paper [7], proving the existence of global-in-time, spatially inhomogeneous, and L^1-renormalized solutions for the SRS system, in the case that there is no mass transfer and no alteration of molecular diameters.

In the SRS kinetic theory, the molecules behave as if they were single mass points and the elastic and reactive collisions are assumed of hard-sphere type. Moreover, reactive collisions occur when the collision is sufficiently energetic, that is when the kinetic energy of the colliding molecules exceeds the activation energy of the chemical reaction. This theory incorporates some important aspects that renders the SRS kinetic model so interesting. For example, reactive and elastic collisions are treated in equal pair, contrary to those other models that consider the reactive terms as a small perturbation of the elastic ones (see, for example, [8–10]). As a consequence, the SRS theory results to be appropriate to deal with processes considerably influenced by chemical reactions. Also, the dynamics of the reaction is fully integrated into the model, so that the detailed balance condition and the microscopic reversibility principle are built in the model. Additionally, when the chemical reactions are turned off, the SRS kinetic model reduces to the Boltzmann kinetic system of four species.

All these aspects motivate the authors to investigate some mathematical and physical problems for the SRS kinetic theory.

Continuing with the work presented in paper [2], we consider here a simplified version of the SRS model that accounts for a coloring process with probability α_R described by a reversible law of type $A_1 + A_2 \rightleftharpoons A_3 + A_4$. Since this process does not modify the mass or chemical binding energy of the reactants, the reactive terms become straightforward and the corresponding SRS kinetic system results much simpler than the one of the general model considered in [2]. This allows to obtain a more tractable linearized SRS system and a better representation of the elastic and reactive kernels. The main purpose of the present paper is to use this representation of the kernels to prove the compactness of some of the elastic and reactive linearized operators, in particular, operators $\mathscr{Q}^{(1)}$, $\mathscr{Q}^{(2)}$, $\mathscr{T}^{(2)}$, and $\mathscr{R}^{(2)}$, in the notation of Sect. 6. On the other hand, additional techniques are required for the remaining operators, $\mathscr{Q}^{(3)}$, $\mathscr{T}^{(3)}$ and $\mathscr{R}^{(3)}$, and work is in progress to complete the proof of compactness for these operators. Once the compactness of these linearized operators is achieved, we plan to provide full spectral analysis of these operators, and this will be used to obtain expressions for the transport coefficients, the asymptotic behavior of the evolution operator associated with the linearized SRS system, and ultimately hydrodynamics limits for these systems.

The paper is organized as follows. In the next section, with reference to the considered coloring process, we describe the SRS kinetic model and state its consistency properties. Section 3 is devoted to the equilibrium solutions and trend to equilibrium. In Sect. 4 we present the linearized SRS system and state its main properties. In Sect. 5 we provide the explicit representation of the kernels of the linearized elastic and reactive operators and state some symmetry properties. Finally, in Sect. 6 we prove the compactness of two elastic and one reactive linearized operators.

2 SRS Kinetic System

With reference to the simple reacting spheres (SRS) model [4–6], we consider a dilute gas mixture of four constituents, whose particles behave as if they were single mass points. Internal degrees of freedom for the gas particles, such as vibrational and rotational energies, are not taken into account. Particles undergo binary elastic collisions and reactive collisions, both of hard-sphere type. Reactive collisions obey the chemical law

$$A + B \rightleftharpoons A^* + B^*, \tag{1}$$

where A, B, A^* and B^* indicate the constituents of the mixture. We use the indices $1, 2, 3, 4$ for the constituents A, B, A^* and B^*, respectively. Furthermore, m_i, d_i and E_i denote the molecular mass, the molecular diameter and the chemical binding energy of each i-constituent. In general, the chemical reaction results in a redistribution of masses among the constituents, in such a way that the law of mass conservation holds in the form

$$m_1 + m_2 = m_3 + m_4.$$

At the same time, the chemical reaction implies a rearrangement of energies (kinetic and binding) so that a conservation law of the total energy holds. The balance of chemical binding energy of products and reactants of the chemical reaction defines the reaction heat, namely $Q_R = E_3 + E_4 - E_1 - E_2$. If Q_R is positive, the reverse chemical reaction, $A^* + B^* \rightharpoonup A + B$, is exothermic.

In our paper [2], we described the SRS kinetic system for a general bimolecular chemical reaction. In particular, we introduced in detail the mathematical aspects of the SRS system and the relevant properties of the collisional operators for what concerns the mathematical and physical consistency of the model. Our main result in that paper is the representation of the SRS system linearized around the equilibrium and the explicit representation of the kernels of the reactive operators.

In the present paper we consider a simplified version of the SRS system, in which the reaction law (1) describes a *coloring* process with a fixed probability α_R. We assume that

$$m_3 = m_1, \quad m_4 = m_2, \qquad d_3 = d_1, \quad d_4 = d_2, \qquad E_3 = E_1, \quad E_4 = E_2. \tag{2}$$

When a hard-sphere A collides with another hard-sphere B, there is a probability α_R that they change their colors in the process $A + B \rightharpoonup A^* + B^*$, without modifying their molecular masses or diameters, and without altering their chemical binding energies. The same probability rule is applied to the reverse reaction $A^* + B^* \rightharpoonup A + B$.

This dynamical model represents a simplified version of the SRS system that is appropriate to describe a coloring process.

2.1 Collisional Dynamics

An elastic collision between particles from constituents i and s, with velocities v and w, respectively, results in a change of velocities, say $(v, w) \to (v', w')$, with $i, s = 1, ..., 4$. Since the linear momentum and kinetic energy of the colliding particles are conserved, we have

$$m_i v + m_s w = m_i v' + m_s w', \tag{3a}$$

$$m_i v^2 + m_s w^2 = m_i v'^2 + m_s w'^2. \tag{3b}$$

We consider elastic cross sections of hard-spheres type, given by

$$\sigma_{is}^2 = \frac{1}{4}(d_i + d_s)^2. \tag{4}$$

Conditions (3) and (4), imply that the elastic post-collisional velocities are given by

$$v' = v - 2\frac{\mu_{is}}{m_i}\langle \varepsilon, v - w\rangle \varepsilon \qquad \text{and} \qquad w' = w + 2\frac{\mu_{is}}{m_s}\langle \varepsilon, v - w\rangle \varepsilon, \tag{5}$$

where ε is a unit vector along the line passing through the centres of the colliding particles at the moment of the impact, and μ_{is} is the reduced mass of the colliding particles,

$$\varepsilon \in \mathbb{S}_+^2, \text{ with } \mathbb{S}_+^2 = \left\{\varepsilon \in \mathbb{R}^3 : \|\varepsilon\| = 1, \langle \varepsilon, v - w\rangle > 0\right\}, \text{ and } \mu_{is} = \frac{m_i m_s}{m_i + m_s}. \tag{6}$$

On the other hand, a reactive collision in our model is a coloring event in which the participating molecules only change, besides their velocities, their color. Thus the reactive collision between particles from constituents i and j with velocities v and w, respectively, results in a transition into constituents k and l with different colors from i and j, and a consequent change of velocities to $v^{\odot}$ and $w^{\odot}$. The indices (i, j, k, l) can take the values $(1, 2, 3, 4)$, $(2, 1, 4, 3)$, $(3, 4, 1, 2)$ and $(4, 3, 2, 1)$. The conservation laws of linear momentum and total energy of the colliding particles are represented by

$$m_1 v + m_2 w = m_3 v^{\odot} + m_4 w^{\odot}, \tag{7a}$$

$$m_1 v^2 + m_2 w^2 = m_3 (v^{\odot})^2 + m_4 (w^{\odot})^2, \tag{7b}$$

where the molecular masses satisfy conditions (2). The chemical reactive cross sections associated to the coloring process are given by

$$\sigma_{12}^{*2} = \alpha_R \sigma_{12}^2 \quad \text{and} \quad \sigma_{34}^{*2} = \alpha_R \sigma_{34}^2 \tag{8}$$

for the direct and inverse reaction, respectively. The post-collisional velocities associated to the direct reaction $A + B \rightharpoonup A^* + B^*$, for the constituents k and l, are given by

$$v^{\odot} = v - 2\frac{\mu_{ij}}{m_k}\langle \varepsilon, v - w\rangle \varepsilon \qquad \text{and} \qquad w^{\odot} = w + 2\frac{\mu_{ij}}{m_l}\varepsilon \langle \varepsilon, v - w\rangle. \tag{9}$$

where $(i, j, k, l) \in \{(1, 2, 3, 4), (2, 1, 4, 3), (3, 4, 1, 2), (4, 3, 2, 1)\}$. The post-collisional velocities associated to the backward reaction $A^* + B^* \rightharpoonup A + B$, for the constituents i and j, are given by the same expressions (9), thanks to the assumptions (2) on the molecular masses.

2.2 Kinetic System

Within the kinetic theory for rarefied gases, the state of the mixture is described by the one-particle distribution functions $f_i(t, x, v)$, $i = 1, 2, 3, 4$, such that $f_i(t, x, v)$ represents the probability density of finding, at time t, a particle of the A_i-constituent occupying the position x with velocity v. Here, the time-space evolution of the functions $f_i(t, x, v)$ is given by the SRS kinetic system formulated for the coloring problem, in the form

$$\frac{\partial f_i}{\partial t} + v \cdot \nabla_x f_i = J_i^E + J_i^R, \quad i = 1, \ldots, 4, \tag{10}$$

where J_i^E is the elastic collisional operator, given by

$$\begin{aligned} J_i^E = & \sum_{s=1}^{4} \sigma_{is}^2 \int_{\mathbb{R}^3} \int_{\mathbb{S}_+^2} \Big[f_i(t, x, v') f_s(t, x, w') - f_i(t, x, v) f_s(t, x, w) \Big] \langle \varepsilon, v - w\rangle d\varepsilon dw \\ & - \alpha_R \sigma_{ij}^2 \int_{\mathbb{R}^3} \int_{\mathbb{S}_+^2} \Big[f_i(t, x, v') f_j(t, x, w') - f_i(t, x, v) f_j(t, x, w) \Big] \langle \varepsilon, v - w\rangle d\varepsilon dw \end{aligned} \tag{11}$$

with $(i, j) \in \{(1, 2), (2, 1), (3, 4), (4, 3)\}$, and J_i^R is the reactive collisional operator, given by

$$J_i^R = \alpha_R \sigma_{ij}^2 \int_{\mathbb{R}^3} \int_{\mathbb{S}_+^2} \Big[f_k(t, x, v^{\odot}) f_l(t, x, w^{\odot}) - f_i(t, x, v) f_j(t, x, w) \Big] \langle \varepsilon, v - w\rangle d\varepsilon dw, \tag{12}$$

with $(i, j, k, l) \in \{(1, 2, 3, 4), (2, 1, 4, 3), (3, 4, 1, 2), (4, 3, 2, 1)\}$. Velocities v', w' and $v^{\odot}$, $w^{\odot}$ are given by Eqs. (5) and (9), respectively. Moreover, the second term in the expression (11) of J_i^E, with $0 \leq \alpha_R \leq 1$ in front of it, singles out those collisions that result in a coloring process and prevent double counting of the events in the elastic collisional integrals.

The collisional operators J_i^E and J_i^R satisfy some mathematical properties, which assure the physical consistency of the model, at least from the formal point of view.

Proposition 1 *The collisional operators are such that*

(a) $\int_{\mathbb{R}^3} J_i^E dv = 0, \; i = 1, 2, 3, 4;$

(b) $\int_{\mathbb{R}^3} J_1^R dv = \int_{\mathbb{R}^3} J_2^R dv = -\int_{\mathbb{R}^3} J_3^R dv = -\int_{\mathbb{R}^3} J_4^R dv;$

(c) $\sum_{i=1}^{4} \int_{\mathbb{R}^3} \psi_i (J_i^E + J_i^R) dv = 0$, *where* $\psi = (\psi_1, \psi_2, \psi_3, \psi_4)$ *is alternatively given by* $\psi = (1, 0, 1, 0)$, $\psi = (1, 0, 0, 1)$, $\psi = (0, 1, 1, 0)$, *or by* $\psi_i = m_i v_1$, $\psi_i = m_i v_2$, $\psi_i = m_i v_3$, *with* v_1, v_2, v_3 *being the spatial components of the velocity* v, *and* $\psi_i = \frac{1}{2} m_i v^2$. □

Property (a) assures that elastic collisions do not modify the number of particles of each color.
Property (b) states the correct exchange rates of all constituents, in agreement with the chemical law (1) describing the coloring process.
Property (c) implies seven independent macroscopic conservation laws, for the total number of particles of the reactant-product pairs of the form A_1–A_3, A_1–A_4 and A_2–A_3, as well as for the three momentum components and kinetic energy of the whole mixture. Proper linear combinations of the former three functions ψ in (c) reproduce conservation of the particle number density and of total mass of the mixture. Note that, in contrast to a non-reactive mixture, there is no mass conservation for the individual constituents in our simple model.
The conservation laws involve macroscopic quantities defined, as usual, as appropriate velocity moments of the distribution functions $f_i(t, x, v)$. Paper [2] includes a rather complete definition of the macroscopic variables for the more general SRS system with bimolecular chemical reaction. The paper also includes the set of balance equations and conservation laws of the system. Here, for sake of brevity, we omit the analysis of the macroscopic framework of the model.

3 Equilibrium Solutions and H-Theorem

Definition 1 The equilibrium solutions of system (10), (11) and (12) are distribution functions $f_i(t, x, v)$ satisfying

$$J_i^E + J_i^R = 0, \qquad i = 1, \ldots, 4. \tag{13}$$

The above definition of equilibrium solutions is motivated by the following result:

Proposition 2 *For* $n_i(t, x)$, $u(t, x)$, *and* $T(t, x)$ *with* $f_i(t, x, v) \geq 0$, *the following statements are equivalent:*

1. $M_i = n_i \left(\frac{m_i}{2\pi kT}\right)^{3/2} \exp\left(-\frac{m_i(v-u)^2}{2kT}\right)$, $i = 1, \ldots, 4$, *and* $n_1 n_2 = n_3 n_4$*;*

2. $J_i^E(\{M_i\}) = 0$ *and* $J_i^R(\{M_i\}) = 0$, $i = 1, \ldots, 4$*;*

3. $\sum_{i=1}^{4} \int_{\mathbb{R}^3} \left[J_i^E(\{M_i\}) + J_i^R(\{M_i\})\right] \log\left(M_i/\mu_{ij}\right) dv = 0.$

The notations $J_i^E(\{M_i\})$ *and* $J_i^R(\{M_i\})$ *signify the fact that for* $i = 1, \ldots, 4$*, the collisional operators depend on the set one-particle distribution functions,* M_1*,* M_2*,* M_3*, and* M_4*. Also* $\mu_{ij} = m_i m_j/(m_i + m_j)$ *with* $(i, j) = (1, 2), (2, 1), (3, 4), (4, 3)$*, and*

$$n_i = \int_{\mathbb{R}^3} f_i(t, x, v) dv, \qquad i = 1, \ldots, 4, \tag{14a}$$

$$u = \sum_{i=1}^{4} \int_{\mathbb{R}^3} m_i v f_i(t, x, v) dv \Big/ \sum_{i=1}^{4} \int_{\mathbb{R}^3} m_i f_i(t, x, v) dv, \tag{14b}$$

$$T = \frac{1}{3k} \sum_{i=1}^{4} \int_{\mathbb{R}^3} m_i (v-u)^2 f_i(t, x, v) dv \Big/ \sum_{i=1}^{4} n_i, \tag{14c}$$

where k is the Boltzmann constant. □

Equation $n_1 n_2 = n_3 n_4$ appearing in item 1. of Proposition 2 represents the mass action law for our coloring system. The proof of Proposition 2 follows the same line of arguments as the proof of Proposition 3.2 in [7]. See also the discussion in Refs. [2, 11].
The Maxwellian distribution functions satisfy the properties stated in the next two lemmas, which will be used in the next section.

Lemma 1 *The Maxwellian distribution functions* M_i *appearing in Proposition 2 with the number densities* n_i *are such that*

$$M_i(t, x, v')M_s(t, x, w') = M_i(t, x, v)M_s(t, x, w), \quad i, s = 1, 2, 3, 4, \tag{15}$$

where v and w are pre-collisional velocities of species A_i *and* A_s*, respectively, and* v' *and* w' *are the corresponding elastic post-collisional velocities given by (5).* □

Lemma 2 *The Maxwellian distribution functions* M_i *with number densities* n_i *constrained by the mass action law are such that*

$$M_k(t, x, v^{\odot})M_l(t, x, w^{\odot}) = M_i(t, x, v)M_j(t, x, w), \tag{16}$$

where $(i, j, k, l) \in \{(1, 2, 3, 4), (2, 1, 4, 3), (3, 4, 1, 2), (4, 3, 2, 1)\}$, *$v$ and w are the pre-collisional velocities of constituents A_i and A_j, respectively, and $v^{\odot}$ and $w^{\odot}$ are the corresponding reactive post-collisional velocities of constituents A_k and A_l, respectively, given by (9).* □

Lemma 1 follows from the energy conservation of elastic events (3b) and Lemma 2 follows from the law of mass action and from the energy conservation of reactive events (7b).

The trend to equilibrium of the reactive mixture and the macroscopic irreversibility of the coloring process are assured by an $\mathscr{H}$-theorem. In fact, for a nonnegative smooth solution $f_i(t, x, v)$ of our kinetic system (10), (11) and (12), the convex function $H(t)$ defined by

$$H(t) = \sum_{i=1}^{4} \int_{\Omega} \int_{\mathbb{R}^3} f_i(t, x, v) \log \left[f_i(t, x, v)/\mu_{ij} \right] dv dx, \tag{17}$$

is a Lyapunov functional (see for example, [7]). Here, μ_{ij} is the reduced mass defined in Proposition 2, and $(i, j) = (1, 2), (2, 1), (3, 4), (4, 3)$. Moreover, $H(t)$ attains its minimum when the velocity distribution functions are Maxwellians given in Proposition 2. Here, Ω is the physical evolution domain of the mixture. We assume that Ω can be either the whole space $\mathbb{R}^3$ or a box with boundary conditions of periodic type or boundary conditions of specular reflection at the walls.

4 Linearized SRS Kinetic System

We assume that the evolution regime of the gas mixture corresponds to a small deviation of the thermodynamical equilibrium. We then expand the distribution function $f_i(t, x, v)$ around the Maxwellian $M_i(t, x, v)$ with zero drift velocity $u = 0$ and densities satisfying the mass action law. We introduce a new unknown function $h_i(t, x, v)$ connected to $f_i(t, x, v)$ through the expansions

$$f_i(t, x, v) = M_i(t, x, v) + M_i^{1/2}(t, x, v)\, h_i(t, x, v), \qquad i = 1, \ldots, 4. \tag{18}$$

For each $i = 1, \ldots, 4$, the term $M_i^{1/2} h_i$ represents the deviation of the distribution function f_i from the equilibrium. The factors $M_i^{1/2}$ in front of h_i, $i = 1, \ldots, 4$, make the linearized operator symmetric in L^2 space without a need to introduce additional weigh functions, see [12]. Inserting expansions (18) into Eqs. (10–12), and using the properties (15) and (16), we obtain the linearized SRS kinetic system, as stated in the next proposition. For sake of brevity, we use the notations $\underline{h} = (h_1, h_2, h_3, h_4)^{\mathrm{T}}$, $h_i(v)$ instead of $h_i(t, x, v)$, and $\widehat{h_i} = M_i^{1/2} h_i$.

Proposition 3 *The linearized SRS kinetic system for our model is given by*

$$\frac{\partial \widehat{h}_i}{\partial t} + v \cdot \nabla_x \widehat{h}_i = \mathscr{L}_i^E(\underline{\widehat{h}}) + \mathscr{L}_i^R(\underline{\widehat{h}}) \equiv \mathscr{L}_i(\underline{\widehat{h}}), \qquad i = 1, \ldots, 4, \tag{19}$$

with

$$\begin{aligned}
\mathscr{L}_i^E(\underline{\widehat{h}}) &= \sum_{s=1}^{4} \sigma_{is}^2 \int_{\mathbb{R}^3} \int_{\mathbb{S}_+^2} \Big[M_s(w) M_i^{1/2}(v) M_i^{-1/2}(v') \widehat{h}_i(v') \\
&\quad + M_s(w) M_i^{1/2}(v) M_s^{-1/2}(w') \widehat{h}_s(w') - M_s(w) \widehat{h}_i(v) \\
&\quad - M_s^{1/2}(w) M_i^{1/2}(v) \widehat{h}_s(w) \Big] \langle \varepsilon, v - w \rangle \, d\varepsilon \, dw \\
&\quad - \alpha_R \sigma_{ij}^2 \int_{\mathbb{R}^3} \int_{\mathbb{S}_+^2} \Big[M_j(w) M_i^{1/2}(v) M_i^{-1/2}(v') \widehat{h}_i(v') \\
&\quad + M_j(w) M_i^{1/2}(v) M_j^{-1/2}(w') \widehat{h}_j(w') - M_j(w) \widehat{h}_i(v) \\
&\quad - M_j^{1/2}(w) M_i^{1/2}(v) \widehat{h}_j(w) \Big] \langle \varepsilon, v - w \rangle \, d\varepsilon \, dw,
\end{aligned} \tag{20}$$

$$\begin{aligned}
\mathscr{L}_i^R(\underline{\widehat{h}}) &= \alpha_R \sigma_{ij}^2 \int_{\mathbb{R}^3} \int_{\mathbb{S}_+^2} \Big[M_j(w) M_i^{1/2}(v) M_k^{-1/2}(v^{\odot}) \widehat{h}_k(v^{\odot}) \\
&\quad + M_j(w) M_i^{1/2}(v) M_l^{-1/2}(w^{\odot}) \widehat{h}_l(w^{\odot}) - M_j(w) \widehat{h}_i(v) \\
&\quad - M_j^{1/2}(w) M_i^{1/2}(v) \widehat{h}_j(w) \Big] \langle \varepsilon, v - w \rangle \, d\varepsilon \, dw,
\end{aligned} \tag{21}$$

where $(i, j, k, l) \in \{(1, 2, 3, 4), (2, 1, 4, 3), (3, 4, 1, 2), (4, 3, 2, 1)\}$. □

We introduce the linearized collision operator $\mathscr{L}$ with components $\mathscr{L}_i = \mathscr{L}_i^E + \mathscr{L}_i^R$, $i = 1, 2, 3, 4$, where $\mathscr{L}_i^E$ and $\mathscr{L}_i^R$ are defined above. Such operator possesses fundamental properties that are stated below. At this end, we consider the Hilbert space $Y = L^2(\mathbb{R}^3) \times L^2(\mathbb{R}^3) \times L^2(\mathbb{R}^3) \times L^2(\mathbb{R}^3)$ endowed with the inner product defined by

$$\langle \underline{F}, \underline{G} \rangle = \sum_{i=1}^{4} \int_{\mathbb{R}^3} F_i(v) G_i(v) dv, \tag{22}$$

for $\underline{F} = (F_1, F_2, F_3, f_4) \in Y$ and $\underline{G} = (G_1, G_2, G_3, G_4) \in Y$, with F_i, G_i $(i = 1, 2, 3, 4)$ real-valued functions. Then we have the following result.

Proposition 4 *The linearized collisional operator* $\mathscr{L}$ *is symmetric and non-positive semi-definite, that is*

(a) $\langle \mathscr{L}(\underline{F}), \underline{G} \rangle = \langle \underline{F}, \mathscr{L}(\underline{G}) \rangle$, *for all* $\underline{F}, \underline{G} \in Y$;

(b) $\langle \mathscr{L}(\underline{F}), \underline{F} \rangle \leq 0$, *for all* $\underline{F} \in Y$. *Additionally,* $\langle \mathscr{L}(\underline{F}), \underline{F} \rangle = 0$ *if and only if* $\underline{F}$ *is a collisional invariant weighted by* $\underline{M}^{1/2} = (M_1^{1/2}, M_2^{1/2}, M_3^{1/2}, M_4^{1/2})$, *that is the null space of the operator* $\mathscr{L}$ *is 7-dimensional and is spanned by the* *collisional invariants (weighted by $\underline{M}^{1/2}$) associated to the conservation laws of the partial number densities $n_1 + n_3$, $n_1 + n_4$, $n_2 + n_3$, momentum components and kinetic energy of the mixture.* □

5 Kernels of the Linearized Operators

In the case of a gas mixture, even if without chemical reaction, the representation of the kernels is much more involved than in the case of a single inert gas. In fact, the complexity of the collisional terms increase significantly when collisions between particles of different constituents and different molecular masses are considered, especially when chemical reactions are also allowed. The computation of the kernels of the linearized collisional operators requires much more work and therefore the explicit representation of the kernels becomes, in our opinion, a very difficult task.

In the last years, we have been interested in this topic and some results have been obtained in this direction. In particular, the explicit representation of the kernels of the SRS linearized operators, both elastic and reactive, is presented in [11], for the general case of a quaternary mixture with reversible bimolecular reaction and arbitrary molecular masses. The expressions are, in general, quite long and rather complicated, because some of the transformations used to compute the kernels do not retain the symmetry of the integral operators. A better representation has been obtained in paper [2], for a similar chemically reactive mixture but considering that the constituents have equal molecular masses. The expressions are more tractable but those of the reactive operators are still intricate.

For the simplified version of the SRS model considered in the present paper, we were able to obtain a rather simple representation of the kernels. In particular, the expressions of both elastic and reactive kernels keep the symmetry property of the corresponding integral operators. In this section, we give the kernels of the operators and briefly explain the computations. We also state some symmetry properties of the kernels useful in the next Sect. 6.

5.1 Explicit Representation of the Kernels

In view of the explicit computations, we first work on the expression of the linearized operator $\mathscr{L}_i$ by manipulating expressions (20) and (21) of $\mathscr{L}_i^E$ and $\mathscr{L}_i^R$. The last two contributions of $\mathscr{L}_i^E$ and $\mathscr{L}_i^R$ are equal with opposite signs and thus they cancel out. Then, using the energy conservation (3b) of elastic events we modify those contributions appearing in the expression of $\mathscr{L}_i^E$ with the perturbation function $\widehat{h}$ evaluated in the post collisional velocities v' and w'.

Analogously, we use the energy conservation (7b) of the reactive events and the mass action law to work on those contributions in $\mathscr{L}_i^R$ with $\widehat{h}$ evaluated in the post collisional velocities $v^{\odot}$ and $w^{\odot}$. Using the same notations $\mathscr{L}_i^E$ and $\mathscr{L}_i^R$ for the elastic and reactive contributions in the resulting expression of $\mathscr{L}_i$, we obtain

$$\begin{aligned}\mathscr{L}_i^E(\widehat{\underline{h}}) &= \sum_{s=1}^{4}\sigma_{is}^2\int_{\mathbb{R}^3}\int_{\mathbb{S}_+^2}\Big[M_s^{1/2}(w)M_s^{1/2}(w')\widehat{h}_i(v') + M_s^{1/2}(w)M_i{}^{1/2}(v')\widehat{h}_s(w')\\ &\qquad - M_s(w)\widehat{h}_i(v) - M_s^{1/2}(w)M_i^{1/2}(v)\widehat{h}_s(w)\Big]\langle\varepsilon, v-w\rangle\, d\varepsilon\, dw\\ &\quad -\alpha_R\sigma_{ij}^2\int_{\mathbb{R}^3}\int_{\mathbb{S}_+^2}\Big[M_j^{1/2}(w)M_j^{1/2}(w')\widehat{h}_i(v')\\ &\qquad + M_j^{1/2}(w)M_i^{1/2}(v')\widehat{h}_j(w')\Big]\langle\varepsilon, v-w\rangle\, d\varepsilon\, dw,\end{aligned} \tag{23}$$

$$\begin{aligned}\mathscr{L}_i^R(\widehat{\underline{h}}) &= \alpha_R\sigma_{ij}^2\int_{\mathbb{R}^3}\int_{\mathbb{S}_+^2}\Big[M_j^{1/2}(w)M_l^{1/2}(w^{\odot})\widehat{h}_k(v^{\odot})\\ &\qquad + M_j^{1/2}(w)M_k^{1/2}(v^{\odot})\widehat{h}_l(w^{\odot})\Big]\langle\varepsilon, v-w\rangle\, d\varepsilon\, dw,\end{aligned} \tag{24}$$

where $(i, j, k, l) \in \{(1, 2, 3, 4), (2, 1, 4, 3), (3, 4, 1, 2), (4, 3, 2, 1)\}$.

5.1.1 Kernels of the Elastic Operators

We split the elastic linearized operator $\mathscr{L}_i^E$ given by (23) in several contributions, as follows

$$\mathscr{L}_i^E(\underline{h}) = -\nu_i(v)\widehat{h}_i(v) - \mathscr{Q}_i^{(1)}(\underline{h}) + \mathscr{Q}_i^{(2)}(\underline{h}) + \mathscr{Q}_i^{(3)}(\underline{h}) - \mathscr{T}_i^{(2)}(\underline{h}) - \mathscr{T}_i^{(3)}(\underline{h}), \tag{25}$$

with

$$\nu_i(v) = \sum_{s=1}^{4}\sigma_{is}^2\int_{\mathbb{R}^3}\int_{\mathbb{S}_+^2} M_s(w)\langle\varepsilon, v-w\rangle\, d\varepsilon dw, \tag{26}$$

$$\mathscr{Q}_i^{(1)}(\widehat{\underline{h}}) = \sum_{s=1}^{4}\sigma_{is}^2\int_{\mathbb{R}^3}\int_{\mathbb{S}_+^2} M_i^{1/2}(v)M_s^{1/2}(w)\widehat{h}_s(w)\langle\varepsilon, v-w\rangle\, d\varepsilon dw, \tag{27}$$

$$\mathscr{Q}_i^{(2)}(\hat{\underline{h}}) = \sum_{s=1}^{4}\sigma_{is}^2\int_{\mathbb{R}^3}\int_{\mathbb{S}_+^2} M_s^{1/2}(w)M_s^{1/2}(w')\widehat{h}_i(v')\langle\varepsilon, v-w\rangle\, d\varepsilon dw, \tag{28}$$

$$\mathscr{Q}_i^{(3)}(\widehat{\underline{h}}) = \sum_{s=1}^{4}\sigma_{is}^2\int_{\mathbb{R}^3}\int_{\mathbb{S}_+^2} M_s^{1/2}(w)M_i{}^{1/2}(v')\widehat{h}_s(w')\langle\varepsilon, v-w\rangle\, d\varepsilon dw, \tag{29}$$

$$\mathscr{T}_i^{(2)}(\widehat{\underline{h}}) = \alpha_R\sigma_{ij}^2\int_{\mathbb{R}^3}\int_{\mathbb{S}_+^2} M_j^{1/2}(w)M_j^{1/2}(w')\widehat{h}_i(v')\langle\varepsilon, v-w\rangle\, d\varepsilon dw, \tag{30}$$

$$\mathscr{T}_i^{(3)}(\underline{\widehat{h}}) = \alpha_R \sigma_{ij}^2 \int_{\mathbb{R}^3} \int_{\mathbb{S}_+^2} M_j^{1/2}(w) M_i^{1/2}(v') \widehat{h}_j(w') \langle \varepsilon, v - w \rangle \, d\varepsilon dw, \quad (31)$$

where $\nu_i(v)$ represents a generalized collision frequency.
We work separately on each operator. It is easy to extract the kernel of each operator $\mathscr{Q}_i^{(1)}$ and $\mathscr{Q}_i^{(2)}$. We obtain

$$\mathscr{Q}_i^{(1)}(\underline{\widehat{h}}) = \sum_{s=1}^{4} \int_{\mathbb{R}^3} K_{is}^{(1)}(v, w) \widehat{h}_s(w) \, dw, \quad (32)$$

$$\mathscr{Q}_i^{(2)}(\underline{\widehat{h}}) = \sum_{s=1}^{4} \int_{\mathbb{R}^3} K_{is}^{(2)}(v, w) \widehat{h}_i(w) \, dw, \quad (33)$$

with the kernels $K_{is}^{(1)}$ and $K_{is}^{(2)}$ $(i, s = 1, \ldots, 4)$ given by

$$K_{is}^{(1)}(v, w) = \pi \sigma_{is}^2 \, \|v - w\| \sqrt{n_i n_s} \left(\frac{m_i m_s}{(2\pi k T)^2} \right)^{3/4} \exp\left(-\frac{m_i v^2 + m_s w^2}{4kT} \right), \quad (34)$$

$$K_{is}^{(2)}(v, w) = \sigma_{is}^2 n_s \left(\frac{m_s}{2\pi k T} \right)^{1/2} \left(\frac{m_i}{2\mu_{is}} \right)^2 \frac{1}{\|v - w\|} \times \exp\left[-\frac{m_s}{8kT} \frac{\left(v^2 - w^2\right)^2}{\|v - w\|^2} - \frac{m_s}{8kT} \left(\frac{m_i}{m_s} \right)^2 (v - w)^2 \right]. \quad (35)$$

The computation of the kernel of $\mathscr{Q}_i^{(3)}$ requires more work. It results convenient to split the operator $\mathscr{Q}_i^{(3)}$ as a summation according to the molecular masses m_s and m_i. Thus, we introduce the following sets of indices

$$I = \{1, 2, 3, 4\},$$

$$I^0 = \{s \in I : \; m_s = m_i\}, \quad I^+ = \{s \in I : \; m_s > m_i\}, \quad I^- = \{s \in I : \; m_s < m_i\},$$

and split the operator $\mathscr{Q}_i^{(3)}$ into three contributions, namely

$$\mathscr{Q}_i^{(3)}(\underline{\widehat{h}}) = \mathscr{Q}_i^{(3^0)}(\underline{\widehat{h}}) + \mathscr{Q}_i^{(3^+)}(\underline{\widehat{h}}) + \mathscr{Q}_i^{(3^-)}(\underline{\widehat{h}}), \quad (36)$$

where

$$\mathscr{Q}_i^{(3^0)}(\underline{\widehat{h}}) = \sum_{s \in I^0} \int_{\mathbb{R}^3} K_{is}^{(3^0)}(v, w) \widehat{h}_s(w) \, dw, \quad (37)$$

$$\mathscr{Q}_i^{(3^+)}(\widehat{h}) = \sum_{s\in I^+} \int_{\mathbb{R}^3} K_{is}^{(3^+)}(v,w)\widehat{h}_s(w)\,dw, \tag{38}$$

$$\mathscr{Q}_i^{(3^-)}(\widehat{h}) = \sum_{s\in I^-} \int_{\mathbb{R}^3} K_{is}^{(3^-)}(v,w)\widehat{h}_s(w)\,dw. \tag{39}$$

Then we obtain the kernels of the operators (37–39) in the form

$$K_{is}^{(3^0)}(v,w) = \sigma_{is}^2\,(n_i n_s)^{1/2}\left(\frac{m_i}{2\pi kT}\right)^{1/2}\frac{1}{\|v-w\|} \tag{40}$$
$$\times \exp\left[-\frac{m_i}{8kT}\frac{(v^2-w^2)^2}{\|v-w\|^2}-\frac{m_i}{8kT}(v-w)^2\right],\quad i=1,2,3,4,\quad s\in I^0,$$

$$K_{is}^{(3^+)}(v,w) = \frac{2\pi\sigma_{is}^2}{\left(\frac{m_s-m_i}{m_s+m_i}\right)^2}(n_i n_s)^{1/2}\left(\frac{m_i m_s}{(2\pi kT)^2}\right)^{3/4}\|v-w\|$$
$$\times \exp\left[-\frac{1}{4kT}\frac{(m_s w-m_i v)^2}{m_i+m_s}\right] \tag{41}$$
$$\times \int_0^{\pi/2}\cos^2\theta\exp\left[-\frac{\mu_{is}}{kT}\left(\left(\frac{\|v+w\|}{2}-\frac{\|v-w\|\cos\theta}{\frac{m_s-m_i}{m_s+m_i}}\right)^2\right.\right.$$
$$\left.\left.+\frac{\|v-w\|\,\|v+w\|\cos\theta\,(1-\cos\theta^\star)}{\frac{m_s-m_i}{m_s+m_i}}\right)\right]d\theta,\quad i=1,2,3,4,\quad s\in I^+,$$

$$K_{is}^{(3^-)}(v,w) = \frac{2\pi\sigma_{is}^2}{\left(\frac{m_s-m_i}{m_s+m_i}\right)^2}(n_i n_s)^{1/2}\left(\frac{m_i m_s}{(2\pi kT)^2}\right)^{3/4}\|v-w\|$$
$$\times \exp\left[-\frac{1}{4kT}\frac{(m_s w-m_i v)^2}{m_i+m_s}\right] \tag{42}$$
$$\times \int_{\pi/2}^{\pi}\cos^2\theta\exp\left[-\frac{\mu_{is}}{kT}\left(\left(\frac{\|v+w\|}{2}-\frac{\|v-w\|\cos\theta}{\frac{m_s-m_i}{m_s+m_i}}\right)^2\right.\right.$$
$$\left.\left.+\frac{\|v-w\|\,\|v+w\|\cos\theta\,(1-\cos\theta^\star)}{\frac{m_s-m_i}{m_s+m_i}}\right)\right]d\theta,\quad i=1,2,3,4,\quad s\in I^-.$$

In the above Eqs. (41) and (42), $\theta^\star$ is the angle defined by the vectors $v+w$ and ε. Observe that, in (41), we have $m_s > m_i$ because $s\in I^+$, whereas in (42), we have $m_s < m_i$, because $s\in I^-$.
Now we proceed with the operators $\mathscr{T}_i^{(\alpha)}$. We obtain

$$\mathscr{T}_i^{(2)}(\underline{\widehat{h}}) = \int_{\mathbb{R}^3} N_{ij}^{(2)}(v,w)\widehat{h}_i(w)\,dw, \tag{43}$$

$$\mathscr{T}_i^{(3)}(\underline{\widehat{h}}) = \int_{\mathbb{R}^3} N_{ij}^{(3)}(v,w)\widehat{h}_j(w)\,dw, \tag{44}$$

and the computations are similar to those developed for $\mathscr{Q}_i^{(2)}$ and $\mathscr{Q}_i^{(3)}$. For sake of brevity, we omit their expressions here.

5.1.2 Kernels of the Reactive Operators

The calculation of the reactive kernels are similar. We start by splitting the linearized reactive collisional operator (24) as follows

$$\widehat{\mathscr{L}}_i^R(\underline{\widehat{h}}) = \mathscr{R}_i^{(2)}(\underline{\widehat{h}}) + \mathscr{R}_i^{(3)}(\underline{\widehat{h}}), \tag{45}$$

with

$$\mathscr{R}_i^{(2)}(\underline{\widehat{h}}) = \alpha_R\sigma_{ij}^2 \int_{\mathbb{R}^3}\int_{\mathbb{S}_+^2} M_j^{1/2}(w)M_l^{1/2}(w^{\odot})\widehat{h}_k(v^{\odot})\langle\varepsilon, v-w\rangle\,d\varepsilon dw, \tag{46}$$

$$\mathscr{R}_i^{(3)}(\underline{\widehat{h}}) = \alpha_R\sigma_{ij}^2 \int_{\mathbb{R}^3}\int_{\mathbb{S}_+^2} M_j^{1/2}(w)M_k^{1/2}(v^{\odot})\widehat{h}_l(w^{\odot})\langle\varepsilon, v-w\rangle\,d\varepsilon dw. \tag{47}$$

We obtain

$$\mathscr{R}_i^{(2)}(\underline{\widehat{h}}) = \int_{\mathbb{R}^3} \Upsilon_{ij}^{(2)}(v,w)\widehat{h}_k(w)\,dw, \tag{48}$$

$$\mathscr{R}_i^{(3)}(\underline{\widehat{h}}) = \int_{\mathbb{R}^3} \Upsilon_{ij}^{(3)}(v,w)\widehat{h}_l(w)\,dw. \tag{49}$$

with the kernels given as follows. For the operator $\mathscr{R}_i^{(2)}$ the calculations are simple and

$$\Upsilon_{ij}^{(2)}(v,w) = \alpha_R\sigma_{ij}^2\sqrt{n_jn_l}\left(\frac{m_j}{2\pi kT}\right)^{1/2}\left(\frac{m_i}{2\mu_{ij}}\right)^2\frac{1}{\|v-w\|} \times \exp\left[-\frac{m_j}{8kT}\frac{\left(v^2-w^2\right)^2}{\|v-w\|^2} - \frac{m_j}{8kT}\left(\frac{m_i}{m_j}\right)^2(v-w)^2\right]. \tag{50}$$

Concerning the kernel of the operator $\mathscr{R}_i^{(3)}$, similarly to what we have done before for the elastic operator $\mathscr{Q}_i^{(3)}$, we consider three different cases. If $m_j = m_i$, and therefore all constituents have equal molecular masses, then the kernel is given by

$$\Upsilon_{ij}^{(3^0)}(v,w)=\alpha_R\sigma_{ij}^2\,(n_jn_k)^{1/2}\left(\frac{m_i}{2\pi kT}\right)^{1/2}\frac{1}{\|v-w\|}$$
$$\times\exp\left[-\frac{m_i}{8kT}\frac{(v^2-w^2)^2}{\|v-w\|^2}-\frac{m_i}{8kT}(v-w)^2\right]. \tag{51}$$

If $m_j > m_i$, the kernel is given by

$$\Upsilon_{ij}^{(3^+)}(v,w)=\frac{2\pi\alpha_R\sigma_{ij}^2}{\left(\frac{m_j-m_i}{m_i+m_j}\right)^2}\,(n_jn_k)^{1/2}\left(\frac{m_im_j}{(2\pi kT)^2}\right)^{3/4}\|v-w\|$$
$$\times\exp\left[-\frac{1}{4kT}\frac{(m_jw\;-\;m_iv)^2}{m_i\;+\;m_j}\right] \tag{52}$$
$$\times\int_0^{\pi/2}\cos^2\theta\exp\left[-\frac{\mu_{ij}}{kT}\left(\left(\frac{\|v+w\|}{2}-\frac{\|v-w\|\cos\theta}{\frac{m_j-m_i}{m_i+m_j}}\right)^2\right.\right.$$
$$\left.\left.+\frac{\|v-w\|\,\|v+w\|\cos\theta\,(1-\cos\theta^\star)}{\frac{m_j-m_i}{m_i+m_j}}\right)\right]d\theta.$$

If $m_j < m_i$, the kernel $\Upsilon_{ij}^{(3^-)}$ is given by a similar expression to (52), but with the integral in θ extended to $[\pi/2,\pi]$ instead of $[0,\pi/2]$. In expressions (50), (51) and (52), the indices i, j, k, l are from the set $\{(1,2,3,4), (2,1,4,3), (3,4,1,2), (4,3,2,1)\}$.

5.2 Properties of the Kernels

From the explicit representation of the kernels given in the previous section, it is easy to state the properties given below.

Proposition 5 *For any* $i, s = 1,2,3,4$, *the following properties of the elastic kernels* $K_{is}^{(\alpha)}$, $\alpha = 1,2,3$, *hold for all* v, w *a.e. in* $\mathbb{R}^3$:

(a) $K_{is}^{(1)}(v,w) = K_{si}^{(1)}(w,v)$; *(c)* $K_{is}^{(3^0)}(v,w) = K_{si}^{(3^0)}(w,v)$;
(b) $K_{is}^{(2)}(v,w) = K_{is}^{(2)}(w,v)$; *(d)* $K_{is}^{(3^+)}(v,w) = K_{si}^{(3^-)}(w,v)$. □

The properties stated in Proposition 5 assure the symmetry of the elastic integral operators $\mathscr{Q}_i^{(\alpha)}$ ($\alpha = 1,2,3$). See expressions (32), (33), (37), (38) and (39). Similar properties hold for the kernels $N_{is}^{(\alpha)}$ of the elastic operators $\mathscr{T}_i^{(\alpha)}$ ($\alpha = 2,3$), assuring the symmetry of these operators.

Proposition 6 *For any* $(i, j) \in \{(1,2), (2,1), (3,4), (4,3)\}$, *the following properties of the reactive kernels* $\Upsilon_{ij}^{(\alpha)}$, $\alpha = 2, 3$, *hold for all* v, w *a.e. in* $\mathbb{R}^3$*:*

(a) $\Upsilon_{ij}^{(2)}(v, w) = \Upsilon_{ij}^{(2)}(w, v)$; *(c)* $\Upsilon_{ij}^{(3^+)}(v, w) = \Upsilon_{ji}^{(3^-)}(w, v)$.

(b) $\Upsilon_{ij}^{(3^0)}(v, w) = \Upsilon_{ji}^{(3^0)}(w, v)$; □

The properties stated in Proposition 6 assure the symmetry of the reactive integral operators $\mathscr{R}_i^{(\alpha)}$ ($\alpha = 2, 3$). See expressions (48) and (49).

6 Compactness of the Linearized Collisional Operator

In this section, we study the compactness of the elastic and reactive linearized operators, as operators from $\left(L^2(\mathbb{R}^3)\right)^4$ to $\left(L^2(\mathbb{R}^3)\right)^4$. We introduce the notation

$$\begin{aligned}
\mathscr{Q}^{(\alpha)} &= (\mathscr{Q}_1^{(\alpha)}, \mathscr{Q}_2^{(\alpha)}, \mathscr{Q}_3^{(\alpha)}, \mathscr{Q}_4^{(\alpha)}), \quad \alpha = 1, 2, 3,\\
\mathscr{T}^{(\alpha)} &= (\mathscr{T}_1^{(\alpha)}, \mathscr{T}_2^{(\alpha)}, \mathscr{T}_3^{(\alpha)}, \mathscr{T}_4^{(\alpha)}), \quad \alpha = 2, 3,\\
\mathscr{R}^{(\alpha)} &= (\mathscr{R}_1^{(\alpha)}, \mathscr{R}_2^{(\alpha)}, \mathscr{R}_3^{(\alpha)}, \mathscr{R}_4^{(\alpha)}), \quad \alpha = 2, 3.
\end{aligned} \tag{53}$$

We extend to our mixture of reactive gases the techniques introduced by Grad (see [12, 13]) and then revisited by Cercignani et al. (see [14]), for one single inert gas. However, these techniques seem to be not enough for the operators $\mathscr{Q}^{(3)}$, $\mathscr{T}^{(3)}$ and $\mathscr{R}^{(3)}$, due to the cumbersome expressions of their terms when the molecular masses are different. See expressions (41), (42) and (52). For these operators, our idea is to adapt to our model the methodology recently proposed by Boudin et al. (see [15]) for a mixture of non-reactive gases. In paper [15], the authors consider the non-reactive Boltzmann equation for gas mixtures and propose a new approach to treat the terms of the linearized operators associated with particles with different masses.
Work is in progress to extend the approach of paper [15] to the similar terms appearing in our model. In the present paper we focus on the compactness of both the elastic operator $\mathscr{Q}^{(1)}$ and the reactive operator $\mathscr{R}^{(2)}$. Operators $\mathscr{Q}^{(2)}$ and $\mathscr{T}^{(2)}$ can be treated in a similar way to $\mathscr{R}^{(2)}$.

6.1 Compactness of the Operator $\mathscr{Q}^{(1)}$

We consider the elastic kernel $K_{is}^{(1)}$ of the operator $\mathscr{Q}^{(1)}$ defined by expression (34). We start with the estimate for for $K_{is}^{(1)}$ stated in the following lemma.

Lemma 3 *For each* $i, s = 1, 2, 3, 4$, *the elastic kernel* $K_{is}^{(1)}$ *is square integrable with respect to* w. *We have*

$$\int_{\mathbb{R}^3} \left(K_{is}^{(1)}\right)^2 (v, w)\, dw \leq C(1 + \|v\|^2) \exp\left(-\frac{m_i v^2}{2kT}\right), \tag{54}$$

where C depends only on m_i, n_i, σ_{is}, and T.

Proof For $i, s = 1, 2, 3, 4$, we have

$$\begin{aligned}
\int_{\mathbb{R}^3} \left(K_{is}^{(1)}\right)^2 (v, w)\, dw &= \pi^2 \sigma_{is}^4 n_i n_s \left(\frac{m_i m_s}{(2\pi kT)^2}\right)^{\frac{3}{2}} \exp\left(-\frac{m_i v^2}{2kT}\right) \\
&\quad \times \int_{\mathbb{R}^3} |v - w|^2 \exp\left(-\frac{m_s w^2}{2k\mathscr{T}}\right) dw \\
&\leq 2\pi^2 \sigma_{is}^4 n_i n_s \left(\frac{m_i m_s}{(2\pi kT)^2}\right)^{\frac{3}{2}} \exp\left(-\frac{m_i v^2}{2kT}\right) \\
&\quad \times \int_{\mathbb{R}^3} (|v|^2 + |w|^2) \exp\left(-\frac{m_s w^2}{4kT}\right) dw \\
&\leq C(1 + |v|^2) \exp\left(-\frac{m_i v^2}{2kT}\right),
\end{aligned}$$

and the proof is complete. □

Then we can prove the compactness of $\mathscr{Q}^{(1)}$, as follows.

Theorem 1 *The linearized elastic operator $\mathscr{Q}^{(1)}$ is a compact operator from $(L^2(\mathbb{R}^3))^4$ to $(L^2(\mathbb{R}^3))^4$.*

Proof It is enough to show that each $\mathscr{Q}_i^{(1)}$ $(i = 1, 2, 3, 4)$ is a compact operator from $(L^2(\mathbb{R}))^4$ to $L^2(\mathbb{R})$. To this end, we notice that, for fixed i and s, the operator A with the kernel $K_{is}^{(1)}$ given by (34) is a compact operator from $L^2(\mathbb{R}^3)$ to $L^2(\mathbb{R}^3)$. Indeed, condition (54) implies that $K_{is}^{(1)} \in L^2(\mathbb{R}^3 \times \mathbb{R}^3)$, showing that the operator A, as a Hilbert-Schmidt operator, is compact from $L^2(\mathbb{R}^3)$ to $L^2(\mathbb{R}^3)$. Since this is true for each $s = 1, 2, 3, 4$, the operator $\mathscr{Q}_i^{(1)}$ $(i = 1, 2, 3, 4)$ is a compact operator from $(L^2(\mathbb{R}^3))^4$ to $L^2(\mathbb{R}^3)$, and this completes the proof. □

6.2 Compactness of the Operator $\mathscr{R}^{(2)}$

Similarly to what we have done in the previous subsection, we consider the reactive kernel $\Upsilon_{ij}^{(2)}$ of the operator $\mathscr{R}^{(2)}$ defined by expression (50). We first state the following result.

Proposition 7 *For each* $i, j = 1, 2, 3, 4$, *the reactive kernel* $\Upsilon_{ij}^{(2)}$ *is square integrable with respect to w. Moreover, we have*

$$\|\Upsilon_{ij}^{(2)}(v, \cdot)\|_{L^2(\mathbb{R}^3)} \leq C, \tag{55}$$

where C is independent of v.

Proof From

$$\begin{aligned}
\Upsilon_{ij}^{(2)}(v, w) &= \alpha_R\, \sigma_{ij}^2 \sqrt{n_j n_l} \left(\frac{m_j}{2\pi kT}\right)^{1/2} \left(\frac{m_i}{2\mu_{ij}}\right)^2 \frac{1}{\|v - w\|} \\
&\quad \times \exp\left[-\frac{m_j}{8kT}\frac{\left(v^2 - w^2\right)^2}{\|v - w\|^2} - \frac{m_j}{8kT}\left(\frac{m_i}{m_j}\right)^2 (v - w)^2\right] \\
&\leq \alpha_R\, \sigma_{ij}^2 \sqrt{n_j n_l} \left(\frac{m_j}{2\pi kT}\right)^{1/2} \left(\frac{m_i}{2\mu_{ij}}\right)^2 \frac{1}{\|v - w\|} \\
&\quad \times \exp\left[-\frac{m_j}{8kT}\left(\frac{m_i}{m_j}\right)^2 (v - w)^2\right],
\end{aligned} \tag{56}$$

and for any $M > 0$, we obtain

$$\begin{aligned}
\int_{\mathbb{R}^3} \left(\Upsilon_{ij}^{(2)}\right)^2 (v, w)\, dw &\leq C_1 \int_{|w| \leq M} \frac{1}{\|v - w\|^2}\, dw \\
&\quad + C_2 \int_{|w| > M} \exp\left[-\frac{m_j}{8kT}\left(\frac{m_i}{m_j}\right)^2 (v - w)^2\right] dw < C.
\end{aligned} \tag{57}$$

The first term on the right hand side of (57) is bounded since $1/\|v - w\|^2$ is integrable on any bounded set containing v, while the second term is bounded because $\exp(-cx^2)$ is integrable over $\mathbb{R}^3$ for any $c > 0$. The constant C does not depend on v since the expression in (57) is translational invariant. □

We also have the following estimate for $\Upsilon_{ij}^{(2)}$.

Proposition 8 *For each* $i, j = 1, 2, 3, 4$, *the reactive kernel* $\Upsilon_{ij}^{(2)}$ *is integrable with respect to w, and we have*

$$\int_{\mathbb{R}^3} \Upsilon_{ij}^{(2)}(v, w)\, dw \leq \frac{C}{(1 + \|v\|^2)^{1/2}}, \tag{58}$$

where C is independent of v.

Proof See [12] or [14]. □

Then we can prove the compactness of $\mathscr{R}^{(2)}$, as follows.

Theorem 2 *The linearized reactive operator* $\mathscr{R}^{(2)}$ *is a compact operator from* $(L^2(\mathbb{R}^3))^4$ *to* $(L^2(\mathbb{R}^3))^4$.

Proof It is enough to show that each $\mathscr{R}_i^{(2)}$ $(i=1,2,3,4)$ is a compact operator from $(L^2(\mathbb{R}^3))^4$ to $L^2(\mathbb{R}^3)$. For fixed i and j, consider the operator A with the kernel $\Upsilon_{ij}^{(2)}$ given by (50).
For $M>0$, define χ_M to be the characteristic function of the set $\{v\in\mathbb{R}^3 : |v|\le M\}$. Consider the identity $Af=\chi_M Af+(1-\chi_M)Af$ for $f\in L^2(\mathbb{R}^3)$. Proposition 7 implies that for each $M>0$, the operator $\chi_M A$, as a Hilbert-Schmidt operator, is compact from $L^2(\mathbb{R}^3)$ to $L^2(\mathbb{R}^3)$. Now, if $\|\cdot\|_{B(L^2(\mathbb{R}^3))}$ denotes the operator norm in the space of bounded operators on $L^2(\mathbb{R}^3)$, then using Schwartz inequality, Propositions 7 and 8, the operator $(1-\chi_M)A$ has the property (see [14], Theorem 7.2.4),

$$
\begin{aligned}
\|(1-\chi_M)\ A\|^2_{B(L^2(\mathbb{R}^3))} &= \sup_{\|f\|_{L^2(\mathbb{R}^3)}\le 1} \int_{\mathbb{R}^3}\Big(1-\chi_M(v)\Big)\Big[\int_{\mathbb{R}^3}\Upsilon_{ij}^{(2)}(v,w)f(w)\,dw\Big]^2 dv\\
&\le \sup_{\|f\|_{L^2(\mathbb{R}^3)}\le 1} \int_{\mathbb{R}^3}\Big(1-\chi_M(v)\Big)\Big[\int_{\mathbb{R}^3}\Upsilon_{ij}^{(2)}(v,w)\,dw\Big]\\
&\qquad\times\Big[\int_{\mathbb{R}^3}\Upsilon_{ij}^{(2)}(v,w)(f(w))^2\,dw\Big]dv\\
&\le C\sup_{\|f\|_{L^2(\mathbb{R}^3)}\le 1} \int_{\mathbb{R}^3}\Big(1-\chi_M(v)\Big)\Big(1+\|v\|^2\Big)^{-1/2}\\
&\qquad\times\Big[\int_{\mathbb{R}^3}\Upsilon_{ij}^{(2)}(v,w)(f(w))^2\,dw\Big]dv\\
&\le C\Big(1+M^2\Big)^{-1/2}\sup_{\|f\|_{L^2(\mathbb{R})}\le 1}\int_{\|v\|>M}\Big[\int_{\mathbb{R}^3}\Upsilon_{ij}^{(2)}(v,w)(f(w))^2\,dw\Big]dv\\
&\le C(1+M^2)^{-1/2}\sup_{\|f\|_{L^2(\mathbb{R}^3)}\le 1}\int_{\mathbb{R}^3}\Big[\int_{\mathbb{R}^3}\Upsilon_{ij}^{(2)}(v,w)(f(w))^2\,dw\Big]dv\\
&\le C\Big(1+M^2\Big)^{-1/2}\sup_{\|f\|_{L^2(\mathbb{R}^3)}\le 1}\int_{\mathbb{R}^3}(f(w))^2\,dw\ \le\ C\Big(1+M^2\Big)^{-1/2}.
\end{aligned}
$$

Therefore, $(1-\chi_M)A\to 0$ in the operator norm as $M\to\infty$. This also yields that $\chi_M A\to A$ in the operator norm as $M\to\infty$. Since for each $M>0$, $\chi_M A$ is a compact operator from $L^2(\mathbb{R}^3)$ to $L^2(\mathbb{R}^3)$, and the set of compact operators is closed in the space of bounded operators with the operator norm $\|\cdot\|_{B(L^2(\mathbb{R}^3))}$, we conclude that A is compact as an operator from $L^2(\mathbb{R}^3)$ to $L^2(\mathbb{R}^3)$. Since this is true for each $j = 1, 2, 3, 4$, the operator $\mathscr{R}_i^{(2)}$ ($i = 1, 2, 3, 4$) is a compact operator from $(L^2(\mathbb{R}^3))^4$ to $L^2(\mathbb{R}^3)$. Thus, the operator $\mathscr{R}^{(2)}$ is compact from $(L^2(\mathbb{R}^3))^4$ to $(L^2(\mathbb{R}^3))^4$ and the proof is complete. □

Acknowledgments The paper is partially supported by the Research Centre of Mathematics of the University of Minho, through the National Funds from the "Fundação para a Ciência e a Tecnologia", Project PEstOE/MAT/UI0013/2014.

References

1. Polewczak, J., Soares, A.J.: Kinetic theory of simple reacting spheres I. In: AIP Conference Proceedings of 27th International Symposium on Rarefied Gas Dynamics, 2010, vol. 1333, pp. 117–122. Pacific Grove (2011)
2. Carvalho, F., Polewczak, J., Soares, A.J.: On the kinetic systems for simple reacting spheres: modeling and linearized equations. In: Particle Systems and Partial Differential Equations, 2012, Braga. Springer Proceedings in Mathematics & Statistics, vol. 75, pp. 251–267 (2014)
3. Marron, M.T.: Simple collision theory of reactive hard spheres. J. Chem. Phys. **52**, 4060–4061 (1970)
4. Xystris, N., Dahler, J.S.: Mass and momentum transport in dilute reacting gases. J. Chem. Phys. **68**, 354–373 (1978)
5. Xystris, N., Dahler, J.S.: Kinetic theory of simple reacting spheres. J. Chem. Phys. **68**, 387–401 (1978)
6. Dahler, J.S., Qin, L.: The kinetic theory of a simple reacting fluid: scattering functions and relaxation processes. J. Chem. Phys. **103**, 725–750 (1995)
7. Polewczak, J.: The kinetic theory of simple reacting spheres: I. Global existence result in a dilute-gas case. J. Stat. Phys. **100**, 327–362 (2000)
8. Prigogine, I., Xhrouet, E.: On the perturbation of Maxwell distribution function by chemical reaction in gases. Physica **XV**, 913–932 (1949)
9. Prigogine, I., Mahieu, M.: Sur La Perturbation De La Distribution De Maxwell Par Des Réactions Chimiques En Phase Gazeuse. Physica **XVI**, 51–64 (1950)
10. Shizgal, B., Karplus, M.: Nonequilibrium contributions to the rate of reaction. I. Perturbation of the velocity distribution function. J. Chem. Phys. **52**, 4262–4278 (1970)
11. Carvalho, F.: Mathematical methods for the Boltzmann equation in the context of chemically reactive gases. Doctoral thesis, Repositorium of the University of Minho (2012). http://hdl.handle.net/1822/24430
12. Grad, H.: Asymptotic theory of the Boltzmann Equation II. In: Laurmann J.A. (ed.) Proceedings of 3rd International Symposium on Rarefied Gas Dynamics, vol. 1, pp. 26–59. Academic Press, New York (1963)
13. Grad, H.: Asymptotic theory of the Boltzmann equation. Phys. Fluids **6**, 147–181 (1963)
14. Cercignani, C., Illner, R., Pulvirenti, M.: The Mathematical Theory of Dilute Gases. Springer, New York (1994)
15. Boudin, L., Grec, B., Pavić, M., Salvarani, F.: Diffusion asymptotics of a kinetic model for gaseous mixtures. Kinet. Relat. Models **6**, 137–157 (2013)

Can Translation Invariant Systems Exhibit a Many-Body Localized Phase?

Wojciech De Roeck and François Huveneers

Abstract We review some recent works related to the exploration of Many-Body Localization in the absence of quenched disorder. We stress that, for systems where not all eigenstates of the Hamiltonian are expected to be localized, as it is generically the case for translation invariant systems with short range interactions, some rare large ergodic spots constitute a possible mechanism for thermalization, even though such spots occur just as well in systems with strong quenched disorder, where all eigenstates are localized. Nevertheless, we show that there is a regime of asymptotic localization for some translation invariant Hamiltonians.

Keywords Statistical physics · Condensed matter · Interacting particles systems · Anderson localization · Thermal transport

1 Introduction

Anderson localization refers usually to the single-particle wave function of an electron being confined in some limited region of space for all times [1]. In perturbative regimes, being for example that of strong disorder, this phenomenon is by now well understood at both physical and mathematical levels [11]. The topic benefited recently from a huge revival of interest, when it was argued that localization could persist in presence of electron-electron interactions [3]. Since in an interacting system, there is no appropriate analogue for one-particle wave functions, the phenomenon of Many-Body Localization (MBL) has to be characterized in other ways, than spatial decay of wave functions. One of the most appealing ways is to contrast MBL with an ergodic phase, where the latter can be defined as a phase in which the system

W. De Roeck
Instituut voor Theoretische Fysica, KU Leuven, Leuven, Belgium
e-mail: Wojciech.DeRoeck@fys.kuleuven.be

F. Huveneers (✉)
Ceremade, Université Paris Dauphine, Paris, France
e-mail: huveneers@ceremade.dauphine.fr; fhuvenee@ceremade.dauphine.fr

P. Gonçalves and A.J. Soares (eds.), *From Particle Systems to Partial Differential Equations II*, Springer Proceedings in Mathematics & Statistics 129,
DOI 10.1007/978-3-319-16637-7_5

thermalizes if initially prepared out of equilibrium. Several indicators can serve to quantify the "absence of thermalization" and characterize the MBL phase:

1. The existence of a complete set of integrals of motion,
2. An area law for the entanglement entropy of eigenstates,
3. The breakdown of the Eigenstate Thermalization Hypothesis (ETH),
4. The vanishing of transport coefficients (e.g. thermal conductivity), etc....

These characterizations are not equivalent (e.g. while the first one fully describes the structure of the many-body Hilbert space, the last one implies only that relaxation times are sub-diffusive, strictly speaking); we refer to the recent review [19] for more explanations.

Both one and many body localization can be understood as the result of frequency mismatch, or lack of resonances, between the different states of the system, in a similar way as the persistence of KAM tori can be understood in classical mechanics. From a mathematical point of view, the study of MBL can be regarded as the investigation of the fate of KAM-like phenomena in the thermodynamic limit, where the volume is sent to infinity while the temperature is kept constant and positive. Recently, a mathematical approach based on the KAM scheme was indeed proposed to show the existence of MBL in one-dimensional spin chains [17].

1.1 MBL Without Disorder

As Anderson localization is naturally associated to an inhomogeneous medium, MBL has mostly been investigated for quenched disordered systems. Nevertheless, in a many-body set-up, interaction between particles could supply the needed frequency mismatch, as it is the case for the finite dimensional systems in the regime where KAM theory applies. The possibility of finding an MBL phase in translation invariant systems was recently explored by several authors [7, 13–15, 18, 21].

To illustrate this, let us consider two classical models of coupled oscillator chains. First, the one-dimensional pinned disordered harmonic chain, with Hamiltonian

$$H(q,p) = \frac{1}{2}\sum_x \{p_x^2 + \omega_x^2 q_x^2 + \mathsf{J}(q_{x+1}-q_x)^2\}, \qquad \omega_x \text{ i.i.d.}, \qquad \mathsf{J} > 0, \qquad (1)$$

is a well known example of perfect thermal insulator, as a consequence of Anderson localization [4]. Because the chain is harmonic, this system can indeed be seen as an example of one-particle localization rather than MBL.[1] This, however, is as such not important for the point we want to make here: to compare this system with the

[1] In fact, one believes that only quantum systems, e.g. like in [17], exhibit MBL in genuinely interacting systems.

rotor chain, described by the Hamiltonian (with the angles $q_x \in \mathbb{T} = \mathbb{R}/(2\pi\mathbb{Z})$ and the angular momenta $\omega_x \in \mathbb{R}$ canonical conjugates),

$$H(\omega, q) = \frac{1}{2}\sum_x \{\omega_x^2 + \mathsf{J}\cos(q_{x+1} - q_x)\}. \tag{2}$$

In contrast to the disordered harmonic chain, ω_x are now dynamical variables rather than frozen degrees of freedom. Nevertheless, they can be regarded as i.i.d. random variables with respect to the Gibbs state at positive temperature. Moreover, stability results for Hamiltonian mechanics at finite volume suggest that ω_x could be stable for very long or even infinite times. Therefore, one might try to compare the high-temperature regime to the large disorder regime for disordered systems, and be led to the somehow provocative conjecture that thermal fluctuations can generate MBL.

1.2 Mobile Bubbles

Despite this suggestive analogy, the replacement of a fixed external disordered field by dynamical variables can in principle deeply modify the long time behavior of the system. Following [8, 15], it is the first purpose of this article to explore a mechanism for delocalization by means of rare mobile thermal bubbles that do not appear in quenched disorder, fully MBL systems (having all their eigenstates localized), but that seem to be unavoidable in translation invariant systems with short range interactions.

We hope that these findings can also shed some light on the localization-delocalization transition in quenched disordered systems. First, in the cases where a transition in function of the energy density is expected [3], the argument developed here suggests in fact that there is no MBL phase at all (except possibly at zero temperature) in the true thermodynamic limit. Second, the absence of an MBL phase in classical disordered anharmonic systems could be explained by a mechanism of rare mobile ergodic bubbles too [2, 20].

At the moment of writing, the validity of our theoretical scenario is being investigated more thoroughly and also tested numerically [9]. Indeed, what our argument ultimately shows is that an overwhelming majority of the states in the Hilbert space are connected via a sequence of resonant transitions (see below). Strictly speaking, delocalization does not follow at once; we refer to [8] for a more detailed discussion on this point.

1.3 Asymptotic Localization

However, though true MBL can fail for the systems considered in this work, some asymptotic localization effects are surely expected, comparable to Nekhoroshev estimates at finite volume. We review here several recent mathematical works, where such estimates have in some sense been extended in the thermodynamic limit, predicting a non-analytic behavior of the thermal conductivity near the (trivial) critical point for various systems, as classical anharmonic disordered chains [16], the rotor and DNLS chain [6], or a quantum particle system analogous to the Bose-Hubbard model [7].

1.4 Organization of the Paper

The main quantum model studied in this paper is introduced in Sect. 2, where the principal difference between quenched and thermal disorder is discussed. In Sect. 3, a first version of the mechanism for delocalization, based on perturbation theory, is presented. In Sect. 4, it is shown that non-perturbative considerations lead to the same conclusion, through a slightly less explicit but probably more robust way. Results on asymptotic localization are reviewed in Sect. 5. Some technical issues are gathered in the appendices.

2 Quenched and Thermal Disorder

We study quantum lattice systems defined in a large volume $V \subset \mathbb{Z}^d$ in the thermodynamic limit $|V| \to \infty$. We think of each lattice site x as containing a variable number of particles $\eta_x \in \{0, \dots, \mathsf{N}\}$, where $\mathsf{N} \geq 1$ is the maximal occupation number per site. There is thus a preferred product basis in the many-body Hilbert space, consisting of classical configurations $|\eta\rangle = |(\eta_x)_x\rangle$. The total Hamiltonian is the sum of local operators:

$$H = \sum_{A \subset \mathbb{Z}^d, |A| \leq R} H_A \qquad \text{for some} \qquad R \geq 1, \tag{3}$$

where A are connected and H_A acts inside A, i.e.,

$$\langle \eta' | H_A | \eta \rangle = 0 \quad \text{if} \quad \eta'_x \neq \eta_x \quad \text{for} \quad x \notin A. \tag{4}$$

We assume that the total number of particles is conserved: $[H, N] = 0$ with $N = \sum_x b_x^* b_x$, where b_x and b_x^* are respectively bosonic annihilation and creation operators with cut-off N: $b_x|\dots, \eta_x, \dots\rangle = \sqrt{\eta_x}|\dots, \eta_x - 1, \dots\rangle$ if $1 \leq \eta_x \leq \mathsf{N}$ and $b_x|\eta\rangle = 0$ if $\eta_x = 0$. Moreover, we consider the case where H is a small perturbation of a Hamiltonian $H^{(0)}$ which is manifestly MBL, for example $H^{(0)}$ of the type (3) and diagonal in the $\{|\eta\rangle\}$ basis,

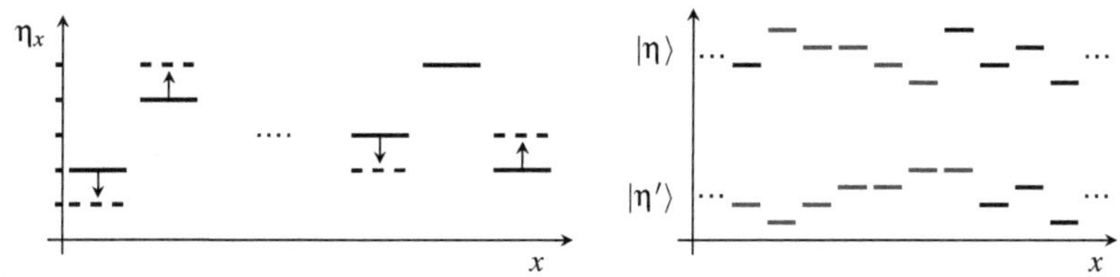

Fig. 1 First order resonances for the Hamiltonian (9). *Left panel* Non-resonant nearest-neighbor hopping on the *left*; resonant second neighbor hopping on the *right*. *Right panel* As shown in Appendix 1, the classical state $|\eta\rangle$ is connected to $|\eta'\rangle$ through a sequence of resonant transitions. The naive resonant spot in $|\eta\rangle$, indicated in *blue*, appears thus as part of a larger one seen in $|\eta'\rangle$

$$H = H^{(0)} + \mathsf{J}H^{(1)}, \qquad \mathsf{J} \ll 1. \tag{5}$$

More precisely, we require that the matrix elements of $\mathsf{J}H^{(1)}$ are smaller than the level spacing of the local Hamiltonians H_A.

Following [8], we give two examples. In the first example, translation invariance is broken by an external field (quenched disorder). At small enough coupling $\mathsf{J} > 0$, resonances, that are potentially responsible for delocalization in the considered perturbative regime, are rare and their location is determined by the external field. In the second example, translation invariance is restored. Although, for small values of J, resonances are equally rare in the second example as in the first one, they can appear everywhere and therefore they can be possibly part of a giant cluster in configuration space (see the right panel in Fig. 1). This is the main difference between these two types of systems.

2.1 Quenched Disordered Hamiltonian

We let $d = 1$, $\mathsf{N} = 1$ (i.e. we have hard-core bosons) and we define a Hamiltonian of the form (5) by

$$H = \sum_x \omega_x b_x^* b_x + \mathsf{J} \sum_x \left\{ b_x^*(b_{x+1} + b_{x+2}) + b_x(b_{x+1}^* + b_{x+2}^*) \right\}, \qquad \omega_x \text{ i.i.d.} \tag{6}$$

(second neighbor hopping is introduced to break integrability). We simply refer to $H^{(0)}$ as the uncoupled energy, and to $H^{(1)}$ as the hopping term. To facilitate the comparison with the translation invariant case described below, we assume that the values of ω_x are uniformly distributed in a finite set $\{\alpha_0, \dots, \alpha_{\mathsf{N}'}\}$ for some $\mathsf{N}' \gg 1$. We think of the numbers $\alpha_0 < \dots < \alpha_{\mathsf{N}'}$ as picked at random in an interval of size 1, so that $\alpha_{k+1} - \alpha_k \sim 1/\mathsf{N}'$. We assume $\mathsf{J} \ll 1/\mathsf{N}'$.

Consider now two classical states $|\eta\rangle$ and $|\eta'\rangle$ connected in the first order in perturbation in J: $\langle \eta' | H^{(1)} | \eta \rangle \neq 0$. These two states only differ from each other by one particle that hopped one or two sites away. Thanks to the properties of the random field $(\omega_x)_x$ described above, at most of the places, the change in energy due to the hopping cannot compensate the resulting change in uncoupled energy:

$$\mathsf{J}\big|\langle\eta'|H^{(1)}|\eta\rangle\big| \ll \big|H^{(0)}(\eta') - H^{(0)}(\eta)\big| = |\omega_y - \omega_x|, \qquad (7)$$

assuming that the particle hops from x to y, with $|y - x| \leq 2$ (here and below, the convention $A(\phi) = \langle\phi|A|\phi\rangle$ is used whenever ϕ is an eigenstate of A). As a consequence, it is possible to remove these transitions via a perturbative procedure [17]. However, in some rare places, the hopping does not produce any change in the uncoupled energy:

$$H^{(0)}(\eta') - H^{(0)}(\eta) = \omega_y - \omega_x = 0. \qquad (8)$$

In these cases, despite the fact that $\mathsf{J} \ll 1$, the hopping becomes the dominant effect and cannot longer be treated perturbatively. These latter transitions are called resonant. However, the most important point seen from (8), is that the location of resonances is determined by the external field $(\omega_x)_x$ alone: the second equality in (8) only depends on the points x and y, and not on the configurations $|\eta\rangle$ and $|\eta'\rangle$. Moreover, the probability with respect to the distribution of $(\omega_x)_x$, of finding a resonance somewhere is of the order of $1/\mathsf{N}'$, so that even for $d > 1$, they form small isolated clusters in physical space. This is the reason why they entail no delocalization.

2.2 Translation Invariant Hamiltonian

We let $d = 1$, $\mathsf{N} \gg 1$ and we consider the Bose-Hubbard Hamiltonian with the cut-off N on the number of particles per site, and with second neighbor hopping:

$$H = \sum_x (b_x^* b_x)^2 + \mathsf{J}\sum_x \big\{b_x^*(b_{x+1} + b_{x+2}) + b_x(b_{x+1}^* + b_{x+2}^*)\big\}. \qquad (9)$$

This Hamiltonian is of the form (5): $H^{(0)}$creates a repulsion between particles, and is referred to as the interaction energy, while $H^{(1)}$ allows again for particles to hop.

First order resonances in J are defined as in the disordered case: two classical states $|\eta\rangle$ and $|\eta'\rangle$ such that $\langle\eta'|H^{(1)}|\eta\rangle \neq 0$, enter in resonance if the inequality in (7) is violated. Thanks to the strong anharmonicity in the interaction, coming from the square in $(b_x^* b_x)^2$, and thanks to the assumption $\mathsf{J} \ll 1$, a resonance between the states $|\eta\rangle$ and $|\eta'\rangle$, due to the hopping of a particle from x to y, with $|x - y| \leq 2$, only occurs if

$$\eta_x' = \eta_y = \eta_x + 1 = \eta_y' + 1 \quad \text{or} \quad \eta_x' = \eta_y = \eta_x - 1 = \eta_y' - 1. \qquad (10)$$

This is illustrated on the left panel of Fig. 1.

For the disordered Hamiltonian considered above, the distribution of the disorder allowed to quantify how likely it was for a hopping between two given sites to induce a resonant transition. This comes as a little surprise on second thought: the inequality in (7) involves two states $|\eta\rangle$ and $|\eta'\rangle$ so that we would naturally expect the nature of the transition (resonant or not) to depend on these states. In general, the Gibbs state can serve to quantify the occurrence of resonances. Here for example, thanks to (10), we can tell how likely it is for a resonance to occur due to a hop between x and y, for a typical classical state $|\eta\rangle$ in the Gibbs state. To simplify the discussion, we consider the Gibbs state at infinite temperature, such that, in the basis of classical configurations, it can be viewed as a bonafide probability measure that gives equal weight to each configuration, allowing to define unambiguously the probability of a resonance. However, as the number of particles is conserved, we can and will fix the density ρ of particles per site. By definition, $0 \leq \rho \leq \mathsf{N}$ is the average number of particles per site.

We distinguish two regimes. First, at very low densities[2] ($\rho \ll 1$), most of the particles are isolated, so that, according to (10), they can typically hop via resonant transitions. An ergodic phase is expected, though, since $d = 1$, atypical clusters of particles can slow down thermalization very much [5]. Second, at densities close to $\mathsf{N}/2$, we deduce from (10) that the probability of finding a resonance somewhere becomes of the order of $1/\mathsf{N}$, precisely as in the quenched disordered case, with N' there being N here. This observation is a possible starting point to address the question of Many-Body Localization [7, 8, 21]. However, due to translation invariance, resonant spots have no preferred location. Therefore, there is no reason why the picture of small isolated clusters would remain valid. Contrary to what a naive look at typical classical configurations suggests, resonant spots might be just the "visible part of the iceberg", a giant cluster in configuration space, as illustrated on the right panel of Fig. 1.

3 Rare Bubbles Due to Perturbative Effects

Following [8], we here study more systematically the effect illustrated on the right panel of Fig. 1, though we switch to $d = 2$ as that allows for a more transparent discussion. More precisely, we show that, in the thermodynamic limit, the overwhelming majority of pairs of classical configurations are connected via a sequence of resonant transitions in the first order in J, as long as some obvious conservation laws are satisfied (resonant transitions preserve the number of sites with a given occupation number). Therefore, we conclude that the choice of a typical classical state as initial state, does not suffice for the picture of small resonant localized islands to survive.

To achieve this, we observe that, in a typical classical state at some density $\rho \sim \mathsf{N}/2$, arbitrary large low density bubbles are found in the thermodynamic limit. Since, as noticed in the previous section, low density states are expected to be ergodic

[2] An analogous conclusion actually also holds at very large densities due to the cut-off N.

for the Hamiltonian (11) below, it should be possible to displace these bubbles across the system through a sequence of resonant transitions in the first order in J, at least if this phenomenon is already visible in a first order approximation. We prove below that this is indeed the case, by showing that resonant transitions allow the bubble to absorb and expel particles from its surrounding. By continuously absorbing and expelling, it can thus move across the system and carry particles all over the places, providing a mechanism to connect all states satisfying the above mentioned conservation laws.[3]

3.1 The Resonant Hamiltonian

Let $d = 2$ and $\mathsf{N} \gg 1$. We consider the Bose-Hubbard Hamiltonian with the cut-off N on the number of particles per site:

$$H = H^{(0)} + \mathsf{J}H^{(1)} = \sum_x (b_x^* b_x)^2 + \frac{\mathsf{J}}{2} \sum_{x,y:x\sim y} \{b_x^* b_y + b_x b_y^*\}, \qquad (11)$$

where $x \sim y$ means $|x-y|_1 = 1$ (nearest neighbors). Transitions that are non-resonant in the first order in J can be removed perturbatively, up to second order corrections [17]. Therefore, if these corrections are neglected in a first approximation, the dynamics is dominated by the resonant Hamiltonian H_{res} defined by

$$\begin{aligned} \langle \eta' | H_{res} | \eta \rangle &= \langle \eta' | H^{(0)} | \eta \rangle + \frac{\mathsf{J}}{2} \delta\left(H^{(0)}(\eta') - H^{(0)}(\eta)\right) \langle \eta' | H^{(1)} | \eta \rangle \\ &= \langle \eta' | H^{(0)} | \eta \rangle + \mathsf{J} \sum_{x,y:x\sim y} \delta(\eta_y - \eta_x - 1) \langle \eta' | b_x^* b_y | \eta \rangle, \end{aligned} \qquad (12)$$

with $\delta(\cdot)$ the Kronecker delta function. Let us make some remarks:

1. The fact that two classical states $|\eta\rangle$ and $|\eta'\rangle$ are connected via a sequence of resonant transitions, can now simply be expressed by saying that $|\eta\rangle$ and $|\eta'\rangle$ are connected via H_{res}, meaning

$$\langle \eta' | \mathrm{e}^{-itH_{res}} | \eta \rangle \neq 0 \quad \text{for some} \quad t \geq 0. \qquad (13)$$

2. By definition of H_{res}, if two states are connected via H_{res}, then they have the same value of $H^{(0)}$. Therefore, the term $\langle \eta' | H^{(0)} | \eta \rangle$ in (12) is constant in every

[3] One could wonder whether localization could emerge at some larger scale. As all involved transitions are resonant, there is surely no obvious reason to think that bubbles are localized. Nevertheless, we will see that a long specific sequence of transitions is required for a bubble to move over a distance of order one, as, for a random sequence of the same length, the bubble would rather shrink to a minimal size than start moving. This feature could suggest that the bubbles are ultimately localized, though no conclusion in that direction can be drawn from this observation alone.

subspace left invariant by H_{res}, and could thus be removed. This confirms that resonant transitions are non-perturbative even though $\mathsf{J} \ll 1$.

3. If the dynamics generated by H_{res} is delocalized, it is hard to conceive that higher orders could restore localization. On the other hand, should the dynamics generated by H_{res} be localized, we would need to check the effect of higher orders before we conclude that there is localization.

3.2 The Bubble

Let us consider a classical initial state $|\eta\rangle$ that is typical at the density $\rho \sim \mathsf{N}/2$ in the thermodynamic limit, and let $\mathscr{B} \subset V$ be a connected region corresponding to a bubble with density $\tilde{\rho}$ so small that $\tilde{\rho} \ll 1$. Since $\mathsf{N} \gg 1$, the number of particles sitting on a given site in $\mathscr{B}$, is approximately given by an exponential distribution with mean $\tilde{\rho}$. We assume that $|\mathscr{B}|$ is large enough so that it is likely to find at least one site occupied by N particles in $\mathscr{B}$ ($|\mathscr{B}| \gtrsim \mathrm{e}\mathsf{d}^{\mathsf{N}/\tilde{\rho}}$ is thus needed). Let us denote by n_k be the number of sites in $\mathscr{B}$ occupied by k particles ($0 \leq k \leq \mathsf{N}$). There exists a number $p \sim \mathrm{e}\mathsf{d}^{1/\tilde{\rho}} \gg 1$ so that $n_k \leq (1/p)n_{k-1}$ for $1 \leq k \leq \mathsf{N}$.

For convenience, let us assume that $\mathscr{B}$ has a rectangular shape, as depicted on Fig. 2. Let us first show that, inside the bubble, it is possible to move the particles

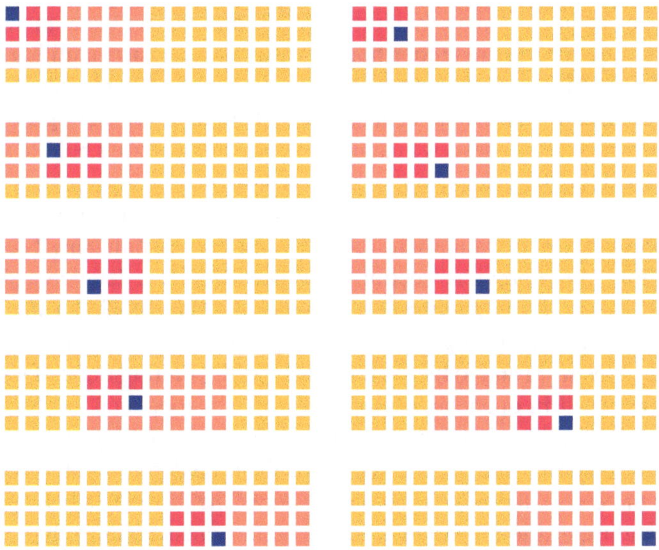

Fig. 2 Possible motion of particles inside a bubble. *Blue square* site with 3 particles, *dark pink square* site with 2 particles, *light pink square* site with 1 particle, *yellow square* site with no particle. From *left* to *right* and from *top* to *bottom* as an example, we use a systematic procedure to show that a classical state where the site occupied by 3 particles is in the *upper left corner*, is connected via H_{res} to a classical state where this site is in the *lower right corner*

all over the places via a sequence of resonant transitions. To go directly to the point, let us consider an initial classical state $|\eta\rangle$ such as depicted on Fig. 2, with high density sites in the upper left corner (our argument shows that we could well have started from a more typical configuration inside the bubble). On Fig. 2, we show a systematic strategy to connect this state via a sequence of resonant transitions (see (13)) to another classical state $|\eta'\rangle$ where the high density sites are in the lower right corner. This is exemplary: it is seen that the sites with N particles (the blue square on Fig. 2) could first be placed everywhere, and that sites with $\mathsf{N}-1$ particles (the dark pink squares) could then in turn be placed at (almost) any of the remaining places, etc...[4]

As a consequence, starting from a classical configuration $|\eta\rangle$ inside the bubble, it is possible to populate any given site on the inside border of the bubble by any given number of particles, via resonant transitions only. Therefore, it is also possible to absorb a particle on a site on the outside border inside the bubble via a sequence of resonant transitions, or to expel a particle outside. We conclude that the bubble can move across the system, and carry particles from any site to any site, via a sequence of resonant transitions.

4 Rare Bubbles Due to Non-perturbative Effects

A drawback of the analysis of Sect. 3 is that it relies on fine details of the model, there given by (11). It is actually shown in [8], and we will see another example below, that if only nearest neighbor hopping was present in (9), all but a vanishing proportion of the eigenstates of the corresponding first order resonant Hamiltonian would be localized. It could thus seem conceivable that, by restricting our attention to a well-chosen class of models, we cannot find any delocalizing bubble at any order, and hence conclude the existence of a localized phase.

We aim to show here that this is not the case, once non-perturbative effects are taken into account. Once again, we follow [8] (a similar reasoning has been developed by [15]). While in the previous section we proceeded via a detailed analysis at low orders, we now assume that, for the full dynamics, ETH is satisfied inside low density bubbles. When their size is large enough, these bubbles can act as baths: they absorb and eject particles from and to the system. It is worth pointing out that such baths are eventually observed in quenched disordered systems too, such as (6), due to large deviations in the disorder that appear unavoidably at large enough volume, no matter how strong the disorder is. In both cases, delocalization does not follow at once since, due to their finite size, these baths are not perfect. The difference between quenched disordered and translation invariant systems comes once again from the fact that, in the latter case, the bubbles can be located everywhere and can therefore move across

[4] It is well possible that, due to geometrical constraints, not precisely all configurations can be reached that way, but this is not relevant for the discussion.

the system by interacting with the rest of the system, while in the former case, their location can be determined a priori if the disorder is strong enough.

We now proceed to a more precise description. Though we study a rather specific case, where our argument can be stated neatly, the described phenomenon is very general. In particular, we expect it to show up for all the Hamiltonians of the type (3–5). In contrast to the cases considered until now, in models (9) and (11), we do not longer take $H^{(0)}$ to be uniformly localized. We know indeed that in these examples, H cannot be such, and we wish to incorporate this feature into $H^{(0)}$. Below, we first describe $H^{(0)}$ in more details, and then show how arbitrarily small perturbations force all eigenstates of $H^{(0)}$ to hybridize to form the eigenstates of H.

4.1 The Hamiltonian $H^{(0)}$

We let $d = 2$ and $\mathsf{N} \gg 1$. The Hamiltonian $H^{(0)}$ is itself the sum of a diagonal interaction term and a hopping term: $H^{(0)} = H_i^{(0)} + \lambda H_h^{(0)}$. We assume $\mathsf{J} \ll \lambda \ll 1$. A natural choice for $H_h^{(0)}$ would be given by $H_h^{(0)} = \frac{1}{2}\sum_{x\sim y}\{b_x^* b_y + b_x b_y^*\}$ but, since non-resonant transitions can be removed perturbatively at first order in λ as pointed out in the previous section, let us consider a caricature, similar to (12), where only resonant hoppings appear:

$$\langle\eta'|H^{(0)}|\eta\rangle = \langle\eta'|H_i^{(0)}|\eta\rangle + \frac{\lambda}{2}\,\delta\left(H_i^{(0)}(\eta') - H_i^{(0)}(\eta)\right)\sum_{x,y:x\sim y}\langle\eta'|b_x^* b_y + b_x b_y^*|\eta\rangle. \quad (14)$$

Let us specify the interaction energy $H_i^{(0)}$. We set

$$H_i^{(0)} = \sum_{A\subset\mathbb{Z}^2:|A|=2} H_A^{(0)} \qquad (A \text{ connected}). \quad (15)$$

The operators $H_A^{(0)}$ are diagonal in the $\{|\eta\rangle\}$ basis, and are such that, for any translation τ_x by x, it holds that $H_{\tau_x A}^{(0)}(\eta) = H_A^{(0)}(\tau_{-x}\eta)$, so that $H_i^{(0)}$ is translation invariant. Therefore, we only need to specify $H_A^{(0)}$ for A being $A_1 = \{(0,0),(1,0)\}$ or $A_2 = \{(0,0),(0,1)\}$, so two functions $f_1, f_2 : \{0,\dots,\mathsf{N}\}^2 \to \mathbb{R}$ such that

$$H_{A_1}^{(0)}(\eta) = f_1\left(\eta_{(0,0)},\eta_{(1,0)}\right), \qquad H_{A_2}^{(0)}(\eta) = f_2\left(\eta_{(0,0)},\eta_{(0,1)}\right).$$

In analogy with the Bose-Hubbard Hamiltonian, we require that typically $H_{A_j}^{(0)}(\eta) \sim \mathsf{N}^2$ $(j = 1, 2)$. Moreover, we impose a genericity condition on f_1, f_2: these functions are injective, the set of values of f_1 has no intersection with the set of values of f_2, and the values $\alpha_1, \ldots, \alpha_{2(\mathsf{N}+1)^2} \in f_1(\{0, \ldots N\}^2) \cup f_2(\{0, \ldots N\}^2)$ are not in any rational combination of each others:

$$k_1\alpha_1 + \cdots + k_{2(\mathsf{N}+1)^2}\alpha_{2(\mathsf{N}+1)^2} \neq 0 \qquad \forall\ (k_1, \ldots, k_{2(\mathsf{N}+1)^2}) \in \mathbb{Z}^{2(\mathsf{N}+1)^2} \backslash \{0\}. \quad (16)$$

We can think of the values $\alpha_1, \ldots, \alpha_{2(\mathsf{N}+1)^2}$ as $2(\mathsf{N}+1)^2$ numbers picked at random in an interval of size 1. In particular, $H_i^{(0)}$ is not invariant under lattice rotations.

The study of the localized and delocalized phases of the Hamiltonian $H^{(0)}$ defined through (14 and 15) boils down to a percolation analysis in $\mathbb{Z}^2$. This is a consequence of the following property, shown in Appendix 2. Let $|\eta\rangle$ be an initial classical state, and let $|\phi(t)\rangle = \mathrm{ed}^{-itH^{(0)}}|\eta\rangle$ be its time evolution after a time $t \geq 0$. Consider four distinct points x, y, z, w forming a square of nearest neighbors:

$$|x-y|_1 \;=\; |y-z|_1 \;=\; |z-w|_1 \;=\; |w-x|_1 \;=\; 1.$$

If it holds that

$$|\eta_x - \eta_y| \;\geq\; 2, \quad |\eta_y - \eta_z| \;\geq\; 2, \quad |\eta_z - \eta_w| \;\geq\; 2, \quad |\eta_w - \eta_x| \;\geq\; 2, \qquad (17)$$

then, for all $t \geq 0$, it holds that

$$\phi_x(t) = \eta_x, \quad \phi_y(t) = \eta_y, \quad \phi_z(t) = \eta_z, \quad \phi_w(t) = \eta_w. \qquad (18)$$

The following picture emerges. At very low density, for a typical initial classical configuration $|\eta\rangle$, particles are few and far between. Under the dynamics generated by $H^{(0)}$, given by (14 and 15), isolated particles evolve like on a billiard: they move freely with the additional constraint that two particles can never come at a distance less or equal to one from each other. Therefore, though a cluster of frozen particles appears here and there, for example because some particles are in a configuration satisfying (17 and 18), we observe a low density gas of interacting particles. Ergodic behavior is expected.

At higher density instead, above some threshold, for a typical initial classical state $|\eta\rangle$, the sites included in a square of nearest neighbors where (17 and 18) is satisfied, form a giant cluster in the lattice, so that the dynamics is frozen. The system is in the MBL phase. Nevertheless, the localization length is not uniform: it becomes arbitrarily large on some islands, whose locations are determined by the initial state $|\eta\rangle$. Indeed, in the thermodynamic limit, low density bubbles of arbitrary large size are found in a typical classical configuration. Since normal thermal behavior is again expected inside the bubbles, the localization length becomes there of the size of the bubble.

In view of the above, despite the rather specific, and perhaps artificial, definition of $H^{(0)}$ given by (14 and 15), we hope that this model furnishes a reasonable cartoon of the MBL phase in translation invariant systems, so that our conclusions will ultimately not depend on the particular choice of this model.

4.2 Hybridization of Eigenstates

Let us assume that the density is large enough so that $H^{(0)}$ is in the localized phase described above. Moreover, we assume that $H^{(1)}$ allows for nearest and second neighbor hopping. Let us give two classical states $|\eta\rangle$ and $|\eta'\rangle$. To simplify the discussion let us assume that the volume V is large but finite, and that $|\eta\rangle$ and $|\eta'\rangle$ contain only one thermal bubble, located at precisely the same place, while all the other sites are frozen. The state $|\eta'\rangle$ is distinct from $|\eta\rangle$ in that a single particle on a frozen site on the outside border of the bubble in $|\eta\rangle$ has moved inside the bubble in $|\eta'\rangle$; we have $\langle\eta'|H^{(0)}|\eta\rangle = 0$ but $\langle\eta'|H^{(1)}|\eta\rangle \neq 0$. We denote by $|\psi\rangle$ any eigenstate of $H^{(0)}$ such that $\langle\eta|\psi\rangle \neq 0$, and by $|\psi'\rangle$ any eigenstate of $H^{(0)}$ such that $\langle\eta'|\psi'\rangle \neq 0$.

We show below that, as soon as the size of the bubble is big enough, given any state $|\psi\rangle$, we can find at least one (but actually much more than one) state $|\psi'\rangle$ such that

$$\left| \frac{\mathrm{J}\langle\psi'|H^{(1)}|\psi\rangle}{H^{(0)}(\psi') - H^{(0)}(\psi)} \right| \gg 1. \tag{19}$$

This means that the extraction of a particle from the surrounding to the inside of the bubble can be achieved via a resonant transition. As this is then true for the reverse process too, we conclude, as in the previous section, that the bubble can eventually move everywhere by absorbing and expelling particles.

In [8], (19) is shown to follow from ETH inside the bubble. In the hope of being more explicit, we here prove (19) under stronger hypotheses (Berry conjecture, see [22]). Let us denote by $\mathcal{N}$ the dimension of the Hilbert space spanned by all the classical states that are connected to $|\eta\rangle$ via a sequence of resonant transitions (see (13)), and by $|\mathcal{B}|$ the physical size of the bubble. It holds that $\mathcal{N} \sim \mathrm{e}^{s|\mathcal{B}|}$ for some entropy density $s > 0$, that depends on the density of particles inside the bubble. Obviously, the values $\mathcal{N}$ and $\mathcal{B}$ would be slightly different for $|\eta\rangle$ replaced by $|\eta'\rangle$ (i.e. the values change slightly before and after the absorption of a particle from the surrounding), but we neglect this as it is irrelevant for the present computation. Let us fix $|\psi\rangle$.

We first estimate the numerator in (19),

$$\langle\psi'|H^{(1)}|\psi\rangle = \sum_{\tilde\eta,\tilde\eta'} \langle\psi'|\tilde\eta'\rangle\langle\tilde\eta|\psi\rangle\langle\tilde\eta'|H^{(1)}|\tilde\eta\rangle,$$

where the sum ranges over the classical states $\tilde\eta$ and $\tilde\eta'$ such that $\langle\tilde\eta|\mathrm{e}^{-itH^{(0)}}|\eta\rangle \neq 0$ for some $t \geq 0$, and $\langle\tilde\eta'|\mathrm{e}^{-it'H^{(0)}}|\eta'\rangle \neq 0$ for some $t' \geq 0$ (the sum is thus over $\mathcal{N} \times \mathcal{N}$ states). Since the bubble is ergodic, we assume, inspired by [22], that

$$\langle\psi'|\tilde\eta'\rangle \sim \frac{\mathrm{e}^{i\theta(\psi',\tilde\eta')}}{\sqrt{\mathcal{N}}}, \qquad \langle\eta|\tilde\psi\rangle \sim \frac{\mathrm{e}^{-i\theta(\psi,\tilde\eta)}}{\sqrt{\mathcal{N}}} \tag{20}$$

for some phases $\theta(\psi', \tilde{\eta}')$ and $\theta(\psi, \tilde{\eta})$. Next, for any state $|\tilde{\eta}\rangle$, there are at most $|\mathscr{B}|$ states $|\tilde{\eta}'\rangle$ such that $\langle\tilde{\eta}'|H^{(1)}|\tilde{\eta}\rangle \neq 0$. Moreover, following again [22], we assume that the phases in (20) are distributed randomly. Therefore, by a central limit argument, we find

$$\mathsf{J}\left|\langle\psi'|H^{(1)}|\psi\rangle\right| \lesssim \mathsf{J}\frac{|\mathscr{B}|\sqrt{\mathscr{N}}}{\mathscr{N}} = \frac{\mathsf{J}|\mathscr{B}|}{\sqrt{\mathscr{N}}}.$$

Let us then look at the denominator in (19). Since there are $\mathscr{N}$ possible states $|\psi'\rangle$, the denominator takes $\mathscr{N}$ possible values depending on the choice of $|\psi'\rangle$, all sitting in an interval of length $|\mathscr{B}|$. Assuming them to be approximately equidistributed, we conclude that, for some $|\psi'\rangle$, the denominator is of size $|\mathscr{B}|/\mathscr{N}$. Altogether we conclude that

$$\left|\frac{\mathsf{J}\langle\psi'|H^{(1)}|\psi\rangle}{H^{(0)}(\psi') - H^{(0)}(\psi)}\right| \sim \frac{\mathsf{J}\mathscr{N}}{\sqrt{\mathscr{N}}} \gg 1$$

for a large enough bubble. This is (19).

5 Asymptotic Localization

For this last section, we consider both classical and quantum systems, and we restrict ourselves to $d = 1$ though this is probably not at all necessary. We assume that $V = \{-(|V|-1)/2, \dots, (|V|-1)/2\}$ for some odd integer $|V| \geq 1$. We still take a Hamiltonian of the form (3), but write it now more simply as $H = \sum_{x\in V} H_x$, where H_x only acts on the variables at the sites x and $x+1$.[5] We also assume that H_x is of the type $H_x = H_x^{(0)} + \mathsf{J}H_x^{(1)}$ with $\{H_x^{(0)}, H_y^{(0)}\} = 0$ for all $x, y \in V$, where $\{\cdot,\cdot\}$ denotes the Poisson bracket in classical mechanics, or the commutator in quantum mechanics. Given a site $a \in V$, we can define the energy current $G_{a,a+1}$ across the bond $(a, a+1)$ by

$$\begin{aligned}\mathsf{J}G_{a,a+1} &= \Big\{H, \sum_{x>a} H_x\Big\} = \{H_a, H_{a+1}\} \\ &= \mathsf{J}\left(\{H_a^{(0)}, H_{a+1}^{(1)}\} + \{H_a^{(1)}, H_{a+1}^{(0)}\}\right) + \mathsf{J}^2\{H_a^{(1)}, H_{a+1}^{(1)}\}.\end{aligned}$$

The Green-Kubo thermal conductivity is defined as the space-time variance of the current in equilibrium at temperature T:

$$\kappa(\mathsf{J}, T) = \lim_{t\to\infty}\lim_{|V|\to\infty} \frac{\mathsf{J}^2}{T^2}\left\langle\left|\frac{1}{\sqrt{t}}\int_0^t \frac{1}{\sqrt{|V|}}\sum_{a\in V} G_{a,a+1}(s)\,\mathrm{d}s\right|^2\right\rangle_T, \tag{21}$$

[5] $H_x = H_{\{x,x+1\}}$ in the notation introduced in (4).

$\langle\cdot\rangle_T$ being the Gibbs state at temperature T, and $G_{a,a+1}(s)$ being the time evolution of $G_{a,a+1}$ at time s under the action of the Hamiltonian.

We aim to understand the behavior of $\kappa(\mathsf{J}, T)$ in the integrable limit, being most notably the limit $\mathsf{J} \to 0$ for fixed $T \sim 1$, though it can also correspond to the limit $T \to 0$ or $T \to \infty$ at fixed $\mathsf{J} \sim 1$ for some of the examples considered below. Rigorous results on the conductivity of non-integrable Hamiltonian dynamics are most likely too hard to get at the present time. Nevertheless, in the examples below, we make the idea of asymptotic localization mathematically precise, by introducing some cut-off, possibly modeled by a energy-conserving noise, in the time integral in (21). We expect this cut-off to furnish a reasonable approximation to obtain an upper bound on κ, in both the scenarios where the dynamics is truly MBL or diffusive.

5.1 Classical Disordered Anharmonic Chain

We consider a quenched disordered system as a first example. Despite the fact that translation invariance is broken, no true MBL phase is expected [2, 10, 20], and it is possible that rare mobile chaotic spots constitute the main delocalization mechanism. Though we could allow for much more generality, let us take as an example the Hamiltonian

$$H(q,p) = \sum_{x\in V}\left(p_x^2 + \omega_x^2 q_x^2 + \mathsf{J}(q_{x+1}-q_x)^2 + \mathsf{J}q_x^4\right), \tag{22}$$

with the convention $q_{(|V|+1)/2} = q_{(|V|-1)/2}$, meaning that free boundary conditions are taken on both ends. We assume the frequencies ω_x to be i.i.d., with smooth compactly supported density, bounded away from zero. Let $n \geq 1$ be some integer (we are interested in the case $n \gg 1$). As a cut-off in time, we perturb the Hamiltonian dynamics by an energy preserving noise that becomes relevant on time scales of order J^{-n}, generating a very slow diffusive behavior. The generator of the full dynamics is given by

$$\begin{aligned} \mathscr{L} &= L_H + \mathsf{J}^n S \quad \text{with} \\ L_H &= \{H,\cdot\} \quad \text{and} \quad Sf(q,p) = \sum_{x\in V}\big(f(q,\dots,-p_x,\dots) - f(q,p)\big). \end{aligned} \tag{23}$$

Let $\tilde{\kappa}(\mathsf{J}, T, n)$ be the Green-Kubo conductivity defined for this noisy dynamics. In [16], we show

Theorem 1 *For given $T > 0$, and for any integer $n \geq 1$, there exists a constant $C_{T,n}$ such that, for almost all realizations of the frequencies ω_x, it holds that $\tilde{\kappa}(\mathsf{J}, T, n) \leq C_{T,n}\mathsf{J}^n$.*

This result suggests that $\kappa(\mathsf{J}, T) = \mathscr{O}(\mathsf{J}^n)$ for any $n \geq 1$ as $\mathsf{J} \to 0$. We expect that the theorem remains true if $\mathsf{J}(q_{x+1}-q_x)^2$ is replaced by $(q_{x+1}-q_x)^2$ in (22), since all the eigenmodes of the chain are localized in the absence of nonlinearity (in $d = 1$ or even in $d > 1$ at strong enough disorder). In that case, as the scaling $\kappa(\mathsf{J}, T) = \kappa(\mathsf{J}/r, rT)$ holds then for any $r > 0$, we conclude that asymptotic localization at small coupling implies also asymptotic localization at low temperature.

5.2 Classical Rotors and DNLS Chain

In [6], Theorem 1 is extended to the translation invariant rotors chain (2) with the same noise as in (23). In this case, the scaling for the true conductivity is given by $\kappa(\mathsf{J}, T) = \frac{1}{r}\kappa(r^2\mathsf{J}, r^2T)$ for any $r > 0$, so that, this time, asymptotic localization at small coupling implies asymptotic localization at large temperature, in agreement with the observations of [12].

The Discrete Non-Linear Schrödinger (DNLS) chain, with Hamiltonian

$$H(\psi, \overline{\psi}) = \sum_{x \in V} \left(|\psi_x|^4 + \mathsf{J}|\psi_{x+1} - \psi_x|^2 \right), \qquad \psi_x \in \mathbb{C} \tag{24}$$

with again $\psi_{(|V|+1)/2} = \psi_{(|V|-1)/2}$, displays a thermal behavior similar to that of the rotors chain in the small coupling limit. Nevertheless, the introduction of an energy preserving noise is non-obvious. We adopt therefore a different, less physical, type of time cut-off. For an integer $n \geq 1$, we define

$$\hat{\kappa}(\mathsf{J}, T, n) = \lim_{|V| \to \infty} \frac{\mathsf{J}^2}{T^2} \left\langle \left| \frac{1}{\sqrt{\mathsf{J}^{-n}}} \int_0^{\mathsf{J}^{-n}} \frac{1}{\sqrt{|V|}} \sum_{a \in V} G_{a,a+1}(s)\, \mathrm{d}s \right|^2 \right\rangle_T ,$$

and in [6], we prove

Theorem 2 *For given $T > 0$, and for the Hamiltonian dynamics generated by the Hamiltonian* (24)*, it holds that* $\lim_{\mathsf{J} \to 0} \mathsf{J}^{-(n-1)} \hat{\kappa}(\mathsf{J}, T, n) = 0$ *for any integer $n \geq 1$.*

Again, this result suggests that $\kappa(\mathsf{J}, T) = \mathscr{O}(\mathsf{J}^n)$ for any $n \geq 1$ as $\mathsf{J} \to 0$. It is observed that asymptotic localization is not to be attributed to any statistical interplay with a second conserved quantity (total momentum for the rotors chain, total ℓ^2-norm for the DNLS chain). The conservation of these quantities could be broken by an additional perturbation. We conjecture that this result stays valid in higher dimensions and for many more similar Hamiltonians, as for example $H(q, p) = \sum_x \{p_x^2 + q_x^4 + \mathsf{J}(q_{x+1} - q_x)\}$. This has however not been shown at the present time.

5.3 Quantum Chain Analogous to the Bose-Hubbard Chain

In [7], we show that a quantum chain, analogous to the Bose-Hubbard chain, becomes asymptotically MBL in the limit $T \to \infty$.[6] The Hamiltonian is given by

$$H = \sum_{x \in V} \left((a_x^* a_x)^q + \mathsf{J}(a_x^* a_{x+1} + a_x a_{x+1}^*) \right), \qquad q > 2,$$

where a_x and a_x^* are respectively bosonic annihilation and creation operators (without cut-off). The need for the constraint $q > 2$ is seen as follows.[7] The typical energy per site is $N_x^q \sim T$ (with $N_x = a_x^* a_x$), so that the typical energy difference due to hopping is given by $(N_x + 1)^q - N_x^q \sim N_x^{q-1} \sim T^{1-1/q}$. The coupling is of order $\mathsf{J} N_x \sim \mathsf{J} T^{1/q}$. Therefore, the ratio between coupling and energy difference is given by $\mathsf{J} T^{2/q-1}$. This ratio needs to go to zero as $T \to \infty$ for resonances to become sparse in this limit, imposing $q > 2$.

For any integer $n \geq 1$, let us define

$$\hat{\kappa}(\mathsf{J}, T, n) = \lim_{|V| \to \infty} \frac{\mathsf{J}^2}{T^2} \left\langle \left| \frac{1}{\sqrt{T^n}} \int_0^{T^n} \frac{1}{\sqrt{|V|}} \sum_{a \in V} G_{a,a+1}(s)\, \mathrm{d}s \right|^2 \right\rangle_T .$$

The following theorem is shown in [7].

Theorem 3 *There exists $c < +\infty$ such that, for the Hamiltonian dynamics generated by the Hamiltonian (5), and for fixed $\mathsf{J} > 0$, it holds that* $\lim_{T \to \infty} T^{n-c} \hat{\kappa}(\mathsf{J}, T, n) = 0$ *for any integer $n \geq 1$.*

Let us notice that the conductivity depends separately on the temperature T and the coupling strength J. Our result does not imply asymptotic localization as $\mathsf{J} \to 0$ for a fixed temperature $T \sim 1$. Actually, especially for $d \geq 2$, the considerations of Sect. 3 lead us to think that the conductivity behaves as a power law in J in the limit $\mathsf{J} \to 0$.

Acknowledgments We thank D. Huse, J. Imbrie, R. Nandkishore, M. Müller and M. Schiulaz for discussions.

---

[6] This limit could be compared to the limit $\mathsf{N} \to \infty$, while keeping the temperature infinite, in the set-up of Sect. 2.

[7] A regime of asymptotic MBL could still be expected for $q = 2$ if both $T \to \infty$ and $\mathsf{J} \to 0$.

Appendix 1

We show a sequence of possible steps to connect $|\eta\rangle$ to $|\eta'\rangle$ via resonant transitions on Fig. 1. We start from $|\eta\rangle$ in the upper left corner, go from left to right and from top to bottom, and end up with $|\eta'\rangle$ in the lower right corner. Occupation numbers marked in red are the ones that will get swapped.

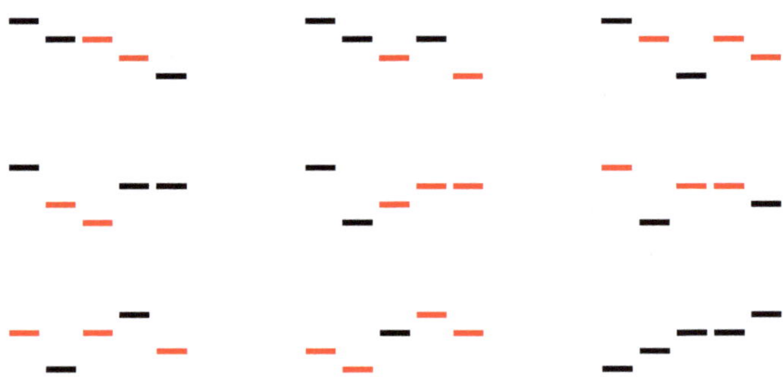

Appendix 2

We prove (17 and 18). Consider an initial classical state $|\eta\rangle$ and another classical state $|\eta'\rangle$ such that $\langle\eta'|H^{(0)}|\eta\rangle \neq 0$. There are thus points x, y satisfying $|x-y|_1 = 1$ and such that $\eta'_x = \eta_x + 1$ and $\eta'_y = \eta_y - 1$, while $\eta'_z = \eta_z$ for all $z \neq x, y$. For concreteness, let us assume that $x = (x_1, x_2)$ and $y = (x_1, x_2 + 1)$ (other cases analogous), let us write $\eta_x = a$, $\eta_y = b$ and let us denote by $r, s, \ldots, w$ the occupation numbers on the neighboring sites. The transition from $|\eta\rangle$ to $|\eta'\rangle$ is represented as

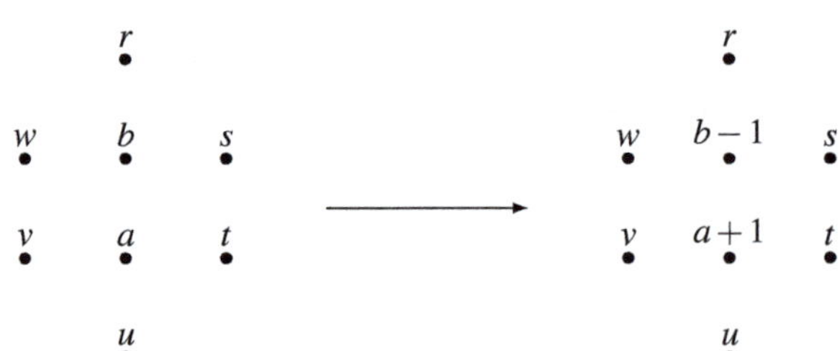

Because transitions must preserve the interaction energy according to (14), and because of the genericity condition (16), the states $|\eta\rangle$ and $|\eta'\rangle$ must contain "the same oriented nearest neighbors pairs with the same multiplicity", meaning that all the patterns (w, b), (b, s), (v, a), (a, t), $\binom{r}{b}$, $\binom{b}{a}$, $\binom{a}{u}$ must be found in η'. This imposes strong constraints on $|\eta\rangle$ for the transition to be possible. First we find that we need to have $b = a + 1$, and then that $|\eta\rangle$ must actually be a configuration of one of the following two types

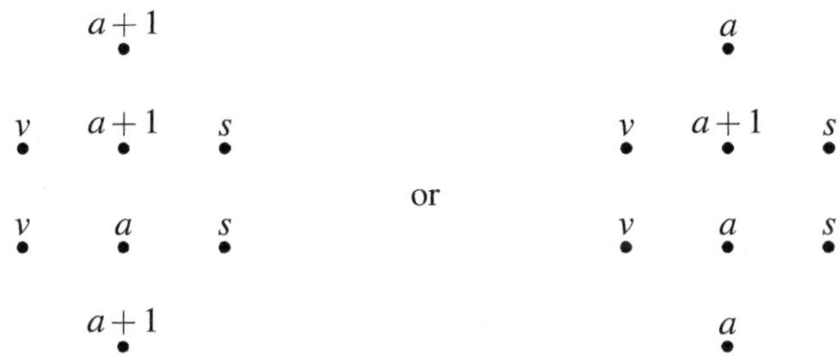

for some s and v. Now, for $|\eta\rangle$ satisfying (17), we find a configuration depicted as

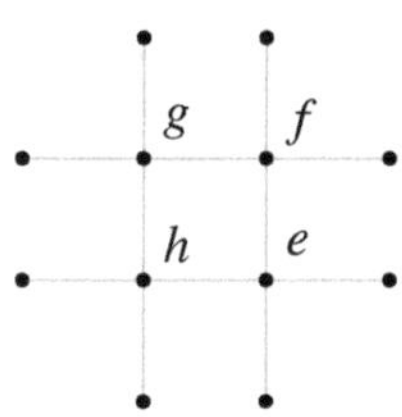

$$|g-f| \geq 2,\ |f-e| \geq 2,\ |e-h| \geq 2,\ |h-g| \geq 2.$$

In order for $|\eta\rangle$ to be connected by $H^{(0)}$ to a state $|\eta'\rangle$ where at least one of the values e, f, g or h has changed, a particle must hop along one of the twelve bonds appearing on the last figure. By the above, this is not possible. This shows (18).

References

1. Anderson, P.W.: Absence of diffusion in certain random lattices. Phys. Rev. **109**(5), 1492–1505 (1958)
2. Basko, D.M.: Weak chaos in the disordered nonlinear Schrödinger chain: destruction of Anderson localization by Arnold diffusion. Ann. Phys. **326**(7), 1577–1655 (2011)
3. Basko, D.M., Aleiner, I.L., Altshuler, B.L.: Metal-insulator transition in a weakly interacting many-electron system with localized single-particle states. Ann. Phys. **321**(5), 1126–1205 (2006)
4. Bernardin, C., Huveneers, F.: Small perturbation of a disordered harmonic chain by a noise and an anharmonic potential. Probab. Theor. Relat. Fields **157**, 301–331 (2013)
5. Carleo, G., Becca, F., Schiró, M., Fabrizio, M.: Localization and glassy dynamics of many-body quantum systems. Sci. Rep. **2**, 243 (2012)
6. De Roeck, W., Huveneers, F.: Asymptotic localization of energy in non-disordered oscillator chains, to appear in Communications on Pure and Applied Mathematics (2013). arXiv:1305.5127
7. De Roeck, W., Huveneers, F.: Asymptotic quantum many-body localization from thermal disorder. Commun. Math. Phys. (2014). doi:10.1007/s00220-014-2116-8

8. De Roeck, W., Huveneers, F.: Scenario for resonant delocalization in translation invariant systems (2014). arXiv:1405.3279
9. De Roeck, W., Huveneers, F., Müller, M., Schiulaz, M.: In preparation
10. Dhar, A., Lebowitz, J.L.: Effect of phonon-phonon interactions on localization. Phys. Rev. Lett. **100**(13), 134301 (2008)
11. Fröhlich, J., Spencer, T.: Absence of diffusion in the Anderson tight binding model for large disorder or low energy. Commun. Math. Phys. **88**(2), 151–184 (1983)
12. Giardinà, C., Livi, R., Politi, A., Vassalli, M.: Finite thermal conductivity in 1D lattices. Phys. Rev. Lett. **84**(10), 2144–2147 (2000)
13. Grover, T., Fisher, M.: Quantum disentangled liquids, to appear in J. Stat. Mech.: Theor. Exp. (2013). arXiv:1307.2288
14. Hickey, J.M., Genway, S., Garrahan, J.P.: Signatures of many-body localisation in a system without disorder and the relation to a glass transition (2014). arXiv:1405.5780
15. Huse, D.A., Nandkishore, R.: Private communication
16. Huveneers, F.: Drastic fall-off of the thermal conductivity for disordered lattices in the limit of weak anharmonic interactions. Nonlinearity **26**(3), 837–854 (2013)
17. Imbrie, J.Z.: On many-body localization for quantum spin chains (2014). arXiv:1403.7837
18. Kagan, Y., Maksimov, L.A.: Localization in a system of interacting particles diffusing in a regular crystal. Zhurnal Eksperimental'noi i Teoreticheskoi Fiziki **87**, 348–365 (1984)
19. Nandkishore, R., Huse, D.A.: Many body localization and thermalization in quantum statistical mechanics, to appear in Annual Reviews of Condensed Matter Physics **6**(2015). arXiv:1404.0686 (2014)
20. Oganesyan, V., Pal, A., Huse, D.: Energy transport in disordered classical spin chains. Phys. Rev. B **80**(11), 115104 (2009)
21. Schiulaz, M., Müller, M.: Ideal quantum glass transitions: many-body localization without quenched disorder. AIP Conf. Proc. **1610**, 11 (2014)
22. Srednicki, M.: Chaos and quantum thermalization. Physical Review E **50**(2), 888–901 (1994)

Stability of Non-deterministic Systems

Pedro Duarte and Maria Joana Torres

Abstract A space of non-deterministic dynamical systems of Markov type on compact manifolds is considered. This is a natural space for stochastic perturbations of maps. For such systems, both the combinatorial stability, of the periodic attractors, and the spectral stability, of the invariant measures, are characterized and its genericity established.

Keywords Stochastic dynamical systems · Combinatorial stability · Spectral stability

1 Introduction

Given a state space X, any function f that associates to each state $x \in X$ a state probability transition f_x on X will be called a (discrete time) stochastic dynamical system or, simply, a Markov system. Deterministic dynamical systems correspond to such functions when each value $f_x = \delta_{f(x)}$ is a Dirac measure sitting on some point $f(x) \in X$.

In the present work we will consider a space $\mathscr{H}(X)$ of stochastic dynamical systems defined on a compact Riemannian manifold X, with volume measure m, which is large and natural to make stochastic perturbations in continuous deterministic dynamical systems. Our main goal is to study and compare, for generic systems $f \in \mathscr{H}(X)$, the combinatorial-topological stability of the limit behaviour of the nondeterministic system $\varphi_f : x \mapsto supp(f_x)$, with the spectral stability of the

P. Duarte
CMAF, Departamento de Matemática, Faculdade de Ciências da, Universidade de Lisboa, Campo Grande, 1749-016 Lisboa, Portugal
e-mail: pedromiguel.duarte@gmail.com

M.J. Torres (✉)
CMAT, Departamento de Matemática e Aplicações, Universidade do Minho, Campus de Gualtar, 4700-057 Braga, Portugal
e-mail: jtorres@math.uminho.pt

P. Gonçalves and A.J. Soares (eds.), *From Particle Systems to Partial Differential Equations II*, Springer Proceedings in Mathematics & Statistics 129,
DOI 10.1007/978-3-319-16637-7_6

linear operator $\mathscr{L}_f : \mu \mapsto f_*\mu$, that to each probability distribution μ associates the μ-conditional probability distribution in the next instant, also known as the Perron-Frobenius operator.

Given a point-set map $\varphi : X \to \mathscr{P}(X)$, any sequence $x_0, x_1, \cdots, x_n$ such that $x_i \in \varphi(x_{i-1})$ for $i = 1, \cdots, n$ will be called an *orbit* of φ, and we shall say that x_n is an *iterate* of the state x_0. If there is $\varepsilon > 0$ such that $d(x_i, \varphi(x_{i-1})) < \varepsilon$ for all $i = 1, \ldots, n$ then $x_0, x_1, \cdots, x_n$ will be called an *ε-pseudo orbit* of φ and we shall say that x_n is an *ε-pseudo iterate* of x_0. We shall say that x_n is a *pseudo iterate* of x_0 if for every $\varepsilon > 0$, x_n is an ε-pseudo iterate of x_0.

The *recurrent set* of φ, denoted by $\Omega(\varphi)$, is the set of all states $x \in X$ such that x is an iterate of x. Analogously, the *chain recurrent set* of φ, denoted by $R(\varphi)$, is the set of all states $x \in X$ such that x is a pseudo iterate of x.

The point-set map φ has two limit sets: the *Ω-final*, or *final recurrent set*, denoted by $\Omega_{\text{final}}(\varphi)$, of all states $x \in \Omega(\varphi)$ such that every iterate of x still has some iterate which comes back to x, and the *R-final*, or *final chain recurrent set*, denoted by $R_{\text{final}}(\varphi)$, of all states $x \in R(\varphi)$ such that every pseudo iterate of x still has some pseudo iterate which comes back to x. These limit sets contain all the asymptotic dynamical behaviour of the point-set map φ.

A point-set map φ is called *open* when its graph

$$\text{graph}\,(\varphi) = \{(x, y) \in X \times X \,:\, y \in \varphi(x),\ \forall x \in X\}$$

is an open set in $X \times X$, and $\varphi(x)$ is connected for all $x \in X$.

Theorem
If φ is an open map then both limit sets $\Omega_{\text{final}}(\varphi)$ and $R_{\text{final}}(\varphi)$ decompose into a finite number of connected pieces which are permuted by φ.

The restriction of φ to each of these pieces is, in some sense, irreducible.

The system φ will be called *combinatorially stable* when this topological decomposition is stable under perturbations. Since any iterate is also a pseudo iterate, there is a natural relation between the connected pieces of $\Omega_{\text{final}}(\varphi)$ and those of $R_{\text{final}}(\varphi)$. We shall see that for a generic system φ, with respect to some natural topology, this relation is bijective, and that whenever this happens the system is combinatorially stable, thus obtaining:

Theorem
There is an open and dense set of combinatorially stable systems.

Given $f \in \mathscr{H}(X)$, both topological decompositions for the limit sets of φ_f, correspond to the unit circle spectral decomposition of the linear operator $\mathscr{L}_f$ acting on the Banach space $L^1(X, m)$. There is an $\mathscr{L}_f$-invariant spectral decomposition $L^1(X, m) = E_0(f) \oplus E_1(f)$ which corresponds to the spectrum partition $\sigma\left(\mathscr{L}_f\right) = \sigma_0\left(\mathscr{L}_f\right) \cup \sigma_1\left(\mathscr{L}_f\right)$, where

$$\sigma_0\left(\mathscr{L}_f\right) = \{\lambda \in \sigma\left(\mathscr{L}_f\right) \ : \ |\lambda| < 1\},$$

and

$$\sigma_1\left(\mathscr{L}_f\right) = \{\lambda \in \sigma\left(\mathscr{L}_f\right) \ : \ |\lambda| = 1\}.$$

This second component of the spectrum consists of a finite number of eigenvalues, all with finite multiplicity. Hence $\dim E_1(f) < +\infty$, and the operator $\mathscr{L}_f$ is quasi-compact.

The system f will be called *spectrally stable* if there is some $0 < k < 1$ such that for every small perturbation g of f:

(a) $\mathscr{L}_f\,|_{E_1(f)}$ is conjugated to $\mathscr{L}_g\,|_{E_1(g)}$;
(b) $\left\|\mathscr{L}_g\,|_{E_0(g)}\right\| \leq k$.

In our setting the operator $\mathscr{L}_f$ depends continuously on f and, therefore, so does the spectrum $\sigma\left(\mathscr{L}_f\right)$. The spectral stability of f relates with the fact that no eigenvalue can enter, or leave, the unit circle.

In Sect. 5 we establish the spectral stability for a generic system f, with respect to some natural metric in $\mathscr{H}(X)$.

Theorem
There is an open and dense set of systems $f \in \mathscr{H}(X)$ for which φ_f is combinatorially stable and f is spectrally stable.

We are naturally led to consider finite state Markov chains when trying to approximate a continuous dynamical system by discretizing the manifold X. Finite state Markov chains are the stochastic, or random dynamical systems on a finite state space. One may think that these dynamical systems are what we actually see when running computer simulations of deterministic dynamical systems. Each such dynamical system is specified by a *stochastic matrix* with the state probability transitions. The stochastic matrix is the Perron operator of this finite state system. The Markov chain also determines an *oriented graph*, encapsulating some qualitative aspects of the system behaviour. The theory of finite state Markov chains establishes a correspondence between spectral properties of the stochastic matrix on one side, and combinatorial properties of the corresponding graph on the other hand. See, e.g., [1].

For measurable spaces and Markov systems satisfying the *Doeblin-condition* similar spectral results were obtained in a more general setting by Doob [2], but the stability problem is never addressed. As far as we know the topological-combinatorial approach of our work is also new (see [3, 4]).

In the theory of smooth deterministic hyperbolic systems $f : X \to X$ the *spectral decomposition theorem* states that there is a decomposition of the non-wandering set $\Omega(f)$ into a finite number of hyperbolic basic sets which are permuted by f. The dynamics of f partially orders the basic set components of $\Omega(f)$, the minimal, or final, elements being the *attractors* of f. Analogously, in our setting, there are partially ordered decompositions of the recurrent and chain recurrent sets, $\Omega(\varphi)$ and

$R(\varphi)$, respectively. The final connected components in $\Omega_{\text{final}}(\varphi)$, and $R_{\text{final}}(\varphi)$, are the *attractors*' equivalent in $\Omega(\varphi)$, and $R(\varphi)$, respectively. As the name indicates this decomposition relates with the spectral decomposition of the linear operator which describes the action of f on the tangent vector fields to X. The ergodic theory of these systems is also well studied. It is well known that in each attractor there is a unique ergodic stochastically stable measure, called the *physical measure* of the attractor. Furthermore, almost every point $x \in X$ lies in the basin of attraction of one of these physical measures. The concept of stochastic stability is usually attributed to Kolmogorov. Roughly, a measure μ is said stochastically stable if it is stable under small stochastic perturbations of the system f. In general it has been conjectured by Palis [6] that for a dense set of dynamics, *the system has a finite set of transitive, stochastically stable attractors whose basins of attraction cover almost every point in* X. This conjecture suggested the main motivation for the present study: to understand stochastic stability in the realm of stochastic dynamical systems, at least in a class of Markov systems which is suitable for stochastic perturbations of continuous maps.

2 Topological Semigroups of Open Maps

Several semigroups of point-set maps are defined, namely open, continuous and Lipschitz point-set maps. The key concept of topological semigroup of open maps is introduced.

Throughout this work X will denote a compact Riemannian manifold of dimension n, d will be the geodesic distance on X and m will be the corresponding normalized ($m(X) = 1$) Riemannian volume. Similar notation will be used on $X \times X$, where d will stand for the metric $d((x_1, y_1), (x_2, y_2)) = \max\{d(x_1, x_2), d(y_1, y_2)\}$.

Let $\mathscr{S}(X)$ be the space of all point-set maps on X, that is, $\varphi \in \mathscr{S}(X)$ if and only if φ is a map from X into the power set of X, i.e. $\varphi : X \to \mathscr{P}(X)$. For a point-set map $\varphi \in \mathscr{S}(X)$ and a subset $A \subseteq X$ the *image* $\varphi(A) \in \mathscr{P}(X)$ is defined by $\varphi(A) = \cup_{x \in A}\varphi(x)$. We say that $A \subseteq X$ is *φ-invariant* when $\varphi(A) \subseteq A$. Analogously, we say that A is *fully φ-invariant* if $\varphi(A) = A$. For two point-set maps $\varphi, \psi \in \mathscr{S}(X)$ the usual composition product $\varphi \circ \psi : X \to \mathscr{P}(X)$ of φ and ψ at x is defined by

$$(\varphi \circ \psi)(x) = \varphi(\psi(x)) = \cup_{y \in \psi(x)}\varphi(y) \ .$$

Clearly, $\mathscr{S}(X)$ with this composition product is a semigroup.

We define $\mathscr{O}(X)$ to be the space of all open point-set maps φ in $\mathscr{S}(X)$. See the definition of open map in the introduction.

Proposition 1 ([3, Proposition 4.1.]) *Given $\varphi \in \mathscr{O}(X)$ and an open set $C \subseteq X$, if C is connected then $\varphi(C)$ is open and connected.*

From the previous proposition we easily have the following.

Proposition 2 ([3, Proposition 4.2.]) *$\mathscr{O}(X)$ is a subsemigroup of $\mathscr{S}(X)$.*

Given sets $U, V \subseteq X$ we will use the following notation: $B_r(U)$, respectively $\overline{B}_r(U)$, denotes the set of all points whose distance d to U is less than, respectively less than or equal to r, and $\rho_H(U, V) = \inf \{ r > 0 \ : U \subseteq B_r(V) \text{ and } V \subseteq B_r(U) \}$ denotes the Hausdorff distance between U and V. Furthermore, $\overline{U}$ is the closure of U in X and U° is the interior of U in X. Similar notation will be used on $X \times X$.

We say that a point-set map is *Lipschitz* if and only if the map $x \mapsto \varphi(x)$ is Lipschitz with respect to the Hausdorff distance ρ_H, i.e. there is $K > 0$ such that $\rho_H(\varphi(x), \varphi(y)) \leq Kd(x, y)$ for every $y \in X$. We shall denote by $\mathrm{Lip}(\varphi)$ the greatest lower bound of all Lipschitz constants K for this map. We denote by $\mathscr{O}_{\mathrm{Lip}}(X)$ the subset of all Lipschitz point-set maps in $\mathscr{O}(X)$.

We recall the following continuity concepts. A point-set map $\varphi : X \to \mathscr{P}(X)$ with non-empty values is called *lower semi-continuous*, respectively *upper semi-continuous*, if for every $x \in X$ and $\varepsilon > 0$ there is a neighborhood N_x of x in X such that for any $y \in N_x$, $\varphi(x) \subseteq B_\varepsilon(\varphi(y))$, respectively $\varphi(y) \subseteq B_\varepsilon(\varphi(x))$. It is called *continuous* if $\varphi : X \to \mathscr{P}(X)$ is both lower semi-continuous and upper semi-continuous. This means continuity with respect to the Hausdorff distance. We denote by $\mathscr{O}_{\mathrm{Cont}}(X)$ the subset of all continuous point-set maps in $\mathscr{O}(X)$.

Proposition 3 ([3, Proposition 4.6.]) *$\mathscr{O}_{\mathrm{Lip}}(X)$ and $\mathscr{O}_{\mathrm{Cont}}(X)$ are both subsemigroups of $\mathscr{O}(X)$.*

Given $\varphi \in \mathscr{O}(X)$ we define $\overline{\varphi}$, $\widehat{\varphi} : X \to \mathscr{P}(X)$ by setting

$$\mathrm{graph}\,(\overline{\varphi}) = \overline{\mathrm{graph}\,(\varphi)} \quad \text{and}$$

$$\mathrm{graph}\,(\widehat{\varphi}) = \left(\overline{\mathrm{graph}\,(\varphi)}\right)^\circ ,$$

We call the open map $\widehat{\varphi}$, whose graph is the interior of the closure of graph (φ), the *regularization* of φ. The following proposition is easily proved.

Proposition 4 ([3, Proposition 4.7.])

(1) *All maps φ in $\mathscr{O}(X)$ are lower semi-continuous.*
(2) *For every map $\varphi \in \mathscr{O}(X)$, $\overline{\varphi}$ is upper semi-continuous.*
(3) *Every map $\varphi \in \mathscr{O}_{\mathrm{Lip}}(X)$ is continuous.*

Because there are several natural non-equivalent ways of endowing $\mathscr{O}(X)$, and its sub-semigroups, with some topology we give the following abstract definition. Consider any sub-semigroup of open maps $\mathscr{O}_1 \subseteq \mathscr{O}(X)$, endowed with some topology.

Definition 1 We say that $\mathscr{O}_1$ is a *topological semigroup of open maps* if

(1) the Hausdorff distance between open map graphs is continuous;
(2) for each $\varphi \in \mathscr{O}_1$, there is a family of open maps $\{\tilde{\varphi}_\varepsilon\}_{\varepsilon>0}$ in $\mathscr{O}_1$ such that

(a) $\overline{\text{graph}(\varphi)} = \bigcap_{\varepsilon>0} \text{graph}(\tilde{\varphi}_\varepsilon)$;

(b) for all $\varepsilon_1, \varepsilon_2$, if $\varepsilon_1 > \varepsilon_2 > 0$ then $\overline{\text{graph}\left(\tilde{\varphi}_{\varepsilon_2}\right)} \subseteq \text{graph}\left(\tilde{\varphi}_{\varepsilon_1}\right)$; and

(c) $\lim_{\varepsilon\to 0^+} \tilde{\varphi}_\varepsilon = \varphi$ w.r.t. $\mathscr{O}_1$ topology;

(3) given $\varepsilon > 0$, an integer $N \in \mathbb{N}$, and non-empty open subsets $U, V \subseteq X$ such that $\overline{U \times V} \subseteq \text{graph}\left(\varphi^N\right)$, there is a neighborhood $\mathscr{N}$ of φ in $\mathscr{O}_1$ such that for all $\psi \in \mathscr{N}$ and $x \in \overline{U}$, $m(V \setminus \widehat{\psi^N}(x)) < \varepsilon$, where $\widehat{\psi}$ denotes the regularization of ψ.

Condition (2) above is an outer continuity assumption that says every open map φ can be well approximated from above within the topology. Condition (3) expresses a kind of inner, or lower, continuity.

Identifying each $\varphi \in \mathscr{O}(X)$ with its graph we can see $\mathscr{O}(X)$ as a subset of the space of all non-empty connected open subsets of $X \times X$. Therefore, we can consider on $\mathscr{O}(X)$ and its sub-semigroups, topologies induced from general (topological) spaces of open sets. See [5] for an overview on topological spaces of sets. We shall now topologize $\mathscr{O}(X)$ with a topology that is natural to address the subtle concept of *combinatorial stability* for continuous deterministic dynamical systems (see [3, Sect. 6]). First let $\mathscr{U}(X)$ denote the space of all non-empty connected open subsets of X. We define the following pseudo-metric ρ in $\mathscr{U}(X)$. Given $U, V \in \mathscr{U}(X)$,

$$\rho(U, V) = \max\left\{\rho_H(U, V), \rho_H(U^e, V^e)\right\},$$

where ρ_H stands for the Hausdorff distance and U^e denotes the exterior of U in X. Now consider $\mathscr{O}(X)$ as a subset of $\mathscr{U}(X \times X)$ and let ρ be the induced pseudo-metric, which is given by

$$\rho(\varphi, \psi) = \rho\left(\text{graph}(\varphi), \text{graph}(\psi)\right).$$

Proposition 5 *With the topology associated to ρ, $\mathscr{O}(X)$ is a topological semigroup of open maps.*

To prove Proposition 5, we first introduce the following two families of open maps. Given $\varphi \in \mathscr{O}(X)$, define φ_ε^* by

$$\text{graph}\left(\varphi_\varepsilon^*\right) = B_\varepsilon(\text{graph}(\varphi)),$$

and define $\varphi_\varepsilon^\circ$ setting $\varphi_\varepsilon^\circ(x)$ to be the largest connected component of the open set

$$\left\{ y \in X \, : \, d\left((x, y), \text{graph}(\varphi)^c\right) > \varepsilon \right\},$$

where $\text{graph}(\varphi)^c$ denotes the complement of $\text{graph}(\varphi)$ in $X \times X$. Then $\varphi_\varepsilon^* \in \mathscr{O}(X)$, and $\varphi_\varepsilon^\circ \in \mathscr{O}(X)$ for all small enough $\varepsilon > 0$.

Next we provide the following characterization of the ε-ball for the pseudo-metric ρ. Given open maps $\varphi, \psi : X \to \mathscr{P}(X)$, we will write $\varphi \prec \psi$ to mean that $\overline{\text{graph}(\varphi)} \subseteq \text{graph}(\psi)$.

Proposition 6 *Given $\varepsilon > 0$, for every $\varphi, \psi \in \mathscr{O}(X)$,*

$$\rho(\varphi, \psi) < \varepsilon \implies \psi_\varepsilon^\circ \prec \widehat{\varphi} \prec \psi_\varepsilon^* .$$

Proof First $\rho_H(\varphi, \psi) < \varepsilon$ implies that

$$\overline{\operatorname{graph}(\widehat{\varphi})} \subseteq \operatorname{graph}(\overline{\varphi}) \subseteq B_\varepsilon(\operatorname{graph}(\psi)) = \operatorname{graph}\left(\psi_\varepsilon^*\right).$$

On the other hand $\rho_H(\operatorname{graph}(\varphi)^e, \operatorname{graph}(\psi)^e) < \varepsilon$ implies that

$$\operatorname{graph}(\varphi)^e \subseteq B_\varepsilon(\operatorname{graph}(\psi)^e) \subseteq B_\varepsilon(\operatorname{graph}(\psi)^c),$$

which in turn implies that $\psi_\varepsilon^\circ \prec \widehat{\varphi}$. □

We prove now Proposition 5.

Proof It is clear that Definition 1(1) holds. To prove Definition 1(2) we just need to take $\tilde{\varphi}_\varepsilon \equiv \varphi_\varepsilon^*$. To prove Definition 1(3), let $U, V \subseteq X$ be non-empty open sets such that $\overline{U \times V} \subseteq \overline{\operatorname{graph}\left(\varphi^N\right)}$. Taking $\delta > 0$ small enough we have $\overline{U \times V} \subseteq \operatorname{graph}\left((\varphi_\delta^\circ)^N\right)$. Consider the $\delta-$neighbourhood $\mathscr{N} = B_\delta(\varphi)$ with respect to the pseudo-metric ρ. If $\psi \in \mathscr{N}$ then, by Proposition 6, $\varphi_\delta^\circ \prec \widehat{\psi}$, implying that

$$\overline{U \times V} \subseteq \operatorname{graph}\left((\varphi_\delta^\circ)^N\right) \subseteq \operatorname{graph}\left((\widehat{\psi})^N\right) \subseteq \operatorname{graph}\left(\widehat{\psi^N}\right).$$

Therefore, $m(V \setminus \widehat{\psi^N}(x)) = 0$ for all $x \in \overline{U}$. □

3 Combinatorial Stability of Open Maps

Combinatorial stability of open maps is defined and characterized. Its genericity is proved.

Let us briefly recall the main dynamical concepts for open maps (see [3]). Given $\varphi \in \mathscr{O}(X)$, a sequence $x_0, x_1, \cdots, x_n$ such that $x_i \in \varphi(x_{i-1})$ for all $i = 1, \cdots, n$ is called a $\varphi-$*orbit*, and we say that x_n is a $\varphi-$*iterate* of x_0. If for every $\varepsilon > 0$, y is a φ_ε^*-iterate of x, where φ_ε^* is the open map whose graph is an $\varepsilon-$radius ball of graph (φ), we say that y is a φ*–pseudo-iterate* of x. The *recurrent* and *chain-recurrent sets* of φ are defined respectively by $\Omega(\varphi) = \{x \in X : x \text{ is a } \varphi\text{–iterate of } x\}$ and $R(\varphi) = \{x \in X : x \text{ is a } \varphi\text{–pseudo-iterate of } x\}$. Both these sets split into equivalence classes, each class being formed by states which are accessible from each other. The set of all these classes is then partially ordered by the dynamics of φ. At the bottom of this hierarchy are two special limit sets: the *final recurrent* and the *final chain-recurrent sets*, denoted respectively by $\Omega_{\text{final}}(\varphi)$ and $R_{\text{final}}(\varphi)$, of all states $x \in \Omega(\varphi)$ ($x \in R(\varphi)$) such that every iterate (pseudo-iterate) of x still has some iterate (pseudo-iterate) which comes back to x. These limit sets contain all the asymptotic dynamical behaviour of φ. They both decompose into a finite number of equivalence classes, called respectively $\Omega-$final and $R-$final classes. We denote by $\Lambda^{\Omega}_{\text{final}}(\varphi)$ respectively $\Lambda^{R}_{\text{final}}(\varphi)$ the set of all equivalence classes of the limit sets $\Omega_{\text{final}}(\varphi)$ and $R_{\text{final}}(\varphi)$. Each $\Omega-$final and $R-$final class decomposes into a finite number of connected pieces, called respectively *$\Omega-$final* and *$R-$final components*, which are permuted by φ. See Theorems 5.1 and 5.2 of [3]. The restriction of φ to each of these pieces is in some sense irreducible. We call period of a final class to the number of its connected components. The period of a connected component is the period of its class. We denote by $\Sigma^{\Omega}_{\text{final}}(\varphi)$ respectively $\Sigma^{R}_{\text{final}}(\varphi)$ the set of connected pieces of the limit sets $\Omega_{\text{final}}(\varphi)$ and $R_{\text{final}}(\varphi)$. Thus, each open map $\varphi \in \mathscr{O}(X)$ induces a permutation π_φ on the set $\Sigma^{\Omega}_{\text{final}}(\varphi)$ of $\Omega-$final components.

Definition 2 Let $\varphi, \psi \in \mathscr{O}(X)$. We say that φ is *combinatorially equivalent* to ψ, and write $\varphi \bowtie \psi$, if and only if the permutations π_φ and π_ψ are conjugated, that is, there is a bijective map $h : \Sigma^{\Omega}_{\text{final}}(\varphi) \to \Sigma^{\Omega}_{\text{final}}(\psi)$ such that the following diagram commutes:

$$\begin{array}{ccc} \Sigma^{\Omega}_{\text{final}}(\varphi) & \xrightarrow{h} & \Sigma^{\Omega}_{\text{final}}(\psi) \\ \pi_\varphi \downarrow & & \downarrow \pi_\psi \\ \Sigma^{\Omega}_{\text{final}}(\varphi) & \xrightarrow{h} & \Sigma^{\Omega}_{\text{final}}(\psi) \end{array} .$$

Definition 3 Given a topological subsemigroup $\mathscr{O}_1 \subseteq \mathscr{O}(X)$, we say that $\varphi \in \mathscr{O}(X)$ is *combinatorially stable in* $\mathscr{O}_1$ if and only if there is a neighbourhood $\mathscr{U}$ of φ in $\mathscr{O}_1$ such that all $\psi \in \mathscr{U}$ are combinatorially equivalent to φ.

Theorem 1 (Combinatorial stability characterization) *For any $\varphi \in \mathscr{O}(X)$, φ is combinatorially stable in $(\mathscr{O}(X), \rho)$ if and only if φ satisfies the following combinatorial stability condition: φ induces the same permutation on $\Sigma^{\Omega}_{\text{final}}(\varphi)$ and $\Sigma^{R}_{\text{final}}(\varphi)$, or, equivalently, $\left|\Lambda^{\Omega}_{\text{final}}(\varphi)\right| = \left|\Lambda^{R}_{\text{final}}(\varphi)\right|$ and $\left|\Sigma^{\Omega}_{\text{final}}(\varphi)\right| = \left|\Sigma^{R}_{\text{final}}(\varphi)\right|$.*

Proof In [3, Theorem 5.3.]) we have proved that given any topological semigroup of open maps $\mathscr{O}_1$, φ is combinatorially stable in $\mathscr{O}_1$ if and only if φ satisfies the combinatorial stability condition. Thus the proof follows immediately in view of Proposition 5. □

Theorem 2 (Genericity of combinatorial stability) *The set of combinatorially stable systems is open and dense in $(\mathscr{O}(X), \rho)$.*

Proof In [3, Theorem 5.4.] we have proved that for any topological semigroup of open maps $\mathscr{O}_1$, the set of $\mathscr{O}_1$-combinatorially stable maps is open and dense in the semigroup $\mathscr{O}_1$. Thus the proof follows immediately in view of Proposition 5. □

4 Topological Semigroups of Markov Systems

A semigroup $\mathscr{H}(X)$ of Markov systems is defined. The key concept of topological semigroup of Markov systems is introduced. For each $f \in \mathscr{H}(X)$, the Perron-Frobenius operator $\mathscr{L}_f$ is recalled. Invariant measures are defined.

We denote by $\mathscr{M}_{\text{prob}}(X)$ the space of all Borel probability measures on the compact manifold X. This is a subset of the Banach space $\mathscr{M}(X)$ of all finite Borel real measures on X, with the usual total variation norm $\|\mu\|$. $\mathscr{M}(X)$ is the dual of the Banach space of continuous real-valued functions on X, denoted here by $C^0(X)$, endowed with the uniform proximity norm $\|\cdot\|_\infty$. The space $\mathscr{M}_{\text{prob}}(X)$ is a compact and convex subset of $\mathscr{M}(X)$ with respect to the *weak-$*$* topology, which is the weak topology of $\mathscr{M}(X)$ as dual of $C^0(X)$. We will call here *Markov system* to any weak-$*$ continuous mapping $p : X \to \mathscr{M}_{\text{prob}}(X)$. The probability measure $p(x) = p_x$ is referred as the transition probability at state $x \in X$. We denote by $\mathscr{M}\mathscr{S}(X)$ the set of all Markov systems. A Markov system $p : X \to \mathscr{M}_{\text{prob}}(X)$ will also be referred as a stochastic dynamical system. A Markov system is called deterministic if for some continuous mapping $f : X \to X$, we have $p(x) = \delta_{f(x)}$ for every $x \in X$, where $\delta_{f(x)}$ denotes the Dirac measure sitting at the point $f(x)$. The Perron-Frobenius operator of a Markov system $p : X \to \mathscr{M}_{\text{prob}}(X)$ is the linear operator $\mathscr{L}_p : \mathscr{M}(X) \to \mathscr{M}(X)$, defined by

$$\mathscr{L}_p(\mu) = \int_X p(x)\, d\,\mu(x)\,, \qquad \text{for every } \mu \in \mathscr{M}(X)\,.$$

The integral of the measure-valued function p is well defined, in a sense that can be found, for instance, in [7]. The adjoint operator $\mathscr{L}_p^* : C^0(X) \to C^0(X)$, is given by

$$\mathscr{L}_p^*(\psi)(x) = \int_X \psi(y)\, d\,p_x(y)\,, \qquad \text{for every } \psi \in C^0(X)\,.$$

Both $\mathscr{L}_p$ and $\mathscr{L}_p^*$ are bounded linear operators with norms less or equal than 1. The convolution of two Markov systems $p, q \in \mathscr{M}\mathscr{S}(X)$ is $p * q : X \to \mathscr{M}_{\text{prob}}(X)$, where

$$(p * q)(x) = \mathscr{L}_p(q_x) = \mathscr{L}_p(\mathscr{L}_q(\delta_x)) \qquad \text{for every } x \in X\,.$$

The space $(\mathscr{M}\mathscr{S}(X), *)$ is a semigroup with identity, where the identity is the deterministic Markov system $x \mapsto \delta_x$. The map $p \mapsto \mathscr{L}_p$ is a semigroup homomorphism taking $\mathscr{M}\mathscr{S}(X)$ into the algebra of bounded linear operators on the Banach space $\mathscr{M}(X)$. We will say that a measure $\mu \in \mathscr{M}(X)$ is $p-$invariant when $\mathscr{L}_p\mu = \mu$, and that a measurable set $A \subseteq X$ is $p-$invariant when $\mathscr{L}_p^*\chi_A = \chi_A$, where χ_A denotes the characteristic function $\chi_A(x) = \begin{cases} 1 & \text{if } x \in A \\ 0 & \text{if } x \in X - A \end{cases}$.

We denote by $L^1(X, m)$ the Banach space of m-integrable functions on X with the usual L^1-norm, $\|h\|_1 = \int_X |h(x)|\, dm(x)$. This space is isometrically embedded in $\mathscr{M}(X)$ through the inclusion map $L^1(X, m) \hookrightarrow \mathscr{M}(X)$, $h \mapsto hm$. We say that a Markov system $p : X \to \mathscr{M}(X)$ is *absolutely continuous with respect to* m if $p_x = f_x\, m$, with $f_x \in L^1(X, m)$, for every $x \in X$. Absolutely continuous Markov systems are defined by *stochastic transition functions* $f : X \times X \to \mathbb{R}$ such that:

(a) $f(x, y)$ is measurable on $X \times X$,
(b) $f(x, y) \geq 0$, for every $(x, y) \in X \times X$,
(c) $\int_X f(x, y)\, dm(y) = 1$, for every $x \in X$,
(d) the real valued function $X \to \mathbb{R}$, $x \mapsto \int f(x, y)\, \psi(y)\, dm(y)$, is continuous for every test function $\psi \in C^0(X)$.

A function $f : X \times X \to \mathbb{R}$ satisfying (a), (b), (d) and

(c') $\int_X f(x, y)\, dm(y) \leq 1$, for every $x \in X$,

is called a *sub-stochastic transition function*.

The subset of all absolutely continuous Markov systems forms a sub-semigroup, without identity, of $\mathscr{M}\mathscr{S}(X)$. Given two transition functions $f, g : X \times X \to \mathbb{R}$, the convoluted Markov system is defined by the usual function convolution

$$(f * g)(x, z) = \int_X f(x, y) g(y, z)\, dm(y)\ .$$

From now on we shall identify each absolutely continuous Markov system with its probability transition function $f : X \times X \to \mathbb{R}$. Given any such absolutely continuous Markov system f, the operator $\mathscr{L}_f$ takes $\mathscr{M}(X)$ into $L^1(X, m)$ and its restriction to $L^1(X, m)$ is given by

$$\mathscr{L}_f(q)(y) = \int_X q(x) f(x, y)\, dm(x) \qquad q \in Ł^1(X, m)\ .$$

The adjoint action on $L^\infty(X, m)$ is given by

$$\mathscr{L}_f^*(g)(x) = \int_X f(x, y)\, g(y)\, dm(y) \qquad g \in Ł^\infty(X, m)\ .$$

Given a Markov system $p : X \to \mathscr{M}_{\text{prob}}(X)$, $\sigma(\mathscr{L}_p)$ will denote the spectrum of the Perron-Frobenius operator $\mathscr{L}_p$. The *spectral radius* of $\mathscr{L}_p$, i.e. the lowest upper bound for absolute values of elements in $\sigma(\mathscr{L}_f)$, will be denoted by $r(\mathscr{L}_p)$. Of course $r(\mathscr{L}_p) = 1$. The *discrete spectrum* of $\mathscr{L}_p$, i.e. the set of all eigenvalues in $\sigma(\mathscr{L}_p)$ that are isolated and have finite multiplicity, will be denoted by $\sigma_{\text{disc}}\left(\mathscr{L}_p\right)$. The complement of $\sigma_{\text{disc}}\left(\mathscr{L}_p\right)$ in $\sigma(\mathscr{L}_p)$ is called the *essential spectrum* of $\mathscr{L}_p$, and denoted by $\sigma_{\text{ess}}\left(\mathscr{L}_p\right)$. The *essential spectral radius* of $\mathscr{L}_p$, i.e. the lowest upper bound for absolute values of elements in $\sigma_{\text{ess}}\left(\mathscr{L}_f\right)$, is denoted by $r_{\text{ess}}(\mathscr{L}_p)$. It is well known, see for instance [8], that the Perron operator $\mathscr{L}_f$ of any absolutely continuous Markov system f is a weakly compact operator. In particular, $r_{\text{ess}}\left(\mathscr{L}_f\right) = 0$ and, therefore, the spectrum $\sigma(\mathscr{L}_f)$ is at most countable. All spectrum points in $\sigma(\mathscr{L}_f) - \{0\}$ are isolated eigenvalues with finite multiplicity. Given an absolutely continuous Markov system f, we can decompose the spectrum of $\mathscr{L}_f$ as:

$$\sigma(\mathscr{L}_f) = \sigma_0(\mathscr{L}_f) \cup \sigma_1(\mathscr{L}_f),$$

where $\sigma_0(\mathscr{L}_f) = \{\lambda \in \sigma(\mathscr{L}_f) : |\lambda| < 1\}$, and $\sigma_1(\mathscr{L}_f) = \sigma(\mathscr{L}_f) - \sigma_0(\mathscr{L}_f)$. Of course $\sigma_1(\mathscr{L}_f)$ is finite while $\sigma_0(\mathscr{L}_f)$ is at most countable but closed for the complex plane topology. Consequently, $\sigma_0(\mathscr{L}_f)$ and $\sigma_1(\mathscr{L}_f)$ are disjoint compact sets and, therefore, there is an associated decomposition of $L^1(X, m)$ into two $\mathscr{L}_f$-invariant subspaces:

$$L^1(X, m) = E_0(f) \oplus E_1(f) ,$$

where $E_1(f)$ has finite dimension. We shall denote by $r_{\text{int}}(\mathscr{L}_f)$ the *interior spectral radius* of $\mathscr{L}_f$, i.e. the lowest upper bound of all absolute values of elements in $\sigma_0(\mathscr{L}_f)$.

Given any absolutely continuous Markov system f a sequence $x_0, x_1, \cdots, x_n$ such that $f(x_{i-1}, x_i) > 0$ for all $i = 1, \cdots, n$ is called an f*−orbit*, and we say that x_n is an f*−iterate* of x_0. An absolutely continuous Markov system is called *irreducible* if for almost all points $x, y \in X$ there is some $n \in \mathbb{N}$ such that the probability transition density from x to y in n iterates is positive, i.e. $f^n(x, y) > 0$. A *recurrence time* is any integer $n \in \mathbb{N}$ such that the set $E_n = \{x \in X : f^n(x, x) > 0\}$ has positive measure. Given an absolutely continuous irreducible Markov system f the greatest common divisor d of all recurrence times $n \in \mathbb{N}$ is called the period of f. An irreducible Markov system f is called *acyclic* if it has period one. The state space X of an irreducible Markov system f of period d can be decomposed into a finite union of f^d-invariant subsets $X = X_0 \cup \cdots \cup X_{d-1}$ such that each restriction $(f^d)_{X_i} : X_i \times X_i \to \mathbb{R}$, is an irreducible acyclic Markov system on X_i.

We shall denote by f_R the restriction to $R \times R$ of a given function $f : X \times X \to \mathbb{R}$, for any subset $R \subseteq X$. If f is stochastic transition function then:

1. f_R is a sub-stochastic transition function.
2. f_R is a stochastic transition function $\Leftrightarrow$ R is f-invariant.

Let $\mathscr{H}(X)$ be the set of all absolutely continuous Markov systems (i.e. probability transition functions) $f : X \times X \to \mathbb{R}$ satisfying the following extra conditions:

(1) f is bounded on $X \times X$;
(2) f is lower semi-continuous on $X \times X$;
(3) for each $x \in X$, the open set $\varphi_f(x) = \{y \in X : f(x, y) > 0\}$ is connected.

The space $\mathscr{H}(X)$ is a convolution sub-semigroup of $\mathscr{MS}(X)$. Item (2) in the definition of $\mathscr{H}(X)$ ensures that $\varphi_f \in \mathscr{O}(X)$. Thus, this semigroup carries a natural homomorphism $\varphi : \mathscr{H}(X) \to \mathscr{O}(X)$.

Given $f \in \mathscr{H}(X)$ and an open φ_f−invariant set $R \subseteq X$, let

$$\begin{aligned}\tau_f(R) &= \frac{1}{2} \sup_{x,z\in R} \int_R |f(x,y) - f(z,y)| \, dm(y) \\ &= 1 - \min_{x,z\in R} \int_R f(x,y) \wedge f(z,y) \, dm(y) \; . \end{aligned} \tag{1}$$

and

$$\tau_f^*(R) = \inf_{n\geq 1} \left[\tau_{f^n}(R) \right]^{1/n} \; .$$

The quantity $-\ln\left(\tau_f^*(R)\right)$ is a kind of *mixing rate* for the action of $\mathscr{L}_f$ on $\mathscr{M}_{\text{prob}}(R)$, which measures how fast the $\mathscr{L}_f$-iterates of any probability distribution on R will converge to the unique $\mathscr{L}_f$-invariant measure supported in R. Next, we make some trivial remarks on this concept:

1. $\tau_{f^n}(R) = 0 \Leftrightarrow$ the transition probabilities $f_x^n(\cdot) = f^n(x,\cdot)$ do not depend on x, for x over R.
2. If for some pair of points $x, y \in R$, the transition probabilities f_x^n and f_y^n have disjoint supports, then $\tau_{f^n}(R) = 1$.
3. If $\tau_f^*(R) < 1$ then the restriction Markov system f_R on R is irreducible and acyclic.

Under the same invariance assumption on $R \subseteq X$, $\varphi_f(R) \subseteq R$, we define

$$\begin{aligned}\beta_f(R) &= 1 - \min_{x\in X} \int_R f(x,y) \, dm(y) \\ &= \sup_{x\in X} \int_{R^c} f(x,y) \, dm(y)\end{aligned} \tag{2}$$

and

$$\beta_f^*(R) = \inf_{n\geq 1} \left[\beta_{f^n}(R) \right]^{1/n} \; .$$

The quantity $-\ln(\beta_f^*(R))$ is a kind of *escape rate*, which measures how fast the restriction to R^c of the $\mathscr{L}_f$-iterates of any probability distribution on X will tend to zero. We also make some obvious remarks on this concept:

1. $\beta_{f^n}(R) = 0 \Leftrightarrow \varphi_{f^n}(X) = (\varphi_f)^n(X) \subseteq R$.
2. If for some point $x \in X$, the transition probability f_x^n has support disjoint from R, then $\beta_{f^n}(R) = 1$.
3. If $\beta_f^*(R) < 1$ then for every $x \in X$ the probability density $(f_x^n)_{R^c}$ converges to zero in $L^1_{R^c}$, as $n \to +\infty$.

We shall say that an open φ_f−invariant set $R \subseteq X$ is an *acyclic spectral attractor* for $f \in \mathscr{H}(X)$ if and only if R is connected and $\tau_f^*(R) < 1$. When the set R splits as a disjoint union of d connected sets,

$$R = R_0 \cup \varphi_f(R_0) \cup \cdots \cup \varphi_f^d(R_0),$$

such that R_0 is an acyclic spectral attractor for f^d we say that R is a periodic spectral attractor of period d. We shall say that an open φ_f−invariant set $R \subseteq X$ is *spectrally attractive* for $f \in \mathscr{H}(X)$ if and only if $\beta_f^*(R) < 1$.

We can extract some spectral information on $\mathscr{L}_f$ from the combinatorics of φ_f.

Proposition 7 ([4, Proposition 5.9.]) *Given $f \in \mathscr{H}(X)$, each Ω−final class of period d is a periodic spectral attractor of period d for f.*

Proposition 8 ([4, Proposition 5.10.]) *Given $f \in \mathscr{H}(X)$, $\Omega_{\text{final}}(\varphi_f)$ is spectrally attractive for f.*

Corollary 1 ([4, Corollary 5.11.]) *Given $f \in \mathscr{H}(X)$, let $\Sigma_{\text{final}}^{\Omega}(\varphi_f) = \{R_1, R_2, \cdots, R_s\}$. Let κ_f be the maximum between $\beta_f^*(R)$ and $\tau_f^*(R_i)$, for $i = 1, \cdots, s$. Then*

$$r_{\text{int}}(\mathscr{L}_f) \leq \kappa_f.$$

Corollary 2 ([4, Corollary 5.12.]) *Given $f \in \mathscr{H}(X)$, let $\Sigma_{\text{final}}^{\Omega}(\varphi_f) = \{R_1, R_2, \cdots, R_s\}$, where each component R_i is $\varphi_f^{d_i}$−invariant for some power $d_i \geq 1$. Then there is a f^{d_i}-invariant measure supported on R_i, $\mu_i = \mathscr{L}_{f^{d_i}}\mu_i$, for each $i = 1, \cdots, s$, such that:*

1. *The sum $E_1(f)$ of all generalized eigenspaces associated with eigenvalues in the unit circle is the s−dimensional space spanned by the measures $\mu_1, \cdots, \mu_s$.*
2. *The action of $\mathscr{L}_f$ on the invariant subspace $E_1(f)$ w.r.t. the basis $\{\mu_1, \cdots, \mu_s\}$ is represented by the permutation matrix associated with the permutation π_{φ_f}.*
3. *The eigenvalues of $\mathscr{L}_f$ in the unit circle are, with multiplicity, the d−unity roots $\mathbb{U}^d = \{\lambda \in \mathbb{C} : \lambda^d = 1\}$, counted for every cycle of length d in permutation π_{φ_f}.*
4. *The operator induced by $\mathscr{L}_f$ on the quotient $\mathscr{M}_{\text{prob}}(X)/E_1(f)$ is contractive, i.e., it has norm less than one.*

Consider now any sub-semigroup of Markov systems $\mathscr{H}_1 \subseteq \mathscr{H}(X)$, endowed with some topology.

Definition 4 We say that $\mathscr{H}_1$ is a *topological semigroup of Markov systems* over a topological semigroup of open maps $\mathscr{O}_1$ if and only if for any $f \in \mathscr{H}_1$:

(1) $\varphi_f \in \mathscr{O}_1$;
(2) The map $f \mapsto \varphi_f$ is continuous for the topology of $\mathscr{O}_1$;
(3) $\mathscr{H}_1$ admits outer approximations in the sense that given $f \in \mathscr{H}_1$, for every neighborhood $\mathscr{N}$ of f in $\mathscr{H}_1$ there is $g \in \mathscr{N}$ such that $\varphi_f \prec \varphi_g$;
(4) $\lim_{g \to f} \left\| \mathscr{L}_f^* \varphi - \mathscr{L}_g^* \varphi \right\|_\infty = 0$ for all $\varphi \in C^0(X)$;
(5) The quantities $\tau_f(R)$ and $\beta_f(R)$, defined in (1) and (2), vary upper semicontinuously with f, for any set $R \subseteq X$.

We now topologize the semigroup $\mathscr{H}(X)$ turning it into a topological semigroup of Markov systems. Consider

$$d_\infty(f, g) = \max_{(x,y)\in X\times X} |f(x, y) - g(x, y)|$$

and

$$d_1(f, g) = \max_{x\in X} \int_X |f(x, y) - g(x, y)|\, dm(y).$$

Define

$$\rho_\infty(f, g) = \max\{d_\infty(f, g), \rho(\varphi_f, \varphi_g)\}$$

and

$$\rho_1(f, g) = \max\{d_1(f, g), \rho(\varphi_f, \varphi_g)\}.$$

Proposition 9 *With the topology associated to any of the metrics ρ_∞ and ρ_1, $\mathscr{H}(X)$ is a topological semigroup of Markov systems.*

Proof In [4, Proposition 6.2.], we have proved the following result. Consider any sub-semigroup $\mathscr{O}_1 \subseteq \mathscr{O}(X)$ with a topology defined by some metric ρ which makes it a topological semigroup of open maps. Consider the metrics ρ_∞ and ρ_1 as defined above. Then $\mathscr{H}_1 = \{f \in \mathscr{H}(X) : \varphi_f \in \mathscr{O}_1\}$ with the topology associated with any of the metrics ρ_∞ and ρ_1 is a topological semigroup of Markov systems over $\mathscr{O}_1$. Thus the proof follows immediately in view of Proposition 5. □

5 Spectral Stability of Markov Systems

Spectral stability of Markov systems is defined and characterized. Its genericity is proved. Continuity of the invariant measures.

The main goal of this section is to relate, for generic systems $f \in \mathscr{H}(X)$, the combinatorial stability of φ_f with the spectral stability of f, defined below. The novelty here with respect to finite state Markov system theory is that in this context, because we are dealing with continuous systems, it makes sense to define stability. Assume $\mathscr{H}_1 \subseteq \mathscr{H}(X)$ is a sub-semigroup endowed with some topology.

Definition 5 We say that $f \in \mathscr{H}(X)$ is *spectrally stable in* $\mathscr{H}_1$ if and only if there is a neighborhood $\mathscr{U}$ of f in $\mathscr{H}_1$ and there is $0 < k < 1$ such that for all $g \in \mathscr{U}$:

(1) there is a linear map $h_g : E_1(f) \to E_1(g)$ that conjugates $\mathscr{L}_f |_{E_1(f)}$ to $\mathscr{L}_g |_{E_1(g)}$;
(2) the map h_g depends continuously on f w.r.t. the topology in $\mathscr{H}_1$, in the sense that for any $\varphi \in C^0(X)$, $\lambda_\varphi \circ h_g$ converges to λ_φ as g tends to f in $\mathscr{H}_1$, where $\lambda_\varphi : L^1(X, m) \to \mathbb{R}$ is defined by $\lambda_\varphi(\mu) = \int \varphi\, d\mu$;
(3) $\sigma_0(\mathscr{L}_g) \cap \{\lambda \in \mathbb{C} : k < |\lambda| < 1\} = \emptyset$.

We note that item (2) above is equivalent to say that the invariant measures of $\mathscr{L}_f$ vary continuously with f w.r.t. the weak-$*$ topology. The fixed points of this linear operator are precisely the system invariant measures. The spectral stability of f relates with the fact that no eigenvalues can enter, or leave, the unit circle.

Theorem 3 (Spectral stability characterization) *For any $f \in \mathscr{H}(X)$, f is spectrally stable in any of the spaces $(\mathscr{H}(X), \rho_\infty)$ and $(\mathscr{H}(X), \rho_1)$ if and only if φ_f satisfies the combinatorial stability condition.*

Proof In [4, Theorem A]) we have proved that given any topological semigroup of Markov systems $\mathscr{H}_1$, f is spectrally stable in $\mathscr{H}_1$ if and only if φ_f satisfies the combinatorial stability condition. Thus the proof follows immediately in view of Proposition 9. □

Theorem 4 (Genericity of spectral stability) *The set of spectrally stable systems is open and dense in any of the spaces $(\mathscr{H}(X), \rho_\infty)$ and $(\mathscr{H}(X), \rho_1)$.*

Proof In [4, Theorem B]) we have proved that for any topological semigroup of Markov systems $\mathscr{H}_1$, the set of $\mathscr{H}_1$-spectrally stable Markov systems is open and dense in the semigroup $\mathscr{H}_1$. Thus the proof follows immediately in view of Proposition 9. □

Acknowledgments The first author was supported by "Fundação para a Ciência e a Tecnologia" through the Program POCI 2010 and the Project "Randomness in Deterministic Dynamical Systems and Applications" (PTDC-MAT-105448-2008). The second author was partially supported by the Research Centre of Mathematics of the University of Minho with the Portuguese Funds from the "Fundação para a Ciência e a Tecnologia" through the Project PEstOE/MAT/UI0013/2014.

References

1. Billingsley, P.: Probability and Measure. Wiley Interscience, New York (1985)
2. Doob, J.L.: Stochastic Processes. Wiley, New York (1953)
3. Duarte, P., Torres, M.J.: Combinatorial stability of non-deterministic systems. Ergodic Theory Dyn. Syst. **26**, 93–128 (2006)
4. Duarte, P., Torres, M.J.: Spectral stability of Markov systems. Nonlinearity **21**, 381–397 (2008)
5. Nadler Jr, S.: Hyperspaces of sets. In: Pure and Applied Mathematics. A Series of Monographs and Textbooks. Marcel Dekker, New York (1978)
6. Palis, J.: A global view of dynamics and a conjecture on the denseness of finitude of attractors. Astérisque **261**, 335–347 (2000)
7. Rudin, W.: Functional Analysis. Tata McGraw-Hill Publishing Company LTD, New Delhi (1982)
8. Weis, L.: Decompositions of positive operators and some of their applications. In: Biersted, K.D., Fuchssteiner, B. (eds.) Functional Analysis: Surveys and Recent Results III, pp. 95–115. North-Holland, Amsterdam (1983)

Derivation of the Stochastic Burgers Equation from the WASEP

Patrícia Gonçalves

Abstract In these notes we give a simple proof of the second-order Boltzmann-Gibbs Principle, which is the main tool in order to prove that the equilibrium fluctuations of the WASEP are given, in the regime of the critical strength asymmetry, by the Stochastic Burgers equation.

Keywords WASEP · Boltzmann-Gibbs principle · Stochastic Burgers equation · Additive functionals

1 Introduction

The KPZ equation was proposed in [1] as the default stochastic partial differential equation ruling the evolution of the profile of a growing interface. For a time t and a space variable x, if $h_t(x)$ denotes the height of that interface at t and x, then the KPZ equation reads as

$$dh_t = A\Delta h_t dt + B(\nabla h_t)^2 dt + \sqrt{C}\mathscr{W}_t,$$

where A, B, C are constants which depend on the thermodynamical quantities of the interface, $\Delta = \partial_x^2$, $\nabla = \partial_x$ and $\mathscr{W}_t$ is a space-time white noise. This equation is ill-posed since its solutions are similar, locally, to the Brownian motion and the main problem comes from the nonlinear term $(\nabla h_t)^2$, which makes no sense. For a very detailed exposition on the KPZ equation we refer to [2]. A way to solve this equation is to consider its Cole-Hopf solution, namely, $u_t(x) = e^{\frac{B}{A}h_t(x)}$. The

P. Gonçalves (✉)
Departamento de Matemática, PUC-RIO, Rua Marquês de São Vicente, No. 225, Rio de Janeiro, RJ 22453-900, Brazil
e-mail: patricia@mat.puc-rio.br

P. Gonçalves
CMAT, Centro de Matemática da Universidade do Minho, Campus de Gualtar, 4710-057 Braga, Portugal

P. Gonçalves and A.J. Soares (eds.), *From Particle Systems to Partial Differential Equations II*, Springer Proceedings in Mathematics & Statistics 129,
DOI 10.1007/978-3-319-16637-7_7

Cole-Hopf solution $u_t(x)$ solves the linearized version of the KPZ equation, namely, the stochastic heat equation (SHE) with multiplicative noise given by

$$du_t = A\Delta u_t dt + \frac{BC}{A} u_t \mathscr{W}_t.$$

Since the equation is now linear, its solutions can be constructed and characterized by means of the constants A, B, C. In [3] it was proved that these Cole-Hopf solutions can be obtained as a scaling limit of the weakly asymmetric simple exclusion process (wasep). Their approach consists in taking a microscopic analogue of the Cole-Hopf solutions, that is, they exponentiate the underling microscopic dynamics and use this new process to characterize the solutions of the SHE by means of martingale problems. The advantage of this new process is that its martingale decomposition has a compensator which is "linear", contrarily to what happens when dealing with the original process.

More recently, there has been many advances in characterizing the KPZ universality class. The approach used is to describe several functionals of asymmetric conservative systems in terms of determinantal formulas that can be solved by using the machinery of random matrix theory. This approach allows to obtain very much detailed information about the solutions of the KPZ equation, nevertheless, it depends highly on the properties of the underlying model. For more details on this we refer to [4] and references therein.

Another stochastic partial differential equation related to the KPZ equation is the stochastic Burgers equation (SBE) which can be obtained from the KPZ equation, at least formally, by taking $\mathscr{Y}_t = \nabla h_t$. In this case, $\mathscr{Y}_t$ satisfies

$$d\mathscr{Y}_t = A\Delta \mathscr{Y}_t dt + B\nabla \mathscr{Y}_t^2 dt + \sqrt{C}\nabla \mathscr{W}_t.$$

In [5] the SBE was obtained by a tightness argument and by showing that all limit points are energy solutions of the SBE equation, see Definition 1. This was done in [5] for general exclusion processes in equilibrium and in [6] for a general class of models in a non-equilibrium scenario. If uniqueness of energy solutions is proved, then convergence would follow. The approach of [5, 6] is to work directly with the given microscopic dynamics and to characterize the solutions by means of a martingale problem. The difficulty in this approach is to make sense to the non-linear term in the SBE. To overcome that, they used a second order Boltzmann-Gibbs Principle which allows to identify the non-linear term in the SBE from the underlying dynamics. This principle is the most difficult step to achieve in this approach and the purpose of this paper is to give a simple proof of it in the case of the most classical weakly asymmetric interacting particle system, namely, the wasep. We consider the process evolving on $\mathbb{Z}$ but we remark that all the results presented here are also true when considering the process evolving on the one-dimensional torus $\mathbb{T}$.

The outline of this paper is as follows. In Sect. 2 we give the statement of results, we describe the model, we give the notion of energy solutions of the SBE equation and we introduce the density fluctuation field. In Sect. 3, we characterize the limiting points of the sequence of density fields by means of a martingale problem. In Sect. 4 we study the non-linear term in the SBE, we state and prove the second order Boltzmann-Gibbs Principle and we generalize it to any local function of the dynamics.

2 Statement of Results

2.1 The Model

In this section we introduce the most classical weakly asymmetric interacting particle system, namely, the wasep. Its dynamics can informally be described as follows. Fix a parameter $a > 0$. In the one-dimensional lattice, we allow at most one particle per site and at each bond we associate an exponential clock. Clocks associate to different bonds are independent. When one of these clocks ring, the occupation variables at the vertices of the bond are interchanged with a certain rate. More precisely, a particle at x jumps to $x+1$ (resp. $x-1$) at rate $p_n := 1/2 + a/2\sqrt{n}$ (resp. $q_n := 1 - p_n$) if and only if the destination site is empty. If the destination site is occupied, then nothing happens and the clocks restart (Fig. 1).

Formally, we denote by $\{\eta_t := \eta_{tn^2} \,:\, t \geq 0\}$, the speeded-up, one-dimensional weakly asymmetric simple exclusion process with state space $\Omega = \{0,1\}^{\mathbb{Z}}$. The configurations of the state space are denoted by the symbol η, so that $\eta(x) = 1$ if the site x is occupied and $\eta(x) = 0$ if the site x is empty. Its infinitesimal generator is denoted by $n^2\mathscr{L}_n$ where $\mathscr{L}_n$ acts on functions $f : \Omega \to \mathbb{R}$ as

$$\mathscr{L}_n f(\eta) = \sum_{x\in\mathbb{Z}} \big\{p_n\eta(x)\big(1-\eta(x+1)\big) + q_n\eta(x+1)\big(1-\eta(x)\big)\big\}\nabla_{x,x+1}f(\eta),$$

where $\nabla_{x,x+1}f(\eta) = f(\eta^{x,x+1}) - f(\eta)$ and for $x \in \mathbb{Z}$,

$$\eta^{x,x+1}(y) = \eta(x+1)\mathbf{1}_{y=x} + \eta(x)\mathbf{1}_{y=x+1} + \eta(y)\mathbf{1}_{y\neq x,x+1}.$$

Let $\rho \in [0,1]$ and denote by ν_ρ the Bernoulli product measure on Ω with density ρ, which is defined as the unique measure in Ω such that the variables $\{\eta(x) : x \in \mathbb{Z}\}$ are independent and such that $\nu_\rho\{\eta(x) = 1\} = \rho$ for any $x \in \mathbb{Z}$. The measures

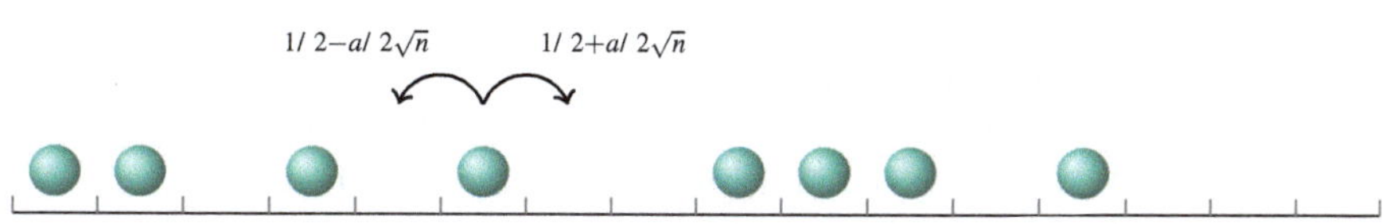

Fig. 1 The one-dimensional weakly asymmetric simple exclusion process

$\{\nu_\rho : \rho \in [0, 1]\}$ are invariant, ergodic and reversible when $a = 0$, but for $a \neq 0$ they are no longer reversible.

We start by remarking some microscopic functions of the model that will be important in what follows. For that purpose, from now on up to the rest of this article, we fix a density $\rho \in (0, 1)$ and a positive time T. We will consider the process η_t with initial distribution ν_ρ and we denote by E_ρ the expectation with respect to ν_ρ. We denote by $\mathbb{P}_\rho$ the distribution of $\{\eta_t : t \in [0, T]\}$ in the space of càdlàg trajectories $\mathscr{D}([0, T], \Omega)$ and we denote by $\mathbb{E}_\rho$ the expectation with respect to $\mathbb{P}_\rho$.

Fix $x \in \mathbb{Z}$. A simple computation shows that $\mathscr{L}_n(\eta(x)) = j_{x-1,x} - j_{x,x+1}$, where for a bond $\{x, x+1\}$, $j_{x,x+1}$ denotes the instantaneous current through that bond, that is, the difference between the jump rate from x to $x+1$ and the jump rate from $x+1$ to x:

$$j_{x,x+1}(\eta) = \Big(\frac{1}{2} + \frac{a}{2\sqrt{n}}\Big)\eta(x)(1 - \eta(x+1)) - \Big(\frac{1}{2} - \frac{a}{2\sqrt{n}}\Big)\eta(x+1)(1 - \eta(x)).$$

We will use the following decomposition of the microscopic current

$$\begin{aligned} j_{x,x+1}(\eta) &:= j^S_{x,x+1}(\eta) + \frac{1}{\sqrt{n}} j^A_{x,x+1}(\eta), \\ &= \frac{1}{2}(\eta(x) - \eta(x+1)) + \frac{a}{2\sqrt{n}}(\eta(x) - \eta(x+1))^2. \end{aligned} \tag{1}$$

Notice that $j^S_{x,x+1}$ is written as the gradient of the function $D(\eta) = \frac{1}{2}\eta(x)$. This point will be important in what follows.

For a local function f, we denote by $\tilde{f}(\rho)$ its expectation with respect to ν_ρ, that is, $\tilde{f}(\rho) = \int f(\eta)\, d\nu_\rho$. Therefore, $\tilde{j}^S(\rho) = 0$ and $\tilde{j}^A(\rho) = 2a\chi(\rho)\tilde{D}'(\rho)$, where

$$\chi(\rho) = \rho(1 - \rho) \quad \text{and} \quad \tilde{D}(\rho) = \frac{\rho}{2}$$

are, the static compressibility of the system and the diffusivity of the system, respectively.

2.2 *Stochastic Burgers Equation*

Before introducing the SBE we need to set up some notation. Let $\mathscr{S}(\mathbb{R})$ be the Schwarz space of test functions and $\mathscr{S}'(\mathbb{R})$ its topological dual with respect to inner product of $\mathscr{L}^2(\mathbb{R})$. We denote by $\|F\|_2$ the $\mathscr{L}^2(\mathbb{R})$-norm of a function $F : \mathbb{R} \to \mathbb{R}$, that is, $\|F\|_2^2 = \int_{\mathbb{R}} F^2(x)\, dx$. Fix $T > 0$ and let $\mathscr{C}([0, T], \mathscr{S}'(\mathbb{R}))$ be the space of continuous trajectories in $\mathscr{S}'(\mathbb{R})$.

Definition 1 We say that a stochastic process $\{\mathscr{Y}_t : t \in [0, T]\}$ with trajectories in $\mathscr{C}([0, T], \mathscr{S}'(\mathbb{R}))$ is an *energy solution* of the Stochastic Burgers equation

$$d\mathscr{Y}_t = A(\rho)\Delta\mathscr{Y}_t dt + B(\rho)\nabla\mathscr{Y}_t^2 dt + \sqrt{C(\rho)}\nabla\mathscr{W}_t$$

(i) if there exists a constant $\kappa \in (0, \infty)$ such that for any $F \in \mathscr{S}(\mathbb{R})$, any $\varepsilon > \varepsilon' > 0$ and any $t \in [0, T]$ we have

$$\mathbb{E}\big[\big(\mathscr{A}_t^{\varepsilon}(F) - \mathscr{A}_t^{\varepsilon'}(F)\big)^2\big] \leq \kappa t\varepsilon\|\nabla F\|_2^2,$$

where

$$\mathscr{A}_t^{\varepsilon}(F) := \int_0^t \int_{\mathbb{R}} \mathscr{Y}_s(i_\varepsilon(x))^2 F'(x)\,dx\,ds$$

and for $x \in \mathbb{R}$, $i_\varepsilon(x, y) = \varepsilon^{-1}\mathbb{1}_{(x,x+\varepsilon]}(y)$.

(ii) if the $\mathscr{S}'(\mathbb{R})$-valued process $\{\mathscr{A}_t : t \in [0, T]\}$ defined as $\mathscr{A}_t(F) := \lim_{\varepsilon\to 0}\mathscr{A}_t^{\varepsilon}(F)$ for $t \in [0, T]$ and $F \in \mathscr{S}(\mathbb{R})$, has trajectories in $\mathscr{C}([0, T], \mathscr{S}'(\mathbb{R}))$,

(iii) if for any function $F \in \mathscr{S}(\mathbb{R})$, the process

$$\mathscr{M}_t(F) = \mathscr{Y}_t(F) - \mathscr{Y}_0(F) - A(\rho)\int_0^t \mathscr{Y}_s(\Delta F)ds - \frac{1}{2}B(\rho)\mathscr{A}_t(F) \quad (2)$$

is a continuous martingale of quadratic variation $C(\rho)t\|\nabla F\|_2^2$.

We remark here that the uniqueness of energy solutions of the SBE equation, as given above, has not been proved.

2.3 The Density Fluctuation Field

Recall that we have fixed a density $\rho \in (0, 1)$ and we consider the starting measure ν_ρ. The density fluctuation field $\{\mathscr{Y}_t^n : t \in [0, T]\}$ is the linear functional defined on $\mathscr{D}([0, T], \mathscr{S}'(\mathbb{R}))$ and given on $F \in \mathscr{S}(\mathbb{R})$ by

$$\mathscr{Y}_t^n(F) = \frac{1}{\sqrt{n}}\sum_{x\in\mathbb{Z}} T_t F\Big(\frac{x}{n}\Big)(\eta_t(x) - \rho),$$

where $T_t F(x) = F(x - \sqrt{n}(\tilde{j}^A)'(\rho)t)$. Now we explain why we remove the velocity $n^{3/2}(\tilde{j}^A)'(\rho)t = n^{3/2}(1 - 2\rho)at$ in the test function F. It is well known (see [7] and references therein) that, at a first order, the density fluctuation field of asymmetric systems are rigidly transported along the characteristics of the corresponding hydrodynamic equation. In order to see a non trivial temporal evolution, we should look at the density fluctuation field *around the characteristics* of the system. Therefore, we need to look at the density fluctuation field evolving in a time dependent reference frame as given above.

Now we are ready to state our main result. Roughly speaking, it says that, in the limit $n \to \infty$, the density fluctuation field is an energy solution of the stochastic Burgers equation.

Theorem 1 *The sequence of processes* $\{\mathscr{Y}_t^n : t \in [0,T]\}_{n\in\mathbb{N}}$ *is tight with respect to the Skorohod topology of* $\mathscr{D}([0,T], \mathscr{S}'(\mathbb{R}))$ *and all its limit points are energy solutions of the SBE equation*

$$d\mathscr{Y}_t = \tilde{D}'(\rho)\Delta\mathscr{Y}_t dt + \frac{1}{2}(\tilde{j}^A)''(\rho)\nabla\mathscr{Y}_t^2 dt + \sqrt{2\chi(\rho)\tilde{D}'(\rho)}\nabla\mathscr{W}_t, \tag{3}$$

where $\tilde{j}^A(\rho) = 2a\chi(\rho)\tilde{D}'(\rho)$.

The method of the proof of last theorem is classical in the literature and can be done in two steps. First, we prove that the sequence is tight. Second, we characterize its limit points by means of a martingale problem. Given the uniqueness of solutions of the stochastic partial differential equation, convergence follows. For tightness issues we refer the interested reader to [5]. As mentioned above, up to now, uniqueness of energy solutions of the SBE has not been proved. If one proves uniqueness, then convergence would follow. In [8] a new approach for proving uniqueness of solutions of the SBE, and other stochastic partial differential equations, when the equations are taken on $\mathbb{T}$, has emerged. Nevertheless, up to our knowledge, so far, the methods developed by [8] do not extend to solutions evolving on $\mathbb{R}$ and do not give any answer about uniqueness of energy solutions on $\mathbb{T}$ nor $\mathbb{R}$.

3 Characterization of Limits Points

In this section we prove that any limit point of the sequence $\{\mathscr{Y}_t^n : t \in [0,T]\}_{n\in\mathbb{N}}$ is an energy solution of the SBE equation (3). According to Definition 1 we have to check three points. In these notes we will give a simpler proof of (iii) and we refer the reader to [5] for the remaining points. Recall that we want to characterize the limiting field $\mathscr{Y}_t$ as a solution of (3) by means of a martingale problem. For that purpose we fix a test function $F \in \mathscr{S}(\mathbb{R})$ and we use the Dynkin's formula which is a tool that provides a decomposition of functions of a Markov process as a martingale plus a compensator. More precisely, we apply the Dynkin's formula in our setting to the density fluctuation field as follows. For each $n \in \mathbb{N}$,

$$\mathscr{M}_t^n(F) = \mathscr{Y}_t^n(F) - \mathscr{Y}_0^n(F) - \int_0^t (n^2\mathscr{L}_n + \partial_s)\mathscr{Y}_s^n(F)ds \tag{4}$$

is a martingale of quadratic variation given by

$$\int_0^t n^2\mathscr{L}_n(\mathscr{Y}_s^n(F))^2 - 2\mathscr{Y}_s^n(F)n^2\mathscr{L}_n\mathscr{Y}_s^n(F)ds. \tag{5}$$

Now we look to the compensator in (4), that is, the integral term in that equation. Recall that $\mathscr{L}_n(\eta(x)) = j_{x-1,x} - j_{x,x+1}$, therefore, a summation by parts gives

$$\begin{aligned} n^2\mathscr{L}_n(\mathscr{Y}_s^n(F)) &= \frac{n}{\sqrt{n}}\sum_{x\in\mathbb{Z}}\nabla_n T_s F\Big(\frac{x}{n}\Big) j_{x,x+1}(\eta_s) \\ &= \sqrt{n}\sum_{x\in\mathbb{Z}}\nabla_n T_s F\Big(\frac{x}{n}\Big) j^S_{x,x+1}(\eta_s) + \sum_{x\in\mathbb{Z}}\nabla_n T_s F\Big(\frac{x}{n}\Big) j^A_{x,x+1}(\eta_s), \end{aligned}$$

where $\nabla_n T_s F(x) = n(T_s F(x+1) - T_s F(x))$.

Since $j^S_{x,x+1}(\eta)$ is written as the gradient of the function $\frac{1}{2}(\eta(x)-\rho)$, the term on the left hand side of last expression can be rewritten as

$$\frac{1}{2\sqrt{n}}\sum_{x\in\mathbb{Z}}\Delta_n T_s F\Big(\frac{x}{n}\Big)(\eta_s(x)-\rho),$$

where $\Delta_n T_s F(x) = n^2(T_s F(x+1) + T_s F(x-1) - 2T_s F(x))$.

Since $j^A_{x,x+1}(\eta) = \frac{a}{2}(\eta(x)-\eta(x+1))^2 = \frac{a}{2}(\eta(x) - 2\eta(x)\eta(x+1) + \eta(x+1))$ we can write the remaining term as

$$a\sum_{x\in\mathbb{Z}}\nabla_n T_s F\Big(\frac{x}{n}\Big)\Big(\frac{1}{2}\eta_s(x) - \eta_s(x)\eta_s(x+1) + \frac{1}{2}\eta_s(x+1)\Big).$$

On the other hand, $\partial_s\mathscr{Y}_s^n(F)$ equals to

$$-a\sum_{x\in\mathbb{Z}}\partial_x T_s F\Big(\frac{x}{n}\Big)(1-2\rho)(\eta_s(x)-\rho).$$

The sum of the two last terms can be written as

$$\begin{aligned} &a\sum_{x\in\mathbb{Z}}\nabla_n T_s F\Big(\frac{x}{n}\Big)\Big(\frac{\eta_s(x)}{2} - \eta_s(x)\eta_s(x+1) + \frac{\eta_s(x+1)}{2} - (1-2\rho)(\eta_s(x)-\rho)\Big) \\ &+ a\sum_{x\in\mathbb{Z}}\Big(\nabla_n T_s F\Big(\frac{x}{n}\Big) - \partial_x T_s F\Big(\frac{x}{n}\Big)\Big)(1-2\rho)(\eta_s(x)-\rho). \end{aligned}$$

By the Taylor expansion of $T_s F$, a simple computation shows that the $L^2(\mathbb{P}_\rho)$-norm of the time integral of last term vanishes, as $n\to\infty$. Now we look at the first term in the previous expression. Since

$$\frac{\eta_s(x)}{2} - \eta_s(x)\eta_s(x+1) + \frac{\eta_s(x+1)}{2} - (1-2\rho)(\eta_s(x)-\rho)$$
$$= -(\eta_s(s)-\rho)(\eta_s(x+1)-\rho) + \Big(\rho - \frac{1}{2}\Big)(\eta(x)-\eta(x+1)),$$

that term can be written as

$$-a\sum_{x\in\mathbb{Z}} \nabla_n T_s F\Big(\frac{x}{n}\Big)(\eta_s(s)-\rho)(\eta_s(x+1)-\rho)$$
$$+\frac{a}{n}\sum_{x\in\mathbb{Z}} \Delta_n T_s F\Big(\frac{x}{n}\Big)\Big(\rho-\frac{1}{2}\Big)(\eta(x)-\rho).$$

Notice that in all the expressions above we can introduce constants since the terms $\nabla_n T_s F$ and $\Delta_n T_s F$ add up to zero.

A simple computation shows that the $L^2(\mathbb{P}_\rho)$-norm of the time integral of the second term on the previous expression vanishes, as $n \to \infty$. Therefore, we can write the integral part of the martingale in (4) as the sum of the following terms

$$\mathscr{I}_t^n(F) = \int_0^t \frac{1}{2\sqrt{n}} \sum_{x\in\mathbb{Z}} \Delta_n T_s F\Big(\frac{x}{n}\Big)(\eta_s(x)-\rho)ds, \tag{6}$$

$$\mathscr{B}_t^n(F) = -a\int_0^t \sum_{x\in\mathbb{Z}} \nabla_n T_s F\Big(\frac{x}{n}\Big)(\eta_s(s)-\rho)(\eta_s(x+1)-\rho)ds, \tag{7}$$

plus some term $\mathscr{R}_t^n(F)$ whose $L^2(\mathbb{P}_\rho)$-norm vanishes, as $n \to \infty$. Notice that

$$\mathscr{I}_t^n(F) = \frac{1}{2}\int_0^t \mathscr{Y}_t^n(\Delta_n F)\, ds,$$

and when $n \to \infty$, this term will converge to the third term on the right hand side of (2).

Now it remains to analyze the term $\mathscr{B}_t^n(F)$. This is the most difficult task and we postpone it to the next section.

Finally, the quadratic variation of the martingale in (5) can be written as

$$\int_0^t \frac{1}{2n}\sum_{x\in\mathbb{Z}} \Big(\nabla_n T_s F\Big(\frac{x}{n}\Big)\Big)^2 \Big((\eta_s(x)-\eta_s(x+1))^2 + \frac{a}{\sqrt{n}}(\eta_s(x)-\eta_s(x+1))\Big)ds.$$

As a consequence,

$$\mathbb{E}_\rho\Big[\Big(\mathscr{M}_t^n(F)\Big)^2\Big] \leq \int_0^t \frac{1}{2n}\sum_{x\in\mathbb{Z}} \Big(\nabla_n T_s F\Big(\frac{x}{n}\Big)\Big)^2 2\chi(\rho)\, ds + O\Big(\frac{1}{\sqrt{n}}\Big),$$

and $\lim_{n\to\infty}\mathbb{E}_\rho[(\mathscr{M}_t^n(F))^2] = t\chi(\rho)\|\nabla F\|_2^2$. Notice that the limit of the quadratic variation is also equal to $2\chi(\rho)\tilde{D}'(\rho)\|\nabla F\|_2^2$, which matches with the strength of the noise in (3).

Remark 1 Suppose that we are looking at a weakly asymmetric dynamics such that its symmetric part of the current $j^S_{x,x+1}$ is written as the gradient of a function $D(\eta)$ which does not allow to immediately write the corresponding term in the martingale decomposition as a function of the density field, as happens here for the term $\mathscr{I}_t^n(F)$. Then, we cannot close the term as we did above. A way to overcome this problem is to apply the classical Boltzmann-Gibbs principle, see Sect. 4, which allows to do that as long as the function $D(\eta)$ is local. In that case, we would be able to write that term in the martingale decomposition as $\tilde{D}'(\rho)\int_0^t \mathscr{Y}_t^n(\Delta_n F)\,ds$, which is exactly what we have above for the dynamics we consider here.

4 The Non-linear Term in SBE

The main goal of this section is to analyze the term $\mathscr{B}_t^n(F)$ given in (7). Looking to that field one can see that the integrand function is not written in terms of the density fluctuation field in a closed form, as happens for the term $\mathscr{I}_t^n(F)$. Therefore, we need to replace it by some function of the density field. This kind of replacement is known in the literature as the Boltzmann-Gibbs Principle. This Principle was first introduced by [9] and it says that, for any local function $f:\Omega\to\mathbb{R}$, $h:\mathbb{R}\to\mathbb{R}$ of compact support and for any $t>0$,

$$\lim_{n\to\infty}\mathbb{E}_\rho\Big[\int_0^t \frac{1}{\sqrt{n}}\sum_{x\in\mathbb{Z}} h\Big(\frac{x}{n}\Big)\Big\{\tau_x f(\eta_s) - \tilde{f}(\rho) - \tilde{f}'(\rho)(\eta_s(x)-\rho)\Big\}\,ds\Big]^2 = 0.$$

We notice that in our case $\tau_x f(\eta) = (\eta(x)-\rho)(\eta(x+1)-\rho)$, so that $\tilde{f}'(\rho)=0$ and the previous does give any useful information about the limit of the term $\mathscr{B}_t^n(F)$. Moreover, in $\mathscr{B}_t^n(F)$ we do not have the factor $1/\sqrt{n}$ in front of the sum. Nevertheless, the following result is the key point in order to write the term $\mathscr{B}_t^n(F)$ as a quadratic function of the field $\mathscr{Y}_t^n$. Basically it says that, when time averaged with h, we can replace the function $(\eta(x)-\rho)(\eta(x+1)-\rho)$ by its conditional expectation in a box of size εn.

Theorem 2 (Second-order Boltzmann-Gibbs Principle) *For every $t>0$ and any measurable function $h:\mathbb{Z}\times[0,T]\to\mathbb{R}$,*

$$\lim_{\varepsilon\to 0}\lim_{n\to\infty}\mathbb{E}_\rho\Big[\int_0^t \sum_{x\in\mathbb{Z}} h_s\Big(\frac{x}{n}\Big)\Big\{\bar{\eta}_s(x)\bar{\eta}_s(x+1) - E_\rho\Big[\bar{\eta}_s(x)\bar{\eta}_s(x+1)\Big|\eta_s^{\varepsilon n}(x)\Big]\Big\}\,ds\Big]^2 = 0,$$

where $\bar{\eta}(x) = \eta(x) - \rho$ *and*

$$\eta_s^{\varepsilon n}(x) = \frac{1}{\varepsilon n} \sum_{y=x}^{x+\varepsilon n-1} \eta(y).$$

By the previous result, we can write

$$\mathscr{B}_t^n(F) = -a \int_0^t \sum_{x\in\mathbb{Z}} \nabla_n T_s F\Big(\frac{x}{n}\Big) E_\rho\Big[\bar{\eta}_s(x)\bar{\eta}_s(x+1)\Big|\eta_s^{\varepsilon n}(x)\Big]ds,$$

plus some error that vanishes, sending $n \to +\infty$ and then $\varepsilon \to 0$. At this point we compute the conditional expectation appearing above. Notice that the projection of ν_ρ over the space of configurations with a fixed number of particles in the box $\{x, x+1, \cdots, \varepsilon n - 1\}$ is the uniform measure [10], so that

$$E_\rho[\eta(x)\eta(x+1)|\eta^{\varepsilon n}(x) = k] = \frac{C_{k-2}^{\varepsilon n-2}}{C_k^{\varepsilon n}} = \frac{k(k-1)}{\varepsilon n(\varepsilon n - 1)},$$

and by a simple computation we get

$$E_\rho[\bar{\eta}(x)\bar{\eta}(x+1)|\eta^{\varepsilon n}(x)] = (\bar{\eta}^{\varepsilon n}(x))^2 - \frac{1}{\varepsilon n - 1}\chi(\eta^{\varepsilon n}(x)). \tag{8}$$

Since $\sum_{x\in\mathbb{Z}} \nabla_n T_s F\big(\frac{x}{n}\big)$ is equal to zero, we can introduce constants in the summation above and since $(\bar{\eta}_s^{\varepsilon n}(x))^2 = \frac{1}{n}\Big(\mathscr{Y}_s^n(\iota_\varepsilon(x/n))\Big)^2$ we write

$$\begin{aligned}\mathscr{B}_t^n(F) = &- a \int_0^t \frac{1}{n} \sum_{x\in\mathbb{Z}} \nabla_n T_s F\Big(\frac{x}{n}\Big) \Big(\mathscr{Y}_s^n(\iota_\varepsilon(x/n))\Big)^2 ds \\ &+ a \int_0^t \sum_{x\in\mathbb{Z}} \nabla_n T_s F\Big(\frac{x}{n}\Big) \frac{1}{\varepsilon n - 1}\Big\{\chi(\eta_s^{\varepsilon n}(x)) - \chi(\rho)\Big\}\, ds.\end{aligned} \tag{9}$$

Now, by the Cauchy-Schwarz's inequality we have

$$\begin{aligned}&\mathbb{E}_\rho\Big[\int_0^t \sum_{x\in\mathbb{Z}} \nabla_n T_s F\Big(\frac{x}{n}\Big) \frac{1}{\varepsilon n - 1}\Big\{\chi(\eta_s^{\varepsilon n}(x)) - \chi(\rho)\Big\}\, ds\Big]^2 \\ &\leq t^2 \sum_{x,y\in\mathbb{Z}} \nabla_n T_s F\Big(\frac{x}{n}\Big)\nabla_n T_s F\Big(\frac{x}{n}\Big)\frac{1}{\varepsilon^2 n^2}\Big\{\chi(\eta_s^{\varepsilon n}(x)) - \chi(\rho)\Big\}\Big\{\chi(\eta_s^{\varepsilon n}(y)) - \chi(\rho)\Big\}.\end{aligned} \tag{10}$$

Since the functions above correlate for x and y at a distance at most εn, we can bound the previous expression by

$$\begin{aligned} & C\frac{t^2}{\varepsilon n}\sum_{x\in\mathbb{Z}}\Big(\nabla_n T_s F\Big(\frac{x}{n}\Big)\Big)^2 E_\rho\Big[\chi(\eta_s^{\varepsilon n}(x))-\chi(\rho)\Big]^2 \\ & \leq C\frac{t^2}{\varepsilon^2 n}\|\nabla F\|_2^2, \end{aligned} \tag{11}$$

which vanishes, as $n\to\infty$.

4.1 Proof of the Second-Order Boltzmann-Gibbs Principle

The proof of Theorem 2 is divided into two steps. For notational convenience we consider a function h that does not depend on time. We start by showing that we can replace the local function $\bar\eta_s(x)\bar\eta_s(x+1)$ by its conditional expectation on a box of size ℓ.

Lemma 1 (One-block estimate)
For every $t>0$, $\ell\geq 2$ and any measurable function $h:\mathbb{Z}\to\mathbb{R}$,

$$\mathbb{E}_\rho\Big[\int_0^t\sum_{x\in\mathbb{Z}}h(x)V_\ell(\eta_s)ds\Big]^2\leq Ct\frac{\ell^3}{n^2}\sum_{x\in\mathbb{Z}}h^2(x),$$

where $V_\ell(\eta_s)=\bar\eta_s(x)\bar\eta_s(x+1)-E_\rho\Big[\bar\eta_s(x)\bar\eta_s(x+1)\Big|\eta_s^\ell(x)\Big]$.

Proof Before giving the proof we introduce some notation. For two functions f,g in $L^2(\nu_\rho)$ we define the inner product $\langle f,-\mathscr{L}_n g\rangle_\rho=-\int f(\eta)\mathscr{L}_n g(\eta)\,d\nu_\rho$. Let H_1 be the Hilbert space generated by $L^2(\nu_\rho)$ and this inner product. Denote by $\|\cdot\|_1$ the norm induced by this inner product and let $\|\cdot\|_{-1}$ be its dual norm with respect to $L^2(\nu_\rho)$:

$$\|f\|_{-1}^2=\sup_{g\in L^2(\nu_\rho)}\Big\{2\int f(\eta)g(\eta)\,d\nu_\rho-\|g\|_1^2\Big\}. \tag{12}$$

Immediately we see that for every $f\in H_{-1}$, $g\in L^2(\nu_\rho)$ and $A>0$

$$2\int f(\eta)g(\eta)\,d\nu_\rho\leq\frac{1}{A}\|f\|_{-1}^2+A\|g\|_1^2. \tag{13}$$

By Proposition A1.6.1 of [10], the expectation in the statement of the lemma is bounded from above by

$$Ct\Big\|\sum_{x\in\mathbb{Z}}h(x)V_\ell\Big\|_{-1}^2,$$

where C is a constant.

By the variational formula for the H_{-1}-norm (12) the previous expression is equal to

$$Ct\sup_{g\in L^2(\nu_\rho)}\Big\{2\sum_{x\in\mathbb{Z}}h(x)\int V_\ell(\eta)g(\eta)\,d\nu_\rho-n^2<g,-\mathscr{L}_ng>_\rho\Big\}.$$

Now we bound $\int V_\ell(\eta)g(\eta)\,d\nu_\rho$. We notice that by a simple computation one can prove that the adjoint of $\mathscr{L}_n$, denoted by $\mathscr{L}_n^*$ is given on functions $f:\Omega\to\mathbb{R}$ by

$$\mathscr{L}_n^*f(\eta)=\sum_{x\in\mathbb{Z}}\{q_n\eta(x)\big(1-\eta(x+1)\big)+p_n\eta(x+1)\big(1-\eta(x)\big)\}\nabla_{x,x+1}f(\eta).$$

We denote the symmetric part of the infinitesimal generator $\mathscr{L}_n$ by $\mathscr{S}_n$, which is given by

$$\mathscr{S}_n:=\frac{\mathscr{L}_n+\mathscr{L}_n^*}{2}.$$

First, we notice that for any $g\in L^2(\nu_\rho)$ it holds that $\langle g,-\mathscr{L}_ng\rangle_\rho=\langle g,-\mathscr{S}_ng\rangle_\rho$. Since $(p_n+q_n)\eta(x)\big(1-\eta(x+1)\big)+(p_n+q_n)\eta(x+1)\big(1-\eta(x)\big)=\eta(x)+\eta(x+1)-2\eta(x)\eta(x+1)$, we obtain that

$$\mathscr{S}_nf(\eta)=\sum_{x\in\mathbb{Z}}\Big\{\frac{1}{2}\eta(x)\big(1-\eta(x+1)\big)+\frac{1}{2}\eta(x+1)\big(1-\eta(x)\big)\Big\}\nabla_{x,x+1}f(\eta),$$

which is the infinitesimal generator of the symmetric simple exclusion process.

Before proceeding we compute $\langle g,-\mathscr{S}_ng\rangle_\rho$. By definition it equals to

$$-\sum_{x\in\mathbb{Z}}\int g(\eta)\Big\{\frac{1}{2}\eta(x)\big(1-\eta(x+1)\big)+\frac{1}{2}\eta(x+1)\big(1-\eta(x)\big)\Big\}\nabla_{x,x+1}g(\eta)d\nu_\rho.$$

Now we write it as twice its half and for each $x\in\mathbb{Z}$ and in one of the parcels we make the exchange η into $\eta^{x,x+1}$ (for which the measure ν_ρ is invariant) to obtain

$$-\frac{1}{2}\sum_{x\in\mathbb{Z}}\int g(\eta)\Big\{\frac{1}{2}\eta(x)\big(1-\eta(x+1)\big)+\frac{1}{2}\eta(x+1)\big(1-\eta(x)\big)\Big\}\nabla_{x,x+1}g(\eta)d\nu_\rho$$
$$+\frac{1}{2}\sum_{x\in\mathbb{Z}}\int g(\eta^{x,x+1})\Big\{\frac{1}{2}\eta(x+1)\big(1-\eta(x)\big)+\frac{1}{2}\eta(x)\big(1-\eta(x+1)\big)\Big\}\nabla_{x,x+1}g(\eta)d\nu_\rho.$$

Now we organize the terms and we get

$$\langle g,-\mathscr{S}_n g\rangle_\rho=\frac{1}{4}\sum_{x\in\mathbb{Z}}\int\Big(\eta(x)-\eta(x+1)\Big)^2\Big(\nabla_{x,x+1}g(\eta)\Big)^2 d\nu_\rho.$$

For simplicity we write $\langle g,-\mathscr{S}_n g\rangle_\rho$ as $\sum_{x\in\mathbb{Z}}I_{x,x+1}(g)$, where

$$I_{x,x+1}(g)=\frac{1}{4}\int\Big(\eta(x)-\eta(x+1)\Big)^2\Big(\nabla_{x,x+1}g(\eta)\Big)^2 d\nu_\rho.$$

Now for fixed $x\in\mathbb{Z}$ and $\ell\geq 2$, let $\mathscr{S}_{x,\ell}$ be the restriction of $\mathscr{S}_n$ to the set $\{x,x+1,\cdots,x+\ell-1\}$. Since $E_\rho[V_\ell|\eta^\ell(x)]=0$, then V_ℓ belongs to the image of the generator $\mathscr{S}_{x,\ell}$. Therefore, by (13) for each $x\in\mathbb{Z}$ and A_x a positive constant it holds that

$$\int V_\ell(\eta)g(\eta)\,d\nu_\rho\leq\frac{1}{2A_x}\langle V_\ell,(-\mathscr{S}_{x,\ell})^{-1}V_\ell\rangle_\rho+\frac{A_x}{2}\langle g,-\mathscr{S}_{x,\ell}g\rangle_\rho.$$

Now, we notice that by translation invariance it holds that

$$\sum_{x\in\mathbb{Z}}\langle g,-\mathscr{S}_{x,\ell}g\rangle_\rho=\sum_{x\in\mathbb{Z}}\sum_{y=x+1}^{x+\ell}I_{y,y+1}(g)\leq\sum_{x\in\mathbb{Z}}\ell I_{x,x+1}(g)=\ell\langle g,-\mathscr{L}_n g\rangle_\rho.$$

Therefore, taking for each x, $A_x=n^2(h(x))^{-1}\ell^{-1}$, the expectation in the statement of the lemma becomes bounded by

$$Ct\sum_{x\in\mathbb{Z}}h^2(x)\frac{\ell}{n^2}\langle V_\ell,(-\mathscr{S}_{x,\ell})^{-1}V_\ell\rangle_\rho. \tag{14}$$

By the spectral gap inequality [11] we have that $\langle V_\ell,(-\mathscr{S}_{x,\ell})^{-1}V_\ell\rangle_\rho\leq\ell^2 Var_\rho[V_\ell]$, where $Var_\rho[V_\ell]$ denotes the variance of the function V_ℓ with respect to ν_ρ. From (8) it is easy to see that $Var_\rho[V_\ell]\leq C$, from where (14) is bounded by

$$Ct\frac{\ell^3}{n^2}\sum_{x\in\mathbb{Z}}h^2(x),$$

which finishes the proof.

We remark that from the previous estimate we cannot take $\ell = \varepsilon n$, otherwise the error would blow up, when taking $n \to \infty$.

The second step consists replacing the conditional expectation of $\bar{\eta}_s(x)\bar{\eta}_s(x+1)$ in the box of size ℓ by its conditional expectation in a box of size 2ℓ.

Lemma 2 (Renormalization step)
For every $t > 0$, $\ell \geq 2$ and any measurable function $h : \mathbb{Z} \to \mathbb{R}$,

$$\mathbb{E}_\rho\Big[\int_0^t \sum_{x\in\mathbb{Z}} h(x)\tilde{V}_{\ell,2\ell}(\eta_s)ds\Big]^2 \leq Ct\frac{\ell}{n^2}\sum_{x\in\mathbb{Z}} h^2(x),$$

where $\tilde{V}_{\ell,2\ell}(\eta_s) = E_\rho\Big[\bar{\eta}_s(x)\bar{\eta}_s(x+1)\Big|\eta_s^\ell(x)\Big] - E_\rho\Big[\bar{\eta}_s(x)\bar{\eta}_s(x+1)\Big|\eta_s^{2\ell}(x)\Big]$.

Proof Using the same arguments as in the proof of Lemma 1, the expectation becomes bounded by

$$Ct \sup_{g\in L^2(\nu_\rho)} \Big\{2\sum_{x\in\mathbb{Z}} h(x)\int \tilde{V}_{\ell,2\ell}(\eta)g(\eta)\,d\nu_\rho - n^2\langle g, -\mathscr{S}_n g\rangle_\rho\Big\}.$$

Therefore, taking for each x, $A_x = n^2(h(x))^{-1}(2\ell)^{-1}$ together with the spectral gap inequality, last expression becomes bounded by

$$Ct\sum_{x\in\mathbb{Z}} h^2(x)\frac{\ell^3}{n^2}Var_\rho[\tilde{V}_{\ell,2\ell}]. \tag{15}$$

Since,

$$\tilde{V}_{\ell,2\ell}(\eta) = (\bar{\eta}^\ell(x))^2 - \frac{1}{\ell-1}\chi(\eta^\ell(x)) - (\bar{\eta}^{2\ell}(x))^2 + \frac{1}{2\ell-1}\chi(\eta^{2\ell}(x)),$$

then a simple computation shows that $Var_\rho[\tilde{V}_{\ell,2\ell}] \leq \frac{C}{\ell^2}$, from where we get that (15) is bounded by

$$Ct\frac{\ell}{n^2}\sum_{x\in\mathbb{Z}} h^2(x), \tag{16}$$

which finishes the proof.

We notice that instead of doubling the box in the previous lemma, we can also estimate the price for replacing for the conditional expectation in a box of size ℓ by the condition expectation in a box of size L.

Lemma 3 (L-Renormalization step)
For every $t > 0$, $\ell \geq 2$ and any measurable function $h : \mathbb{Z} \to \mathbb{R}$,

$$\mathbb{E}_\rho\Big[\int_0^t \sum_{x\in\mathbb{Z}} h(x)\tilde{V}_{\ell,L}(\eta_s)ds\Big]^2 \leq Ct\frac{L^3}{n^2\ell^2}\sum_{x\in\mathbb{Z}} h^2(x),$$

where $\tilde{V}_{\ell,L}(\eta_s) = E_\rho\Big[\bar{\eta}_s(x)\bar{\eta}_s(x+1)\Big|\eta_s^\ell(x)\Big] - E_\rho\Big[\bar{\eta}_s(x)\bar{\eta}_s(x+1)\Big|\eta_s^L(x)\Big]$.

Proof The only difference with respect to the previous proof is that here we take for each x, $A_x = n^2(h(x))^{-1}L^{-1}$, the spectral gap inequality $\langle \tilde{V}_{\ell,L}, (-\mathscr{S}_{x,L})^{-1}\tilde{V}_{\ell,L}\rangle_\rho \leq L^2 Var_\rho[\tilde{V}_{\ell,L}]$ and in this case $Var_\rho(\tilde{V}_{\ell,L}) \leq \frac{C}{\ell^2}$, from where the result follows.

So far we have been able to replace the local function $\bar{\eta}(x)\bar{\eta}(x+1)$ by its condition expectation in a box of size 2ℓ. Now we want to increase the box in order to get to one of size εn. We notice that so far we cannot do that, otherwise the errors (obtained in the One-block estimate and in the Renormalization step) explode, as $n \to \infty$. In the next lemma we compute the price to go any box.

Lemma 4 (Two-blocks estimate)
For every $t > 0$, $\ell \geq 2$, $\ell \geq \ell_0$ and any measurable function $h : \mathbb{Z} \to \mathbb{R}$,

$$\mathbb{E}_\rho\Big[\int_0^t \sum_{x\in\mathbb{Z}} h(x)\tilde{V}_{\ell_0,\ell}(\eta_s)ds\Big]^2 \leq Ct\frac{\ell}{n^2}\sum_{x\in\mathbb{Z}} h^2(x).$$

Proof We start by giving the proof in the case $\ell = 2^M\ell_0$. In this case we write

$$\tilde{V}_{\ell_0,\ell}(\eta_s) = \sum_{i=0}^{M-1} \tilde{V}_{2^i\ell_0,2^{i+1}\ell_0}(\eta_s).$$

Therefore, by Minkowski's inequality the expectation in the statement of the lemma is bounded from above by

$$\Big\{\sum_{i=1}^{M}\Big(\mathbb{E}_\rho\Big[\int_0^t \sum_{x\in\mathbb{Z}} h(x)\tilde{V}_{2^{i-1}\ell_0,2^i\ell_0}ds\Big]^2\Big)^{1/2}\Big\}^2.$$

From the Renormalization step (Lemma 2) we have

$$\mathbb{E}_\rho\Big[\int_0^t \sum_{x\in\mathbb{Z}} h(x)\tilde{V}_{2^{i-1}\ell_0,2^i\ell_0}ds\Big]^2 \leq Ct\frac{2^i\ell_0}{n^2}\sum_{x\in\mathbb{Z}} h^2(x),$$

from where we get that last expression bounded by

$$Ct\Big\{\sum_{i=1}^{M} 2^{i/2}\Big\}^2 \frac{\ell_0}{n^2}\sum_{x\in\mathbb{Z}} h^2(x) \le Ct\frac{2^M\ell_0}{n^2}\sum_{x\in\mathbb{Z}} h^2(x).$$

This proves the lemma for the case $\ell = 2^M\ell_0$. In the other cases, we choose M such that $2^M\ell_0 \le \ell \le 2^{M+1}\ell_0$. Then,

$$\mathbb{E}_\rho\Big[\int_0^t \sum_{x\in\mathbb{Z}} h(x)\tilde{V}_{\ell_0,\ell}(\eta_s)ds\Big]^2 \le \Big\{\sum_{i=1}^{M}\Big(\mathbb{E}_\rho\Big[\int_0^t \sum_{x\in\mathbb{Z}} h(x)\tilde{V}_{2^{i-1}\ell_0,2^i\ell_0}ds\Big]^2\Big)^{1/2}$$
$$+\Big(\mathbb{E}_\rho\Big[\int_0^t \sum_{x\in\mathbb{Z}} h(x)\tilde{V}_{2^M\ell_0,\ell}(\eta_s)ds\Big]^2\Big)^{1/2}\Big\}^2.$$

From the previous computations, we have that the first term on the right hand side of the previous expression is bounded from above by $Ct\frac{\ell}{n^2}\sum_{x\in\mathbb{Z}} h^2(x)$. To bound the second term, we notice that by the L-Renormalization step (Lemma 3) we get that

$$\mathbb{E}_\rho\Big[\int_0^t \sum_{x\in\mathbb{Z}} h(x)\tilde{V}_{2^M\ell_0,\ell}(\eta_s)ds\Big]^2 \le Ct\frac{\ell^3}{n^2(2^M\ell_0)^2}\sum_{x\in\mathbb{Z}} h^2(x)$$
$$\le Ct\frac{(2^{M+1}\ell_0)^3}{n^2(2^M\ell_0)^2}\sum_{x\in\mathbb{Z}} h^2(x) \le Ct\frac{\ell}{n^2}\sum_{x\in\mathbb{Z}} h^2(x).$$

which ends the proof.

4.1.1 The Proof of Theorem 2

From the previous computations we obtain that the expectation appearing in the statement of the Theorem is bounded by

$$Ct\frac{\varepsilon n}{n^2}\sum_{x\in\mathbb{Z}} h^2\Big(\frac{x}{n}\Big)$$

which converges to $Ct\varepsilon\|h\|_2^2$, as $n\to+\infty$, and then vanishes taking $\varepsilon\to 0$.

4.2 The Boltzmann-Gibbs Principle for Occupation Variables

In this section we make some remarks concerning the Boltzmann-Gibbs Principle. To make it as general as we can let $m\in\mathbb{N}$ and $x\in\mathbb{Z}$ and let $f_m^x(\eta) = \prod_{y=x}^{x+m-1}\bar{\eta}(y)$. A simple computation shows that for any $\sigma\in[0,1]$, $E_\sigma[f_m^x] = (\sigma-\rho)^m$.

Now we establish the Boltzmann-Gibbs Principle (Theorem 2) for the function f_m^x. We have already done it for the case $m = 2$ and now we look to the other cases.

Lemma 5 (One-block estimate)
For every $t > 0$, $\ell \geq 2$, $m \in \mathbb{N}$ and any measurable function $h : \mathbb{Z} \to \mathbb{R}$,

$$\mathbb{E}_\rho\Big[\int_0^t \sum_{x\in\mathbb{Z}} h(x)V_\ell^m(\eta_s)ds\Big]^2 \leq Ct\frac{\ell^3}{n^2}\sum_{x\in\mathbb{Z}} h^2(x),$$

where $V_\ell^m(\eta_s) = f_m^x(\eta_s) - E_\rho\Big[f_m^x(\eta_s)\Big|\eta_s^\ell(x)\Big]$.

In this case the error is the same as above, since $Var_\rho[V_\ell^m] \leq C$. Nevertheless, in the Renormalization step, the higher the value of m the less is the error.

Lemma 6 (Renormalization step)
For every $t > 0$, $\ell \geq 2$, $m \in \mathbb{N}$ and any measurable function $h : \mathbb{Z} \to \mathbb{R}$,

$$\mathbb{E}_\rho\Big[\int_0^t \sum_{x\in\mathbb{Z}} h(x)\tilde{V}_{\ell,2\ell}^m(\eta_s)ds\Big]^2 \leq Ct\frac{\ell^{3-m}}{n^2}\sum_{x\in\mathbb{Z}} h^2(x),$$

where $\tilde{V}_{\ell,2\ell}^m(\eta_s) = E_\rho\Big[f_m^x(\eta_s)\Big|\eta_s^\ell(x)\Big] - E_\rho\Big[f_m^x(\eta_s)\Big|\eta_s^{2\ell}(x)\Big]$.

Proof The only difference with respect to the proof in the case $m = 2$ is that here $Var_\rho[\tilde{V}_{\ell,2\ell}^m] \leq \frac{C}{\ell^m}$, from where the result follows.

In the case of the L-Renormalization step we have that

Lemma 7 (L-Renormalization step)
For every $t > 0$, $\ell \geq 2$, $m \in \mathbb{N}$ and any measurable function $h : \mathbb{Z} \to \mathbb{R}$,

$$\mathbb{E}_\rho\Big[\int_0^t \sum_{x\in\mathbb{Z}} h(x)\tilde{V}_{\ell,L}^m(\eta_s)ds\Big]^2 \leq Ct\frac{L^3}{n^2\ell^m}\sum_{x\in\mathbb{Z}} h^2(x),$$

where $\tilde{V}_{\ell,L}^m(\eta_s) = E_\rho\Big[f_m^x(\eta_s)\Big|\eta_s^\ell(x)\Big] - E_\rho\Big[f_m^x(\eta_s)\Big|\eta_s^L(x)\Big]$.

Lemma 8 (Two-blocks estimate)
For every $t > 0$, $\ell \geq 2$, $m \in \mathbb{N}$ and any measurable function $h : \mathbb{Z} \to \mathbb{R}$,

$$\mathbb{E}_\rho\Big[\int_0^t \sum_{x\in\mathbb{Z}} h(x)\tilde{V}_{\ell_0,\ell}^m(\eta_s)ds\Big]^2 \leq Ct\frac{\ell}{n^2}\sum_{x\in\mathbb{Z}} h^2(x).$$

Proof In this case we notice that the only difference in the proof is that now we have the expectation bounded by

$$Ct\Big\{\sum_{i=1}^{M}2^{i(3-m)/2}\Big\}^2\frac{\ell_0^{3-m}}{n^2}\sum_{x\in\mathbb{Z}}h^2(x).$$

Now, if $m=1$ then last expression is bounded by

$$Ct\Big\{\sum_{i=1}^{M}2^{i}\Big\}^2\frac{\ell_0^{2}}{n^2}\sum_{x\in\mathbb{Z}}h(x)\leq Ct\frac{\ell^2}{n^2}\sum_{x\in\mathbb{Z}}h^2(x),$$

while if $m=3$ it is bounded by

$$Ct\Big\{\sum_{i=1}^{M}1\Big\}^2\frac{1}{n^2}\sum_{x\in\mathbb{Z}}h^2(x)=CtM^2\frac{1}{n^2}\sum_{x\in\mathbb{Z}}h^2(x).$$

Since $\ell=2^M\ell_0$, then last expression equals to $Ct\frac{(\log(\ell))^2}{n^2}\sum_{x\in\mathbb{Z}}h^2(x)$. For the other cases of m, one notices that $\sum_{i=1}^{M}2^{i(3-m)/2}\leq\sum_{i=1}^{\infty}2^{i(3-m)/2}<\infty$ from where we get the bound

$$Ct\frac{1}{n^2}\sum_{x\in\mathbb{Z}}h^2(x).$$

We can summarize the Boltzmann-Gibbs Principle as follows:

Theorem 3 (General Boltzmann-Gibbs Principle)
For every $t>0$, $\ell\geq 2$, $m\in\mathbb{N}$ and any measurable function $h:[0,T]\times\mathbb{Z}\to\mathbb{R}$,

$$\mathbb{E}_\rho\Big[\int_0^t\frac{1}{\theta(n)}\sum_{x\in\mathbb{Z}}h_s\Big(\frac{x}{n}\Big)\Big\{f_m^x(\eta_s)-E_\rho\Big[f_m^x(\eta_s)\Big|\eta_s^\ell(x)\Big]\Big\}\,ds\Big]^2\leq Ct\frac{c_m(\ell)}{n(\theta(n))^2}\|h\|_{2,n}^2,$$

where $\|h\|_{2,n}^2=\frac{1}{n}\sum_{x\in\mathbb{Z}}h^2\Big(\frac{x}{n}\Big)$ and

$$c_m(\ell)=\begin{cases}\ell^2, & \text{if } m=1,\\ \ell, & \text{if } m=2,\\ (\log(\ell))^2, & \text{if } m=3,\\ 1, & \text{otherwise,}\end{cases}\qquad(17)$$

In order to conclude we have the following scheme.

1. $m = 1$
   From the previous estimates we have that for $\theta(n) > \sqrt{n}$

$$\lim_{n\to+\infty} \mathbb{E}_\rho\Big[\Big(\int_0^t \frac{1}{\theta(n)} \sum_{x\in\mathbb{Z}} \nabla_n h\Big(\frac{x}{n}\Big) f_m^x(\eta_s)\, ds\Big)^2\Big] = 0,$$

while for $\theta(n) = \sqrt{n}$, since $E_\rho\Big[f_m^x(\eta_s)\Big|\eta_s^{\varepsilon n}(x)\Big] = \bar{\eta}_s^{\varepsilon n}(x) = \frac{1}{\sqrt{n}}\mathcal{Y}_t^n(\iota_\varepsilon(x/n))$ we have

$$\int_0^t \frac{1}{\sqrt{n}} \sum_{x\in\mathbb{Z}} \nabla_n h\Big(\frac{x}{n}\Big)\bar{\eta}_s(x)\, ds \sim \int_0^t \frac{1}{n} \sum_{x\in\mathbb{Z}} \nabla_n h\Big(\frac{x}{n}\Big)\mathcal{Y}_t^n(\iota_\varepsilon(x/n))\, ds$$

$$\downarrow$$

$$\int_0^t \int_{\mathbb{R}} \nabla h(x)\mathcal{Y}_t(\iota_\varepsilon(0))\, ds$$

2. $m = 2$
   From the previous estimates, for $\theta(n) = n^\delta$, with $\delta > 0$ we have

$$\lim_{n\to+\infty} \mathbb{E}_\rho\Big[\Big(\int_0^t \frac{1}{\theta(n)} \sum_{x\in\mathbb{Z}} \nabla_n h\Big(\frac{x}{n}\Big) f_m^x(\eta_s)\, ds\Big)^2\Big] = 0,$$

while for $\theta(n) = 1$, since $E_\rho\Big[f_m^x(\eta_s)\Big|\eta_s^{\varepsilon n}(x)\Big] = (\bar{\eta}_s^{\varepsilon n}(x))^2 - \frac{1}{\varepsilon n - 1}\chi(\eta^{\varepsilon n}(x))$ and since $(\bar{\eta}_s^{\varepsilon n}(x))^2 = \frac{1}{n}(\mathcal{Y}_t^n(\iota_\varepsilon(x/n)))^2$ we get

$$\int_0^t \sum_{x\in\mathbb{Z}} \nabla_n h\Big(\frac{x}{n}\Big)\bar{\eta}_s(x)\bar{\eta}_s(x+1)\, ds \sim \int_0^t \frac{1}{n} \sum_{x\in\mathbb{Z}} \nabla_n h\Big(\frac{x}{n}\Big)(\mathcal{Y}_t^n(\iota_\varepsilon(x/n)))^2\, ds$$

$$\downarrow$$

$$\int_0^t \int_{\mathbb{R}} \nabla h(x)(\mathcal{Y}_t(\iota_\varepsilon(0)))^2\, ds.$$

4.3 The Boltzmann-Gibbs Principle for General Functions

In this section we rephrase the results of the previous subsection in terms of more general functions. First we recall the notion of the degree of a function from [12].

Definition 2 Let $f : \Omega \to \mathbb{R}$ be a local function and for $\sigma \in [0, 1]$, recall that $\tilde{f}(\sigma) = E_\sigma[f]$. The function f is said to have degree $m \in \mathbb{N}$ if $\tilde{f}^j(\rho) = 0$, for all $j = 0, \cdots, m-1$ and $\tilde{f}^m(\rho) \neq 0$.

Notice that the function f_m^x defined above has degree m. From the previous results we have the following bounds:

Corollary 1 *Let $m \in \mathbb{N}$ and let $f : \Omega \to \mathbb{R}$ be a local function of degree m with support contained in $\{x, x+1, \cdots, x+\ell\}$, with $\ell > m$. For every $t > 0$, $\ell \geq 2$ and any measurable function $h : [0,T] \times \mathbb{Z} \to \mathbb{R}$,*

$$\mathbb{E}_\rho\Big[\int_0^t \frac{1}{\theta(n)} \sum_{x\in\mathbb{Z}} h_s\Big(\frac{x}{n}\Big)\Big\{f(\eta_s) - E_\rho\Big[f(\eta_s)\Big|\eta_s^\ell(x)\Big]\Big\}\, ds\Big]^2 \leq Ct\frac{c_m(\ell)}{n(\theta(n))^2}\|h\|_{2,n}^2,$$

where $c_m(\ell)$ was given in (17).

Proof Recall the function f_m^x given above. For a function f with degree m, we define the auxiliary function

$$\psi(\eta) = f(\eta) - \frac{\tilde{f}^m(\rho)}{m!}\prod_{y=x}^{x+m-1}\bar{\eta}(y),$$

where $\tilde{f}^m(\rho) = \frac{d^m\tilde{f}}{d\rho^m}(\rho)$ which is non zero, since f has degree m. A simple computation shows that

$$\tilde{\psi}(\sigma) = \tilde{f}(\sigma) - \frac{\tilde{f}^m(\rho)}{m!}(\sigma-\rho)^m,$$

from where it follows that the degree of ψ is greater or equal than $m+1$. Therefore, as a consequence of the previous results, by using the inequality $(x+y)^2 \leq 2x^2+2y^2$, the fact that $c_{m+1}(\ell) \leq c_m(\ell)$ and also by writing

$$\begin{aligned}
f(\eta) - E_\rho\Big[f(\eta)\Big|\eta^\ell(x)\Big] &= f(\eta) - \frac{\tilde{f}^m(\rho)}{m!}\prod_{y=x}^{x+m-1}\bar{\eta}(y)\\
&+\frac{\tilde{f}^m(\rho)}{m!}\prod_{y=x}^{x+m-1}\bar{\eta}(y) - E_\rho\Big[\frac{\tilde{f}^m(\rho)}{m!}\prod_{y=x}^{x+m-1}\bar{\eta}(y)\Big|\eta^\ell(x)\Big]\\
&+E_\rho\Big[\frac{\tilde{f}^m(\rho)}{m!}\prod_{y=x}^{x+m-1}\bar{\eta}(y)\Big|\eta^\ell(x)\Big] - E_\rho\Big[f(\eta)\Big|\eta^\ell(x)\Big]\\
&= \psi(\eta) - E_\rho\Big[\psi(\eta)\Big|\eta^\ell(x)\Big]\\
&+\frac{\tilde{f}^m(\rho)}{m!}\prod_{y=x}^{x+m-1}\bar{\eta}(y) - E_\rho\Big[\frac{\tilde{f}^m(\rho)}{m!}\prod_{y=x}^{x+m-1}\bar{\eta}(y)\Big|\eta^\ell(x)\Big],
\end{aligned}$$

the proof ends.

Acknowledgments The author thanks CNPq (Brazil) for support through the research project "Additive functionals of particle systems" 480431/2013-2 and CMAT for support by "FEDER" through the "Programa Operacional Factores de Competitividade COMPETE" and by FCT through the project PEst-C/MAT/UI0013/2011.

References

1. Kardar, M., Parisi, G., Zhang, Y.-C.: Dynamic scaling of growing interfaces. Phys. Rev. Lett. **56**(9), 889–892 (1986)
2. Quastel, J.: Lecture notes of the course introduction to KPZ. http://math.arizona.edu/~mathphys/school_2012/IntroKPZ-Arizona.pdf
3. Bertini, L., Giacomin, G.: Stochastic Burgers and KPZ equations from particle systems. Commun. Math. Phys. **183**(3), 571–607 (1997)
4. Corwin, I.: The Kardar-Parisi-Zhan equation and universality class. Random Matrices: Theor. Appl. **1**, 1130001 (2012)
5. Gonçalves, P., Jara, M.: Nonlinear fluctuations of weakly asymmetric interacting particle systems. Arch. Rational Mech. Anal. **212**(2), 597–644 (2014)
6. Gonçalves, P., Jara, M., Sethuraman, S.: A stochastic Burgers equation from a class of microscopic interactions. Ann. Probab. **43**(1), 286–338 (2015)
7. Gonçalves, P.: Central limit theorem for a tagged particle in asymmetric simple exclusion. Stochast. Processes Appl. **118**(3), 474–502 (2008)
8. Hairer, M.: Solving the KPZ equation. Ann. Math. **178**(2), 559–664 (2013)
9. Brox, T., Rost, H.: Equilibrium fluctuations of stochastic particle systems: the role of conserved quantities. Ann. Probab. **12**(3), 742–759 (1984)
10. Kipnis, C., Landim, C.: Scaling Limits of Interacting Particle Systems. Springer, New York (1999)
11. Quastel, J.: Diffusion of color in the simple exclusion process. Commun. Pure Appl. Math. **45**(6), 623–679 (1992)
12. Sethuraman, S., Xu, L.: A central limit theorem for reversible exclusion and zero-range particle systems. Ann. Probab. **24**, 1871–1902 (1996)

Large Population Asymptotics for Interacting Diffusions in a Quenched Random Environment

Eric Luçon

Abstract We review some recent results on large population behavior of interacting diffusions in a random environment. Emphasis is put on the quenched influence of the environment on the macroscopic behavior of the system (law of large numbers and fluctuations). We address the notion of (non-)self-averaging phenomenon for this class of models. A guiding thread in this survey is the Kuramoto synchronization model which has met in recent years a growing interest in the literature.

Keywords Interacting diffusions · Random environment · Self-averaging · Fluctuations · McKean-Vlasov equation · FitzHugh-Nagumo model · Kuramoto model · Stochastic partial differential equation

1 Introduction

1.1 Diffusions in Mean-Field Interaction

In this paper, we review some recent results [39–41] concerning large population asymptotics of interacting diffusions in a random environment. This class of models generalizes systems of particles in a mean-field interaction that have been intensively studied since McKean [44]. A general instance of such mean-field models may be given as follows: for any fixed $T > 0$, for any $N \geq 1$ and $m \geq 1$, consider the system of N coupled stochastic differential equations in $\mathbf{R}^m$

$$d\theta_{i,t} = c(\theta_{i,t})dt + b(\theta_{i,t}, \nu_{N,t})dt + \sigma(\theta_{i,t}, \nu_{N,t}) \cdot dB_{i,t}, \ t \in [0, T], \ i = 1 \ldots, N, \tag{1}$$

E. Luçon (✉)
Laboratoire MAP 5 (UMR CNRS 8145), Université Paris Descartes,
Sorbonne Paris Cité, 45 rue des Saints-Pères, 75270 Paris cedex 06, France
e-mail: eric.lucon@parisdescartes.fr

P. Gonçalves and A.J. Soares (eds.), *From Particle Systems to Partial Differential Equations II*, Springer Proceedings in Mathematics & Statistics 129,
DOI 10.1007/978-3-319-16637-7_8

where ν_N is the empirical measure of the particles $(\theta_1, \ldots, \theta_N)$:

$$\nu_{N,t} := \frac{1}{N} \sum_{i=1}^{N} \delta_{\theta_{i,t}}, \ t \in [0, T]. \tag{2}$$

For fixed N and t, $\nu_{N,t}$ is an element of $\mathscr{M}_1(\mathbf{R}^m)$, the set of probability measures on $\mathbf{R}^m$. In the following, unless specified otherwise, $\mathscr{M}_1$ is endowed with its weak topology. In (1), $(B_i)_{i=1,\ldots,N}$ is a collection of independent Brownian motions in $\mathbf{R}^p$ $(p \geq 1)$, $(\theta, \nu) \mapsto b(\theta, \nu) \in \mathbf{R}^m$ and $(\theta, \nu) \mapsto \sigma(\theta, \nu) \in \mathbf{R}^{m \times p}$ are (possibly not) regular functions of (θ, ν), where ν belongs to $\mathscr{M}_1(\mathbf{R}^m)$. To fix ideas, the reader may think of the particular case where b or σ are of the form $(\theta, \nu) \mapsto \int \Gamma(\theta, \tilde{\theta})\nu(\mathrm{d}\tilde{\theta})$ where Γ is a regular function. In such a case, the mean-field terms in (1) reduce to $\frac{1}{N} \sum_{j=1}^{N} \Gamma(\theta_i, \theta_j)$.

Mean-field models like (1) have been quite popular in recent years in both physical and mathematical literature, as they provide the most natural way to represent the time evolution of a population of particles in all-to-all interaction. In (1), we have distinguished the local contribution $c(\theta_{i,t})\mathrm{d}t$ to the dynamics (that would govern the dynamics if θ_i had not been connected to the rest of the population) from the mean-field contribution of all other particles $b(\theta_{i,t}, \nu_{N,t})\mathrm{d}t$ modeling the interaction with the rest of the population.

A reason for the interest in such systems is that they also provide natural particle approximations for various partial differential equations appearing in physics (e.g. granular media equation [12, 42], porous media equation [11], Vicsek model for alignment of self-propelled particles [20]), biology (neuronal models [4, 22, 38], Keller-Segel model for chemotaxis [30]) or finance (rank-based models [34, 36]).

The problematic concerns the large population behavior of (1): under mild assumptions on the coefficients c, b and σ, one can show that the empirical measure (2) converges weakly as $N \to \infty$ to a measure-valued solution $t \mapsto \nu_t$ of a nonlinear Fokker-Planck equation (or McKean-Vlasov equation [27, 44]). In that extent, a crucial feature of such mean-field systems is *exchangeability*: if, at $t = 0$, the law of the vector $(\theta_{1,0}, \ldots, \theta_{N,0})$ is invariant under permutation, the same property holds for (1) at any positive time t. It is well understood [53] that, under this assumption, the above convergence is equivalent to the notion of *propagation of chaos*: for any fixed $k \geq 1$, the first k particles $(\theta_1, \ldots, \theta_k)$ converges in law as $N \to \infty$ to k independent copies of a *nonlinear process*, whose law is given by the solution to the McKean-Vlasov equation. More generally, a vast literature addresses the questions of fluctuations and large deviations of the empirical measure around its limit (see [19, 24, 26, 27, 32, 46] and references therein) as well as long-time behavior of such processes (uniform propagation of chaos and concentration inequalities [12, 13]).

1.2 Inhomogeneous Interacting Diffusions

1.2.1 General Framework

The class of models we address in this paper is a generalization of (1) in the presence of a random environment. Namely, the system we consider is the following:

$$\mathrm{d}\theta_{i,t} = c(\theta_{i,t}, \omega_i)\mathrm{d}t + \frac{1}{N}\sum_{j=1}^{N} \Gamma(\theta_{i,t}, \omega_i, \theta_{j,t}, \omega_j)\mathrm{d}t + \mathrm{d}B_{i,t},\ i = 1, \ldots, N,\ t \in [0, T], \quad (3)$$

endowed with an initial condition $(\theta_{i,0})_{i=1,\ldots,N}$ i.i.d. with law ζ on $\mathbf{R}^m$.

The difference between (3) and (1) is that the particles $(\theta_1, \ldots, \theta_N)$ now live in a random environment, that is both the local dynamics $c(\cdot)$ and the mean-field term $\Gamma(\cdot)$ are perturbed by a given sequence $(\omega_i)_{i=1,\ldots,N} \in (\mathbf{R}^n)^N$ $(n \geq 1)$ of i.i.d. random variables, independent of the thermal noise $(B_1, \ldots, B_N)$. This sequence models a local inhomogeneity in the system: particles are similar but not necessarily identical. The common law of the random variables $(\omega_i)_{i\geq 1}$ is denoted as $\mu \in \mathcal{M}_1(\mathbf{R}^n)$.

Remark 1 Note that we do not address the whole generality of the coefficients b and σ as described in (1). In particular, we restrict ourselves for simplicity to the case $p = m$ and $\sigma = I$. But the results exposed here should remain valid for general σ, provided further regularity and non-degeneracy assumptions are made.

From the point of view of statistical physics, the additional randomness $(\omega_i)_{i\geq 1}$ in (3) will be considered as a *disorder*. In this framework, there exist two ways to consider (3): one can either study the *averaged* (or *annealed*) model (where one looks at $(\theta_1, \ldots, \theta_N)$ under the joint law of the noise *and* the disorder) or one can fix once and for all a typical realization of the disorder and consider (3) under the law of the noise only (*quenched model*). In this paper, we focus on the quenched model, which is more realistic from a modeling point of view. Under the same hypotheses, the asymptotic results proven here remain valid in the (technically easier) averaged framework.

The difficulty of working in a quenched environment $(\omega_i)_{i\geq 1}$ lies in the fact that the particles $(\theta_1, \ldots, \theta_N)$ are no longer exchangeable. In particular, the question of propagation of chaos concerning (3) is *a priori* not clear. More precisely, the question we want to address is the influence of a fixed realization of the disorder on the behavior of the empirical measure of the system as $N \to \infty$

$$\nu_{N,t} = \nu_{N,t}^{(\omega)} := \frac{1}{N}\sum_{i=1}^{N} \delta_{(\theta_{i,t}, \omega_i)}, \quad t \in [0, T],\ N \geq 1. \quad (4)$$

As in (4), we will specify when required the dependence of the empirical measure in the specific choice of the disorder $(\omega) := (\omega_i)_{i\geq 1}$.

1.2.2 Synchronization Models and Neural Networks

A first motivation for systems such as (3) comes from synchronization models in physics. The Kuramoto model [1], which will be the guiding thread of this review, is a particular case of (3) where the particles θ_i reduce to one-dimensional oscillators (or rotators) on the circle $\mathbf{S} := \mathbf{R}/2\pi$, within a mean-field sine interaction, perturbed by random frequencies $\omega_i \in \mathbf{R}$:

$$d\theta_{i,t} = \omega_i dt + \frac{K}{N}\sum_{j=1}^{N} \sin(\theta_{j,t} - \theta_{i,t})dt + dB_{i,t}, \quad i = 1, \dots, N, \ t \in [0, T]. \tag{5}$$

In (5), $K > 0$ is the coupling strength between rotators, and the disorder ω_i is the local random frequency of the rotator θ_i, that may differ from one rotator to another. As $N \to \infty$, the system is described by the following nonlinear Fokker-Planck equation (whose solution $(t, \theta, \omega) \mapsto q_t(\theta, \omega)$ is the density of rotators at time t at position θ with frequency ω):

$$\partial_t q_t = \frac{1}{2}\partial_\theta^2 q_t(\theta, \omega) - \partial_\theta \left(q_t(\theta, \omega) \left(\omega + K \int_{\mathbf{S}\times\mathbf{R}} \sin(\tilde{\theta} - \theta) q_t(\tilde{\theta}, \tilde{\omega}) d\tilde{\theta} \mu(d\tilde{\omega}) \right) \right). \tag{6}$$

The Kuramoto model was first introduced in [37] in order to study collective behavior of synchronizing individuals (neurons, social insects, cardiac cells) and has been since the subject of a vast literature, mostly in physics (see [1, 37, 51] and references therein) and more recently in mathematics.

The intuition for the dynamics of the Kuramoto model is simple: in a large population, each rotator θ_i obeys to the influence of its local frequency ω_i which tends to desynchronize the rotators, in contradiction with the mean-field coupling that tends to make the particles rotate together. A striking result, first observed by Kuramoto and Sakaguchi [50], is that (6) exhibits a phase transition: if K is smaller than a critical value K_c, the uniform distribution $q \equiv \frac{1}{2\pi}$ is the only stationary solution of (6) (there is no synchronization), whereas it coexists with nontrivial synchronized profiles for $K > K_c$. Recent results address the question of the long-time stability of such synchronized solutions, in the case without disorder [8, 9, 29] and also with small disorder [28].

A second motivation for this work comes from the modeling of the spiking activity of neurons in a noisy environment. The FitzHugh-Nagumo model, which is a 2-dimensional simplification of the Hodgkin-Huxley model (see [4, 23] for further neurophysiological insights on the subject) is given as follows: $\theta_i := (V_i, w_i)$ and

$$\begin{cases} dV_i(t) &= \left(V_i(t) - \frac{V_i(t)^3}{3} - w_i(t) + I \right) dt, \\ dw_i(t) &= \left(a_i(b_i V_i(t) - w_i(t)) \right) dt, \end{cases} \quad i = 1, \dots, N, \ t \in [0, T], \tag{7}$$

where a_i and b_i are random coefficients. The variable $V_i(t)$ denotes the voltage activity of the neuron, $w_i(t)$ is a recovery variable and I is the exterior input current.

In (7), the disorder $\omega_i = (a_i, b_i) \in \mathbf{R}^2$ plays the role of a random discrimination between inhibited and excited neurons: it is well known [23] that, depending on the values of the parameters (a_i, b_i), the dynamics (7) exhibits either periodic behavior around a limit cycle (spiking activity) or convergence to a fixed point (inhibition of the neuron). If one incorporates this dynamics into (3), the mean-field term $\Gamma(\cdot)$ models connections between neurons through electrical synapses. We refer to [4] for precise details and the exposition of more elaborate mean-field models applied to neuronal activity. Note that one difficulty of the dynamics given by (7) is that it is unbounded and not uniformly Lipschitz-continuous.

1.2.3 Existing Literature on Disordered Interacting Diffusions

The subject of diffusions in random environment has already been addressed in the literature. In particular, one should mention the seminal paper of Dai Pra and den Hollander [18] where an averaged large deviations principle for models similar to (3) is proved, with applications to the Kuramoto model and spin-flip systems. Other interesting applications of (3) may be found in the context of statistical physics (random Curie-Weiss model [16], model of social interactions [17]).

Instead of putting the disorder on the particles, it would also make sense to put the disorder on the connections between particles. There is currently a growing interest in mean-field models with random connectivities (in particular with applications to neuronal models, see [14, 25, 54, 55] and references therein), that is, models of the type

$$d\theta_{i,t} = \sum_{j=1}^{N} J_{i,j} \Gamma(\theta_{i,t}, \theta_{j,t})dt + dB_{i,t}, \quad i = 1, \ldots, N, \ t \in [0, T], \tag{8}$$

where $(J_{i,j})_{i,j=1,\ldots,N}$ is a (possibly symmetric) collection of random variables. Large population asymptotics of such models have been first studied through large deviations techniques by Guionnet and Ben Arous [3, 6] in the context of spin-glass systems. The long time analysis of the associated nonlinear process appears to be significantly more difficult than for the models considered here.

Other mean-field models of neurons with integrate-and-fire dynamics have recently been studied [15, 21, 22]. For such models, existence of a solution for all time to the nonlinear Fokker Planck turns out to be problematic.

1.3 *Organization of the Paper*

We summarize in Sect. 2 the main results of the paper, that is a quenched law of large numbers and a quenched central limit theorem for the empirical measure (4). In Sect. 3, the main lines of proof for the two results are indicated.

2 Main Results

In the remaining of the paper, for any $p \geq 1$, the euclidean norm and scalar product in $\mathbf{R}^p$ are respectively denoted as $|u|$ and $u \cdot v$ ($u, v \in \mathbf{R}^p$). $\mathbf{P}$ is the law of the sequence of Brownian motions and $\mathbb{P}$ is the law of the sequence of the disorder. The corresponding expectations are denoted as $\mathbf{E}$ and $\mathbb{E}$ respectively.

2.1 Quenched Law of Large Numbers

The first result concerns a quenched law of large numbers for the empirical measure ν_N defined in (4).

2.1.1 Assumptions

We assume that the coefficients and initial condition in (3) satisfy:

1. The function $(\theta, \omega) \mapsto c(\theta, \omega)$ is locally Lipschitz-continuous in θ (for fixed ω) and satisfy a one-sided Lipschitz condition:
$$\forall(\theta, \omega), (\bar{\theta}, \bar{\omega}),\ \left(\theta - \bar{\theta}\right) \cdot \left(c(\theta, \omega) - c(\bar{\theta}, \bar{\omega})\right) \leq L\left(\left|\theta - \bar{\theta}\right|^2 + |\omega - \bar{\omega}|^2\right), \tag{9}$$
for some constant L. The function c also satisfies a polynomial bound:
$$\forall(\theta, \omega), |c(\theta, \omega)| \leq C\left(1 + |\theta|^\kappa + |\omega|^\iota\right), \tag{10}$$
for some constant $C > 0$ and where $\kappa \geq 2$ and $\iota \geq 1$.
2. The interaction term Γ is bounded and globally Lipschitz on $(\mathbf{R}^m \times \mathbf{R}^n)^2$.
3. For fixed $\tilde{\theta}, \omega, \tilde{\omega}$, the functions $\theta \mapsto c(\theta, \omega)$ and $\theta \mapsto \Gamma(\theta, \omega, \tilde{\theta}, \tilde{\omega})$ are twice differentiable with continuous derivatives.
4. The initial distribution ζ of the particles in (3) and the law μ of the disorder satisfy the following moment conditions:
$$\int_{\mathbf{R}^m} |\theta|^\kappa\, \zeta(\mathrm{d}\theta) < \infty \text{ and } \int_{\mathbf{R}^n} |\omega|^\iota\, \mu(\mathrm{d}\omega) < \infty, \tag{11}$$
where the constants κ and ι are given in (10).

Remark 2 These assumptions differ from the hypotheses of the original proof in [39]. Indeed, in [39], we were mostly concerned with particles living in a compact space (as in the Kuramoto model (5)), in which case assumptions on c, Γ and ζ (especially the differentiability of c and Γ) can be simplified. We refer to [39], Sect. 2.3.1 for further details. Note that the present assumptions specifically include the case of polynomial coefficients (as in the FitzHugh-Nagumo case (7), set $\kappa = 3$ and $\iota = 1$) which is not covered by [39].

2.1.2 Law of Large Numbers and McKean-Vlasov Equation

Theorem 1 ([39, 41]) *Under the hypotheses of Sect. 2.1.1,*

1. *there is a unique process* $t \mapsto \nu_t$ *in* $\mathscr{C}([0,T], \mathscr{M}_1(\mathbf{R}^m \times \mathbf{R}^n))$ *such that* $\nu_0 = \zeta \times \mu$ *and* $\sup_{t\le T} \int (|\theta|^\kappa \vee |\omega|^\iota)\, \nu_t(\mathrm{d}\theta, \mathrm{d}\omega) < +\infty$*, satisfying the weak McKean-Vlasov equation*

$$\langle \nu_t\,,f\rangle = \langle \nu_0\,,f\rangle + \int_0^t \left\langle \nu_s\,,\, \frac{1}{2}\Delta_\theta f + \nabla_\theta f\cdot([\Gamma, \nu_s] + c)\right\rangle \mathrm{d}s,\ t\in[0,T], \tag{12}$$

where $\langle \nu\,,f\rangle := \int f(\theta,\omega)\nu(\mathrm{d}\theta,\mathrm{d}\omega)$ *and*

$$[\Gamma, m](\theta,\omega) := \int \Gamma(\theta,\omega,\tilde\theta,\tilde\omega)m(\mathrm{d}\tilde\theta,\mathrm{d}\tilde\omega). \tag{13}$$

2. *for almost-every sequence* $(\omega_i)_{i\ge1}$*, the sequence* $(\nu_N^{(\omega)})_{N\ge1}$ *converges as* $N\to\infty$ *in* $\mathscr{C}([0,T], \mathscr{M}_1(\mathbf{R}^m \times \mathbf{R}^n))$ *to the unique solution* ν *of* (12).

Remark 3 Note that if we take a test function f in (12) that does not depend on θ, we obtain that $\langle \nu_0\,,f\rangle = \langle \nu_t\,,f\rangle$, $\forall t\in[0,T]$. In particular, the marginal distribution on ω of ν_t is independent of t and equal to μ.

Since the noise in (3) is non-degenerate, using the regularizing properties of the heat kernel, one can prove that for any measure-valued initial condition in (12), the solution of (12) has a regular density $\nu_t(\mathrm{d}\theta,\mathrm{d}\omega) = q_t(\theta,\omega)\mathrm{d}\theta\mu(\mathrm{d}\omega)$ for all $t>0$. Integrating by parts in (12), we find that q_t is a strong solution to

$$\begin{aligned}\partial_t q_t &= \frac{1}{2}\Delta_\theta q_t(\theta,\omega) - \mathrm{div}_\theta\,(q_t(\theta,\omega)c(\theta,\omega))\\ &\quad - \mathrm{div}_\theta\left(q_t(\theta,\omega)\int \Gamma(\theta,\omega,\tilde\theta,\tilde\omega)q_t(\tilde\theta,\tilde\omega)\mathrm{d}\tilde\theta\mu(\mathrm{d}\tilde\omega)\right),\ t>0.\end{aligned} \tag{14}$$

For a proof of this fact and further details, we refer to [28, Prop. A.1].

A consequence of Theorem 1 is that, at the level of the law of large numbers, the system only depends on the law μ of the disorder, but not on a typical realization of the $(\omega_i)_{i\ge1}$: there is a *self-averaging phenomenon*. The notion of self-averaging (or its absence) is crucial in many disordered models of statistical physics and is deeply related to the influence of the disorder on the phase transition in such systems (see e.g. [2, 48] and references therein).

2.2 Quenched Central Limit Theorem

The second result concerns the influence of a typical realization of the disorder $(\omega_j)_{j\geq 1}$ on the fluctuations of the empirical measure (4) around its McKean-Vlasov limit (12). The question is whether or not self-averaging also holds at the level of the fluctuations, and if not, if it is possible to quantify the dependance in the disorder of the system (3) at the level of fluctuations.

2.2.1 Non-self-averaging Phenomenon in the Kuramoto Model

The motivation comes from the Kuramoto model (5). To fix ideas, consider the case where the frequencies ω_i in (5) are sampled with $\mu = \frac{1}{2}(\delta_{-1} + \delta_1)$: this is simply a random decomposition of $(\theta_1, \ldots, \theta_N)$ between two subpopulations, one naturally rotating clockwise ($\omega_i = +1$) and the second rotating anti-clockwise ($\omega_i = -1$). One can imagine that fluctuations in the finite sample $(\omega_1, \ldots, \omega_N) \in \{\pm 1\}^N$ may lead, for example, to a majority of $+1$ with respect to -1, so that the rotators with positive frequency induce a global rotation of the whole system in the direction of the majority. Direction and speed of rotation depend on this initial configuration of the disorder (see Figs. 1 and 2). This phenomenon, noticed numerically in [5], can be computed through the order parameters $\left(r_{N,t}, \psi_{N,t}\right)$:

$$r_{N,t}e^{i\psi_{N,t}} := \frac{1}{N}\sum_{j=1}^{N} e^{i\theta_{j,t}} = \int_{\mathbf{S}\times\mathbf{R}} e^{i\theta}\,\nu_{N,t}(\mathrm{d}\theta, \mathrm{d}\omega), \quad N \geq 1,\ t \geq 0, \tag{15}$$

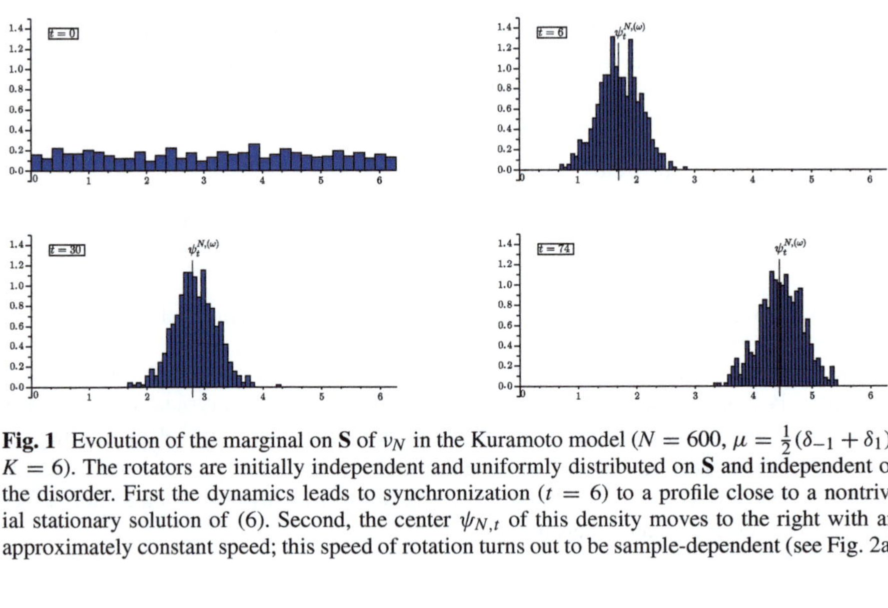

Fig. 1 Evolution of the marginal on $\mathbf{S}$ of ν_N in the Kuramoto model ($N = 600$, $\mu = \frac{1}{2}(\delta_{-1} + \delta_1)$, $K = 6$). The rotators are initially independent and uniformly distributed on $\mathbf{S}$ and independent of the disorder. First the dynamics leads to synchronization ($t = 6$) to a profile close to a nontrivial stationary solution of (6). Second, the center $\psi_{N,t}$ of this density moves to the right with an approximately constant speed; this speed of rotation turns out to be sample-dependent (see Fig. 2a)

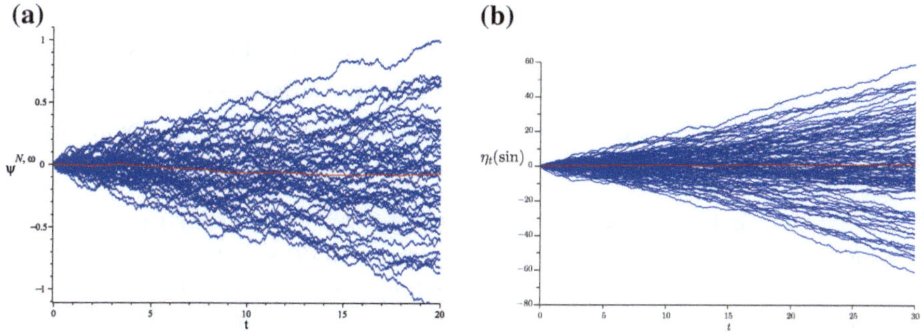

Fig. 2 The *red trajectories* in both figures correspond to the averaged model where there is no disorder-dependent rotation. **a** Trajectories of the center of synchronization ψ_N for different realizations of the disorder $(\omega_1, \dots, \omega_N)$ ($\mu = 1/2\ (\delta_{-0.5} + \delta_{0.5})$, $k = 4$, $n = 400$). Direction and speed of ψ_N depend on the initial configuration of the frequencies. These simulations are compatible with speeds of order $N^{-1/2}$. **b** Trajectories of the process η_t (sin) for different realizations of the mean-value $C(\omega)$ (see 23). Trajectories of the limiting fluctuation process $\eta(\sin)$ are almost linear and compatible with (**a**)

Here $r_{N,t} \in [-1, 1]$ gives a notion of synchronization of the system (5) (e.g. $r_{N,t} = 1$ if the oscillators $\theta_{j,t}$ are all equal) and $\psi_{N,t}$ captures the position of the center of synchronization (see Fig. 1). One can see on Fig. 2a that $t \mapsto \psi_{N,t}$ has an approximately linear behavior whose slope depends on the sample of the disorder. Note that this disorder-induced phenomenon does not happen at the level of the nonlinear Fokker-Planck equation (6), but only at the level of fluctuations (the speed of rotation in Fig. 2a is of order $N^{-1/2}$ which vanishes as $N \to \infty$). Consequently, in order to understand this phenomenon, one needs to make sense to a *quenched central limit theorem* for the empirical measure ν_N.

2.2.2 Weak Quenched Convergence of the Fluctuation Process

We consider, for a fixed realization of the disorder $(\omega) = (\omega_i)_{i\geq 1}$ the fluctuation process $\eta_N = \eta_N^{(\omega)}$ given by

$$\eta_{N,t}^{(\omega)} := \sqrt{N}\left(\nu_{N,t}^{(\omega)} - \nu_t\right),\ t \leq T,\ N \geq 1. \tag{16}$$

For fixed $t \leq T$, (ω) and $N \geq 1$, $\eta_{N,t}^{(\omega)}$ is a random element of $\mathscr{S}'$, the Schwartz space of tempered distributions on $\mathbf{R}^m \times \mathbf{R}^n$.

Remark 4 One observation about (16) is that the convergence of $\eta_N^{(\omega)}$ as $N \to \infty$ cannot hold *for a fixed realization of* (ω): consider the particular case of (3) when $m = 1$, $c(\theta, \omega) = \omega$, $\Gamma = 0$

$$\theta_{i,t} = \omega_i t + B_{i,t},\ t \leq T, i = 1, \dots, N, \tag{17}$$

that is simply Brownian motions with random drifts. Studying the fluctuations of the empirical measure associated to (17) requires to look at functionals of the type $\sqrt{N}\left(\frac{1}{N}\sum_{i=1}^{N}\varphi(\omega_i) - \int \varphi(\omega)\mu(\mathrm{d}\omega)\right)$, for regular functions φ. But *almost-surely* in $(\omega_i)_{i\geq 1}$, the latter quantity does not converge (it only converges *in law* with respect to $(\omega_i)_{i\geq 1}$).

In order to make sense of any possible limit for $\eta_N^{(\omega)}$, one needs to find a *weak formulation of a notion of quenched convergence*. To do so, fix (ω) and denote as $\mathscr{H}_N(\omega)$ the law of the process $\eta_N^{(\omega)}$; $\mathscr{H}_N(\omega)$ belongs to $\mathscr{M}_1(\mathscr{C}([0,T],\mathscr{S}'))$, the set of probability measures on continuous paths with values in $\mathscr{S}'$. As noted in Remark 4, $\mathscr{H}_N(\omega)$ is not likely to converge weakly as $N\to\infty$ for fixed (ω). Instead, consider the random variable:

$$(\omega) \in \left(\mathbf{R}^n\right)^{\mathbf{N}} \mapsto \mathscr{H}_N(\omega) \in \mathscr{M}_1(\mathscr{C}([0,T],\mathscr{S}')). \tag{18}$$

The purpose of Theorem 2 below is precisely to state that this random variable (with values in the big set $\mathscr{M}_1(\mathscr{C}([0,T],\mathscr{S}'))$) converges in law to a random variable $\omega \mapsto \mathscr{H}(\omega) \in \mathscr{M}_1(\mathscr{C}([0,T],\mathscr{S}'))$. The second point of Theorem 2 is to identify $\mathscr{H}(\omega)$ as the law of the solution to a linear stochastic partial differential equation.

Remark 5 It is important to note that this weak notion of quenched convergence differs from an averaged convergence: it is also possible under the same hypotheses to state an averaged central limit theorem which gives a different limit from the one found in (23) (see [39]). The quenched convergence still keeps track of the dependence in the disorder of the particle system as $N\to\infty$ (see Theorem 2), whereas the averaged limit does not.

2.2.3 Assumptions

In addition to the assumptions of Sect. 2.1.1, we suppose that

1. The functions $(\theta,\omega)\mapsto c(\theta,\omega)$ and $(\theta,\omega,\tilde{\theta},\tilde{\omega})\mapsto \Gamma(\theta,\omega,\tilde{\theta},\tilde{\omega})$ are infinitely differentiable. The derivatives of Γ are uniformly bounded and the derivatives of c are bounded in θ and satisfy the same polynomial bound in ω as in (10).
2. The following moment conditions are satisfied:

$$\int_{\mathbf{R}^m} |\theta|^{2\gamma}\,\zeta(\mathrm{d}\theta) < \infty \text{ and } \int_{\mathbf{R}^n} |\omega|^{2(\iota+\gamma)}\,\mu(\mathrm{d}\omega) < \infty, \tag{19}$$

where the constant ι is given by (10), for a sufficiently large constant γ (depending explicitly on m, n, ι).

2.2.4 Quenched Central Limit Theorem

Before stating the result, let us give some definitions: for all $0 \leq s \leq T$, let $\mathscr{L}_s$ be the second order differential operator defined by

$$\mathscr{L}_s f(\theta,\omega) := \frac{1}{2}\Delta_\theta f(\theta,\omega) + \nabla_\theta f(\theta,\omega)\cdot([\Gamma,\nu_s](\theta,\omega)+c(\theta,\omega)) \\ + \int \nabla_\theta f(\tilde{\theta},\tilde{\omega})\cdot \Gamma(\tilde{\theta},\tilde{\omega},\theta,\omega)\nu_s(\mathrm{d}\tilde{\theta},\mathrm{d}\tilde{\omega}).$$

Let W the Gaussian process with covariance (for every $s,t\in[0,T]$):

$$\mathbf{E}(W_t(f_1)W_s(f_2)) = \int_0^{s\wedge t}\int \nabla_\theta f_1(\theta,\omega)\cdot\nabla_\theta f_2(\theta,\omega)\nu_u(\mathrm{d}\theta,\mathrm{d}\omega)\mathrm{d}u. \tag{20}$$

For all f_1, f_2 bounded and continuous on $\mathbf{R}^m\times\mathbf{R}^n$, let

$$\Gamma_1(f_1,f_2) := \int_{\mathbf{R}^n} \mathrm{Cov}_\zeta\,(f_1(\cdot,\omega), f_2(\cdot,\omega))\,\mu(\mathrm{d}\omega), \tag{21}$$
$$= \int_{\mathbf{R}^n}\left\{\int_{\mathbf{R}^m}\left(f_1 - \int_{\mathbf{R}^m} f_1(\cdot,\omega)\mathrm{d}\zeta\right)\left(f_2 - \int_{\mathbf{R}^m} f_2(\cdot,\omega)\mathrm{d}\zeta\right)\mathrm{d}\zeta\right\}\mu(\mathrm{d}\omega),$$

and

$$\Gamma_2(f_1,f_2) = \mathrm{Cov}_\mu\left(\int_{\mathbf{R}^m} f_1\mathrm{d}\zeta, \int_{\mathbf{R}^m} f_2\mathrm{d}\zeta\right), \tag{22}$$
$$= \int_{\mathbf{R}^n}\left(\int_{\mathbf{R}^m} f_1\mathrm{d}\zeta - \int_{\mathbf{R}^m\times\mathbf{R}^n} f_1\mathrm{d}\zeta\mathrm{d}\mu\right)\left(\int_{\mathbf{R}^m} f_2\mathrm{d}\zeta - \int_{\mathbf{R}^m\times\mathbf{R}^n} f_2\mathrm{d}\zeta\mathrm{d}\mu\right)\mathrm{d}\mu.$$

The main theorem is the following:

Theorem 2 ([39]) *Under the assumptions of Sects. 2.1.1 and 2.2.3, the sequence $(\omega)\mapsto\mathscr{H}_N(\omega)$ converges in law to the random variable $\omega\mapsto\mathscr{H}(\omega)$, where $\mathscr{H}(\omega)$ is the law of the Ornstein-Uhlenbeck process η^ω solution in $\mathscr{S}'$ of the following stochastic partial differential equation:*

$$\eta_t^\omega = X + C(\omega) + \int_0^t \mathscr{L}_s^*\eta_s^\omega\mathrm{d}s + W_t, \tag{23}$$

where, $\mathscr{L}_s^*$ is the formal adjoint operator of $\mathscr{L}_s$, X is a centered Gaussian process with covariance Γ_1 and where for fixed ω, $C(\omega)$ is the nontrivial deterministic mean value of the initial condition. As a random variable in ω, $\omega\mapsto C(\omega)$ is a Gaussian process with covariance Γ_2. Moreover, W is independent on the initial value (X, C).

In (23), the linear operator $\mathscr{L}_s^*$ is deterministic ; the only dependence in ω lies in the initial condition $\eta_0^\omega = X + C(\omega)$, through its non trivial mean-value $C(\omega)$. We give here some intuition about the fact that $C(\cdot)$ precisely captures the fluctuations of the disorder in the microscopic model and show how one can understand from (23) the non-self-averaging behavior of the Kuramoto model described in Sect. 2.2.1. For $N \geq 1$ and $f : \mathbf{R}^m \times \mathbf{R}^n \to \mathbf{R}$, the initial fluctuation applied to f is given by

$$\begin{aligned}\eta_{N,0}(f) &= \frac{1}{\sqrt{N}} \sum_{j=1}^{N} \left(f(\theta_{j,0}, \omega_j) - \int_{\mathbf{R}^m \times \mathbf{R}^n} f(\theta, \omega) \zeta(\mathrm{d}\theta) \mu(\mathrm{d}\omega) \right), \\ &= \frac{1}{\sqrt{N}} \sum_{j=1}^{N} \left(f(\theta_{j,0}, \omega_j) - \int_{\mathbf{R}^m} f(\theta, \omega_j) \zeta(\mathrm{d}\theta) \right) \\ &+ \frac{1}{\sqrt{N}} \sum_{j=1}^{N} \left(\int_{\mathbf{R}^m} f(\theta, \omega_j) \zeta(\mathrm{d}\theta) - \int_{\mathbf{R}^m \times \mathbf{R}^n} f(\theta, \omega) \zeta(\mathrm{d}\theta) \mu(\mathrm{d}\omega) \right), \\ &:= X_N(f) + C_N(f).\end{aligned} \tag{24}$$

The process X_N captures the initial fluctuations of the rotators whereas C_N captures the fluctuations of the disorder. It is easy to see that C_N converges in law (w.r.t. the disorder) to the process C with covariance (22). In the framework of the Kuramoto model with binary disorder (recall Sect. 2.2.1), computations show ([40]) that the relevant quantity for the dynamics of (23) is the restriction C_+ of the process C to the component on $+1$:

$$C_+(\psi) := C(\psi \mathbf{1}_{\omega=+1}), \quad \psi : \mathbf{S} \to \mathbf{R}. \tag{25}$$

C_+ is the limit in law of the microscopic process $C_{N,+}$ defined by

$$C_{N,+}(\psi) := \left(\int_{\mathbf{S}} \psi(\cdot) \mathrm{d}\zeta \right) \frac{1}{\sqrt{N}} \sum_{i=1}^{N} \left(\mathbf{1}_{(\omega_i=+1)} - \frac{1}{2} \right) := \left(\int_{\mathbf{S}} \psi(\cdot) \mathrm{d}\zeta \right) \frac{\alpha_N}{\sqrt{N}}. \tag{26}$$

Here, α_N is exactly the (centered) number of frequencies among $(\omega_1, \ldots, \omega_N)$ that are positive, so that $C_{N,+}$ captures the lack of symmetry of the disorder: $\alpha_N > 0$ (resp. $\alpha_N < 0$) represents the case of an asymmetry in favor of positive (resp. negative) frequencies. Hence, Theorem 2 provides a way to study, through the process C, the influence of the asymmetry of the disorder on (3). The main lines of proof of Theorem 2 are given in Sect. 3.2.

2.3 Long-Time Behavior of the Fluctuation Process in the Kuramoto Model

In this paragraph, we restrict the study to the Kuramoto case (5) with binary disorder: $\mu = \frac{1}{2}\left(\delta_{-\omega_0} + \delta_{\omega_0}\right)$ for some $\omega_0 > 0$. One can see numerically (see Fig. 2b) that the initial asymmetry $C(\cdot)$ propagates from $t = 0$ to positive times and provide analogous trajectories to the ones observed in the microscopic model (Fig. 2a). One can make this observation rigorous, at least when ω_0 is small:

Theorem 3 ([40]) *In the Kuramoto model (5) in the small-disorder regime, the solutions of (23) are asymptotically linear and disorder dependent. More precisely, for all* $K > 1$, *there exist an Hilbert space of distributions* $\mathbf{H}$ *and* $\omega^* > 0$ *such that for all* $0 < \omega_0 < \omega^*$ *and for fixed initial value* $\eta_0^\omega = X + C(\omega)$, *there exists* $V(\omega) \in \mathbf{H}$ *such that*

$$\frac{\eta_t^\omega}{t} \xrightarrow{\text{in law}} V(\omega), \quad \text{as } t \to +\infty. \tag{27}$$

Moreover, $\omega \mapsto V(\omega)$ *is a Gaussian process in* $\mathbf{H}$ *with explicit (in terms of* $C(\omega)$*) covariance.*

The proof of Theorem 3 relies on a spectral analysis (based on perturbation arguments from the case without disorder, [8]) of the linear operator $\mathscr{L}_s^*$ governing (23). Following the intuition of the finite-dimensional Jordan block $\binom{x}{y}' = \left(\begin{smallmatrix} 0 & 1 \\ 0 & 0 \end{smallmatrix}\right) \cdot \binom{x}{y}$ (whose solutions are obviously linear), one of the key ingredients to the proof of Theorem 3 is to prove a similar property for the unbounded operator $\mathscr{L}_s^*$: there exist u and v such that $\mathscr{L}_s^* u = 0$ and $\mathscr{L}_s^* v = u$. We refer to [40] for more details.

2.4 Conclusion and Perspectives

Disordered mean-field models such as (3) (and especially the Kuramoto model) are not self-averaging at the level of fluctuations: the dynamics of the quenched fluctuations of (4) is disorder-dependent. However, in order to derive rigorously the exact speed of rotation of synchronized solutions in the Kuramoto model described in Fig. 2, it would be necessary to study (5) on larger time scales, as in [9]. This is currently under investigation.

Another difficulty is that, although both law of large numbers and central limit theorem are valid in a rather general setting, investigating the long-time behavior of the limiting objects ν and η is often very difficult. In that sense, one of the reasons for the popularity of the Kuramoto model is that the stationary solutions of the nonlinear Fokker-Planck equation (6) are explicitly computable [50]. A key point in the analysis of the stability of such solutions [8, 9, 29] is that the Kuramoto model without disorder is *reversible* [8], whereas reversibility is lost for many other neuronal models (e.g. FitzHugh-Nagumo [7]). The results presented here should be applicable to other models of disordered diffusions, provided sufficient information is known about characterization and linear stability of stationary states (see for example [38]).

There is currently a growing interest in generalizations of (3) where the interactions depend on the topology of non-mean-field networks or on the distance between particles. The motivation comes from the biological observation that neurons do not interact in a mean-field way (see [55] and references therein). As far as weighted interactions are considered, models of *moderately interacting diffusions* [35, 46] have been one of the first attempts to go beyond pure mean-field models. Several papers in physics [31, 43, 47] have also considered models of oscillators within P-nearest neighbors or with power-law interaction. Those models are known to exhibit anomalous dynamical properties (*chimera states*). The mean-field limit in this framework has been rigorously shown in [41], showing in particular anomalous speed of convergence to the nonlinear Fokker-Planck equation. The exact central limit theorem in this case is currently under investigation.

Generalizing the result of Dai Pra and den Hollander concerning possible *quenched large deviations* for (3) is of course a natural perspective for this work and is the object of an ongoing project.

3 Sketches of Proofs

3.1 Law of Large Numbers

We give in this paragraph the main lines of the proof of Theorem 1. The difficulty here is that we allow the coefficient c in (3) to have polynomial growth (recall Sect. 2.1.1); in particular, one needs to have *a priori* controls on the moments of any accumulation point of (4). The well-posedness of the McKean-Vlasov equation (12) under the hypotheses of Sect. 2.1.1 can be seen as a consequence of [41]. The existence of a least one solution ν (satisfying the required moment conditions) to (12) is established in [41] via a fixed-point procedure on the nonlinear process associated to (12) using arguments from Sznitman [53] and the uniqueness comes from the fact that for any such solution ν, the empirical measure ν_N necessarily converges to ν as $N \to \infty$, in terms of an adequate Wasserstein metric. Note that we cannot directly use this last result for the quenched convergence of ν_N, since the convergence in [41] is averaged w.r.t. the disorder.

Remark 6 The framework of [41] concerns the more general case of diffusions within weighted spatial interactions (P-nearest neighbor with parameter $R \in]0, 1]$ and power law with exponent $\alpha \geq 0$, see [41], Sect. 1.2.2 for detailed definitions of these models). It is easy to see that one retrieves the full mean-field setting for $R = 1$ or $\alpha = 0$.

As far as the quenched convergence of the empirical measure (4) is concerned, one can proceed as follows. The result for the initial condition is clear: since $(\theta_{i,0}, \omega_i)_{1 \leq i \leq N}$ are i.i.d. random variables with law $\zeta \otimes \mu$, the initial empirical measure $\nu_{N,0}$ converges almost surely in (ω) to $\nu_0(\mathrm{d}\theta, \mathrm{d}\omega) = \zeta(\mathrm{d}\theta)\mu(\mathrm{d}\omega)$, as $N \to \infty$.

An application of Ito's formula to (3) (for any regular function $(\theta, \omega) \mapsto f(\theta, \omega)$) leads to the following semi-martingale representation for ν_N:

$$\langle \nu_{N,t}, f \rangle = \langle \nu_{N,0}, f \rangle + \int_0^t \left\langle \nu_{N,s}, \frac{1}{2}\Delta_\theta f + \nabla_\theta f \cdot c \right\rangle \mathrm{d}s + \int_0^t \left\langle \nu_{N,s}, \nabla_\theta f \cdot \int \Gamma(\cdot, \cdot, \tilde{\theta}, \tilde{\omega}) \nu_{N,s}(\mathrm{d}\tilde{\theta}, \mathrm{d}\tilde{\omega}) \right\rangle \mathrm{d}s + M_{N,t}(f), \tag{28}$$

where $M_{N,t}(f) := \frac{1}{N}\sum_{j=1}^N \int_0^t \nabla_\theta f(\theta_{j,s}, \omega_j) \cdot \mathrm{d}B_{j,s}$ is a martingale. Using usual tightness criteria [45, 49] based on Aldous criterion [10] for real valued continuous processes, it is easy to derive from (28) the tightness of (ν_N) in $\mathscr{C}([0,T], (\mathscr{M}_1(\mathbf{R}^m \times \mathbf{R}^n), v))$, where v is the vague topology (i.e. the coarsest topology that makes the evaluations $\nu \mapsto \langle \nu, f \rangle$ continuous for every f continuous with compact support). Using the one-sided Lipschitz continuity of c, one can prove that, almost surely in (ω), $\sup_{N\geq 1}\sup_{1\leq i\leq N} \mathbf{E}\left(|\theta_{i,t}|^\kappa\right) < \infty$. Using this estimate and a localization argument ([39], Lemma 3.4) one obtains that for any accumulation point $\tilde{\nu}$ of $(\nu_N)_{N\geq 1}$,

$$\int_{\mathbf{R}^m\times\mathbf{R}^n} \left(|\theta|^\kappa \vee |\omega|^\iota\right) \tilde{\nu}(\mathrm{d}\theta, \mathrm{d}\omega) < +\infty. \tag{29}$$

The tightness of $(\nu_N)_{N\geq 1}$ in the weak topology and its convergence to (12) follows from the a priori estimate (29) and the fact that $M_{N,t}(f)$ in (28) vanishes as $N \to \infty$.

3.2 *Fluctuations*

We now turn to the proof of Theorem 2. The principal tool used here is Hilbertian techniques for measure-valued processes developed by Fernandez, Méléard and Jourdain [26, 35] for similar models without disorder.

An application of Ito's formula to (3) leads to a semi-martingale decomposition of the fluctuation process $\eta_N^{(\omega)}$ of the following form: for all regular functions f, for every sequence (ω) and for all $t \leq T$:

$$\left\langle \eta_{N,t}^{(\omega)}, f \right\rangle = \left\langle \eta_{N,0}^{(\omega)}, f \right\rangle + \int_0^t \left\langle \eta_{N,s}^{(\omega)}, \mathscr{L}_{N,s}(f) \right\rangle \mathrm{d}s + W_{N,t}^{(\omega)}(f), \tag{30}$$

where $\mathscr{L}_{N,s}$ is an unbounded linear operator defined by

$$\mathscr{L}_{N,s}f(\theta,\omega) := \frac{1}{2}\Delta_\theta f(\theta,\omega) + \nabla_\theta f(\theta,\omega) \cdot ([\Gamma, \nu_{N,s}](\theta,\omega) + c(\theta,\omega)) + \int \nabla_\theta f(\tilde{\theta}, \tilde{\omega}) \cdot \Gamma(\tilde{\theta}, \tilde{\omega}, \theta, \omega) \nu_s(\mathrm{d}\tilde{\theta}, \mathrm{d}\tilde{\omega}), \ \theta \in \mathbf{R}^m, \omega \in \mathbf{R}^n. \tag{31}$$

and $W_{N,t}^{(\omega)}(f)$ is a real continuous martingale with quadratic variation process

$$\left\langle W_N^{(\omega)}(f) \right\rangle_t = \int_0^t \left\langle \nu_{N,s}^{(\omega)}, |\nabla_\theta f(\theta,\omega)|^2 \right\rangle \mathrm{d}s.$$

Equation (30) is nothing else but a discrete version of the SPDE (23). The natural procedure is then to show that $(\omega) \mapsto \mathscr{H}_N(\omega)$ is tight and identify the limit. Since the identification of the limit is standard, we focus on the main difficulty, that is the tightness result.

3.2.1 The Nonlinear Process

The core of the proof consists in introducing the nonlinear process [35, 42, 52, 53] associated to the McKean-Vlasov equation (12), that is the diffusion $\bar{\theta}$ whose finite dimensional laws are precisely given by $\nu_t, t \geq 0$. This notion was first introduced by Sznitman [53] for systems without disorder. In the context of disordered diffusions, the nonlinear process may be defined as the solution $(\bar{\theta}_t, \omega)_{t\in[0,T]}$ to

$$\begin{cases} \bar{\theta}_t = \theta_0 + \int_0^t c(\bar{\theta}_s, \omega)\mathrm{d}s + \int_0^t [\Gamma, \nu_s](\bar{\theta}_s, \omega)\mathrm{d}s + B_t, \\ \omega \sim \mu, \\ \nu_t = \mathscr{L}(\bar{\theta}_t, \omega), \forall t \in [0, T]. \end{cases} \tag{32}$$

Note that in (32), $\bar{\theta}$ depends on ν, which itself is the law of $\bar{\theta}$, so the existence of such a nonlinear process is unclear.

Proposition 1 ([39]) *There is pathwise existence and uniqueness in (32).*

Proof The proof follows ideas from Sznitman [53], Th 1.1, p. 172. The point is to use a Picard iteration in the space of probabilities on $\mathscr{C}([0, T], \mathbf{R}^m \times \mathbf{R}^n)$ endowed with an appropriate Wasserstein metric. □

The key point in the proof of the fluctuation theorem (see [26]) is to build a coupling between the particle-system $(\theta_i)_{1\leq i\leq N}$ given by (3) and a well-chosen collection of independent nonlinear processes. Namely, for all $i = 1, \dots, N$, consider the nonlinear process $\bar{\theta}_i$ defined by (32), with the same initial value as θ_i, with the same inhomogeneity ω_i, driven by the same Brownian motion B_i.

Proposition 2 ([39]) *Under the assumptions made in Sect. 2.2.3,*

$$\sup_{1\leq i\leq N} \mathbf{E}\left[\sup_{t\leq T} \left|\theta_{i,t} - \bar{\theta}_{i,t}\right|^2\right] \leq Z_N(\omega_1, \dots, \omega_N), \quad N \geq 1, \tag{33}$$

where the random variable $(\omega) \mapsto Z_N(\omega)$ is such that

$$\lim_{A\to\infty} \limsup_{N\to\infty} \mathbb{P}\left(N Z_N(\omega) > A\right) = 0. \tag{34}$$

Proof By (3) and (32) and using the one-sided Lipschitz continuity of c, one has

$$\mathbf{E}\left[\sup_{s\le t}\left|\theta_{i,s}-\bar{\theta}_{i,s}\right|^2\right] \le C\left(\int_0^t \sup_{1\le j\le N}\mathbf{E}\left[\sup_{u\le s}\left|\theta_{j,u}-\bar{\theta}_{j,u}\right|^2\right]\mathrm{d}s \right.$$
$$\left. + \int_0^t \mathbf{E}\left[\left|[\Gamma,\bar{\nu}_{N,s}-\nu_s](\bar{\theta}_{i,s},\omega_i)\right|^2\right]\mathrm{d}s\right),$$

where $\bar{\nu}_N$ is the empirical measure of the nonlinear processes $\bar{\nu}_{N,t} := \frac{1}{N}\sum_{i=1}^N \delta_{(\bar{\theta}_{i,t},\omega_i)}$. Define $T_{i,j,s} := \Gamma(\bar{\theta}_{i,s},\omega_i,\bar{\theta}_{j,s},\omega_j) - \int \Gamma(\theta_{i,s},\omega_i,\tilde{\theta},\tilde{\omega})\nu_s(\mathrm{d}\tilde{\theta},\mathrm{d}\tilde{\omega})$, so that

$$\mathbf{E}\left(\left|[\Gamma,\bar{\nu}_{N,s}-\nu_s](\bar{\theta}_{i,s},\omega_i)\right|^2\right) = \frac{1}{N^2}\mathbf{E}\left(\sum_{j=1}^N\left|T_{i,j,s}\right|^2 + \sum_{k\neq l} T_{i,k,s}\cdot T_{i,l,s}\right)$$
$$\le \frac{1}{N^2}\left(CN + \mathbf{E}\left(\sum_{\substack{k\neq i, l\neq i\\ k\neq l}} T_{i,k,s}\cdot T_{i,l,s}\right)\right),$$

where we used that Γ is bounded. Suppose for a moment that there is no disorder. Taking conditional expectation w.r.t. $(\bar{\theta}_r, r\neq l)$ in the last term and using the exchangeability of the particles, we obtain that this term is zero, leading, by Gronwall's Lemma, to an upper bound in (33) of the form C/N (see [26], Lemma 3.2). Since we work here in a frozen environment (i.e. we do not integrate w.r.t. the disorder and lose the exchangeability of the particles), this additional term becomes nontrivial and not bounded *for fixed* (ω). This fact precisely motivates the weak formulation of the quenched convergence introduced in Sect. 2.2.2. Proof of (34) can be found in [39]. □

3.2.2 Weighted Sobolev Spaces

Proposition 2 is the key result in order to show that the random variable $\mathscr{H}_N$ defined in Sect. 2.2.2 is tight. The second tool we use is the introduction of weighted-Sobolev norms that are specifically adapted to the analysis of (30) (see [26] for a previous similar approach). Namely, for every integer j, $\alpha \ge 0$, we consider the space of all real functions f defined on $\mathbf{R}^m\times\mathbf{R}^n$ differentiable up to order j such that

$$\|f\|_{j,\alpha} := \left(\sum_{|k_1|+|k_2|\le j}\int_{\mathbf{R}^m\times\mathbf{R}^n}\frac{\left|D_\theta^{k_1}D_\omega^{k_2}f(\theta,\omega)\right|^2}{(1+|\theta|^\alpha)^2(1+|\omega|^\alpha)^2}\mathrm{d}\theta\mathrm{d}\omega\right)^{1/2} < \infty,$$

where, if $k = (k_1, \dots, k_m)$ and $\theta = (\theta^{(1)}, \dots, \theta^{(m)})$, we define $|k| := \sum_{i=1}^{m} k_i$ and $D_\theta^k h(\theta) := \partial_{\theta^{(1)}}^{k_1} \dots \partial_{\theta^{(m)}}^{k_m} h(\theta)$. Let $W_0^{j,\alpha}$ be the completion of $\mathscr{C}_c^\infty(\mathbf{R}^m \times \mathbf{R}^n)$ for this norm; $(W_0^{j,\alpha}, \|\cdot\|_{j,\alpha})$ is a Hilbert space and $W_0^{-j,\alpha}$ is its dual space. The only thing that differs from the usual Sobolev norm is the weight which is here to control the possibly unbounded coefficient c in (3) and the unbounded disorder. In this framework, using Proposition 2, one can show that under the assumptions of Theorem 2,

Proposition 3 *There exist well-chosen Sobolev indices (j_1, α_1) and (j_2, α_2) such that for fixed (ω) and $N \geq 1$, for all $T > 0$,*

1. *the imbedding $W_0^{-j_1,\alpha_1} \hookrightarrow W_0^{-j_2,\alpha_2}$ is of Hilbert-Schmidt type,*
2. *the process $\eta_N^{(\omega)}$ is bounded in $W_0^{-j_1,\alpha_1}$:*

$$\sup_{t\leq T} \mathbf{E}\left[\left\| \eta_{N,t}^{(\omega)} \right\|_{-j_1,\alpha_1}^2\right] \leq A_N(\omega_1, \dots, \omega_N), \tag{35}$$

3. *the linear operator $\mathscr{L}_{N,s}$ defined in (31) is continuous from $W_0^{j_2,\alpha_2}$ to $W_0^{j_1,\alpha_1}$,*
4. *the process $\eta_N^{(\omega)}$ is bounded, uniformly in time, in $W_0^{-j_2,\alpha_2}$:*

$$\mathbf{E}\left[\sup_{t\leq T} \left\| \eta_{N,t}^{(\omega)} \right\|_{-j_2,\alpha_2}^2\right] \leq B_N(\omega_1, \dots, \omega_N). \tag{36}$$

Moreover, as random variables in (ω), A_N and B_N satisfy

$$\lim_{A\to\infty} \limsup_{N\to\infty} \mathbb{P}(A_N > A) = 0, \text{ and } \lim_{A\to\infty} \limsup_{N\to\infty} \mathbb{P}(B_N > A) = 0.$$

3.2.3 Tightness Result

In the case without disorder (or in the averaged case), the random variables A_N and B_N would simply be replaced by constants. Using the estimates of Proposition 3 and standard tightness criteria for continuous processes with values in Hilbert spaces (see Joffe and Métivier [33]), proving that the fluctuation process is tight is straightforward [26]. In the quenched model, A_N and B_N are not bounded for fixed (ω). We need to work harder to prove that the random variable $(\omega) \mapsto \mathscr{H}_N(\omega)$ defined in (18) is tight in $\mathscr{M}_1(\mathscr{C}([0,T], \mathscr{S}'))$.

Theorem 4 *For all $\varepsilon > 0$, there exists a relatively compact subset K_ε in the set $\mathscr{M}_1(\mathscr{C}([0,T], \mathscr{S}'))$ such that*

$$\limsup_{N\to\infty} \mathbb{P}\left(\left\{(\omega)\ ;\ \mathscr{H}_N(\omega) \in K_\varepsilon^c\right\}\right) \leq \varepsilon. \tag{37}$$

The idea of the proof of Theorem 4 is that, based on Proposition 3, for all $\varepsilon > 0$, one can choose $A(\varepsilon)$ sufficiently large so that the quantities in (35) and (36) are smaller than ε with large probability. K^{ε} is then constructed as a set of elements in $\mathscr{M}_1\left(\mathscr{C}([0,T],\mathscr{S}')\right)$ satisfying a Joffe and Métivier criterion (hence relatively compact) with parameters (depending on ε) chosen in such a way that $\mathscr{H}_N(\omega)$ belongs to K_{ε} with high probability. We leave the technical details to [39], Theorem 4.10.

Acknowledgments One of the articles reviewed here [41] was written at the Technische Universität of Berlin and was supported by the BMBF, FKZ 01GQ 1001B.

References

1. Acebrón, J.A., Bonilla, L.L., Pérez Vicente, C.J, Ritort, F., Spigler R.: The Kuramoto model: a simple paradigm for synchronization phenomena. Rev. Mod. Phys. **77**, 137–185 (2005)
2. Aizenman, M., Lebowitz, J.L., Ruelle, D.: Some rigorous results on the Sherrington-Kirkpatrick spin glass model. Comm. Math. Phys. **112**(1), 3–20 (1987)
3. Arous, G.B., Guionnet, A.: Large deviations for Langevin spin glass dynamics. Probab. Theory Rel. **102**(4), 455–509 (1995)
4. Baladron, J., Fasoli, D., Faugeras, O., Touboul, J.: Mean-field description and propagation of chaos in networks of Hodgkin-Huxley and FitzHugh-Nagumo neurons. J. Math. Neurosci. **2**, 10 (2012)
5. Balmforth, N.J., Sassi, R.: A shocking display of synchrony. Phys. D: Nonlinear Phenom. **143**, 21–55 (2000)
6. Ben Arous, G., Guionnet, A.: Symmetric Langevin spin glass dynamics. Ann. Probab. **25**(3), 1367–1422 (1997)
7. Berglund, N., Landon, D.: Mixed-mode oscillations and interspike interval statistics in the stochastic Fitzhugh-Nagumo model. Nonlinearity **25**, 2303 (2012)
8. Bertini, L., Giacomin, G., Pakdaman, K.: Dynamical aspects of mean field plane rotators and the Kuramoto model. J. Statist. Phys. **138**, 270–290 (2010)
9. Bertini, L., Giacomin, G., Poquet, C.: Synchronization and random long time dynamics for mean-field plane rotators. Probab. Theory Rel. 1–61 (2013)
10. Billingsley, P.: Convergence of probability measures, Wiley Series in Probability and Statistics: Probability and Statistics, 2nd edn. Wiley, New York (1999)
11. Blanchard, P., Röckner, M., Russo, F.: Probabilistic representation for solutions of an irregular porous media type equation. Ann. Probab. **38**, 1870–1900 (2010)
12. Bolley, F., Gentil, I., Guillin, A.: Uniform convergence to equilibrium for granular media. Arch. Ration. Mech. Anal. 1–17 (2012)
13. Bolley, F., Guillin, A., Villani, C.: Quantitative concentration inequalities for empirical measures on non-compact spaces. Probab. Theory Rel. **137**, 541–593 (2007)
14. Cabana, T., Touboul, J.: Large deviations, dynamics and phase transitions in large stochastic heterogeneous neural networks (2013), arXiv:1302.6951
15. Cáceres, M.J., Carrillo, J.A., Perthame, B.: Analysis of nonlinear noisy integrate & fire neuron models: blow-up and steady states. J. Math. Neurosci. **1**, 33 (2011)
16. Collet, F., Dai Pra, P.: The role of disorder in the dynamics of critical fluctuations of mean field models. Electron. J. Probab. **17**(26), 40 (2012)
17. Collet, F., Dai Pra, P., Sartori, E.: A simple mean field model for social interactions: dynamics, fluctuations, criticality. J. Stat. Phys. **139**, 820–858 (2010)
18. Dai Pra, P., den Hollander, F.: McKean-Vlasov limit for interacting random processes in random media. J. Statist. Phys. 184735–184772 (1996)

19. Dawson, D.A., Gärtner, J.: Large deviations from the McKean-Vlasov limit for weakly interacting diffusions. Stochastics **20**, 247–308 (1987)
20. Degond, P., Frouvelle, A., Liu, J.-G.: Macroscopic limits and phase transition in a system of self-propelled particles. J. Nonlin. Sci. 1–30 (2012)
21. Delarue, F., Inglis, J., Rubenthaler, S., Tanré, E.: Global solvability of a networked integrate-and-fire model of McKean-Vlasov type (2012), arXiv:1211.0299
22. Delarue, F., Inglis, J., Rubenthaler, S., Tanré, E.: Particle systems with a singular mean-field self-excitation, Application to neuronal networks (2014), arXiv:1406.1151
23. Ermentrout, G.B., Terman, D.H.: Mathematical foundations of neuroscience. Interdisciplinary Applied Mathematics, vol. 35. Springer, New York (2010)
24. Fathi, M.: A gradient flow approach to large deviations for diffusion processes (2014), arXiv:1405.3910
25. Faugeras O., MacLaurin, J.: A large deviation principle for networks of rate neurons with correlated synaptic weights (2013), arXiv:1302.1029
26. Fernandez, B., Méléard, S.: A Hilbertian approach for fluctuations on the McKean-Vlasov model. Stochast. Process. Appl. **71**, 33–53 (1997)
27. Gärtner, J.: On the McKean-Vlasov limit for interacting diffusions. Math. Nachr. **137**, 197–248 (1988)
28. Giacomin, G., Luçon, E., Poquet, C.: Coherence stability and effect of random natural frequencies in populations of coupled oscillators. J. Dynam. Differ. Equ. **26**, 333–367 (2014)
29. Giacomin, G., Pakdaman, K., Pellegrin, X.: Global attractor and asymptotic dynamics in the Kuramoto model for coupled noisy phase oscillators. Nonlinearity **25**, 1247 (2012)
30. Godinho, D., Quininao, C.: Propagation of chaos for a sub-critical Keller-Segel model. Ann. Inst. H. Poincaré Probab. Statist. (2013), arXiv:1306.3831
31. Gupta, S., Potters, M., Ruffo, S.: One-dimensional lattice of oscillators coupled through power-law interactions: Continuum limit and dynamics of spatial fourier modes. Phys. Rev. E **85**, 066201 (2012)
32. Hitsuda, M., Mitoma, I.: Tightness problem and stochastic evolution equation arising from fluctuation phenomena for interacting diffusions. J. Multivar. Anal. **19**, 311–328 (1986)
33. Joffe, A., Métivier, M.: Weak convergence of sequences of semimartingales with applications to multitype branching processes. Adv. Appl. Probab. **18**, 20–65 (1986)
34. Jourdain, B., Malrieu, F.: Propagation of chaos and Poincaré inequalities for a system of particles interacting through their CDF. Ann. Appl. Probab. **18**, 1706–1736 (2008)
35. Jourdain, B., Méléard, S.: Propagation of chaos and fluctuations for a moderate model with smooth initial data. Ann. Inst. H. Poincaré Probab. Statist. **34**, 727–766 (1998)
36. Jourdain, B., Reygner, J.: Capital distribution and portfolio performance in the mean-field atlas model, to appear in Annals of Finance (2014), arXiv:1312.5660
37. Kuramoto, Y.: Self-entrainment of a population of coupled non-linear oscillators. In: International Symposium on Mathematical Problems in Theoretical Physics (Kyoto Univ., Kyoto). Lecture Notes in Physics, vol. 39, pp. 420–422. Springer, Berlin (1975)
38. Louca, S., Atay, F.M.: Stationary states in infinite networks of spiking oscillators with noise. SIAM J. Appl. Dyn. Syst. **12**, 415–449 (2013)
39. Luçon, E.: Quenched limits and fluctuations of the empirical measure for plane rotators in random media. Electr. J. Prob. **16**, 792–829 (2011)
40. Luçon, E.: Large time asymptotics for the fluctuation SPDE in the Kuramoto synchronization model. J. Funct. Anal. **266**, 6372–6417 (2014)
41. Luçon, E., Stannat, W.: Mean field limit for disordered diffusions with singular interactions. Ann. Appl. Probab. **24**, 1946–1993 (2014)
42. Malrieu, F.: Convergence to equilibrium for granular media equations and their Euler schemes. Ann. Appl. Probab. **13**, 540–560 (2003)
43. Maródi, M., d'Ovidio, F., Vicsek, T.: Synchronization of oscillators with long range interaction: phase transition and anomalous finite size effects. Phys. Rev. E, **66**, 011109 (2002)
44. McKean Jr, H.P.: Propagation of chaos for a class of non-linear parabolic equations. In: Stochastic Differential Equations (Lecture Series in Differential Equations, Session 7, Catholic University (1967) pp. 41–57. Air Force Office Sci. Res, Arlington, Va (1967)

45. Méléard, S., Roelly, S.: Sur les convergences étroite ou vague de processus à valeurs mesures. C. R. Acad. Sci. Paris Sér. I Math. 3**17**, 785–788 (1993)
46. Oelschläger, K.: A law of large numbers for moderately interacting diffusion processes. Z. Wahrsch. Verw. Gebiete **69**, 279–322 (1985)
47. Omelchenko, I., Maistrenko, Y., Hövel, P., Schöll, E.: Loss of coherence in dynamical networks: spatial chaos and chimera states. Phys. Rev. Lett. **106**, 234102 (2011)
48. Pastur, L., Shcherbina, M.: Absence of self-averaging of the order parameter in the Sherrington-Kirkpatrick model. J. Stat. Phys. **62**, 1–19 (1991)
49. Roelly-Coppoletta, S.: A criterion of convergence of measure-valued processes: application to measure branching processes. Stochastics **17**, 43–65 (1986)
50. Sakaguchi, H., Shinomoto, S., Kuramoto, Y.: Phase transitions and their bifurcation analysis in a large population of active rotators with mean-field coupling. Progr. Theoret. Phys. **79**, 600–607 (1988)
51. Strogatz, S.H., Mirollo, R.E.: Stability of incoherence in a population of coupled oscillators. J. Statist. Phys. **63**, 613–635 (1991)
52. Sznitman, A.-S.: Nonlinear reflecting diffusion process, and the propagation of chaos and fluctuations associated. J. Funct. Anal. **56**, 311–336 (1984)
53. Sznitman, A.-S.: Topics in propagation of chaos. In: École d'Été de Probabilités de Saint-Flour XIX–1989, Lecture Notes in Mathematics, vol. 146, pp. 4165–4251. Springer, Berlin (1991)
54. Touboul, J.: Limits and dynamics of stochastic neuronal networks with random heterogeneous delays. J. Stat. Phys. **149**, 569–597 (2012)
55. Wainrib G., Touboul, J.: Topological and dynamical complexity of random neural networks. Phys. Rev. Lett. **110**, 118101 (2013)

Shock Structure and Temperature Overshoot in Macroscopic Multi-temperature Model of Binary Mixtures

Damir Madjarević

Abstract The present study deals with the shock wave profiles in the macroscopic multi-temperature model of binary gaseous mixtures. For that purpose we adopt the hyperbolic model developed within the framework of extended thermodynamics. It is assumed that the mass difference between the constituents has the most prominent influence on the shock structure. Simplicity of the model enables systematic analysis of the results, using a large set of values for parameters, with special regard to the temperature overshoot (TO) of the heavier constituent. We found that TO varies non-monotonically with mass ratio of the constituents. In the context of the previous research, the influence of the different types of dissipation on the shock structure is considered by extending the original hyperbolic system with diffusion terms. It has been observed that TO continued to exist even in the presence of additional dissipative mechanisms.

Keywords Binary mixtures · Shock structure · Extended thermodynamics · Diffusion in gases · Heat transfer

1 Introduction

In macroscopic continuum modeling shock structure problem represents a standard test problem for studying features of rarefied gas flows and validating accuracy of the model [28, 29]. The beginnings of the shock wave structure analysis in binary gas mixtures was related with the work of Cowling [9] which takes into account the diffusion between the components as the only dissipative mechanism. Dyakov [10] and Sherman [26] included viscosity and thermal conductivity. In the framework of the kinetic theory many different models were developed. Goldman and Sirovich [14] in the model included the possibility that each component has its own velocity and temperature.

D. Madjarević (✉)
Faculty of Technical Sciences, University of Novi Sad,
Trg Dositeja Obradovića, Novi Sad, Serbia
e-mail: damirm@uns.ac.rs

P. Gonçalves and A.J. Soares (eds.), *From Particle Systems to Partial Differential Equations II*, Springer Proceedings in Mathematics & Statistics 129,
DOI 10.1007/978-3-319-16637-7_9

Harris and Bienkovski [17] used the method of moments for Maxwell molecules in the kinetic theory of gases. Until then, it was the most complete hydrodynamic model applicable to the analysis of strong shock waves with no restriction on system parameters. They analyzed the equations for the different cases of the mass ratios and densities of components in front of the shock wave. The results are compared with experiments [16] without definitive conclusion about results. Another very important paper [12] is devoted to shock waves in binary gas mixtures in the case where the mass of the components significantly differ. They have developed a method and a set of equations which were not limited to the assumption of Maxwell molecules. In the paper [11] model was developed in the form of moments equations in relaxation approximation. The results were compared with the results of the direct simulation Monte Carlo for the shock structure in a binary mixture of argon and xenon and for the diffusion processes in a binary mixture of argon and helium. Only qualitative agreement in the results has been shown.

Most of existing models due to its complexity do not provide systematic analysis of specific phenomena and reduce to the analysis of particular cases instead.

Nevertheless, these approaches revealed certain important features of shock structure in mixtures: the velocities and the temperatures of the constituents have different values within the profile of the shock wave. Moreover, for certain values of parameters there appears a temperature overshoot (TO) of the heavier constituent—a region within a profile where the temperature increases above the terminal temperature of the mixture [1, 2, 18, 23].

The aim of this paper is to give a systematic analysis of the shock structure problem in binary multi-temperature (MT) mixture of gases using macroscopic hyperbolic MT model [24] in the context of temperature overshoot of the heavier component. Ascribing to each component its own temperature, along with their own velocities, one gets deeper insight into non-equilibrium processes in the mixture. Moreover, it is the simplest way, within the macroscopic framework, to quantitatively describe the mutual exchange of internal energy between the components. Using the macroscopic model over kinetic one also allows us to use simpler low cost numerical schemes. At the same time, model provide accurate solutions for macroscopic quantities and capture important phenomena like TO.

In recent research we deal with the hyperbolic MT model where diffusion between components is only present. We found very good agreement with experimental results for weak shock waves [19]. Our calculation confirmed existence of TO even in that simple case. In this paper we proved that existence of TO was evident even after introducing additional dissipation in the form of viscosity and heat conduction. Systematic analysis of TO in terms of the mass ratio of the constituents reveals its non-monotonic behavior in both cases.

The paper is organized as follows. In Sect. 2 we give a description of the MT model for binary mixtures of gases. Shock structure equation in inviscid and viscous case is given in Sects. 2.1 and 2.3, respectively. A detailed analysis of the temperature overshoot in terms of the parameters of both models is presented in Sect. 3. The paper is closed with conclusions.

2 Multi-temperature Models of Mixtures

In general, kinetic theory of mixtures treats phenomena on a molecular scale. It has a great potential for description of non-equilibrium processes in rarefied gas flows by means of numerical solution of the system of Boltzmann equations.

On the other hand, kinetic models are connected with numerical difficulties resulting from the complex nature of the collision integral of the Boltzmann equation. In most cases, the complexity of numerical schemes prevent massive calculations and detailed study of the shock structure problem.

Besides solving all sorts of kinetic models, another possibility for obtaining solutions of the Boltzmann equation lies in Direct Simulation Monte Carlo (DSMC) [3]. However, like any statistical techniques, the DSMC method suffers from the statistical noise which becomes particularly severe for low-speed flows and weak shock waves. This limits the application of the DSMC on rarefied flow conditions. Alternatively, one can use, direct, deterministic numerical finite-difference method for solving Boltzmann equation whose evaluation requires calculating multidimensional integrals which is very computationally intensive [18, 23].

Between the kinetic theory on one side and the classical thermodynamics of irreversible processes (TIP) on the other side there is a macroscopic theory called extended thermodynamics (ET). Bridging the gap between two theories, ET actually links two scales—macro and meso scale.

The way in which ET achieves that goal is through extension of the set of state variables needed for description of the non-equilibrium processes. This calls for introduction of additional set of balance laws for the new state variables, and treatment of the entropy inequality in a broader sense. This approach, thoroughly described in the book of Müller and Ruggeri [21], will be the one we will stick to in our study. It provides a systematic way for derivation of extended set of field equations for non-equilibrium processes, which is in accordance with the basic physical principles—Galilean invariance and the entropy principle.

Although macroscopic models are widely appreciated, they also have some inherent drawbacks. Physically, they suffer from the paradox of infinite speed of propagation of disturbances, and their predictions are relevant only for a class of processes which occur in the neighborhood of local equilibrium state.

The paradox of infinite speed of diffusion was resolved by Müller [20] through introduction of velocity fields for each constituent as state variables. This model was extended even more by adjoining to each constituent its own temperature [24].

Recently, the hyperbolic MT model was tested against experimentally determined shock structure in Helium-Argon mixture [19]. Even though the model was simplified, since viscosity and heat conductivity were neglected, a good agreement was obtained for available experimental data [16].

2.1 Inviscid Model—Mixture of Euler Fluids

Starting with the hyperbolic MT model established in [24], in this section we shall restrict the analysis to a non-reacting binary mixture of ideal gases which are neither viscous, nor heat conducting. The neglect of viscous dissipation and heat conduction has twofold motivation; for the processes not far from the local equilibrium state it is justifiable to neglect these effects; dissipation will enter through relaxation processes due to mutual exchange of momentum and internal energy between the constituents.

Consider the MT model of binary (two-component) mixtures whose behavior is described by equations:

$$\begin{aligned}
&\partial_t \rho_\alpha + \operatorname{div}(\rho_\alpha \mathbf{v}_\alpha) = r_\alpha, \\
&\partial_t (\rho_\alpha \mathbf{v}_\alpha) + \operatorname{div}(\rho_\alpha \mathbf{v}_\alpha \otimes \mathbf{v}_\alpha - \mathbf{t}_\alpha) = \mathbf{m}_\alpha, \\
&\partial_t \left(\frac{1}{2}\rho_\alpha v_\alpha^2 + \rho_\alpha \varepsilon_\alpha\right) + \operatorname{div}\left\{\left(\frac{1}{2}\rho_\alpha v_\alpha^2 + \rho_\alpha \varepsilon_\alpha\right)\mathbf{v}_\alpha - \mathbf{t}_\alpha \mathbf{v}_\alpha + \mathbf{q}_\alpha\right\} = e_\alpha
\end{aligned} \tag{1}$$

where $\alpha = 1, 2$. Apart from standard notation for mass, momentum and energy densities of the constituents, as well as non-convective fluxes, we have also the production terms. Namely, r_α is the mass rate of change of constituent α due to chemical reaction; $\mathbf{m}_\alpha$ is the momentum rate of change due to mechanical interaction; e_α is the energy rate of change due to exchange of energy between the constituents. The behaviour of the mixture is determined by the global conservation law:

$$\begin{aligned}
&\partial_t \rho + \operatorname{div}(\rho \mathbf{v}) = 0, \\
&\partial_t (\rho \mathbf{v}) + \operatorname{div}(\rho \mathbf{v} \otimes \mathbf{v} - \mathbf{t}) = \mathbf{0}, \\
&\partial_t \left(\frac{1}{2}\rho v^2 + \rho \varepsilon\right) + \operatorname{div}\left\{\left(\frac{1}{2}\rho v^2 + \rho \varepsilon\right)\mathbf{v} - \mathbf{t}\mathbf{v} + \mathbf{q}\right\} = 0.
\end{aligned} \tag{2}$$

State quantities and non-convective fluxes which are defined for the whole mixture by the following relations are called global field variables:

$$\begin{aligned}
&\rho = \sum_{\alpha=1}^{2} \rho_\alpha, \quad \mathbf{v} = \frac{1}{\rho}\sum_{\alpha=1}^{2} \rho_\alpha \mathbf{v}_\alpha, \quad \mathbf{u}_\alpha = \mathbf{v}_\alpha - \mathbf{v}, \\
&\varepsilon_I = \frac{1}{\rho}\sum_{\alpha=1}^{2} \rho_\alpha \varepsilon_\alpha, \quad \varepsilon = \varepsilon_I + \frac{1}{2\rho}\sum_{\alpha=1}^{2} \rho_\alpha u_\alpha^2, \\
&\mathbf{t} = \sum_{\alpha=1}^{2} (\mathbf{t}_\alpha - \rho_\alpha \mathbf{u}_\alpha \otimes \mathbf{u}_\alpha), \quad \mathbf{q} = \sum_{\alpha=1}^{2} \left\{\mathbf{q}_\alpha + \rho_\alpha\left(\varepsilon_\alpha + \frac{1}{2}u_\alpha^2\right)\mathbf{u}_\alpha - \mathbf{t}_\alpha \mathbf{u}_\alpha\right\},
\end{aligned} \tag{3}$$

where $\mathbf{u}_\alpha$ are diffusion velocities of the components and ε_I is the density of the intrinsic internal energy where diffusion term is not included. The source terms must satisfy the following conservation conditions:

$$\sum_{\alpha=1}^{2} r_\alpha = 0, \quad \sum_{\alpha=1}^{2} \mathbf{m}_\alpha = \mathbf{0}, \quad \sum_{\alpha=1}^{2} e_\alpha = 0. \tag{4}$$

Below we assume that both components of the mixture are ideal gases described by thermal and caloric equations of state:

$$p_\alpha = \rho_\alpha \frac{k_\mathrm{B}}{m_\alpha} T_\alpha, \quad \varepsilon_\alpha = \frac{k_\mathrm{B} T_\alpha}{m_\alpha(\gamma_\alpha - 1)} = c_{V_\alpha} T_\alpha, \tag{5}$$

where k_B is the Boltzmann constant, c_{V_α} are the specific heats of the components at constant volume, m_α are atomic masses, γ_α stands for the ratios of specific heats and T_α refers to the temperatures of the components.

The average temperature of the mixture T is defined by an intrinsic internal energy in a state of local equilibrium [25]:

$$(\rho_1 c_{V_1} + \rho_2 c_{V_2}) T = \rho_1 c_{V_1} T_1 + \rho_2 c_{V_2} T_2. \tag{6}$$

It should be noted that this definition is not the only one that can be found in the literature. Average temperature can be defined by Dalton law [30], and also, it can rely on the complete expression for the internal energy density of the mixture. As our analysis is limited to monatomic gases where $\gamma_1 = \gamma_2 = \gamma = 5/3$ total pressure of the mixture p and the density of the intrinsic internal energy can be written in the following form:

$$p = p_1 + p_2 = \rho \frac{k_\mathrm{B}}{m} T, \qquad \rho\varepsilon_I = \rho_1\varepsilon_1 + \rho_2\varepsilon_2 = \rho \frac{k_\mathrm{B}}{(\gamma - 1)m} T \tag{7}$$

where m is the average atomic mass of the mixture.

In the analysis of binary mixtures it is convenient to introduce the mass concentration of each component, namely:

$$c_1 = \frac{\rho_1}{\rho} = c, \quad c_2 = \frac{\rho_2}{\rho} = 1 - c. \tag{8}$$

and on this basis density of the components can be expressed:

$$\rho_1 = \rho c, \quad \rho_2 = \rho(1 - c).$$

Now, the average atomic mass of the mixture and the average temperature of the mixture can be written as follows:

$$\frac{1}{m} = \frac{c}{m_1} + \frac{1 - c}{m_2}, \qquad T = c\frac{m}{m_1} T_1 + (1 - c)\frac{m}{m_2} T_2. \tag{9}$$

As noted before, we will assume in this part of the analysis that the effects of viscosity and heat conduction are negligible. This is reflected in the structure of the expression for the stress tensor and heat flow:

$$\mathbf{t}_\alpha = -p_\alpha \mathbf{I}, \quad \mathbf{q}_\alpha = \mathbf{0}. \tag{10}$$

Neglecting these dissipative mechanisms we implicitly assumed that the analysis is restricted to weak shock waves.

Structure of the source terms is determined in accordance with the basic principles of thermodynamics. In this paper the source terms which describe the influence of chemical reactions will be ignored:

$$r_\alpha = \hat{r}_\alpha = 0.$$

Therefore, the structure of the remaining source terms will be:

$$\mathbf{m}_1 = \hat{\mathbf{m}}_1 = -\mathbf{m}_2, \quad e_1 = \hat{e}_1 + \hat{\mathbf{m}}_1 \cdot \mathbf{v} = -e_2, \tag{11}$$

The velocity-independent part of the source terms is determined taking into account one possible solution of the residual inequality in the entropy balance law:

$$\hat{\mathbf{m}}_1 = -\psi_{11}\left(\frac{\mathbf{u}_1}{T_1} - \frac{\mathbf{u}_2}{T_2}\right), \quad \hat{e}_1 = -\theta_{11}\left(-\frac{1}{T_1} + \frac{1}{T_2}\right). \tag{12}$$

The phenomenological coefficients ψ_{11} and θ_{11} can be expressed in terms of the state variables and relaxation times for the diffusion τ_D and for the temperature τ_T

$$\psi_{11} = \frac{1}{\tau_D}\frac{\rho_1\rho_2}{\rho}T, \qquad \theta_{11} = \frac{1}{\tau_T}\frac{\rho_1 c_{V1}\rho_2 c_{V2}}{\rho_1 c_{V1} + \rho_2 c_{V2}}T^2. \tag{13}$$

In the case of binary mixtures, the relaxation time τ_D and τ_T measures of the speed of convergence of the state variables to their equilibrium values. In MT model equations the relaxation time will appear in the equations of change of momentum and energy for the lighter component (20).

Models developed in the framework of kinetic theory have source terms linear with respect to state variables [5, 6]. In our model, however, the source terms are non-linear, which is expected to play an important role in the analysis of non-equilibrium processes.

In the study of non-equilibrium processes in binary mixtures is very useful from the point of evaluation to use quantities such as diffusion flux vector $\mathbf{J}$ and diffusion temperature Θ:

$$\mathbf{J} = \rho_1\mathbf{u}_1 = -\rho_2\mathbf{u}_2, \quad \Theta = T_2 - T_1. \tag{14}$$

Note that they are genuine non-equilibrium variables, i.e. $\mathbf{J}$ vanish when the velocities of the constituents are equal ($\mathbf{v}_1 = \mathbf{v}_2$) and Θ vanish when the temperatures of the constituents are the same ($T_1 = T_2$). Moreover, the temperatures of the constituents can be expressed in terms of T, Θ and concentration c

$$T_1 = T - f_1(c)\Theta, \quad T_2 = T + f_2(c)\Theta, \tag{15}$$

where

$$f_1(c) = (1-c)\frac{m}{m_2}, \quad f_2(c) = c\frac{m}{m_1}, \tag{16}$$

with average mass $m = m(c)$ introduced in (9).

2.1.1 Shock Structure Equations

Problem of the shock wave structure in one-dimensional case involves the continuous solution in the form of a traveling wave, which is moving at a constant speed s. This assumption involves change of independent variables $\xi = x - st$ and introduction of the relative velocity $u = v_x - s$, where v_x is the component of the velocity vector $\mathbf{v}$ in x-direction. The new independent variable ξ binds observer (moving reference frame) on the wave front where the problem becomes stationary. Based on this, the system of Eqs. (1)–(2) is transformed to the following set of ordinary differential equations:

$$\begin{aligned}
&\frac{d}{d\xi}(\rho u) = 0, \\
&\frac{d}{d\xi}\left(\rho u^2 + p + \frac{J^2}{\rho c(1-c)}\right) = 0, \\
&\frac{d}{d\xi}\left\{\left(\frac{1}{2}\rho u^2 + \rho\varepsilon + p\right)u + \left(\frac{uJ}{\rho c(1-c)} + \frac{1}{\beta}\right)J\right\} = 0, \\
&\frac{d}{d\xi}(\rho c u + J) = 0, \\
&\frac{d}{d\xi}\left\{\rho c u^2 + \frac{J^2}{\rho c} + 2uJ + p_1\right\} = \hat{m}_1, \\
&\frac{d}{d\xi}\left\{\left(\frac{1}{2}\rho c\left(u + \frac{J}{\rho c}\right)^2 + \rho c\varepsilon_1 + p_1\right)\left(u + \frac{J}{\rho c}\right)\right\} = \hat{m}_1 u + \hat{e}_1.
\end{aligned} \tag{17}$$

Equations $(17)_{1-3}$ are conservation laws for the mixture, while $(17)_{4-6}$ are balance laws for the first component. J and $\hat{m}_1$ are labeled components of the diffusion flux vector and source term in x-direction.

Let $\mathbf{U} = (\rho, u, T, c, J, \Theta)^T$ denotes the column vector of state variables. Assuming that the wave front moves from the right to the left at speed $s < 0$, boundary conditions can be rewritten as follows:

$$\lim_{\xi\to-\infty} \mathbf{U}(\xi) = \mathbf{U}_0, \quad \lim_{\xi\to\infty} \mathbf{U}(\xi) = \mathbf{U}_1,$$
$$\lim_{\xi\to\pm\infty} \mathbf{U}'(\xi) = \mathbf{0}. \tag{18}$$

The structure of the shock wave can now be treated as a continuous solution of the system (17) which asymptotically tends to the equilibrium states $\mathbf{U}_0$ and $\mathbf{U}_1$, in front of and behind shock wave. According with the theory of dynamical systems [13, 15] this continuous solution represents a heteroclinic orbit connecting the stationary points $\mathbf{U}_0$ and $\mathbf{U}_1$.

The inviscid shock structure problem (17)–(18) will be addressed and analyzed numerically as an initial problem [27]. The structure of the shock wave is defined by heteroclinic orbit on unlimited domain, $-\infty < \xi < \infty$. We will be restricted to a domain that is large enough to ensure that the terminal value of the state variables $\mathbf{U}(\xi_{0,1})$ can be found in the vicinity of the stationary point $\mathbf{U}_{0,1}$.

In order to solve the problem in dimensionless form we will introduce dimensionless variables:

$$\tilde{\rho} = \frac{\rho}{\rho_0}, \quad \tilde{u} = \frac{u}{a_0}, \quad \tilde{T} = \frac{T}{T_0}, \quad \tilde{J} = \frac{J}{\rho_0 a_0}, \quad \tilde{\Theta} = \frac{\Theta}{T_0}, \quad \tilde{\xi} = \frac{\xi}{\lambda_0}, \quad M_0 = \frac{u_0}{a_0}, \tag{19}$$

where λ_0 is a reference length, $a_0 = \{\gamma (k_B/m_0) T_0\}^{1/2}$ represents the speed of sound, $m_0 = m(c_0)$ is the average atomic mass of the mixture and M_0 is the Mach number at steady state in front of the shock wave. Independent variable ξ and the newly dimensionless state variables were scaled relative to the equilibrium state variables in front of the shock marked with index 0. Dimensionless form of equations (17) reads:

$$
\begin{aligned}
&\frac{d}{d\xi}(\rho u) = 0,\\
&\frac{d}{d\xi}\left(\rho u^2 + \frac{1}{\gamma}\frac{m_0}{m}\rho T + \frac{J^2}{\rho c(1-c)}\right) = 0,\\
&\frac{d}{d\xi}\left\{\left(\frac{1}{2}\rho u^2 + \frac{1}{\gamma-1}\frac{m_0}{m}\rho T + \frac{J^2}{2\rho c(1-c)}\right)u + \left(\frac{uJ}{\rho c(1-c)} + \frac{1}{\beta}\right)J\right\} = 0,\\
&\frac{d}{d\xi}(\rho c u + J) = 0, \qquad\qquad (20)\\
&\frac{d}{d\xi}\left\{\rho c u^2 + \frac{J^2}{\rho c} + 2uJ + \frac{1}{\gamma}\frac{m_0}{m_1}\rho c\,(T - f_1(c)\Theta)\right\}\\
&\quad = -\frac{\lambda_0}{\tau_D a_0} m_\mu(T, c, \Theta) J,\\
&\frac{d}{d\xi}\left\{\left(\frac{1}{2}\rho c\left(u + \frac{J}{\rho c}\right)^2 + \frac{1}{\gamma-1}\frac{m_0}{m_1}\rho c\,(T - f_1(c)\Theta)\right)\left(u + \frac{J}{\rho c}\right)\right\}\\
&\quad = -\frac{\lambda_0}{\tau_D a_0} m_\mu(T, c, \Theta) J u - \frac{\lambda_0}{\tau_T a_0} e_\mu(\rho, T, c, \Theta)\Theta.
\end{aligned}
$$

For the sake of simplicity tilde mark will be dropped in the sequel. Auxiliary functions in the source terms reads:

$$m_\mu(T, c, \Theta) = \frac{T + [(1-c)f_2(c) - cf_1(c)]\Theta}{[T - f_1(c)\Theta][T + f_2(c)\Theta]} T,$$

$$e_\mu(\rho, T, c, \Theta) = \frac{1}{\gamma(\gamma - 1)} \frac{m_0}{m_1} \frac{m}{m_2} \frac{\rho c(1-c)T^2}{[T - f_1(c)\Theta][T + f_2(c)\Theta]},$$

Dimensionless thermal inertia has the following form:

$$\frac{1}{\beta} = \frac{1}{\gamma - 1}\left[\frac{m_0}{m_1}(1-\mu)T - \frac{m_0}{m_1}\frac{m}{m_2}\Theta\right] + \frac{J^2}{2\rho^2}\left[\frac{1}{c^2} - \frac{1}{(1-c)^2}\right],$$

where μ is the ratio of masses of the constituents:

$$\mu = \frac{m_1}{m_2}, \quad 0 < \mu \leq 1.$$

Shock structure equations (20) can be written in the following form:

$$\frac{d\mathbf{F}(\mathbf{U})}{d\xi} = \mathbf{f}(\mathbf{U}), \tag{21}$$

where $\mathbf{F}(\mathbf{U})$ is the vector of fluxes and $\mathbf{f}(\mathbf{U})$ is the vector of production terms, whose structure is obvious and will be skipped for the sake of brevity. Equilibrium states are determined by the condition $\mathbf{f}(\mathbf{U}) = \mathbf{0}$. For a given equilibrium state $\mathbf{U}_0$ in front of the shock wave, the steady state $\mathbf{U}_1$ behind it is determined as solution of the system $\mathbf{F}(\mathbf{U}_1, M_0) = \mathbf{F}(\mathbf{U}_0, M_0)$. In dimensionless form it reads:

$$\mathbf{U}_0 = \begin{bmatrix} 1 \\ M_0 \\ 1 \\ c_0 \\ 0 \\ 0 \end{bmatrix}, \ \mathbf{U}_1 = \begin{bmatrix} \frac{4M_0^2}{3+M_0^2} \\ \frac{3+M_0^2}{4M_0} \\ \frac{1}{16}\left(14 - \frac{3}{M_0^2} + 5M_0^2\right) \\ c_0 \\ 0 \\ 0 \end{bmatrix}. \tag{22}$$

Note that the order of the system (20) can be reduced due to the presence of conservation laws (first integrals), as it was done e.g. in [6]. We shall, however, retain the complete system, which certainly does not affect final results.

2.2 Parameters of the Model

The main quality of the dimensionless analysis is in reduced number of relevant parameters of the system. In our Eq. (20) there is only two dimensionless parameters:

$$\Pi_D = \frac{\lambda_0}{\tau_D a_0} \quad \text{and} \quad \Pi_T = \frac{\lambda_0}{\tau_T a_0}. \tag{23}$$

The values of the relaxation times for the diffusion τ_D, temperature τ_T and the reference length λ_0 will be determined in the following using the results of the kinetic theory of gases [7] in the case of hard spheres gas model. In the case of monatomic gases it is possible to establish a relationship between the relaxation times in the form:

$$\frac{\tau_T}{\tau_D} = \frac{m_1 + m_2}{c\, m_2 + (1-c)\, m_1}. \tag{24}$$

On the other hand relaxation times can be related to the transport coefficient. More specifically, the relaxation time for diffusion τ_D is proportional to the diffusivity of the binary mixtures of D_{12}:

$$\tau_D = \frac{n m_1 m_2}{\rho k_B T} D_{12}, \tag{25}$$

where $n = \rho/m$ represent the number density of the mixture:

$$n = n_1 + n_2 = \frac{\rho_1}{m_1} + \frac{\rho_2}{m_2}. \tag{26}$$

Based on the well-known classical approaches to the kinetic theory [4, 8] diffusivity of binary mixtures of ideal gases D_{12} can be expressed as follows:

$$D_{12} = \frac{3}{8 n d_{12}^2} \left(\frac{k_B T}{2\pi} \frac{m_1 + m_2}{m_1 m_2} \right)^{1/2}, \tag{27}$$

where $d_{12} = (d_1 + d_2)/2$ is average atomic diameter of the mixture.

Finally, using (24)–(26), the relaxation times can be expressed as a function of diffusivity:

$$\tau_D = \frac{c m_2 + (1-c) m_1}{k_B T} D_{12}, \qquad \tau_T = \frac{m_1 + m_2}{k_B T} D_{12}. \tag{28}$$

In the kinetic theory of gases the mean free path of molecules i.e. atoms λ_0 is closely associated with the viscosity of gas mixture. Since we start with the assumption that the viscosity of the components is negligible, this method of determining the average length of free path of atoms λ_0 is not adequate. To overcome this difficulty, we have used the fact that the λ_0 can be expressed through the properties of components of mixtures, atomic mass, atomic radius and number density [4]:

$$\lambda_0 = \frac{n_1}{n}(\lambda_1)_0 + \frac{n_2}{n}(\lambda_2)_0, \tag{29}$$

where $(\lambda_\alpha)_0$ represents average mean free path of the components:

$$(\lambda_\alpha)_0 = \frac{1}{\pi d_{12}^2}\left[n_1(1+\frac{m_\alpha}{m_1})^{1/2} + n_2(1+\frac{m_\alpha}{m_2})^{1/2}\right]^{-1}. \tag{30}$$

Index 0 denotes the equilibrium value of in front of the shock wave.

2.3 Viscous Model

In Sect. 2.1, dedicated to the analysis of a hyperbolic model in inviscid approximation, we have seen that due to limitations inherent in hyperbolic systems, a more detailed analysis is possible only for a limited set of system parameters that correspond to weak shock waves. By including an additional dissipative mechanism in the existing model structure of shock wave can be determined for higher values of the Mach number, which corresponds to a strong shock wave. In other words, it means that the analysis is extended to the cases where the transport phenomena can not be ignored. However, the analysis will focus primarily on determining the effect of the mass of the components on the structure of shock waves and comparison with the results obtained in inviscid approximation.

Mathematical model of multi-temperature mixture in viscous approximation contains the stress tensor and heat flux vector, representing the dissipative effects described by the constitutive relations of Navier-Stokes-Fourier type (diffusion type phenomenological relations). In this way, the model gets a mixed character, containing the law of conservation of mass, equation of change of momentum and energy in which dissipative effects occur in two ways—as a source terms of the relaxation type and as of a diffusion type. This model will be, as before, put into a dimensionless form. Also, as in the previous case system of ordinary differential equations that describes the structure of the shock wave will be established.

The basic assumption of the viscous approximation of the multi-temperature binary mixtures is that the stress tensor and heat flow vector components no longer have the structure given by Eq. (10). Now, they include terms in which the effects of viscosity and thermal conductivity are described by constitutive relations Navier-Stokes-Fourier type:

$$\begin{aligned} \mathbf{t}_\alpha &= -p_\alpha \mathbf{I} + \boldsymbol{\sigma}_\alpha, \quad \boldsymbol{\sigma}_\alpha = 2\mu_\alpha \mathbf{D}_\alpha, \quad \mathbf{D}_\alpha = \frac{1}{2}\left(\nabla \mathbf{v}_\alpha + (\nabla \mathbf{v}_\alpha)^T\right), \\ \mathbf{q}_\alpha &= -\kappa_\alpha \nabla T_\alpha, \quad (\alpha = 1, 2) \end{aligned} \tag{31}$$

where μ_α is viscosity, and κ_α thermal conductivity of the components.

The equations which describe the thermodynamic processes in mixture under the above assumptions, have the following structure:

$$
\begin{aligned}
&\frac{\partial \rho}{\partial t} + \operatorname{div}(\rho \mathbf{v}) = 0; \\
&\frac{\partial \rho \mathbf{v}}{\partial t} + \operatorname{div}\left(\rho \mathbf{v} \otimes \mathbf{v} + p\mathbf{I} - \boldsymbol{\sigma}^* + \frac{1}{\rho c(1-c)}\mathbf{J} \otimes \mathbf{J}\right) = \mathbf{0}; \\
&\frac{\partial \left(\frac{1}{2}\rho v^2 + \rho \varepsilon\right)}{\partial t} + \operatorname{div}\left\{\left(\frac{1}{2}\rho v^2 + \rho \varepsilon + p\right)\mathbf{v} - \boldsymbol{\sigma}^* \mathbf{v} + \mathbf{q}^* \right. \\
&\qquad \left. - \left(\frac{\boldsymbol{\sigma}_1}{\rho c} - \frac{\boldsymbol{\sigma}_2}{\rho(1-c)}\right)\mathbf{J} + \left(\frac{\mathbf{v} \cdot \mathbf{J}}{\rho c(1-c)} + \frac{1}{\beta}\right)\mathbf{J}\right\} = 0; \\
&\frac{\partial (\rho c)}{\partial t} + \operatorname{div}(\rho c \mathbf{v} + \mathbf{J}) = 0; \qquad\qquad (32)\\
&\frac{\partial (\rho c \mathbf{v} + \mathbf{J})}{\partial t} + \operatorname{div}\left\{\rho c \mathbf{v} \otimes \mathbf{v} + \frac{1}{\rho c}\mathbf{J} \otimes \mathbf{J} + \mathbf{v} \otimes \mathbf{J} + \mathbf{J} \otimes \mathbf{v} + p_1 \mathbf{I} - \boldsymbol{\sigma}_1\right\} = \mathbf{m}_1; \\
&\frac{\partial \left(\frac{1}{2}\rho c \left(\mathbf{v} + \frac{\mathbf{J}}{\rho c}\right)^2 + \rho c \varepsilon_1\right)}{\partial t} \\
&\qquad + \operatorname{div}\left\{\left(\frac{1}{2}\rho c \left(\mathbf{v} + \frac{\mathbf{J}}{\rho c}\right)^2 + \rho c \varepsilon_1 + p_1\right)\left(\mathbf{v} + \frac{\mathbf{J}}{\rho c}\right) - \boldsymbol{\sigma}_1 \left(\mathbf{v} + \frac{\mathbf{J}}{\rho c}\right) + \mathbf{q}_1\right\} = e_1,
\end{aligned}
$$

where:

$$\boldsymbol{\sigma}^* = \boldsymbol{\sigma}_1 + \boldsymbol{\sigma}_2, \quad \mathbf{q}^* = \mathbf{q}_1 + \mathbf{q}_2.$$

Equation (32) do not represent a closed system of equations. They must be associated with constitutive relations (31).

2.3.1 Shock Structure Equations

Assuming a solution in the form of the normal traveling waves in Eqs. (31)–(32) we get a system of equations that describes the structure of the shock wave. One will consist of ten ordinary differential equations of the first order by an unknown state variables $\mathbf{U} = (\rho, u, T, c, J, \Theta, \sigma_1, \sigma_2, q_1, q_2)$.

Dimensionless stresses and heat flow rates were determined by the following relations:

$$\tilde{\sigma}_\alpha = \frac{\sigma_\alpha}{\rho_0 \frac{k_B}{m_\alpha} T_0}, \quad \tilde{q}_\alpha = \frac{q_\alpha}{\rho_0 \frac{k_B}{m_\alpha} T_0 a_0}. \tag{33}$$

As in the inviscid case, the tilde will be omitted when writing dimensionless equations.

Equations for the shock wave structure in dimensionless form are as follows:

$$\begin{aligned}
&\frac{d}{d\xi}(\rho u) = 0,\\
&\frac{d}{d\xi}\left(\rho u^2 + \frac{1}{\gamma}\frac{m_0}{m}\rho T - \frac{1}{\gamma}\sigma^* + \frac{J^2}{\rho c(1-c)}\right) = 0,\\
&\frac{d}{d\xi}\left\{\left(\frac{1}{2}\rho u^2 + \frac{1}{\gamma-1}\frac{m_0}{m}\rho T + \frac{J^2}{2\rho c(1-c)}\right)u\right.\\
&\qquad \left. - \frac{1}{\gamma}\sigma^* u + \frac{1}{\gamma}q^* + \left[\frac{uJ}{\rho c(1-c)} - \frac{1}{\gamma}\left(\frac{\sigma_1}{\rho c} - \frac{\sigma_2}{\rho(1-c)}\right) + \frac{1}{\beta}\right]J\right\} = 0,\\
&\frac{d}{d\xi}(\rho c u + J) = 0,\\
&\frac{d}{d\xi}\left\{\rho c u^2 + \frac{J^2}{\rho c} + 2uJ + \frac{1}{\gamma}\frac{m_0}{m_1}\rho c\left(T - f_1(c)\Theta - \frac{1}{\gamma}\sigma_1\right)\right\}\\
&\qquad = -\frac{\lambda_0}{\tau_D a_0} m_\mu(T, c, \Theta) J,\\
&\frac{d}{d\xi}\left\{\left(\frac{1}{2}\rho c\left(u + \frac{J}{\rho c}\right)^2 + \frac{1}{\gamma-1}\frac{m_0}{m_1}\rho c\,(T - f_1(c)\Theta)\right)\left(u + \frac{J}{\rho c}\right)\right.\\
&\qquad \left. - \frac{1}{\gamma}\sigma_1\left(u + \frac{J}{\rho c}\right) + \frac{1}{\gamma}q_1\right\} = -\frac{\lambda_0}{\tau_D a_0} m_\mu(T, c, \Theta) J u - \frac{\lambda_0}{\tau_T a_0} e_\mu(\rho, T, c, \Theta)\Theta.
\end{aligned} \tag{34}$$

They are also associated with constitutive relations which are written in dimensionless form:

$$\begin{aligned}
\frac{d}{d\xi}\left(u - \frac{J}{\rho c}\right) &= \frac{3}{4}\frac{\rho_0 a_0 \lambda_0}{\gamma \mu_{10}}\frac{\sigma_1}{\sqrt{T - f_1(c)\Theta}},\\
\frac{d}{d\xi}\left(u - \frac{J}{\rho(1-c)}\right) &= \frac{3}{4}\frac{\rho_0 a_0 \lambda_0}{\gamma \mu_{20}}\frac{\sigma_2}{\sqrt{T + f_2(c)\Theta}},\\
\frac{d}{d\xi}(T - f_1(c)\Theta) &= -\frac{\rho_0 a_0^3 \lambda_0}{\gamma T_0 \kappa_{10}}\frac{q_1}{\sqrt{T - f_1(c)\Theta}},\\
\frac{d}{d\xi}(T + f_2(c)\Theta) &= -\frac{\rho_0 a_0^3 \lambda_0}{\gamma T_0 \kappa_{20}}\frac{q_2}{\sqrt{T + f_2(c)\Theta}}.
\end{aligned} \tag{35}$$

In further analysis, viscosity and thermal conductivity components, which appear in previous equations, will be expressed using the relation of kinetic theory of gases for model of hard spheres:

$$\begin{aligned}
\mu_{10} &= \frac{5}{16}\frac{a_0}{\sqrt{\pi}d_1^2}\sqrt{\frac{m_0 m_1}{\gamma}}, \quad \mu_{20} = \frac{5}{16}\frac{a_0}{\sqrt{\pi}d_2^2}\sqrt{\frac{m_0 m_2}{\gamma}},\\
\kappa_{10} &= \frac{15}{4}\frac{k}{m_1}\mu_{10}, \quad \kappa_{20} = \frac{15}{4}\frac{k}{m_2}\mu_{20},
\end{aligned} \tag{36}$$

where, as we mention before, the d_1 and d_2 are atomic diameters of the mixture components. Hence, there are two important consequences. First, in the model appear four new dimensionless groups:

$$K_{\mu_1} = \frac{3}{4}\frac{\rho_0 a_0 \lambda_0}{\gamma \mu_{10}}, \quad K_{\mu_2} = \frac{3}{4}\frac{\rho_0 a_0 \lambda_0}{\gamma \mu_{20}}, \quad K_{\kappa_1} = -\frac{\rho_0 a_0^3 \lambda_0}{\gamma T_0 \kappa_{10}}, \quad K_{\kappa_2} = -\frac{\rho_0 a_0^3 \lambda_0}{\gamma T_0 \kappa_{20}}. \tag{37}$$

Second, the presence of viscosity and thermal conductivity in the model implicitly introduced atomic diameters of the components, which are not present in inviscid approximation. So, apart from the influence of the parameters that have been previously analyzed (M_0, c_0, μ) the effect of the atomic diameters on the structure of the shock wave can be introduced into the analysis through the ratio $\mathrm{DR} = d_1/d_2$. This fact entails another very debatable question that should be addressed when analyzing the structure of the shock wave in terms of the different mass of the components. In fact, with the change of the mass ratio the ratio DR also changes. In the general case, this relationship is different from the unit. In our analysis, we have adopted the $\mathrm{DR} = d_{\mathrm{He}}/d_{\mathrm{Ar}} = 0.4366$.

3 Temperature Overshoot

Temperature overshoot (TO) is one of the peculiarities of the shock structure in mixtures whose constituents have disparate masses. It manifests through existence of the region of non-zero width where the temperature of one constituent raises above the terminal, i.e. downstream equilibrium temperature of the mixture. TO has been defined as:

$$\mathrm{TO} = \frac{T_2^{\max} - T_1}{T_1 - T_0}, \tag{38}$$

where $T_2^{\max}$ denotes the maximum temperature of the heavier constituent within the profile, whereas T_0 and T_1 are upstream and downstream equilibrium temperatures of the mixture, respectively.

Available experimental data [16] do not provide enough evidence to support existence of the temperature overshoot. However, authors supposed existence of TO of the heavier component temperature for efficiently small mole fractions of it in the mixture. This phenomenon has been observed in numerical calculations based upon Boltzmann equations for mixtures [1, 6, 18, 22], extended thermodynamics [19], as well as DSMC [3].

In [2] Abe and Oguchi offered physical explanation of this phenomenon. They stated that in the case of vanishingly small mole fraction of heavier component the main structure of the shock wave is determined by the lighter one. This causes the deceleration of heavier component and, at the same time, conversion of kinetic into thermal energy. However, dissipation through conduction is slow process which cannot diffuse thermal energy gained by deceleration. As a consequence, internal energy (temperature) of heavier component is raised above the terminal one.

Despite that, in most studies, due to the nature of the process and limited number of numerical simulations, TO is not pronounced and visible and so difficult to distinguish it from measurement i.e. calculation error. To supply evidence of TO appearance, the vast amount of numerical simulations was carried out and presented in the following subsection.

3.1 Temperature Overshoot in Inviscid Model

In Sect. 2.1, we excluded the influence of viscosity and heat conductivity from the model. We focused only on dissipation caused by mutual exchange of momentum and energy between the constituents, where the most prominent role is played by their mass ratio μ. From this perspective we examined the TO, analyzing its dependence on the mass ratio μ, as well as upstream Mach number M_0 and upstream concentration c_0.

The central result of our study is concerned with the analysis of TO in terms of mass ratio μ and with fixed concentration c_0. It has an outstanding feature, not reported in previous studies, that the *temperature overshoot varies non-monotonically with mass ratio*. Different pattern appears when the Mach number is increased. Namely, there exists a value μ^* of the mass ratio which determines the local minimum of the TO. Further analysis, revealed two typical patterns shown in Fig. 1. The first one, which appears for low Mach numbers, is characterized by the existence of the minimal value of the mass ratio below which the TO does not occur. For mass ratios above this value, TO increases with the increase of μ. The second one appears with the rise of Mach number where TO experiences non-monotonic change. In other words, there is a value μ_{crit} at which TO attains minimum $\text{TO}(\mu_{\text{crit}}) = \text{TO}_{\min} \neq 0$. This very important feature of TO was not observed in earlier studies which dealt only with individual cases.

Temperature overshoot can also be analyzed in terms of upstream concentration c_0, for fixed mass ratio μ (Fig. 2). It is obvious that TO increases with c_0, which corresponds to low fraction of heavier component. This is in sound agreement with the known results obtained by other methods. Interestingly enough, there is also a region of low values of c_0, i.e. high fraction of heavier component, for which TO exhibits a slight increase. For low values of Mach number one can also observe that TO does not have significant variations for a broad range of c_0.

From this point of view, the reason for non-monotonic behavior remains unrevealed. We may found explanation after completing the model by introducing more dissipation.

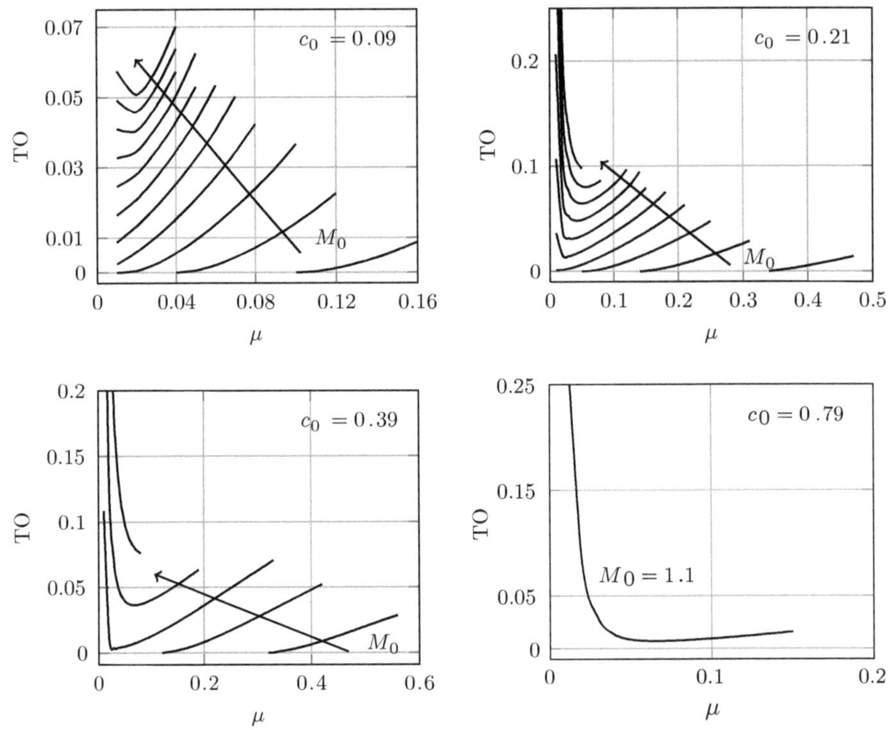

Fig. 1 Dependence of TO on mass ratio μ for fixed mass concentration c_0. Mach number is increased from $M_0 = 1.1$ with an increment 0.1

3.2 Temperature Overshoot in Viscous Model

Our analysis of the viscous model will be primarily concerned with the analysis of TO. On the basis of individual results (Fig. 3) it can be concluded that the introduction of dissipative mechanism is not enough to completely exclude the occurrence of TO. Specifically, in relation to the values of the TO in inviscid case, here the values of TO are considerably lower. This also means that at the same time the minimal value of the Mach number at which TO occurs is increased.

Analysis of the temperature profiles provides new information about the dominant dissipative mechanisms (Fig. 3). Compared to the inviscid case, it is shown that the TO is reduced but still present. Moreover, the newly introduced dissipative mechanisms also increase the width of the shock wave.

At this point we should mention a few characteristics of the temperature profile in the viscous case. The occurrence of the TO, which was observed in the inviscid model for certain (low) values of Mach number, now ceases to exist for certain values of mass ratio μ (Fig. 3a). Namely, it can be seen from Fig. 4a that in viscous case TO is smaller and even vanishes for certain values of mass ratio.

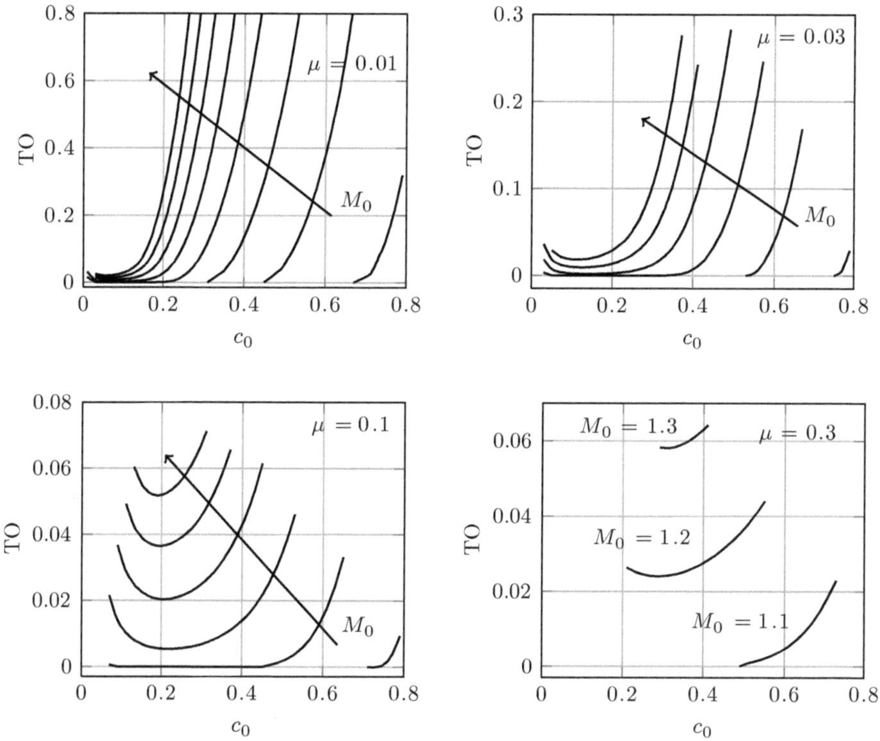

Fig. 2 Dependence of TO on mass concentration c_0 for fixed mass ratio μ. Mach number is increased from $M_0 = 1.1$ with an increment 0.1

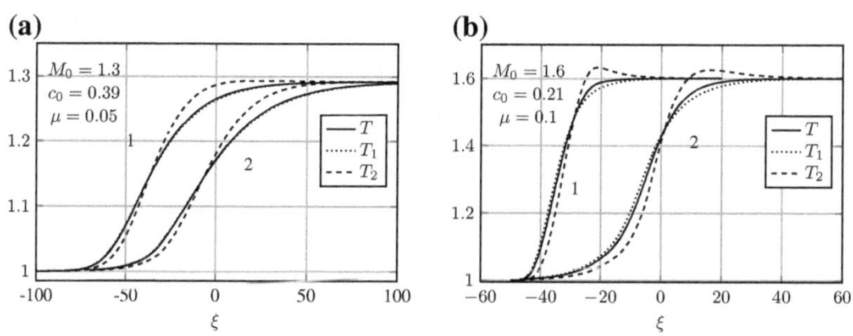

Fig. 3 Temperature profiles: *1* inviscid model, *2* viscous model; T average temperature of the mixture, T_1 temperature of the lighter component, T_2 temperature of the heavier component

On the other hand, there are values of the parameters for which TO exists also in viscous case (Fig. 3b). Detailed analysis shows that TO could exist for all values of μ and also retains its non-monotonic character (Fig. 4b), which confirms the existence of this phenomenon in viscous case. Thus, viscosity cannot attenuate TO, but just make it smaller than in the inviscid case, especially for larger values of μ.

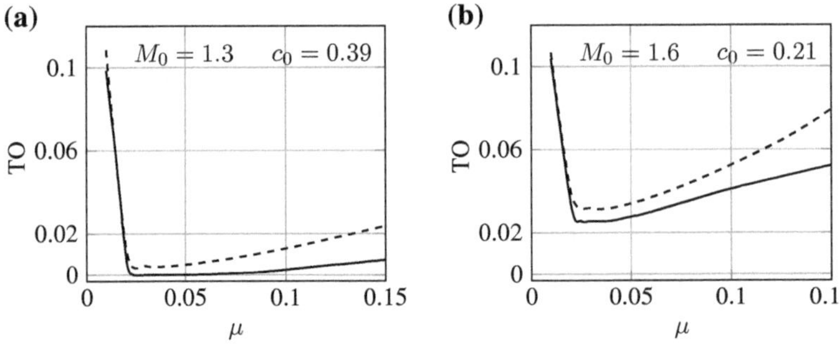

Fig. 4 *TO* comparison of viscous and inviscid model; *dashed line* inviscid model, *solid line* viscous model

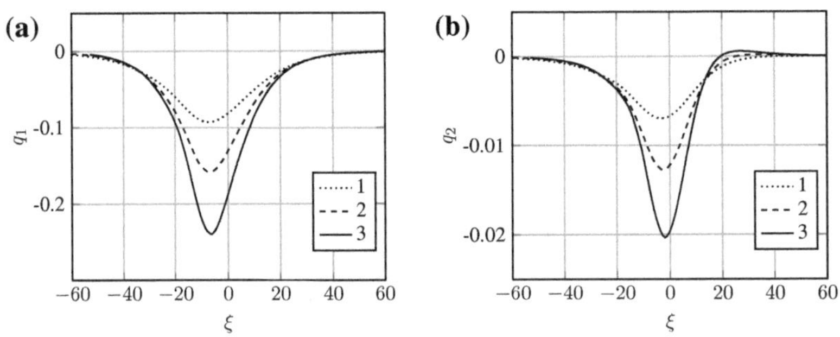

Fig. 5 Heat fluxes q_1 and q_2 for different parameter values: *1* ($M_0 = 1.3$, $c_0 = 0.2$, $\mu = 0.1$), *2* ($M_0 = 1.4$, $c_0 = 0.2$, $\mu = 0.1$), *3* ($M_0 = 1.6$, $c_0 = 0.2$, $\mu = 0.1$)

As well as in the inviscid case, at higher values of the Mach number, there is a noticeable *non-monotonic change of temperature overshoot* depending on the mass ratio, which confirms the existence of this phenomenon (Fig. 4b).

Our results for the heat fluxes support the hypothesis set by Abe and Oguchi [1] that the occurrence of TO is attributed to the insufficient intensity of the heat flow within the heavier component (see Fig. 5). Influence of viscosity and heat conductivity on temperature profile for strong shock waves can be seen in Fig. 5 where heat fluxes increase with increasing of the Mach number.

4 Conclusion

This paper analyzes the structure of shock waves in gas mixtures with special emphasis on the analysis of the temperature overshoot and its existence. We used multi-temperature model for gas mixtures developed within the extended thermodynamics. In order to capture this phenomenon, interaction of the components is modeled with the help of phenomenological coefficients taken from the kinetic theory of gases for the model of hard spheres. Previous studies did not provide much concrete evidence to support existence of TO because they dealt with particular cases only. Our approach allows detailed systematic analysis of TO for the wide range of model parameters. Led by the excellent results of the comparisons with experimental data [19] for the case where the viscosity and thermal conductivity were excluded from the model, we have confirmed the existence of the overshoot. By introducing the additional dissipative mechanism we concluded that the overshoot persists which indicates that it is the essential characteristic of the model. Also, in some cases, we noticed a non-monotonic change of TO with the change of the mass ratio of the components which are not given in previous studies. Explanation for this new behaviour should be found in a comparative analysis with shock thickness and this will be the major task of our future study.

Acknowledgments This work was supported by the Ministry of Education and Science, Republic of Serbia, through the project Mechanics of nonlinear and dissipative systems—contemporary models, analysis and applications, Project No. ON174016.

References

1. Abe, K., Oguchi, H.: Shock wave structures in binary gas mixtures with regard to temperature overshoot. Phys. Fluids **17**, 1333–1334 (1974)
2. Abe, T., Oguchi, H.: A hierarchy kinetic model and its applications. Prog. Astronaut. Aeronaut. **51**, 781–793 (1977)
3. Bird, G.: The structure of normal shock waves in a binary gas mixture. J. Fluid Mech. **31**(4), 657–668 (1968)
4. Bird, G.: Molecular Gas Dynamics and the Direct Simulation of Gas Flows. Carendon Press, Oxford (1994)
5. Bisi, M., Martalò, G., Spiga, G.: Multi-temperature Euler hydrodynamics for a reacting gas from a kinetic approach to rarefied mixtures with resonant collisions. EPL (Europhys. Lett.) 95(5), 55002 (2011)
6. Bisi, M., Martalò, G., Spiga, G.: Shock wave structure of multi-temperature euler equations from kinetic theory for a binary mixture. Acta Applicandae Mathematicae 1–11 (2014)
7. Bose, T.: High Temperature Gas Dynamics. Springer, Berlin (2004)
8. Chapman, S., Cowling, T.G.: The mathematical theory of non-uniform gases: an account of the kinetic theory of viscosity, thermal conduction and diffusion in gases. Cambridge University Press, Cambridge (1991)
9. Cowling, T.: VI. The influence of diffusion on the propagation of shock waves. Philos. Mag. **33**(216), 61–67 (1942)
10. D'yakov, S.: Shock waves in binary mixtures. Zh. Eksperim. i Teor. Fiz **27**, 283–287 (1954)
11. Elizarova, T., Graur, I., Lengrand, J.C.: Macroscopic Equations for a Binary Gas Mixture. Technical report, DTIC Document (2000)
12. Fernandez-Feria, R., De La Mora Fernandez, J.: Shock wave structure in gas mixtures with large mass disparity. J. Fluid Mech. **179**, 21–40 (1987)
13. Gilbarg, D., Paolucci, D.: The structure of shock waves in the continuum theory of fluids. J. Rat. Mech. Anal. **2**(617), 220 (1953)
14. Goldman, E., Sirovich, L.: The structure of shock-waves in gas mixtures. J. Fluid Mech. **35**(3), 575–597 (1969)
15. Guckenheimer, J., Holmes, P.: Nonlinear oscillations, dynamical systems, and bifurcations of vector fields, vol. 42. Springer, New York (1983)

16. Harnet, L.N., Muntz, E.P.: Experimental investigation of normal shock wave velocity distribution functions in mixtures of argon and helium. Phys. Fluids **10**, 565–572 (1972)
17. Harris Sr, W.L., Bienkowski, G.K.: Structure of normal shock waves in gas mixtures. Phys. Fluids **14**, 2652 (1971)
18. Kosuge, S., Aoki, K., Takata, S.: Shock-wave structure for a binary gas mixture: finite-difference analysis of the Boltzmann equation for hard sphere molecules. Eur. J. Mech. B Fluids **17**, 87–126 (2001)
19. Madjarević, D., Simić, S.: Shock structure in helium-argon mixturea comparison of hyperbolic multi-temperature model with experiment. EPL (Europhys. Lett.) **102**(4), 44002 (2013)
20. Müller, I.: A thermodynamic theory of mixtures of fluids. Arch. Ration. Mech. Anal. **28**(1), 1–39 (1968)
21. Müller, I., Ruggeri, T.: Rational Extended Thermodynamics. Springer, New York (1998)
22. Raines, A.: Study of a shock wave structure in gas mixtures on the basis of the Boltzmann equation. Eur. J. Mech. B Fluids **21**(5), 599–610 (2002)
23. Raines, A.: Numerical investigation of the temperature macroparameters in a shock wave in a binary gas mixture using the kinetic Boltzmann equation. Fluid Dyn. **38**(1), 132–142 (2003)
24. Ruggeri, T., Simić, S.: On the hyperbolic system of a mixture of eulerian fluids: a comparison between single-and multi-temperature models. Math. Methods Appl. Sci. **30**(7), 827–849 (2007)
25. Ruggeri, T., Simić, S.: Average temperature and maxwellian iteration in multitemperature mixtures of fluids. Phys. Rev. E 80, 026317 (2009)
26. Sherman, F.S.: Shock-wave structure in binary mixtures of chemically inert perfect gases. J. Fluid Mech. **8**(3), 465–480 (1960)
27. Simić, S.: Shock structure in continuum models of gas dynamics: stability and bifurcation analysis. Nonlinearity **22**, 1337 (2009)
28. Struchtrup, H.: Macroscopic transport equations for rarefied gas flows. Springer, Berlin (2005)
29. Torrilhon, M., Struchtrup, H.: Regularized 13-moment equations: shock structure calculations and comparison to Burnett models. J. Fluid Mech. **513**, 171–198 (2004)
30. Vincenti, W., Krüger, C.: Introduction to physical gas dynamics. Wiley, New York (1965)

Diffusive Limit for the Random Lorentz Gas

Alessia Nota

Abstract We review some recent results concerning the derivation of the diffusion equation and the validation of Fick's law for the microscopic model given by the random Lorentz Gas. These results are achieved by using a linear kinetic equation as an intermediate level of description between our original mechanical system and the diffusion equation. The diffusion coefficient is given by the Green-Kubo formula associated to the generator of the stochastic process dictated by the linear Landau equation and the linear Boltzmann equation respectively, according to the weak-coupling regime and low density regime we are considering.

Keywords Diffusion · Weak-coupling limit · Diffusion limit · Low density limit · Heat equation · Linear Landau equation · Linear Boltzmann equation · Fick's law · Green- Kubo · Lorentz gas

1 Introduction

The problem of deriving macroscopic evolution equations from the microscopic laws of motion governed by Newton's laws of classical mechanics is one of the most important keystones in mathematical physics.

Here we consider a simple microscopic model, namely a gas of non-interacting particles in a fixed random configuration of scatterers. This dynamical system is usually referred to as the Lorentz gas, since it was proposed by H.A. Lorentz in 1905, see [23], to explain the motion of electrons in metals applying the methods of the kinetic theory of gases. Even though this model is quite simple, it is still paradigmatic. Indeed complexities and interesting features come up in the analysis showing new and unexpected macroscopic phenomena.

A. Nota (✉)
Dipartimento di Matematica, Università di Roma La Sapienza,
Piazzale Aldo Moro 5, 00185 Roma, Italy
e-mail: nota@mat.uniroma1.it

P. Gonçalves and A.J. Soares (eds.), *From Particle Systems to Partial Differential Equations II*, Springer Proceedings in Mathematics & Statistics 129,
DOI 10.1007/978-3-319-16637-7_10

The Lorentz gas consists of a particle moving through infinitely heavy, randomly distributed scatterers. The interaction between the Lorentz particle and the scatterers is specified by a central potential of finite range. Hence the motion of the Lorentz particle is defined through the solution of Newton's equation of motion. Lorentz's idea was to view electrons as a gas of light particles colliding with the metallic atoms; neglecting collisions between electrons, Lorentz described the interaction of electrons with the metallic atoms by a collision integral analogous to Boltzmann's. The original system is Hamiltonian, the only stochasticity being that of the positions of the scatterers. This randomness is absolutely necessary to obtain the correct kinetic description. Indeed, for this system, one can prove, under suitable scaling limits, a rigorous validation of linear kinetic equations and, from this, of diffusion equations.

We can argue in terms of stochastic processes. The motion of the Lorentz particle is a stochastic process which is non Markovian. The scaling limit procedure can be understood as a Markovian approximation which leads to a Markov process whose forward equation is a suitable kinetic equation. More precisely the scaling limits we are considering consist of a kinetic scaling of space and time, namely $t \to \varepsilon t$, $x \to \varepsilon x$ and a suitable rescaling of the density of the obstacles and the intensity of the interaction. Accordingly to the resulting frequency of collisions, the mean free path of the particle can have or not macroscopic length and different kinetic equations arise. Typical examples are the linear Boltzmann equation and the linear Landau equation.

The first scaling one could consider is the Boltzmann-Grad limit. The first result in this direction was obtained by Gallavotti in 1969, see [14], who derived the linear Boltzmann equation starting from a random distribution of fixed hard scatterers in the Boltzmann-Grad limit (low density), namely when the number of collisions is small, thus the mean free path of the particle is macroscopic. This result was improved and extended to more general distribution by Spohn [26]. In [4] Boldrighini, Bunimovich and Sinai proved that the limiting Boltzmann equation holds for almost every scatterer configuration drawn from a Poisson distribution. Moreover, for the sake of completeness, we refer to [15, 25] for a rigorous derivation of the nonlinear Boltzmann equation from a system of hard spheres, or a system of Newtonian particles interacting via a short-range potential, in the low density limit. As we already pointed out we remind that the randomness of the distribution of the scatterers is essential in the derivation of the linear Boltzmann equation, in fact for a periodic configuration of scatterers the linear Boltzmann equation fails (see [7]), and the random flight process that emerges in the Boltzmann-Grad limit is substantially more complicated. The first complete proof of the Boltzmann-Grad limit of the periodic Lorentz gas, valid for all lattices and in all space dimensions, can be found in [24]. The mathematical properties of the generalized linear Boltzmann equation derived are analyzed in [8].

Another scaling of interest is the weak coupling limit. The idea of the weak coupling limit is that, by some kind of central limit effect, very many but weak collisions should lead to a diffusion type evolution. The correct kinetic equation which is derived in this scaling limit is the Linear Landau equation

$$(\partial_t + v \cdot \nabla_x) f(x, v, t) = B \Delta_{|v|} f(x, v, t), \tag{1}$$

where $\Delta_{|v|}$ is the Laplace-Beltrami operator on the d-dimensional sphere of radius $|v|$. It is a Fokker-Planck equation for the stochastic process $(V(t), X(t))$, where the velocity process V is a Brownian motion on the (kinetic) energy sphere, and the position X is an additive functional of V. The velocity diffusion follows from the facts that there are many elastic collisions. The diffusion coefficient B is proportional to the variance of the transferred momentum in a single collision and depends on the shape of the interaction potential. The first result in this direction was obtained by Kesten and Papanicolau for a particle in $\mathbb{R}^3$ in a weak mean zero random force field, see [16]. Dürr, Goldstein and Lebowitz proved that in $\mathbb{R}^2$ the velocity process converges in distribution to Brownian motion on a surface of constant speed for sufficiently smooth interaction potentials, see [9].

The linear Landau equation appears also in an intermediate scale between the low density and the weak-coupling regime, namely when the (smooth) interaction potential ϕ rescales according to $\phi \to \varepsilon^{\alpha}\phi$, $\alpha \in (0, 1/2)$ and the density of the obstacles is of order $\varepsilon^{-2\alpha-(d-1)}$ [11, 17]. The limiting cases $\alpha = 0$ and $\alpha = 1/2$ correspond respectively to the low density limit and the weak-coupling limit.

The rigorous derivation of hydrodynamical equations grounds on the heuristic idea that after a few mean free times the Lorentz gas is already very close to the local equilibrium which subsequently evolves according to the diffusion equation. Clearly the only hydrodynamic equation for the Lorentz gas is the diffusion equation since the only conserved quantity is the mass.

The rigorous derivation of the heat equation from the mechanical system given by the Lorentz gas is actually a very difficult and still unsolved problem. In fact we would expect that, under the diffusive scaling, the distribution density of the test particle converges to that of a diffusion process. Bunimovich and Sinai (see [5]) showed that such diffusive limit holds when the scatterers are periodically distributed. This is the most important result in the transition from the microscopic to the macroscopic description.

Nonetheless one can handle this problem by deriving the diffusion equation from the correct kinetic equation which arises, according to the suitable kinetic scaling performed, from the random Lorentz gas. We remark, however, that the hydrodynamics for the Lorentz model is not equivalent to the hydrodynamics for the kinetic equation.

In this direction, in [2], we provide a rigorous derivation of the heat equation from the particle system (the Lorentz model) using the linear Landau equation as a bridge between our original mechanical system and the diffusion equation. It works once having an explicit control of the error in the kinetic limit (see also [10], where the set of bad configurations are explicitly estimated). The diffusive limit can be achieved since the control of memory effects still holds for a longer time scale.

Moreover, since it is well known how important and challenging is the characterization of stationary nonequilibrium states exhibiting transport phenomena in the rigorous approach to nonequilibrium Statistical Mechanics, we are interested in considering the Lorentz model out of equilibrium. Energy or mass transport in non equilibrium macroscopic systems are described phenomenologically by Fourier's and Fick's law respectively. There are very few rigorous results in this direction in the current literature (see for instance [20–22]). A contribution in this direction, discussed in Sect. 4, is the validation of the Fick's law for the Lorentz model in a low density situation which has been recently proven in [3].

2 From Microscopic to Macroscopic Description

We consider a Poisson distribution of fixed scatterers in $\mathbb{R}^2$ and denote by $c_1, \ldots, c_N$ their centers. This means that, given $\mu > 0$, the probability density of finding N obstacles in a bounded measurable set $A \subset \mathbb{R}^2$ is

$$\mathbb{P}(d\mathbf{c}_N) = e^{-\mu|A|}\frac{\mu^N}{N!}dc_1 \ldots dc_N, \tag{2}$$

where $|A| = \text{meas}(A)$ and $\mathbf{c}_N = (c_1, \ldots, c_N)$. The equations of motion for the point particle of unitary mass are

$$\begin{cases} \dot{x} = v \\ \dot{v} = -\sum_{i=1}^{N} \nabla\phi(|x - c_i|), \end{cases} \tag{3}$$

where x and v denote position and velocity of the test particle, t the time and, as usual, $\dot{A} = \frac{dA}{dt}$ indicates the time derivative for any time dependent variable A. Finally $\phi : \mathbb{R}^+ \to \mathbb{R}$ is given by

$$\phi(r) = \begin{cases} 1 & \text{if } r < 1 \\ 0 & \text{otherwise,} \end{cases} \tag{4}$$

namely a circular potential barrier.

This choice for the potential arises from a problem of geometric optics. We are looking at the optical path followed by a light ray traveling in a inhomogeneous medium. More precisely we have a medium, for example water, in which circular drops of a different substance are distributed. These drops are made of a different substance with smaller refractive index, for example air. The analogy between geometric optics and classical mechanics implies that the trajectory of the light ray is the trajectory of a test particle moving in a random distribution of scatterers where each scatterer generates a circular potential barrier.

To outline a kinetic behavior of the particle, we introduce the scale parameter $\varepsilon > 0$, indicating the ratio between the macroscopic and the microscopic variables, and rescale according to

$$x \to \varepsilon x, \; t \to \varepsilon t, \; \phi \to \varepsilon^{\alpha}\phi \tag{5}$$

with $\alpha \in [0, 1/2]$. Then Eq. (3) become

$$\begin{cases} \dot{x} = v \\ \dot{v} = -\varepsilon^{\alpha-1} \sum_i \nabla\phi(\frac{|x-c_i|}{\varepsilon}). \end{cases} \tag{6}$$

We rescale also the intensity μ of the scatterers as $\mu_\varepsilon = \mu\varepsilon^{-\delta}$, where $\delta = 1 + 2\alpha$. Accordingly we denote by $\mathbb{P}_\varepsilon$ the probability density (2) with μ replaced by μ_ε and $\mathbb{E}_\varepsilon$ will be the expectation with respect to the measure $\mathbb{P}_\varepsilon$.

Now let $T^t_{\mathbf{c}_N}(x, v)$ be the Hamiltonian flow solution of Eq. (6) with initial datum (x, v) in a given sample $\mathbf{c}_N = (c_1, \ldots, c_N)$ of obstacles (skipping the ε dependence for notational simplicity). $T^t_{\mathbf{c}_N}(x, v)$ is generated by the Hamiltonian

$$H(x, v, \mathbf{c}_N) = \frac{1}{2}v^2 + \varepsilon^\alpha \sum_j \phi\left(\frac{|x - c_j|}{\varepsilon}\right), \tag{7}$$

where ϕ is given by (4). For this choice of the potential $\nabla\phi$ is not well defined. However the explicit solution of the equation of motion is obtained by solving the single scattering problem using the energy and angular momentum conservation (see Fig. 1).

In Fig. 1 we represent the scattering of a particle entering in the ball

$$B(0, 1) = \{x \text{ s.t. } |x| < 1\}$$

toward a potential barrier of intensity $\phi(x) = \varepsilon^\alpha$.

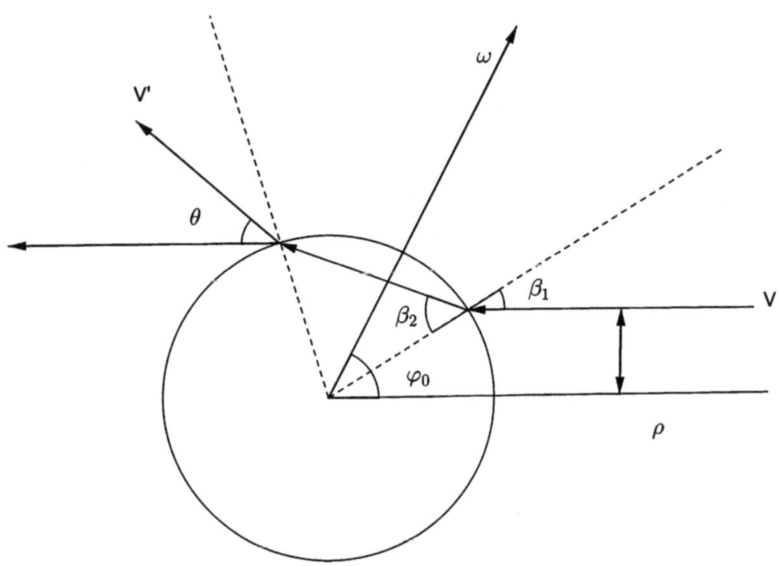

Fig. 1 Scattering by a spherical potential barrier. The particle moves in a straight line which is refracted on entering and leaving the barrier

Using the Snell law of refraction we have an explicit expression for the refractive index, i.e.

$$n_\varepsilon = \frac{\sin \beta_1}{\sin \beta_2} = \frac{|\bar{v}|}{|v|} = \sqrt{1 - \frac{2\varepsilon^\alpha}{v^2}}, \tag{8}$$

where v is the initial velocity, $\bar{v}$ the velocity inside the barrier, β_1 the angle of incidence and β_2 the angle of refraction. The scattering angle is $\theta = \pi - 2\varphi_0 = 2(\beta_2 - \beta_1)$ and the impact parameter is $\rho = \sin \beta_1$.

Remark 1 Formula (8) makes sense if $\frac{2\varepsilon^\alpha}{v^2} < 1$ and $\rho = \sin \beta_1 < \sqrt{1 - \frac{2\varepsilon^\alpha}{v^2}}$. When one of these two inequalities is violated, the outgoing velocity is the one given by the elastic reflection.

A careful computation (see Appendix 1 in [2] for further details) shows that the explicit expression for the scattering angle is given by

$$\theta_\varepsilon(\rho) = \begin{cases} 2\left(\arcsin\left(\frac{\rho}{n_\varepsilon}\right) - \arcsin(\rho)\right) & \text{if } \rho \le n_\varepsilon \\ 2\arccos(\rho) & \text{if } \rho > n_\varepsilon. \end{cases} \tag{9}$$

Here we are not considering possible overlappings of obstacles. The scattering process can be solved in this case as well. However this event is negligible because of the moderate densities we are considering.

Now let $f_0 = f_0(x, v)$ be the initial probability distribution. We are interested in characterizing the asymptotic behavior, under the scaling illustrated above, of the evolved distribution

$$f_\varepsilon(x, v, t) = \mathbb{E}_\varepsilon[f_0(T^{-t}_{\mathbf{c}_N}(x, v))]. \tag{10}$$

We expect that the probability distribution (10), in the limit $\varepsilon \to 0$, solves a linear kinetic equation, more precisely the linear Landau equation. However, due to the particular choice of the interaction potential, new features emerges at a mesoscopic level. The novelty, compared to [10, 17], is that we have a logarithmic divergence of the diffusion coefficient appearing in the Landau equation, due to the lack of smoothness of the potential. This divergence suggests to look at a longer time scale in which a diffusion in space arises. In fact, for a potential of the form (4), the classical formula giving the diffusion coefficient in the Landau equation (1), i.e.

$$B := \lim_{\varepsilon \to 0} \frac{\mu \varepsilon^{-2\alpha}}{2} |v| \int_{-1}^{1} \theta_\varepsilon^2(\rho)\, d\rho, \tag{11}$$

becomes

$$B = \lim_{\varepsilon \to 0} \mu \left[\frac{2\alpha}{|v|^3} |\log(\varepsilon)|\right] = +\infty, \tag{12}$$

where θ_ε is the scattering angle defined in (9). For the detailed computation of the diffusion coefficient we refer to [2], Appendix 2. Roughly speaking, we can state that the asymptotic equation for the density of the Lorentz particle reads as

$$(\partial_t + v \cdot \nabla_x) f(x, v, t) \sim |\log \varepsilon| \tilde{B} \Delta_{|v|} f(x, v, t), \quad \tilde{B} < +\infty. \tag{13}$$

Hence, the asymptotic behavior of the mechanical system we are considering is the same as the Markov process ruled by the linear Landau equation with a diverging factor in front of the collision operator. This is equivalent to consider the limit in the Euler scaling of the linear Landau equation, which is trivial. Therefore we do not get any hydrodynamical equation and the system quickly thermalizes to the local equilibrium. To detect something non trivial we have to look at a longer time scale $t \to |\log \varepsilon| t$ in which the equilibrium starts to evolve. As expected, a diffusion in space arises.

The main results are summarized in the following theorem ([2], Theorem 2.1).

Theorem 1 *Suppose $f_0 \in C_0(\mathbb{R}^2 \times \mathbb{R}^2)$ a continuous, compactly supported initial probability density. Suppose also that $|D_x^k f_0| \leq C$, where D_x is any partial derivative with respect to x and $k = 1, 2$. Assume $\mu_\varepsilon = \varepsilon^{-2\alpha-1}$, with $\alpha \in (0, 1/8)$. Then the following statement holds*

$$\lim_{\varepsilon \to 0} f_\varepsilon(x, v, t) = \langle f_0 \rangle := \frac{1}{2\pi} \frac{1}{|v|} \int_{S_{|v|}} f_0(x, v)\, dv, \tag{14}$$

$\forall t \in (0, T]$, $T > 0$. The convergence is in $L^2(\mathbb{R}^2 \times S_{|v|})$.

Moreover, define $F_\varepsilon(x, v, t) := f_\varepsilon(x, v, t|\log \varepsilon|)$. Then, for all $t \in [0, T)$, $T > 0$,

$$\lim_{\varepsilon \to 0} F_\varepsilon(x, v, t) = \rho(x, t),$$

where ρ solves the following heat equation

$$\begin{cases} \partial_t \rho = D \Delta \rho \\ \rho(x, 0) = \langle f_0 \rangle, \end{cases} \tag{15}$$

with D given by the Green-Kubo formula

$$D = \frac{2}{\mu} |v| \int_{S_{|v|}} v \cdot \big(-\Delta_{|v|}^{-1} \big) v\, dv = \frac{2\pi}{\mu} |v|^2 \int_0^\infty \mathbb{E}\big[v \cdot V(t, v) \big]\, dt, \tag{16}$$

where $V(t, v)$ is the stochastic process generated by $\Delta_{|v|}$ starting from v and $\mathbb{E}[\cdot]$ denotes the expectation with respect to the invariant measure, namely the uniform measure on $S_{|v|}$. The convergence is in $L^2(\mathbb{R}^2 \times S_{|v|})$.

Remark 2 To recover instead the kinetic picture, we can rescale suitably the density of the Poisson process. More precisely, if $\mu_\varepsilon = \frac{\varepsilon^{-2\alpha-1}}{|\log \varepsilon|}$, the microscopic solution f_ε defined by (10) converges to the solution of the linear Landau equation (1) with a renormalized diffusion coefficient $B := \lim_{\varepsilon \to 0} \frac{\mu_\varepsilon}{2} \varepsilon\, |v| \int_{-1}^{1} \theta_\varepsilon^2(\rho)\, d\rho = \frac{2\alpha}{|v|^3} \mu$. This is stated in [2], Theorem 2.2. The explicit expression for the renormalized B can be found in [2], Appendix 2.

We have seen that, according to the particular choice for the potential we are considering, the natural divergence of the diffusion coefficient B leads to a diffusion when we look at the system on a longer time scale. We can wonder if this result can be achieved in presence of a smooth, radial, short-range potential $\tilde{\phi} \in C^2([0,1])$. In the same spirit as in [13, 18], we scale the variables, the density and the potential according to

$$\begin{cases} x \to \varepsilon x \\ t \to \varepsilon^\lambda \varepsilon t, \\ \mu_\varepsilon = \varepsilon^{-(2\alpha+\lambda+1)} \mu \\ \tilde{\phi} \to \varepsilon^\alpha \tilde{\phi}. \end{cases} \tag{17}$$

The naive idea is that the kinetic regime describes the system for kinetic times $O(1)$ (i.e. $\lambda = 0$). One can go further to diffusive times provided that $\lambda > 0$ is not too large. Indeed the distribution function f_ε "almost" solves

$$\left(\varepsilon^\lambda \partial_t + v \cdot \nabla_x\right) f_\varepsilon \approx \varepsilon^{-2\alpha-\lambda} \mathrm{L}_\varepsilon f_\varepsilon \approx \varepsilon^{-\lambda} c\, \Delta_{|v|} f_\varepsilon$$

which is the analogue of (13) above. In other words there is a scale of time for which the system diffuses. However such times should not prevent the Markov property. This gives a constraint on λ. In fact, we can prove that there exists a threshold $\lambda_0 = \lambda(\alpha)$, emerging from the explicit estimate of the set of pathological configurations producing memory effects, s.t. for $\lambda < \lambda(\alpha)$, for $t > 0$ and $\varepsilon \to 0$,

$$f_\varepsilon(x, v, \varepsilon^\lambda t) \to \rho(x,t) \quad \text{in} \quad L^2(\mathbb{R}^2 \times \mathbb{R}^2),$$

solution of the heat equation

$$\begin{cases} \partial_t \rho = D \Delta \rho \\ \rho(x,0) = \langle f_0 \rangle, \end{cases}$$

with D given by the Green-Kubo formula

$$D = \frac{2}{\mu} |v| \int_{S_{|v|}} v \cdot \left(- \Delta_{|v|}^{-1} \right) v\, dv.$$

This is stated in [2], Theorem 6.1.

3 Ideas of the Proof

To give some ideas about how this machinery works we can divide the problem into two steps. The first step concerns the kinetic limit. We analyze how the limiting process, whose Fokker-Planck equation is the linear Landau equation, is obtained from the deterministic time evolution of the mechanical system in the intermediate regime described by Eq. (5). The other step shows how to pass from the kinetic description ruled by the linear Landau equation to the macroscopic picture where a diffusion in the position variable arises. We remind that the heat equation is, in the present case, the correct hydrodynamic equation.

3.1 The Kinetic Description

Following the explicit approach in [10, 11, 14] we will show the asymptotic equivalence of f_ε, defined by (10), and h_ε solution of the following Boltzmann equation

$$(\partial_t + v \cdot \nabla_x) h_\varepsilon(x, v, t) = \mathrm{L}_\varepsilon h_\varepsilon(x, v, t), \tag{18}$$

where

$$\mathrm{L}_\varepsilon h(v) = \mu \varepsilon^{-2\alpha} |v| \int_{-1}^{1} d\rho \{h(v') - h(v)\}. \tag{19}$$

Here $v' = v - 2(\omega \cdot v)\omega$ where $\omega = \omega(\rho, |v|)$ is the unit vector obtained by solving the scattering problem associated to ϕ (see Fig. 1). This allows to reduce the problem to the analysis of a Markov process which is an easier task. In fact, the series expansion defining h_ε (obtained perturbing around the loss term) reads as

$$h_\varepsilon(x, v, t) = e^{-2\varepsilon^{-2\alpha}|v|t} \sum_{Q \geq 0} \mu_\varepsilon^Q \int_0^t dt_Q \dots \int_0^{t_2} dt_1 \int_{-\varepsilon}^{\varepsilon} d\rho_1 \dots \int_{-\varepsilon}^{\varepsilon} d\rho_Q f_0(\bar{\xi}_\varepsilon(-t), \bar{\omega}_\varepsilon(-t)). \tag{20}$$

with

$$\begin{cases} \bar{\xi}_\varepsilon(-t) = x - vt_1 - v_1(t_2 - t_1) \cdots - v_Q(t - t_Q) \\ \bar{\omega}_\varepsilon(-t) = v_Q. \end{cases} \tag{21}$$

We remark that $\bar{\omega}_\varepsilon$ is an autonomous jump process and $\bar{\xi}_\varepsilon$ is an additive functional of $\bar{\omega}_\varepsilon$. Hence Eq. (20) is an evolution equation for the probability density associated to a particle performing random jumps in the velocity variable at random Markov times.

We consider the microscopic solution f_ε defined by (10). For $(x, v) \in \mathbb{R}^2 \times \mathbb{R}^2$, $t > 0$, we have

$$f_\varepsilon(x, v, t) = e^{-\mu_\varepsilon |B_t(x,v)|} \sum_{N \geq 0} \frac{\mu_\varepsilon^N}{N!} \int_{B_t(x,v)^N} d\mathbf{c}_N f_0(T_{\mathbf{c}_N}^{-t}(x, v)), \tag{22}$$

where $T_{\mathbf{c}_N}^t(x, v)$ is the Hamiltonian flow generated by the Hamiltonian (7). Here $B_t(x, v) := B(x, |v|t)$, where $B(x, R)$ denotes the disk of center x and radius R. Thanks to suitable manipulations (see [2] to go into details), Eq. (22) becomes

$$f_\varepsilon(x, v, t) = \sum_{Q \geq 0} \frac{\mu_\varepsilon^Q}{Q!} \int_{B_t(x,v)^Q} d\mathbf{b}_Q \, e^{-\mu_\varepsilon |\mathscr{T}(\mathbf{b}_Q)|} f_0(T_{\mathbf{b}_Q}^{-t}(x, v))$$
$$\chi(\{\text{the } \mathbf{b}_Q \text{ are internal}\}), \tag{23}$$

where here and in the sequel $\chi(\{\dots\})$ is the characteristic function of the event $\{\dots\}$ and the "internal obstacles" are the obstacles of the configuration which, up to time t, influence the motion of the light particle. Moreover $\mathscr{T}(\mathbf{b}_Q)$ is the tube

$$\mathscr{T}(\mathbf{b}_Q) = \{y \in B_t(x, v) \text{ s.t. } \exists s \in (-t, 0) \text{ s.t. } |y - x_\varepsilon(s)| < \varepsilon\}. \tag{24}$$

Here $(x_\varepsilon(s), v_\varepsilon(s)) = T_{\mathbf{c}_N}^s(x, v)$. We introduce

$$\tilde{f}_\varepsilon(x, v, t) = e^{-2\varepsilon^{-2\alpha}|v|t} \sum_{Q \geq 0} \frac{\mu_\varepsilon^Q}{Q!} \int_{B_t(x,v)^Q} d\mathbf{b}_Q$$
$$\chi(\{\text{the } \mathbf{b}_Q \text{ are internal}\}) \chi_1(\mathbf{b}_Q) f_0(T_{\mathbf{b}_Q}^{-t}(x, v)), \tag{25}$$

where χ_1 is the characteristic function of the set of configurations $\mathbf{b}_Q$ for which the particle is outside the range of all scatterers at time 0 and at time $-t$, namely

$$\chi_1(\mathbf{b}_Q) = \chi\{\mathbf{b}_Q \text{ s.t. } b_i \notin B(x, \varepsilon) \text{ and } b_i \notin B(x_\varepsilon(-t), \varepsilon) \text{ for all } i = 1, \dots, Q\}. \tag{26}$$

Since

$$|\mathscr{T}(\mathbf{b}_Q)| \leq 2\varepsilon|v|t, \tag{27}$$

we have

$$f_\varepsilon \geq \tilde{f}_\varepsilon. \tag{28}$$

The key idea of this Markovian approximation is a suitable change of variables which has been introduced by Gallavotti in [14]. We order the obstacles $b_1, \dots, b_Q$ according to the scattering sequence. Let ρ_i and t_i be the impact parameter and the entrance time of the light particle in the protection disk around b_i, i.e. $B(b_i, \varepsilon)$. Hence we perform the following change of variables

$$b_1, \ldots, b_Q \to \rho_1, t_1, \ldots, \rho_Q, t_Q, \tag{29}$$

with

$$0 \leq t_1 < t_2 < \cdots < t_Q \leq t.$$

Conversely, fixed the impact parameters $\{\rho_i\}$ and the hitting times $\{t_i\}$ we construct the centers of the obstacles $b_i = b(\rho_i, t_i)$. By performing the backward scattering we construct a trajectory $(\xi_\varepsilon(s), \omega_\varepsilon(s)), s \in [-t, 0]$. However $(\xi_\varepsilon(s), \omega_\varepsilon(s)) = (x_\varepsilon(s), v_\varepsilon(s))$ (therefore the mapping (29) is one-to-one) only outside the following pathological situations (Fig. 2).

(i) **Overlapping**.
If b_i and b_j are both internal and $B(b_i, \varepsilon) \cap B(b_j, \varepsilon) \neq \emptyset$.
(ii) **Recollisions**.
There exists b_i such that for $\tilde{s} \in (t_j, t_{j+1}), j > i, \xi_\varepsilon(-\tilde{s}) \in B(b_i, \varepsilon)$.
(iii) **Interferences**.

There exists b_i such that $\xi_\varepsilon(-\tilde{s}) \in B(b_i, \varepsilon)$ for $\tilde{s} \in (t_j, t_{j+1}), j < i$.

In order to skip such events we define

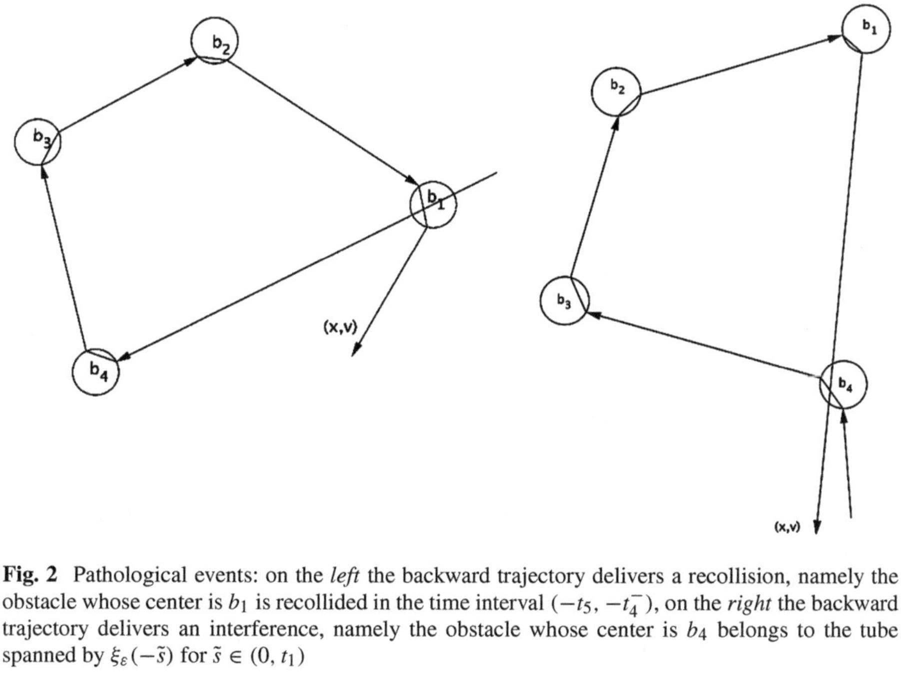

Fig. 2 Pathological events: on the *left* the backward trajectory delivers a recollision, namely the obstacle whose center is b_1 is recollided in the time interval $(-t_5, -t_4^-)$, on the *right* the backward trajectory delivers an interference, namely the obstacle whose center is b_4 belongs to the tube spanned by $\xi_\varepsilon(-\tilde{s})$ for $\tilde{s} \in (0, t_1)$

$$\bar{f}_\varepsilon(x,v,t) = e^{-2\varepsilon^{-2\alpha}|v|t} \sum_{Q\geq 0} \mu_\varepsilon^Q \int_0^t dt_Q \dots \int_0^{t_2} dt_1$$
$$\int_{-\varepsilon}^{\varepsilon} d\rho_1 \dots \int_{-\varepsilon}^{\varepsilon} d\rho_Q \, \chi_1(1-\chi_{ov})(1-\chi_{rec})(1-\chi_{int}) f_0(\xi_\varepsilon(-t), \omega_\varepsilon(-t)), \tag{30}$$

where χ_{ov}, χ_{int} and χ_{rec} are the characteristic functions of the events (i), (ii), (iii) respectively. Moreover we observe that

$$\bar{f}_\varepsilon \leq \tilde{f}_\varepsilon \leq f_\varepsilon.$$

Next we remove $\chi_1(1-\chi_{ov})(1-\chi_{rec})(1-\chi_{int})$ by setting

$$\bar{h}_\varepsilon(x,v,t) = e^{-2\varepsilon^{-2\alpha}|v|t} \sum_{Q\geq 0} \mu_\varepsilon^Q \int_0^t dt_Q \dots \int_0^{t_2} dt_1$$
$$\int_{-\varepsilon}^{\varepsilon} d\rho_1 \dots \int_{-\varepsilon}^{\varepsilon} d\rho_Q f_0(\xi_\varepsilon(-t), \omega_\varepsilon(-t)). \tag{31}$$

Since

$$1-\chi_1(1-\chi_{ov})(1-\chi_{rec})(1-\chi_{int}) \leq (1-\chi_1) + \chi_{ov} + \chi_{rec} + \chi_{int}, \tag{32}$$

from (30) and (31) we have

$$\bar{f}_\varepsilon(t) = \bar{h}_\varepsilon(t) + \varphi_1(\varepsilon,t).$$

with

$$\varphi_1(\varepsilon,t) := e^{-2\mu_\varepsilon \varepsilon t} \sum_{Q\geq 0} (\mu_\varepsilon)^Q \int_0^t dt_Q \dots \int_0^{t_2} dt_1 \int_{-\varepsilon}^{\varepsilon} d\rho_1 \dots \int_{-\varepsilon}^{\varepsilon} d\rho_Q$$
$$\{(1-\chi_1) + \chi_{ov} + \chi_{rec} + \chi_{int}\} f_0(\xi_\varepsilon(-t), \omega_\varepsilon(-t)). \tag{33}$$

We observe that this is the crucial part. In the two-dimensional case the probability of those bad behaviors producing memory effects (correlation between the past and the present) is nontrivial. To control the unphysical trajectories we need an explicit estimate of the set of bad configurations of the scatterers, in other words we have to estimate the error term $\varphi_1(\varepsilon,t)$ showing that it is negligible in the limit. In [2], Sect. 5, we prove that

$$\|\varphi_1(\varepsilon,t)\|_{L^1} \to 0 \quad \text{as} \quad \varepsilon \to 0 \quad \text{for all} \quad t \in [0,T].$$

Moreover the control of memory effects still holds for a longer time scale, namely

$$\|\varphi_1(\varepsilon,t)\|_{L^1} \underset{\varepsilon\to 0}{\longrightarrow} 0 \quad \forall t \in [0, |\log\varepsilon| T], \quad T>0,$$

which implies that we can look at the system on a longer time scale.

Since we are working to achieve the asymptotic equivalence of f_ε and h_ε, we need to compare $\bar{h}_\varepsilon$ with h_ε. This is fulfilled once we consider the collision as instantaneous. More precisely, for the sequence $t_1, \dots, t_Q$ $\rho_1, \dots \rho_Q$ consider the sequence $v_1, \dots, v_Q$ of incoming velocities before the Q collisions. This allows to construct the limiting trajectory $\bar{\xi}_\varepsilon(-t)$, given by (21), which approximates the trajectory $\xi_\varepsilon(-t)$ up to an error vanishing in the limit. Indeed, due to the Lipschitz continuity of f_0, we can assert that

$$\bar{h}_\varepsilon(x,v,t) = h_\varepsilon(x,v,t) + \varphi_2(x,v,t), \tag{34}$$

where

$$\sup_{x,v,t\in[0,T]} |\varphi_2(x,v,t)| \le C\varepsilon^{1-2\alpha}T. \tag{35}$$

3.2 The Diffusive Limit

For the sake of simplicity we set $\eta_\varepsilon = |\log\varepsilon|$. We rewrite the linear Boltzmann equation (18) in the following way

$$\big(\partial_t + v\cdot\nabla_x\big)h_\varepsilon(x,v,t) = \eta_\varepsilon\, \tilde{\mathrm{L}}_\varepsilon h_\varepsilon(x,v,t), \tag{36}$$

where $\tilde{L}_\varepsilon = L/\eta_\varepsilon$, namely

$$\tilde{\mathrm{L}}_\varepsilon f(v) = \mu|v|\frac{\varepsilon^{-2\alpha}}{|\log\varepsilon|}\int_{-1}^{1} d\rho\big[f(v') - f(v)\big]. \tag{37}$$

Performing the limit $\eta_\varepsilon \to \infty$ in (36) we get a trivial result (Eq. (14), Theorem (1)). To obtain something non trivial we look at the solution for times $\eta_\varepsilon t$, thus performing the diffusive scaling for space and time.

We denote by $\tilde{h}_\varepsilon := h_\varepsilon(x,v,\eta_\varepsilon t)$ the solution of the following rescaled linear Boltzmann equation

$$\big(\partial_t + \eta_\varepsilon\, v\cdot\nabla_x\big)\tilde{h}_\varepsilon = \eta_\varepsilon^2\, \tilde{\mathrm{L}}_\varepsilon \tilde{h}_\varepsilon, \tag{38}$$

and we introduce the rescaled Landau equation

$$\big(\partial_t + \eta_\varepsilon\, v\cdot\nabla_x\big)g_{\eta_\varepsilon}(x,v,t) = {\eta_\varepsilon}^2 \mathscr{L} g_{\eta_\varepsilon}(x,v,t), \tag{39}$$

where $\mathscr{L} = \dfrac{\mu}{2}\dfrac{1}{|v|}\Delta_{|v|}$.

Firstly we compare g_{η_ε} with $\tilde{h}_\varepsilon$ to show that they are asymptotically equivalent. We look at the evolution of $\tilde{h}_\varepsilon - g_{\eta_\varepsilon}$, namely

$$\left(\partial_t + \eta_\varepsilon\, v \cdot \nabla_x\right)\left(\tilde{h}_\varepsilon - g_{\eta_\varepsilon}\right) = \eta_\varepsilon^2 \left(\tilde{\mathrm{L}}_\varepsilon \tilde{h}_\varepsilon - \mathscr{L} g_{\eta_\varepsilon}\right),$$

and we obtain

$$\frac{1}{2}\,\partial_t \|\tilde{h}_\varepsilon - g_{\eta_\varepsilon}\| \le \eta_\varepsilon^2 \|\left(\tilde{\mathrm{L}}_\varepsilon - \mathscr{L}\right) g_{\eta_\varepsilon}\|, \tag{40}$$

where $\|\cdot\|$ denotes the L^2–norm. A straightforward computation shows that the collisions are grazing: each collision changes only slightly the velocity of a particle (for the detailed analysis of the scattering angle (9) we refer to [2], Appendix 1). Therefore we perform a Taylor's expansion of g_{η_ε} with respect to the velocity variable and we get

$$\begin{aligned}
\tilde{\mathrm{L}}_\varepsilon g_{\eta_\varepsilon} &= \mu |v| \frac{\varepsilon^{-2\alpha}}{|\log \varepsilon|} \int_{-1}^{1} d\rho \left[g_{\eta_\varepsilon}(x, v', t) - g_{\eta_\varepsilon}(x, v, t) \right] \\
&= \mu |v| \frac{\varepsilon^{-2\alpha}}{|\log \varepsilon|} \Big\{ \frac{1}{2} \Delta_{|v|} g_{\eta_\varepsilon} \int_{-1}^{1} d\rho\, |v' - v|^2 + \int_{-1}^{1} d\rho\, \tilde{R}_{\eta_\varepsilon} \Big\},
\end{aligned}$$

with $\tilde{R}_{\eta_\varepsilon} = O(|v' - v|^4)$. Since

$$\lim_{\varepsilon \to 0} \frac{\varepsilon^{-2\alpha}}{|\log \varepsilon|} \int_{-1}^{1} d\rho\, |v' - v|^2 = \lim_{\varepsilon \to 0} \frac{\varepsilon^{-2\alpha}}{|\log \varepsilon|} \int_{-1}^{1} d\rho \left(4 \sin^2 \frac{\theta_\varepsilon(\rho)}{2} \right) = 2 \frac{\alpha}{|v|^4}$$

and

$$\lim_{\varepsilon \to 0} \frac{\varepsilon^{-2\alpha}}{|\log \varepsilon|} \int_{-1}^{1} d\rho\, \tilde{R}_{\eta_\varepsilon} = \lim_{\varepsilon \to 0} \frac{\varepsilon^{-2\alpha}}{|\log \varepsilon|} \int_{-1}^{1} d\rho\, |v - v'|^4 = \varepsilon^{\alpha} |\log \varepsilon|^{\beta}, \quad -1 < \beta < \frac{5}{2}\alpha - 1,$$

we have

$$\|\left(\tilde{\mathrm{L}}_\varepsilon - \mathscr{L}\right) g_{\eta_\varepsilon}\| \le \varepsilon^{\alpha} |\log \varepsilon|^{\beta} \|\Delta_{|v|}^2 g_{\eta_\varepsilon}\| \le C \varepsilon^{\alpha} |\log \varepsilon|^{\beta}, \qquad C > 0,$$

for some $-1 < \beta < \frac{5}{2}\alpha - 1$ and ε sufficiently small. See [2], Appendix 2, for the details. Consequently, using (40), we have that $\tilde{h}_\varepsilon$, solution of (38), is close to g_{η_ε}, solution of (39), in $L^2(\mathbb{R}^2 \times S_{|v|})$.

Therefore we need to prove that g_{η_ε}, solution of (39), converges to ρ as $\varepsilon \to 0$. The convergence is in $L^2(\mathbb{R}^2 \times S_{|v|})$, uniformly in $t \in (0, T]$. $\rho : \mathbb{R}^2 \times [0, T] \to \mathbb{R}_+$ is the solution of the diffusion equation (15), i.e.

$$\begin{cases} \partial_t \rho = D \Delta \rho \\ \rho(x, 0) = \langle f_0 \rangle, \end{cases}$$

where $\langle f_0 \rangle = \frac{1}{2\pi}\frac{1}{|v|}\int_{S_{|v|}} f_0(x, v)\,dv$ and $D = \frac{2}{\mu}|v|\int_{S_{|v|}} v \cdot \big(-\Delta_{|v|}^{-1}\big) v\,dv.$

The proof relies on a classical tool which is the Hilbert expansion technique. The Hilbert expansion is a formal series, in powers of $\frac{1}{\eta_\varepsilon}$, which allows to write g_{η_ε} in the following way

$$g_{\eta_\varepsilon}(x, v, t) = g^{(0)}(x, t) + \sum_{k=1}^{+\infty} \left(\frac{1}{\eta_\varepsilon}\right)^k g^{(k)}(x, v, t),$$

where the coefficients $g^{(k)}$ are independent of η_ε. The well known idea is to determine them recursively, by imposing that g_{η_ε} is a solution of (39). For the complete statement and the detailed computations we refer to [2], Sect. 4.

We assume that the initial datum of the Cauchy problem associated to (39) depends only on the position variable, namely the initial datum has the form of a local equilibrium, i.e. $g_{\eta_\varepsilon}(x, v, 0) = \langle f_0 \rangle$. Moreover we require $g^{(0)}$ to satisfy the same initial condition as the whole solution g_{η_ε}, namely $g^{(0)}(x, 0) = \langle f_0 \rangle$. We consider the truncated expansion for g_{η_ε} at order η_ε^{-2}, namely

$$g_{\eta_\varepsilon}(x, v, t) = g^{(0)}(x, t) + \frac{1}{\eta_\varepsilon} g^{(1)}(x, v, t) + \frac{1}{\eta_\varepsilon^2} g^{(2)}(x, v, t) + \frac{1}{\eta_\varepsilon} R_{\eta_\varepsilon},$$

where $g^{(i)}, i = 0, 1, 2$ are the first three coefficients of a Hilbert expansion in η_ε, and R_{η_ε} is the reminder. Comparing terms of the same order in η_ε, in (39), we obtain the following equations

$$\begin{aligned}
&(i)\; v \cdot \nabla_x g^{(0)} = \frac{\mu}{2}\frac{1}{|v|}\,\Delta_{|v|} g^{(1)} \\
&(ii)\; \partial_t\, g^{(0)} + v \cdot \nabla_x g^{(1)} = \frac{\mu}{2}\frac{1}{|v|}\,\Delta_{|v|} g^{(2)} \\
&(iii)\; \big(\partial_t + \eta_\varepsilon\, v \cdot \nabla_x\big) R_{\eta_\varepsilon} = \eta_\varepsilon{}^2 \frac{\mu}{2}\frac{1}{|v|}\,\Delta_{|v|} R_{\eta_\varepsilon} - A_{\eta_\varepsilon}(t),
\end{aligned}$$

with $A_{\eta_\varepsilon}(t) = A_{\eta_\varepsilon}(x, v, t) = \partial_t g^{(1)} + \frac{1}{\eta_\varepsilon}\partial_t g^{(2)} + v \cdot \nabla_x g^{(2)}$.

From Eqs. (i), (ii), thanks to suitable computations, we obtain that $g^{(0)}$ is the solution of the heat equation (15). Hence, by showing that the coefficients $g^{(i)} \in L^2(\mathbb{R}^2 \times S_{|v|})$ and that R_{η_ε} is uniformly bounded in L^2, we have that g_{η_ε} converges to $g^{(0)}$ in L^2 as $\eta_\varepsilon \to \infty$. We observe that we have assumed as initial condition for the linear Landau equation a local equilibrium state (independent of v). Strictly speaking this is not necessary since there is an initial regime (the initial layer) in which a general state, depending also on the velocity variable, thermalizes very fast in time and locally in space (see [2], Lemma 4.1).

4 Perspectives: Transport Properties of the Lorentz Gas

Roughly speaking, to consider the system out of equilibrium, we consider the Lorentz gas in a bounded region in the plane and couple the system with two mass reservoirs at the boundaries. We wonder about what to expect in a stationary non equilibrium state. The naive physical intuition tell us that there there exists a stationary state for which

$$J \approx -D\nabla \rho \tag{41}$$

where J is the mass current, ρ is the mass density and $D > 0$ is the diffusion coefficient. Formula (41) is the well known Fick's law which we want to prove in the present context.

In [3] we deal with the validation of the Fick's law of diffusion for the following model. We consider the slice $\Lambda = (0, L) \times \mathbb{R}$ in the plane. In the left half-plane $(-\infty, 0) \times \mathbb{R}$ there is a free gas of light particles at density ρ_1 with correlation functions given by

$$f_j^1(x_1, v_1 \ldots, x_j, v_j) = (\rho_1)^j M(v_1) \ldots M(v_j), \quad j \geq 1,$$

where $M(v_i)$ is the density of the uniform distribution on the unit circle S_1. In the right half-plane $(L, +\infty) \times \mathbb{R}$ there is a free gas of light particles at density ρ_2 with correlation functions given by

$$f_j^2(x_1, v_1 \ldots, x_j, v_j) = (\rho_2)^j M(v_1) \ldots M(v_j) \quad j \geq 1.$$

The two half-planes play the role of mass reservoirs. Inside Λ there is a Poisson distribution of hard core scatterers of diameter ε and intensity μ. We denote by $c_1, \ldots, c_N$ their centers.

A particle in Λ moves freely up to the first instant of contact with an obstacle. Then it is elastically reflected and so on. See Fig. 3.

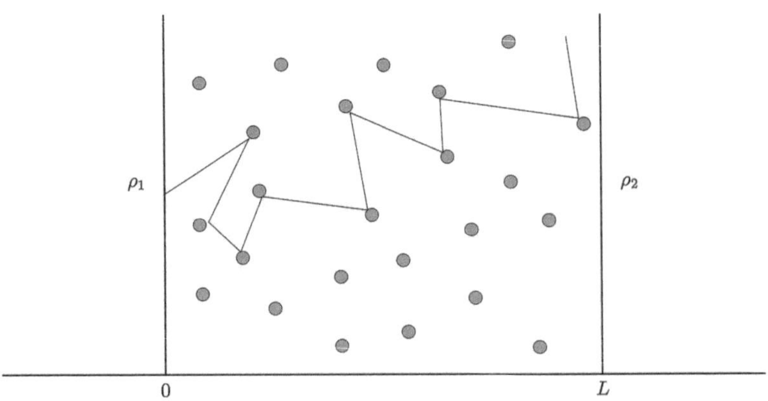

Fig. 3 The two-dimensional strip Λ

Since the modulus of the velocity of the test particle is constant, we assume it to be equal to one, so that the phase space of our system is $\Lambda \times S_1$.

We rescale the intensity μ of the obstacles as $\mu_\varepsilon = \mu\varepsilon^{-1}\eta_\varepsilon$ where, from now on, $\mu > 0$ is fixed and η_ε is slowly diverging as $\varepsilon \to 0$. More precisely we assume that

$$\varepsilon^{\frac{1}{2}}\eta_\varepsilon^6 \to 0 \quad \text{as} \quad \varepsilon \to 0. \tag{42}$$

Hence we are in a low density regime and the scatterer configuration is dilute.

For a given configuration of obstacles $\mathbf{c}_N$, we denote by $T_{\mathbf{c}_N}^{-t}(x, v)$ the (backward) flow with initial datum $(x, v) \in \Lambda \times S_1$ and define $t - \tau$, $\tau = \tau(x, v, t, \mathbf{c}_N)$, as the first (backward) hitting time with the boundary. With $\tau = 0$ we indicate the event such that the trajectory $T_{\mathbf{c}_N}^{-s}(x, v)$, $s \in [0, t]$, never hits the boundary. For any $t \geq 0$ the one-particle correlation function reads

$$f_\varepsilon(x, v, t) = \mathbb{E}_\varepsilon[f_B(T_{\mathbf{c}_N}^{-(t-\tau)}(x, v))\chi(\tau > 0)] + \mathbb{E}_\varepsilon[f_0(T_{\mathbf{c}_N}^{-t}(x, v))\chi(\tau = 0)], \tag{43}$$

where $f_0 \in L^\infty(\Lambda \times S_1)$ and the boundary value f_B is defined by

$$f_B(x, v) := \begin{cases} \rho_1 & \text{if } \; x \in \{0\} \times \mathbb{R}, \quad v_1 > 0, \\ \rho_2 & \text{if } \; x \in \{L\} \times \mathbb{R}, \quad v_1 < 0, \end{cases} \tag{44}$$

with $\rho_1, \rho_2 > 0$. Here we absorbed $M(v) = \frac{1}{2\pi}$, the density of the uniform distribution on S_1, in the definition of the boundary values ρ_1, ρ_2. To deal with the stationary regime we need to introduce the stationary solution of the problem, $f_\varepsilon^S(x, v)$, which solves

$$f_\varepsilon^S(x, v) = \mathbb{E}_\varepsilon[f_B(T_{\mathbf{c}_N}^{-(t-\tau)}(x, v))\chi(\tau > 0)] + \mathbb{E}_\varepsilon[f_\varepsilon^S(T_{\mathbf{c}_N}^{-t}(x, v))\chi(\tau = 0)]. \tag{45}$$

We observe that f_ε^S depends on the space variable only through the horizontal component x_1 since it inherits this feature from the boundary conditions. Moreover we introduce the following observables

$$J_\varepsilon^S(x) = \eta_\varepsilon \int_{S_1} v f_\varepsilon^S(x, v)\, dv, \tag{46}$$

$$\rho_\varepsilon^S(x) = \int_{S_1} f_\varepsilon^S(x, v)\, dv, \tag{47}$$

i.e. the stationary mass flux and the stationary mass density respectively. We define J_ε^S as the total amount of mass flowing through a unit area in a unit time interval and we observe that, although in a stationary problem there is no typical time scale, the factor η_ε appearing in the definition of J_ε^S, is reminiscent of the time scaling necessary to obtain a diffusive limit.

We are interested in the determining the existence and uniqueness of f_ε^S and its asymptotic behavior. We can prove that there exists a unique stationary solution for the microscopic dynamics which converges to the stationary solution of the heat equation, namely to the linear profile of the density. Moreover, in the same regime, the macroscopic current in the stationary state is given by the Fick's law, with the diffusion coefficient determined by the Green-Kubo formula.

The main results are summarized in the following theorems ([3], Theorems 2.1 and 2.2)

Theorem 2 *For ε sufficiently small there exists a unique L^∞ stationary solution f_ε^S for the microscopic dynamics (i.e. satisfying (45)). Moreover, as $\varepsilon \to 0$*

$$f_\varepsilon^S \to \rho^S, \tag{48}$$

where ρ^S is the stationary solution of the heat equation, namely the linear profile of the density

$$\rho^S(x) = \frac{\rho_1(L - x_1) + \rho_2 x_1}{L}. \tag{49}$$

The convergence is in $L^2((0, L) \times S_1)$.

Theorem 3 [Fick's law] *We have*

$$J_\varepsilon^S + D\nabla_x \rho_\varepsilon^S \to 0 \tag{50}$$

as $\varepsilon \to 0$. The convergence is in $\mathscr{D}'(0, L)$ and $D > 0$ is given by the Green-Kubo formula. Moreover

$$J^S = \lim_{\varepsilon \to 0} J_\varepsilon^S(x), \tag{51}$$

where the convergence is in $L^2(0, L)$ and

$$J^S = -D\,\nabla \rho^S = -D\,\frac{\rho_2 - \rho_1}{L}, \tag{52}$$

where ρ^S is the linear profile (49).

As expected by physical arguments, the stationary flux J^S does not depend on the space variable. Furthermore the diffusion coefficient D is determined by the behavior of the system at equilibrium and in particular it is equal to the diffusion coefficient for the time dependent problem.

Note that Theorem 3 is a straightforward consequence of Theorem 2. We observe that in order to prove the convergence of the stationary solutions (48) we can relax the hypothesis on η_ε: it is enough to require η_ε such that $\varepsilon^{\frac{1}{2}} \eta_\varepsilon^5 \to 0$. To prove instead Fick's law we need something more: we require η_ε to satisfy (42).

Our result holds in a low-density regime, hence we can use the linear Boltzmann equation as a bridge between our original mechanical system and the diffusion equation. We exploited this basic idea to prove Theorem 1 by using the linear Landau equation as intermediate level of description (see also [1, 13] where the linear Quantum Boltzmann equation and the linear Boltzmann equation respectively have been used to obtain the heat equation from the particle system in different contexts). This works once one has an explicit control of the error in the kinetic limit, which suggests the scale of times for which the diffusive limit can be achieved. This explains briefly why the above constraint on η_ε emerges. Moreover, since we are using the kinetic picture as an intermediate level of analysis, the diffusion coefficient D appearing in Eqs. (51), (52), is given by the Green-Kubo formula for the linear Boltzmann equation.

Here we have an additional difficulty since we have to deal with a stationary problem. The basic idea is that the explicit solution of the heat equation and the control of the time dependent problem allow us to characterize the stationary solution of the linear Boltzmann equation. This turns out to be the basic tool to obtain the stationary solution of the mechanical system which is the main object in our investigation.

We stress that, to handle the stationary problem, we characterize the stationary solutions in terms of Neumann series, rather than as the long time asymptotics of the time dependent solutions. This trick avoids the problem of controlling the convergence rates, as $t \to \infty$, with respect to the scale parameter ε.

Also in this case the strategy consists in two steps. The first one allows the transition from Boltzmann to the diffusion equation. It is in the same spirit of the one performed in Sect. 3.2. This is the Markov part since we approximate the Brownian motion by the Markov jump process whose generator is the Linear Boltzmann collision operator. We refer to [3], Sect. 4, for a more detailed discussion. Despite the technique is standard, we need an apparently new analysis in L^∞, for the time dependent problem (needed for the control of the Neumann series) and a L^2 analysis for the stationary problem.

The transition from the mechanical system to the Boltzmann equation in a low density regime can be read instead as a Markovian approximation, in the same spirit of the one performed in Sect. 3.1. This allows the transition from the nonmarkovian mechanical system to a Markov process. To reach a diffusive behavior on a longer time scale we need to estimate the set of pathological configurations which produce memory effects. Hence the constructive approach due to Gallavotti is complemented by an explicit analysis of the bad events preventing the Markovianity. The complete analysis is faced in [3], Sect. 5.

Acknowledgments I am indebted to G. Basile, F. Pezzotti and M. Pulvirenti for their collaboration and for the illuminating discussions.

References

1. Bodineau, T., Gallagher, I., Saint-Raymond, L.: The Brownian motion as the limit of a deterministic system of hard-spheres. arxiv: 1305.3397
2. Basile, G., Nota, A., Pulvirenti, M.: A diffusion limit for a test particle in a random distribution of scatterers. J. Stat. Phys. **155**, 1087–1111 (2014)
3. Basile, G., Nota, A., Pezzotti, F., Pulvirenti, M.: Derivation of the Fick's law for the Lorentz model in a low density regime. Commun. Math. Phys. doi:10.1007/s00220-015-2306-z
4. Boldrighini, C., Bunimovich, L.A., Sinai, Y.G.: On the Boltzmann equation for the Lorentz gas. J. Stat. Phys. **32**, 477–501 (1983)
5. Bunimovich, L.A., Sinai, Y.G.: Statistical properties of the Lorentz gas with periodic configuration of scatterers. Commun. Math. Phys. **78**, 478–497 (1981)
6. Cercignani, C., Illner, R., Pulvirenti, M.: The Mathematical Theory of Dilute Gases. Springer, Berlin (1994)
7. Caglioti, E., Golse, F.: The Boltzmann-Grad limit of the periodic Lorentz gas in two space dimensions. C.R. Math. Acad. Sci. Paris **346**, 477–482 (2008)
8. Caglioti, E., Golse, F.: On the Boltzmann-grad limit for the two dimensional periodic lorentz gas. J. Stat. Phys. **141**, 264–317 (2010)
9. Dürr, D., Goldstein, S., Lebowitz, J.: Asymptotic motion of a classical particle in a random potential in two dimensions: Landau model. Commun. Math. Phys. **113**, 209–230 (1987)
10. Desvillettes, L., Pulvirenti, M.: The linear Boltzmann equation for long-range forces: a derivation from particle systems. Models Methods Appl. Sci. **9**, 1123–1145 (1999)
11. Desvillettes, L., Ricci, V.: A rigorous derivation of a linear kinetic equation of Fokker-Planck type in the limit of grazing collisions. J. Stat. Phys. **104**, 1173–1189 (2001)
12. Esposito, R., Pulvirenti, M.: From particles to fluids. In: Hand-book of Mathematical Fluid Dynamics, vol. III, pp. 1–82. North-Holland, Amsterdam (2004)
13. Erdos, L., Salmhofer, M., Yau, H.-T.: Quantum diffusion of the random Schrodinger evolution in the scaling limit. Acta Math. **200**(2), 211–277 (2008)
14. Gallavotti, G.: Grad-Boltzmann Limit and Lorentz's Gas. Statistical Mechanics. A short treatise. Appendix 1.A2. Springer, Berlin (1999)
15. Gallagher, I., Saint-Raymond, L., Texier, B.: From Newton to Boltzmann: hard spheres and short-range potentials. In: Zurich Lectures in Advanced Mathematics (2014)
16. Kesten, H., Papanicolaou, G.: A limit theorem for stochastic acceleration. Commun. Math. Phys. **78**, 19–63 (1981)
17. Kirkpatrick, K.: Rigorous derivation of the Landau Equation in the weak coupling limit. Commun. Pure Appl. Anal. **8**, 1895–1916 (2009)
18. Komorowski, T., Ryzhik, L.: Diffusion in a weakly random Hamiltonian flow. Commun. Math. Phys. **263**, 273–323 (2006)
19. Landau, L.D., Lifshitz, E.M.: Mechanics, Course of Theoretical Physics, vol. 1. Pergamon press, Oxford (1960)
20. Lebowitz, J.L., Spohn, H.: Transport properties of the Lorentz Gas: Fourier's law. J. Stat. Phys. **19**, 633–654 (1978)
21. Lebowitz, J.L., Spohn, H.: Microscopic basis for Fick's law for self-diffusion. J. Stat. Phys. **28**, 539–556 (1982)
22. Lebowitz, J.L., Spohn, H.: Steady state self-diffusion at low density. J. Stat. Phys. **29**, 539–556 (1982)
23. Lorentz, H.A.: The motion of electrons in metallic bodies. Proc. Acad. Amst. **7**, 438–453 (1905)
24. Marklof, J., Strömbergsson, A.: The Boltzmann-Grad limit of the periodic Lorentz gas. Ann. Math. **174**, 225–298 (2011)
25. Pulvirenti, M., Saffirio, C., Simonella, S.: On the validity of the Boltzmann equation for short range potentials. Rev. Math. Phys. **26**, 2 (2014)
26. Spohn, H.: The Lorentz flight process converges to a random flight process. Commun. Math. Phys. **60**, 277–D0290 (1978)

Fractional Boson Gas and Fractional Poisson Measure in Infinite Dimensions

Maria João Oliveira and Rui Vilela Mendes

Abstract As a consequence of Haag's theorem, to obtain a non-trivial theory, one either works with a non-Fock representation or with a Fock representation in a finite volume. Calculations in the Fock representation taking the N,V $\to \infty$ limit with the ratio N/V $= \rho$ fixed, show the equivalence of the free Boson gas and the infinite-dimensional Poisson measure. The N/V limit provides a way to deal with non-trivial infinite systems using the Fock representation. However, by the very nature of the fixed ρ density limit, it is unable to deal with systems with density fluctuations, a shortcoming that is solved by the use of reducible functionals. A particularly interesting reducible functional is the one associated to the infinite-dimensional fractional Poisson measure which we recall in this work.

Keywords Boson gases · Fractional Poisson measure

1 Introduction: Density Fluctuations, Reducible Functionals and Fractional Gases

Systems with an infinite number of degrees of freedom have, in addition to the Fock representation, infinitely many inequivalent representations of the canonical commutation relations. Haag's theorem states that, in a theory with a space-invariant vacuum, any representation equivalent to Fock can only describe a free system. Therefore, to obtain a non-trivial theory, one either works with a non-Fock representation or with a Fock representation in a finite volume. In this latter case one considers N particles in a finite volume V. Calculations are carried out in the Fock representation and in

M.J. Oliveira
Universidade Aberta and Centro de Matemática e Aplicações Fundamentais, Lisbon, Portugal
e-mail: oliveira@cii.fc.ul.pt

R.V. Mendes (✉)
Centro de Matemática e Aplicações Fundamentais, Lisbon, Portugal
e-mail: rvmendes@fc.ul.pt; rvilela.mendes@gmail.com

P. Gonçalves and A.J. Soares (eds.), *From Particle Systems to Partial Differential Equations II*, Springer Proceedings in Mathematics & Statistics 129,
DOI 10.1007/978-3-319-16637-7_11

the end one takes $N, V \to \infty$ with the ratio $N/V = \rho$ being fixed. Thus, the N/V limit provides a way to deal with non-trivial infinite systems using the Fock representation. However, by the very nature of the fixed ρ density limit, this approach is unable to deal with systems with density fluctuations. This shortcoming is solved by the use of the reducible functionals to be described later on.

First, let us recall the connection of the infinite-dimensional Poisson measure to the free Bose gas. A particularly convenient way to establish this connection, and also to explore generalizations, is the framework of non-relativistic current algebra of many-body systems. That is, the basic variables of the many-body system are the smeared currents [1, 2] (see also [3, 4] and references therein)

$$\rho(f) = \int d^3x f(x)\,\rho(x)$$
$$\mathbf{J}(\mathbf{g}) = \int d^3x\,\mathbf{J}(x) \bullet \mathbf{g}(x)$$

f and $\mathbf{g}$ being respectively smooth compact support functions and vector fields. The smeared currents satisfy the infinite-dimensional Lie algebra,

$$\left[\rho(f), \rho(h)\right] = 0$$
$$\left[\rho(f), \mathbf{J}(\mathbf{g})\right] = i\rho(\mathbf{g} \bullet \nabla f)$$
$$\left[\mathbf{J}(\mathbf{g}), \mathbf{J}(\mathbf{k})\right] = i\mathbf{J}(\mathbf{k} \bullet \nabla \mathbf{g} - \mathbf{g} \bullet \nabla \mathbf{k})$$

each particular physical system corresponding to a different Hilbert space representation of this algebra or of the semidirect product group generated by the exponentiated currents

$$U(f) = e^{i\rho(f)}$$
$$V\left(\phi_t^{\mathbf{g}}\right) = e^{i\mathbf{J}(\mathbf{g})}$$

$\phi_t^{\mathbf{g}}$ being the flow of the vector field $\mathbf{g}$

$$\frac{d}{dt}\phi_t^{\mathbf{g}}(x) = \mathbf{g}\left(\phi_t^{\mathbf{g}}(x)\right)$$

For a system of N free bosons in a box of volume V, the normalized ground state is

$$\Omega_{N,V}(x_1, \cdots, x_N) = \left(\frac{1}{\sqrt{V}}\right)^N$$

and the ground state functional

$$L_{N,V}(f) = \left(\Omega_{N,V}, U_{N,V}(f)\,\Omega_{N,V}\right)$$

$$= \frac{1}{V}\left(\int_V d^3x\, e^{if(x)}\right)^N$$

In general, this functional determines not only the representation of $U(f)$ but also that of $V\left(\phi_t^{\mathbf{g}}\right)$, up to a complex phase multiplier.

In the $N \to \infty$ limit with constant average density $\overline{\rho} = \frac{N}{V}$ (also called the N/V limit) one obtains

$$\begin{aligned} L(f) &= \lim_{n\to\infty}\left(1 + \frac{\overline{\rho}}{N}\int\left(e^{if(x)} - 1\right)d^3x\right)^n \\ &= \exp\left(\int\left(e^{if(x)} - 1\right)\overline{\rho}\, d^3x\right) \end{aligned} \tag{1}$$

which one recognizes as the characteristic functional of the infinite-dimensional Poisson measure. Identifying $\overline{\rho}d^3x$ as the measure $d\mu$ in the underlying space M (see Appendix 1), the L functional may also be written as a vacuum expectation functional. Expanding the exponential in (1)

$$L(f) = \sum_{n=0}^{\infty}\frac{e^{-\int_V d\mu}}{n!}\left(\int_V e^{if(x)}d\mu\right)^n \tag{2}$$

one may write

$$L(f) = (\Omega, U(f)\,\Omega)$$

for

$$\Omega = \bigoplus_n e^{-\frac{1}{2}\int_V d\mu}\, 1\!\!1_n$$

$1\!\!1_n$ denotes the identity function in the n-particle space and the $\frac{1}{n!}$ factor in (2) is recovered by the symmetrization operation.

The conclusion is that an infinite dimensional free Boson gas at constant density $\overline{\rho}$ is completely characterized by the infinite-dimensional Poisson measure. However (1) is not the most general consistent representation of the nonrelativistic current algebra, a more general one being [2]

$$L(f) = \int_0^{\infty}\exp\left(\overline{\rho}\int\left(e^{if(x)} - 1\right)d^3x\right)d\mu\left(\overline{\rho}\right) \tag{3}$$

with μ a positive measure on $[0, \infty)$ normalized so that $\int_0^{\infty} d\mu\left(\overline{\rho}\right) = 1$.

Physically this reducible functional represents a Boson gas with density fluctuations. Among the many possible reducible functionals consistent with (3) there is a fractional generalization of (2), namely

$$L_\alpha(f) = \sum_{n=0}^{\infty} \frac{E_\alpha^{(n)}\left(-\int_V d\mu\right)}{n!} \left(\int_V e^{if(x)} d\mu\right)^n \tag{4}$$

$(0 < \alpha \leq 1)$, which corresponds to a vacuum state

$$\Omega_\alpha = \bigoplus_n \sqrt{E_\alpha^{(n)}\left(-\int_V d\mu\right)} 1\!\!1_n$$

$E_\alpha^{(n)}$ denoting the n-th derivative of the Mittag-Leffler function [5]. Ω_α differs from Ω in the weight given to each one of the n-particle spaces.

The measure associated to the functional (4) is called *the infinite-dimensional fractional Poisson measure* and the corresponding physical system *the fractional Boson gas*.

In the same way as the infinite-dimensional Poisson measure completely characterizes the Boson gas, the infinite-dimensional fractional Poisson measure will characterize a "fractional Boson gas" and the remainder of this work is dedicated to recall the main properties of this measure [6]. A detailed analysis of the mathematical properties of the fractional Poisson measure is a precondition for the rigorous formulation of some of the already explored physical implications of fractality (see for example [7, 8]).

2 The Infinite-Dimensional Fractional Poisson Measure

2.1 The Fractional Poisson Process

The Poisson measure π in $\mathbb{R}$ (or $\mathbb{N}$) is

$$\pi(A) = e^{-\sigma} \sum_{n \in A} \frac{\sigma^n}{n!}$$

the parameter σ being called the *intensity*. The Laplace transform of π is

$$l_\pi(\lambda) = \mathbb{E}\left(e^{\lambda \cdot}\right) = e^{-\sigma} \sum_{n=0}^{\infty} \frac{\sigma^n}{n!} e^{\lambda n} = e^{\sigma\left(e^\lambda - 1\right)}$$

For n-tuples of independent Poisson variables one would have

$$l_\pi(\lambda) = e^{\sum \sigma_k \left(e^{\lambda_k} - 1\right)}$$

Continuing λ_k to imaginary arguments $\lambda_k = if_k$, the characteristic function is

$$C_\pi(\lambda) = e^{\sum \sigma_k \left(e^{if_k} - 1\right)} \tag{5}$$

Looked at as a renewal process, $P(X = n) = e^{-\sigma} \frac{\sigma^n}{n!}$ would be the probability of n events occurring in the time interval σ. The survival probability, that is, the probability of no event is

$$\Psi(\sigma) = e^{-\sigma}$$

which satisfies the equation

$$\frac{d}{d\sigma} \Psi(\sigma) = -\Psi(\sigma) \tag{6}$$

Replacing in (6) the derivative $\frac{d}{d\sigma}$ by the (Caputo) fractional derivative

$$D^\alpha \Psi(\sigma) = \frac{1}{\Gamma(1-\alpha)} \int_0^\sigma \frac{\Psi'(\tau)}{(\sigma - \tau)^\alpha} d\tau = -\Psi(\sigma) \quad (0 < \alpha < 1)$$

one has the solution

$$\Psi(\sigma) = E_\alpha\left(-\sigma^\alpha\right)$$

with E_α being the Mittag-Leffler function of parameter α

$$E_\alpha(z) = \sum_{n=0}^{\infty} \frac{z^n}{\Gamma(\alpha n + 1)}, \quad z \in \mathbb{C} \tag{7}$$

($\alpha > 0$). One then obtains a *fractional Poisson process* [9, 10] with the probability of n events

$$P(X = n) = \frac{\sigma^{\alpha n}}{n!} E_\alpha^{(n)}\left(-\sigma^\alpha\right)$$

$E_\alpha^{(n)}$ denoting the n-th derivative of the Mittag-Leffler function. In contrast with the Poisson case ($\alpha = 1$), this process has power law asymptotics rather than exponential, which implies that it is not anymore Markovian. The characteristic function of this process is given by

$$C_\alpha(\lambda) = E_\alpha\left(\sigma^\alpha \left(e^{i\lambda} - 1\right)\right)$$

2.2 The Infinite-Dimensional Fractional Poisson Measure

For the Poisson measure ($\alpha = 1$) an infinite-dimensional generalization is obtained by generalizing (5) to

$$C(\varphi) = e^{\int \left(e^{i\varphi(x)} - 1\right) d\mu(x)}$$

for test functions $\varphi \in \mathscr{D}(M)$, $\mathscr{D}(M)$ being the space of C^{∞}-functions of compact support in a manifold M (fixed from the very beginning), and then using the Bochner-Minlos theorem to show that C is the Fourier transform of a measure on the distribution space $\mathscr{D}'(M)$. Because the Mittag-Leffler function is a natural analytic generalization of the exponential function one conjectures that an infinite-dimensional version of the fractional Poisson measure would have a characteristic functional

$$C_\alpha(\varphi) := E_\alpha\left(\int (e^{i\varphi(x)} - 1)\, d\mu(x)\right), \quad \varphi \in \mathscr{D}(M) \tag{8}$$

with μ a positive intensity measure fixed on the underlying manifold M. However, a priori it is not obvious that this is the Fourier transform of a measure on $\mathscr{D}'(M)$ nor that it corresponds to independent processes because the Mittag-Leffler function does not satisfy the factorization properties of the exponential.

Similarly to the Poisson case, to carry out our construction and analysis in detail we always assume that M is a geodesically complete connected oriented (non-compact) Riemannian C^{∞}-manifold, where we fix the corresponding Borel σ-algebra $\mathscr{B}(M)$, and μ is a non-atomic Radon measure, which we assume to be non-degenerate (i.e., $\mu(O) > 0$ for all non-empty open sets $O \subset M$). Having in mind the most interesting applications, we also assume that $\mu(M) = \infty$.

Theorem 1 *For each* $0 < \alpha \leq 1$ *fixed, the functional* C_α *in Eq. (8) is the characteristic functional of a probability measure* π_μ^α *on the distribution space* $\mathscr{D}'(M)$.

Proof That C_α is continuous and $C_\alpha(0) = 1$ follows easily from the properties of the Mittag-Leffler function. To check the positivity one uses the complete monotonicity of E_α, $0 < \alpha < 1$, which by Appendix 2 (Lemma 1) implies the integral representation

$$E_\alpha(-z) = \int_0^\infty e^{-\tau z} d\nu_\alpha(\tau) \tag{9}$$

for any $z \in \mathbb{C}$ such that $\mathrm{Re}(z) \geq 0$, ν_α being the probability measure (19) (Appendix 2). Hence by (9)

$$\sum_{a,b} C_\alpha(\varphi_a - \varphi_b)\, z_a^* z_b = \int_0^\infty d\nu_\alpha(\tau) \sum_{a,b} e^{-\tau \int_M d\mu(x)\left(1 - e^{i(\varphi_a - \varphi_b)}\right)} z_a^* z_b \tag{10}$$

Each one of the terms in the integrand corresponds to the characteristic function of a Poisson measure. Thus, for each τ the integrand is positive and therefore the spectral integral (10) is also positive. From the Bochner-Minlos theorem it then follows that C_α is the characteristic functional of a probability measure π_μ^α on the measurable space $(\mathscr{D}'(M), \mathscr{C}_\sigma(\mathscr{D}'(M)))$, $\mathscr{C}_\sigma(\mathscr{D}'(M))$ being the σ-algebra generated by the cylinder sets.

For $\alpha = 1$ see e.g. [11]. ■

Introducing the fractional Poisson measure by the above approach yields a probability measure on $(\mathscr{D}'(M), \mathscr{C}_\sigma(\mathscr{D}'(M)))$. The next step is to find an appropriate support for the fractional Poisson measure. Using the analyticity of the Mittag-Leffler function one may informally rewrite (8) as

$$C_\alpha(\varphi) = \sum_{n=0}^{\infty} \frac{E_\alpha^{(n)}\left(-\int d\mu(x)\right)}{n!} \left(\int e^{i\varphi(x)} d\mu(x)\right)^n$$
$$= \sum_{n=0}^{\infty} \frac{E_\alpha^{(n)}\left(-\int d\mu(x)\right)}{n!} \int e^{i(\varphi(x_1)+\varphi(x_2)+\cdots+\varphi(x_n))} d\mu^{\otimes n}$$

For the Poisson case ($\alpha = 1$) instead of $E_\alpha^{(n)}\left(-\int d\mu(x)\right)$ one would have exp $\left(-\int d\mu(x)\right)$ for all n, the rest being the same, cf. Appendix 1. Therefore one concludes that the main difference in the fractional case ($\alpha \neq 1$) is that a different weight is given to each n-particle space, but that a configuration space [12–14] is also the natural support of the fractional Poisson measure. The explicit construction is made below.

Notice however that the different weights, multiplying the n-particle space measures, are physically quite significant in that they have decays, for large volumes, much smaller than the corresponding exponential factor in the Poisson measure.

Using now the spectral representation (9) of the Mittag-Leffler function one may rewrite (8) as

$$C_\alpha(\varphi) = \int_0^\infty \exp\left(\tau \int (e^{i\varphi(x)} - 1)\, d\mu(x)\right) d\nu_\alpha(\tau)$$

with the integrand being the characteristic function of the Poisson measure $\pi_{\tau\mu}$, $\tau > 0$. In other words, the characteristic functional (8) coincides with the characteristic functional of the measure $\int_0^\infty \pi_{\tau\mu}\, d\nu_\alpha(\tau)$. By uniqueness, this implies the integral decomposition

$$\pi_\mu^\alpha = \int_0^\infty \pi_{\tau\mu}\, d\nu_\alpha(\tau)$$

meaning that π_μ^α is an integral (or mixture) of Poisson measures $\pi_{\tau\mu}$, $\tau > 0$. The measure $d\nu_\alpha(\tau)$ corresponds to the measure $d\mu(\overline{\rho})$ in Eq. (3), defining the particular reducible functional that characterizes the fractional Boson gas.

2.3 The Fractional Poisson Measure on Γ

The fractional Poisson measure has support in the configuration space Γ as constructed in Appendix 1.

Given a measure μ on the underlying measurable space $(M, \mathscr{B}(M))$ described as before, consider for each $n \in \mathbb{N}$ the product measure $\mu^{\otimes n}$ on $(M^n, \mathscr{B}(M^n))$. Since $\mu^{\otimes n}(M^n \backslash \widetilde{M^n}) = 0$, one may consider for each $\Lambda \in \mathscr{B}_c(M)$ the restriction of $\mu^{\otimes n}$ to $(\widetilde{\Lambda^n}, \mathscr{B}(\widetilde{\Lambda^n}))$, which is a finite measure, and then the image measure $\mu_\Lambda^{(n)}$ on $(\Gamma_\Lambda^{(n)}, \mathscr{B}(\Gamma_\Lambda^{(n)}))$ under the mapping sym_Λ^n,

$$\mu_\Lambda^{(n)} := \mu^{\otimes n} \circ (\mathrm{sym}_\Lambda^n)^{-1}$$

For $n = 0$ we set $\mu_\Lambda^{(0)} := 1$.[1] Now, for each $0 < \alpha < 1$ one may define a probability measure $\pi_{\mu,\Lambda}^\alpha$ on $(\Gamma_\Lambda, \mathscr{B}(\Gamma_\Lambda))$ by

$$\pi_{\mu,\Lambda}^\alpha := \sum_{n=0}^\infty \frac{E_\alpha^{(n)}(-\mu(\Lambda))}{n!} \mu_\Lambda^{(n)} \tag{11}$$

The family $\{\pi_{\mu,\Lambda}^\alpha : \Lambda \in \mathscr{B}_c(M)\}$ of probability measures yields a probability measure on $(\Gamma, \mathscr{B}(\Gamma))$. In fact, this family is consistent, that is,

$$\pi_{\mu,\Lambda_1}^\alpha = \pi_{\mu,\Lambda_2}^\alpha \circ p_{\Lambda_2,\Lambda_1}^{-1}, \quad \forall\, \Lambda_1, \Lambda_2 \in \mathscr{B}_c(M), \Lambda_1 \subset \Lambda_2$$

and thus, by the version of Kolmogorov's theorem for the projective limit space $(\Gamma, \mathscr{B}(\Gamma))$ [15, Chap. V Theorem 5.1], the family $\{\pi_{\mu,\Lambda}^\alpha : \Lambda \in \mathscr{B}_c(M)\}$ determines uniquely a measure π_μ^α on $(\Gamma, \mathscr{B}(\Gamma))$ such that

$$\pi_{\mu,\Lambda}^\alpha = \pi_\mu^\alpha \circ p_\Lambda^{-1}, \quad \forall\, \Lambda \in \mathscr{B}_c(M)$$

Let us now compute the characteristic functional of the measure $\pi_\mu \alpha$. Given a $\varphi \in \mathscr{D}(M)$ we have $\operatorname{supp}\varphi \subset \Lambda$ for some $\Lambda \in \mathscr{B}_c(M)$, meaning that

$$\langle \gamma, \varphi \rangle = \langle p_\Lambda(\gamma), \varphi \rangle, \quad \forall\, \gamma \in \Gamma$$

Thus

$$\int_\Gamma e^{i\langle \gamma,\varphi\rangle} d\pi_\mu^\alpha(\gamma) = \int_{\Gamma_\Lambda} e^{i\langle \gamma,\varphi\rangle} d\pi_{\mu,\Lambda}^\alpha(\gamma)$$

and the infinite divisibility (11) of the measure $\pi_{\mu,\Lambda}^\alpha$ yields for the right-hand side of the equality

$$\sum_{n=0}^\infty \frac{E_\alpha^{(n)}(-\mu(\Lambda))}{n!} \int_{\Lambda^n} e^{i(\varphi(x_1)+\ldots+\varphi(x_n))} d\mu^{\otimes n}(x) = \sum_{n=0}^\infty \frac{E_\alpha^{(n)}(-\mu(\Lambda))}{n!} \left(\int_\Lambda e^{i\varphi(x)} d\mu(x) \right)^n$$

which corresponds to the Taylor expansion of the function

[1] Of course this construction holds for any Borel set $Y \in \mathscr{B}(M)$. In this case, $\mu_Y^{(n)}(\Gamma_Y^{(n)}) < \infty$ provided $\mu(Y) < \infty$. For more details and proofs see e.g. [16, 17].

$$E_\alpha\left(\int_\Lambda (e^{i\varphi(x)}-1)\,d\mu(x)\right)=E_\alpha\left(\int_M (e^{i\varphi(x)}-1)\,d\mu(x)\right)$$

In other words, the characteristic functional of the measure π_μ^α coincides with the characteristic functional of the probability measure given by Theorem 1 through the Bochner-Minlos theorem.

Similarly to the $\alpha = 1$ case, this shows that the probability measure on $(\mathscr{D}'(M), \mathscr{C}_\sigma(\mathscr{D}'(M)))$ given by Theorem 1 is actually supported on generalized functions of the form $\sum_{x\in\gamma}\delta_x$, $\gamma \in \Gamma$. Thus, each fractional Poisson measure π_μ^α can either be consider on $(\Gamma, \mathscr{B}(\Gamma))$ or on $(\mathscr{D}', \mathscr{C}_\sigma(\mathscr{D}'(M)))$ where, in contrast to Γ, $\mathscr{D}'(M) \supset \Gamma$ is a linear space. Since $\pi_\mu^\alpha(\Gamma) = 1$, the measure space $(\mathscr{D}'(M), \mathscr{C}_\sigma(\mathscr{D}'(M)), \pi_\mu^\alpha)$ can, in this way, be regarded as a linear extension of the fractional Poisson space $(\Gamma, \mathscr{B}(\Gamma), \pi_\mu^\alpha)$.

2.4 *Fractional Poisson Analysis*

2.4.1 Fractional Lebesgue-Poisson Measure and Unitary Isomorphisms

Let us now consider the space of finite configurations

$$\Gamma_0 := \bigsqcup_{n=0}^{\infty} \Gamma_M^{(n)}$$

endowed with the topology of disjoint union of topological spaces, with the corresponding Borel σ-algebra $\mathscr{B}(\Gamma_0)$ and the so-called K-transform [16, 18–22], a mapping which maps functions defined on Γ_0 into functions defined on Γ. By definition, given a $\mathscr{B}(\Gamma_0)$-measurable function G with local support, that is, $G\restriction_{\Gamma_0\setminus\Gamma_\Lambda}\equiv 0$ for some $\Lambda \in \mathscr{B}_c(M)$, the K-transform of G is a mapping $KG : \Gamma \to \mathbb{R}$ defined at each $\gamma \in \Gamma$ by

$$(KG)(\gamma) := \sum_{\substack{\eta\subset\gamma\\ |\eta|<\infty}} G(\eta) \tag{12}$$

Note that for every such function G the sum in (12) has only a finite number of summands different from zero, and thus KG is a well-defined function on Γ. Moreover, if G has support described as before, then the restriction $(KG)\restriction_{\Gamma_\Lambda}$ is a $\mathscr{B}(\Gamma_\Lambda)$-measurable function and $(KG)(\gamma) = (KG)\restriction_{\Gamma_\Lambda}(\gamma_\Lambda)$ for all $\gamma \in \Gamma$.

In terms of the dual operator K^* of the K-transform, this means that the image of a probability measure on Γ under K^* yields a measure on Γ_0. More precisely, given a probability measure ν on $(\Gamma, \mathscr{B}(\Gamma))$ with finite local moments of all orders, that is,

$$\int_\Gamma |\gamma_\Lambda|^n\,d\nu(\gamma) < \infty \quad \text{for all } n \in \mathbb{N} \text{ and all } \Lambda \in \mathscr{B}_c(M)$$

then $K^*\nu$ is a measure on $(\Gamma_0, \mathscr{B}(\Gamma_0))$ defined on each bounded $\mathscr{B}(\Gamma_0)$-measurable set A by

$$(K^*\nu)(A) = \int_\Gamma (K 1\!\!1_A)(\gamma)\, d\nu(\gamma)$$

The measure $K^*\nu$ is called the correlation measure corresponding to ν. In particular, for the Poisson measure π_μ, the correlation measure corresponding to π_μ is called the Lebesgue-Poisson measure

$$\lambda_\mu := \sum_{n=0}^{\infty} \frac{1}{n!}\mu^{(n)}, \quad \mu^{(n)} := \mu^{\otimes n} \circ (\mathrm{sym}_M^n)^{-1}$$

For more details and proofs see e.g. [16].

Theorem 2 *For each* $0 < \alpha < 1$, *the correlation measure corresponding to the fractional Poisson measure* π_μ^α *is the measure on* $(\Gamma_0, \mathscr{B}(\Gamma_0))$ *given by*

$$\lambda_\mu^\alpha := \sum_{n=0}^{\infty} \frac{1}{\Gamma(\alpha n + 1)}\mu^{(n)} \tag{13}$$

In other words, $d\lambda_\mu^\alpha = E_\alpha^{(|\cdot|)}(0)\, d\lambda_\mu$.

In the sequel we call the measure λ_μ^α the *fractional Lebesgue-Poisson measure.*

Proof Let A be a bounded $\mathscr{B}(\Gamma_0)$-measurable set, that is,

$$A \subset \bigsqcup_{n=0}^{N} \Gamma_\Lambda^{(n)}$$

for some $N \in \mathbb{N}_0$ and some $\Lambda \in \mathscr{B}_c(M)$. By the previous considerations, this means that for all $\gamma \in \Gamma$ one has $(K 1\!\!1_A)(\gamma) = (K 1\!\!1_A)(\gamma_\Lambda)$, and thus

$$\begin{aligned}
\int_\Gamma (K 1\!\!1_A)(\gamma)\, d\pi_\mu^\alpha(\gamma) &= \int_{\Gamma_\Lambda} (K 1\!\!1_A)(\gamma)\, d\pi_{\mu,\Lambda}^\alpha(\gamma) \\
&= \sum_{n=0}^{\infty} \frac{E_\alpha^{(n)}(-\mu(\Lambda))}{n!} \int_{\Gamma_\Lambda^{(n)}} (K 1\!\!1_A)(\eta)\, d\mu_\Lambda^{(n)}(\eta) \\
&= \int_{\Gamma_\Lambda} E_\alpha^{(|\eta|)}(-\mu(\Lambda))(K 1\!\!1_A)(\eta)\, d\lambda_\mu(\eta)
\end{aligned}$$

Observe that the latter integral is with respect to the Lebesgue-Poisson measure λ_μ, which properties are well-known (see e.g. [16]). In particular, those yield

$$\int_{\Gamma_\Lambda} E_\alpha^{(|\eta|)}(-\mu(\Lambda))(K 1\!\!1_A)(\eta)\, d\lambda_\mu(\eta)$$
$$= \int_{\Gamma_0} E_\alpha^{(|\eta|)}(-\mu(\Lambda)) \sum_{\xi \subset \eta} 1\!\!1_A(\xi) 1\!\!1_{\Gamma_\Lambda}(\eta \setminus \xi)\, d\lambda_\mu(\eta)$$
$$= \int_{\Gamma_0} 1\!\!1_A(\eta) \left(\int_{\Gamma_0} E_\alpha^{(|\eta \cup \xi|)}(-\mu(\Lambda)) 1\!\!1_{\Gamma_\Lambda}(\xi)\, d\lambda_\mu(\xi) \right) d\lambda_\mu(\eta)$$

where for each $\eta \in \Gamma_0$ fixed, i.e., $\eta \in \Gamma_M^{(m)}$ for some $m \in \mathbb{N}_0$, the integral between brackets is given by

$$\sum_{n=0}^{\infty} \frac{1}{n!} \int_{\Gamma_M^{(n)}} E_\alpha^{(|\eta \cup \xi|)}(-\mu(\Lambda)) 1\!\!1_{\Gamma_\Lambda}(\xi)\, d\mu^{(n)}(\xi)$$
$$= \sum_{n=0}^{\infty} \frac{E_\alpha^{(m+n)}(-\mu(\Lambda))}{n!} (\mu(\Lambda))^n$$
$$= E_\alpha^{(m)}(-\mu(\Lambda) + \mu(\Lambda))$$

As a result,

$$\int_{\Gamma} (K 1\!\!1_A)(\gamma)\, d\pi_\mu^\alpha(\gamma) = \int_{\Gamma_0} 1\!\!1_A(\eta) E_\alpha^{(|\eta|)}(0)\, d\lambda_\mu(\eta)$$

showing that the correlation measure corresponding to π_μ^α is absolutely continuous with respect to the Lebesgue-Poisson measure λ_μ. Moreover, denoting such a correlation measure by λ_μ^α, the density is given by $\frac{d\lambda_\mu^\alpha}{d\lambda_\mu} = E_\alpha^{(|\cdot|)}(0)$.

To conclude, notice that for each $n \in \mathbb{N}_0$ one has

$$E_\alpha^{(n)}(0) = \frac{n!}{\Gamma(\alpha n + 1)}$$

which combined with the definition of the measure λ_μ leads to (13). ■

Throughout this work all L^p-spaces consist of complex-valued functions. For simplicity, the L^p-spaces with respect to a measure ν will be shortly denoted by $L^p(\nu)$ if the underlying measurable space is clear from the context.

Corollary 1 *We have $G \in L^2(\lambda_\mu^\alpha)$ if and only if $G\sqrt{E_\alpha^{(|\cdot|)}(0)} \in L^2(\lambda_\mu)$. Then,*

$$\|G\|_{L^2(\lambda_\mu^\alpha)} = \left\| G\sqrt{E_\alpha^{(|\cdot|)}(0)} \right\|_{L^2(\lambda_\mu)}$$

This result states that there is a unitary isomorphism between the spaces $L^2(\lambda_\mu^\alpha)$ and $L^2(\lambda_\mu)$:

$$I_\alpha : L^2(\lambda_\mu^\alpha) \to L^2(\lambda_\mu)$$
$$I_\alpha(G) := \quad G\sqrt{E_\alpha^{(|\cdot|)}(0)}$$

Hence, through I_α one may extend the unitary isomorphisms defined between the space $L^2(\lambda_\mu)$ and the (Bose or symmetric) Fock space $\mathrm{Exp}L^2(\mu)$ and between the space $L^2(\lambda_\mu)$ and $L^2(\pi_\mu)$ [17] to $L^2(\lambda_\mu^\alpha)$, $0 < \alpha \leq 1$:

$$\begin{array}{ccccccc} L^2(\lambda_\mu^\alpha) & \overset{I_\alpha}{\mapsto} & L^2(\lambda_\mu) & \overset{I_{\lambda\pi}}{\mapsto} & L^2(\pi_\mu) & \overset{I_\pi}{\mapsto} & \mathrm{Exp}L^2(\mu) \\ G & \mapsto & G\sqrt{E_\alpha^{(|\cdot|)}(0)} & \mapsto & \sum_{n=0}^\infty \langle C_n^\mu, g^{(n)}\rangle & \mapsto & \left(g^{(n)}\right)_{n=0}^\infty \end{array}$$

for

$$g^{(n)}(x_1,\ldots,x_n) := \frac{\sqrt{E_\alpha^{(n)}(0)}}{n!}G(\{x_1,\ldots,x_n\}), \quad g^{(0)} := E_\alpha(0)G(\emptyset) = G(\emptyset)$$

and C_n^μ a Charlier kernel.

In particular, the image of a Fock coherent state $e(f) := (\frac{f^{\otimes n}}{n!})_{n=0}^\infty, f \in L^2(\mu)$, under $(I_\pi \circ I_{\lambda\pi})^{-1}$ is the (Lebesgue-Poisson) coherent state $e_\lambda(f) : \Gamma_0 \to \mathbb{C}$ defined for any $\mathscr{B}(M)$-measurable function $f : M \to \mathbb{C}$ by

$$e_\lambda(f,\eta) := \prod_{x\in\eta} f(x), \quad \eta \in \Gamma_0 \setminus \{\emptyset\}, \quad e_\lambda(f,\emptyset) := 1$$

This definition implies that $e_\lambda(f) \in L^p(\lambda_\mu)$ whenever $f \in L^p(\mu)$ for some $p \geq 1$. Moreover, $\|e_\lambda(f)\|^p_{L^p(\lambda_\mu)} = \exp\left(\|f\|^p_{L^p(\mu)}\right)$. For $\alpha \neq 1$, the following result holds.

Proposition 1 *Let $0 < \alpha < 1$ and $p \geq 1$ be given. For all $f \in L^p(\mu)$ we have $e_\lambda(f) \in L^p(\lambda_\mu^\alpha)$ and*

$$\|e_\lambda(f)\|^p_{L^p(\lambda_\mu^\alpha)} = E_\alpha\left(\|f\|^p_{L^p(\mu)}\right)$$

Proof By Theorem 2, given a $f \in L^p(\mu)$ for some $p \geq 1$,

$$\begin{aligned} \|e_\lambda(f)\|^p_{L^p(\lambda_\mu^\alpha)} &= \int_{\Gamma_0} |e_\lambda(f,\eta)|^p E_\alpha^{(|\eta|)}(0)\, d\lambda_\mu(\eta) \\ &= \sum_{n=0}^\infty \frac{1}{\Gamma(\alpha n+1)}\left(\int_M |f(x)|^p\, d\mu(x)\right)^n \end{aligned}$$

which by the Taylor expansion (7) is equal to $E_\alpha\left(\int_M |f(x)|^p\, d\mu(x)\right)$. ∎

According to the latter considerations, the realization of a coherent state $e(f)$, $f \in L^2(\mu)$, in a $L^2(\lambda_\mu^\alpha)$ space is λ_μ-a.e. given by

$$I_\alpha^{-1} e_\lambda(f) = \frac{e_\lambda(f)}{\sqrt{E_\alpha^{(|\cdot|)}(0)}} \tag{14}$$

In addition, given a dense subspace $L \subseteq L^2(\mu)$, the set $\{I_\alpha^{-1} e_\lambda(f) : f \in L\}$ is total in $L^2(\lambda_\mu^\alpha)$. As in the Lebesgue-Poisson case, we define the fractional (Lebesgue-Poisson) coherent state $e_\alpha(f) : \Gamma_0 \to \mathbb{C}$ corresponding to a $\mathscr{B}(M)$-measurable function f by

$$e_\alpha(f, \eta) := \frac{e_\lambda(f, \eta)}{\sqrt{E_\alpha^{(|\eta|)}(0)}}, \quad \forall \eta \in \Gamma_0$$

2.4.2 Annihilation and Creation Operators

The unitary isomorphism between the Fock space and $L^2(\lambda_\mu)$ provides natural operators on the space $L^2(\lambda_\mu)$ by carrying over the standard Fock space operators. In particular, the annihilation and the creation operators, for which the images in $L^2(\lambda_\mu)$ are well-known, see e.g. [23],

$$\left(a_\lambda^-(\varphi)G\right)(\eta) := \int_M G(\eta \cup \{x\})\varphi(x)\, d\mu(x), \quad \eta \in \Gamma_0$$

and

$$\left(a_\lambda^+(\varphi)G\right)(\eta) := \sum_{x\in\eta} G(\eta\backslash\{x\})\varphi(x), \quad \lambda_\mu - a.a.\, \eta \in \Gamma_0$$

Here $\varphi \in \mathscr{D}(M)$ and G is a complex-valued bounded $\mathscr{B}(\Gamma_0)$-measurable function with bounded support, i.e., $G\restriction_{\Gamma_0\backslash\left(\bigsqcup_{n=0}^N \Gamma_\Lambda^{(n)}\right)} \equiv 0$ for some $\Lambda \in \mathscr{B}_c(M)$ and some $N \in \mathbb{N}_0$. In the sequel we denote the space of such functions G by $B_{bs}(\Gamma_0)$.

For more details and proofs see e.g. [17, 24] and the references therein.

Through the unitary isomorphism I_α^{-1}, $0 < \alpha < 1$, the same Fock space operators can naturally be carried over to the space $L^2(\lambda_\mu^\alpha)$.

Proposition 2 *For each $\varphi \in \mathscr{D}(M)$, the following relations hold on $B_{bs}(\Gamma_0)$:*

$$a_\alpha^-(\varphi) := I_\alpha^{-1} a_\lambda^-(\varphi) I_\alpha = \sqrt{\frac{E_\alpha^{(|\cdot|+1)}(0)}{E_\alpha^{(|\cdot|)}(0)}}\, a_\lambda^-(\varphi)$$

and

$$a_\alpha^+(\varphi) := I_\alpha^{-1} a_\lambda^+(\varphi) I_\alpha = \sqrt{\frac{E_\alpha^{(|\cdot|-1)}(0)}{E_\alpha^{(|\cdot|)}(0)}}\, a_\lambda^+(\varphi)$$

Proof One first observes that I_α maps the space $B_{bs}(\Gamma_0)$ into itself. In fact, given a $G \in B_{bs}(\Gamma_0)$, i.e., $G\restriction_{\Gamma_0 \setminus \left(\bigsqcup_{n=0}^{N} \Gamma_\Lambda^{(n)}\right)} \equiv 0$ for some $\Lambda \in \mathscr{B}_c(M)$ and some $N \in \mathbb{N}_0$, one has

$$|(I_\alpha G)(\eta)| = \sqrt{E_\alpha^{(|\eta|)}(0)}|G(\eta)| \leq \max_{0 \leq n \leq N} \sqrt{\frac{n!}{\Gamma(\alpha n + 1)}} \sup_{\eta \in \Gamma_0}(|G(\eta)|), \quad \forall \eta \in \Gamma_0$$

showing that $I_\alpha G$ is bounded. Since the support of $I_\alpha G$ clearly coincides with the support of G, this means that $I_\alpha G \in B_{bs}(\Gamma_0)$.

Hence, given a $G \in B_{bs}(\Gamma_0)$, for all $\eta \in \Gamma_0$ one has

$$\begin{aligned}(a_\lambda^-(\varphi)(I_\alpha G))(\eta) &= \int_M (I_\alpha G)(\eta \cup \{x\})\varphi(x)\, d\mu(x) \\ &= \sqrt{E_\alpha^{(|\eta|+1)}(0)}\, (a_\lambda^-(\varphi)G)(\eta)\end{aligned}$$

which proves the first equality by calculating the image of both sides under I_α^{-1}. A similar procedure applied to $a_\alpha^+(\varphi)$ completes the proof. ■

2.4.3 Second Quantization Operators

Given a contraction operator B on $L^2(\mu)$ one may define a contraction operator Exp$B$ on the Fock space Exp$L^2(\mu)$ acting on coherent states $e(f), f \in L^2(\mu)$, by Exp$B\,(e(f)) = e(Bf)$. In particular, given a positive self-adjoint operator A on $L^2(\mu)$ and the contraction semigroup e^{-tA}, $t \geq 0$, one can define a contraction semigroup Exp$\left(e^{-tA}\right)$ on Exp$L^2(\mu)$ in this way. The generator is the well-known second quantization operator corresponding to A . We denote it by dExp$A$. Through the unitary isomorphism between the Fock space and the space $L^2(\lambda_\mu)$ one may then define the corresponding operator in $L^2(\lambda_\mu)$. We denote the (Lebesgue-Poisson) second quantization operator corresponding to A by H_A^{LP}. The action of H_A^{LP} on coherent states is given by

$$\left(H_A^{LP} e_\lambda(f)\right)(\eta) = \sum_{x \in \eta} (Af)(x) e_\lambda(f, \eta \setminus \{x\}), \quad f \in D(A)$$

Through the unitary isomorphism I_α^{-1}, $0 < \alpha < 1$, the second quantization operator can also be carried over to the space $L^2(\lambda_\mu^\alpha)$:

$$H_A^\alpha := I_\alpha^{-1} H_A^{LP} I_\alpha$$

Proposition 3 *For any* $f \in D(A)$ *we have*

$$\left(H_A^\alpha e_\alpha(f)\right)(\eta) = \sqrt{\frac{E_\alpha^{(|\eta|-1)}(0)}{E_\alpha^{(|\eta|)}(0)}} \sum_{x\in\eta} (Af)(x) e_\alpha(f, \eta\backslash\{x\})$$

Proof According to (14), $I_\alpha e_\alpha(f) = e_\lambda(f)$, and thus for λ_μ-a.a. $\eta \in \Gamma_0$,

$$\left(H_A^{LP}(I_\alpha e_\alpha(f))\right)(\eta) = \left(H_A^{LP} e_\lambda(f)\right)(\eta) = \sqrt{E_\alpha^{(|\eta|-1)}(0)} \sum_{x\in\eta} (Af)(x) e_\alpha(f, \eta\backslash\{x\})$$

leading to the required result by calculating the image of both sides under I_α^{-1}. ■

Appendix 1: The Infinite-Dimensional Poisson Measure and Configuration Spaces

Here a short summary is given of the properties of the infinite-dimensional Poisson measure and its support on configuration spaces [12, 13, 16–19, 24].

An infinite-dimensional generalization of the characteristic function of the Poisson measure is obtained by generalizing (5) to

$$C(\varphi) = e^{\int \left(e^{i\varphi(x)}-1\right) d\mu(x)} \tag{15}$$

for test functions $\varphi \in \mathscr{D}(M)$ in the space of C^∞-functions of compact support in a manifold M. It is easy to prove, using the Bochner-Minlos theorem, that C is indeed the Fourier transform of a measure on the distribution space $\mathscr{D}'(M)$.

A support for this measure is obtained in the *configuration space* $\Gamma := \Gamma_M$ over the manifold M, defined as the set of all locally finite subsets of M (simple configurations)

$$\Gamma := \{\gamma \subset M : |\gamma \cap K| < \infty \text{ for any compact } K \subset M\} \tag{16}$$

Here $|A|$ denotes the cardinality of the set A. As usual one identifies each $\gamma \in \Gamma$ with a non-negative integer-valued Radon measure,

$$\Gamma \ni \gamma \mapsto \sum_{x\in\gamma} \delta_x \in \mathscr{M}(M)$$

where δ_x is the Dirac measure with unit mass at x and $\mathscr{M}(M)$ denotes the set of all non-negative Radon measures on M. In this way the space Γ can be endowed with the relative topology as a subset of the space $\mathscr{M}(M)$ with the vague topology, i.e., the weakest topology on Γ for which the mappings

$$\Gamma \ni \gamma \mapsto \langle \gamma, f \rangle := \int_M f(x) d\gamma(x) = \sum_{x \in \gamma} f(x)$$

are continuous for all real-valued continuous functions f on M with compact support. Denote the corresponding Borel σ-algebra on Γ by $\mathcal{B}(\Gamma)$.

For each $Y \in \mathcal{B}(M)$ let us consider the space Γ_Y of all configurations contained in Y, $\Gamma_Y := \{\gamma \in \Gamma : |\gamma \cap (X \backslash Y)| = 0\}$, and the space $\Gamma_Y^{(n)}$ of n-point configurations,

$$\Gamma_Y^{(n)} := \{\gamma \in \Gamma_Y : |\gamma| = n\}, n \in \mathbb{N}, \quad \Gamma_Y^{(0)} := \{\emptyset\}$$

A topological structure may be introduced on $\Gamma_Y^{(n)}$ through the natural surjective mapping of $\widetilde{Y^n} := \left\{(x_1, ..., x_n) : x_i \in Y, x_i \neq x_j \text{ if } i \neq j\right\}$ onto $\Gamma_Y^{(n)}$,

$$\begin{aligned} \text{sym}_Y^n : \widetilde{Y^n} &\longrightarrow \Gamma_Y^{(n)} \\ (x_1, ..., x_n) &\longmapsto \{x_1, ..., x_n\} \end{aligned}$$

which is at the origin of a bijection between $\Gamma_Y^{(n)}$ and the symmetrization $\widetilde{Y^n}/S_n$ of $\widetilde{Y^n}$, S_n being the permutation group over $\{1, ..., n\}$. Thus, sym_Y^n induces a metric on $\Gamma_Y^{(n)}$ and the corresponding Borel σ-algebra $\mathcal{B}\left(\Gamma_Y^{(n)}\right)$ on $\Gamma_Y^{(n)}$.

For $\Lambda \in \mathcal{B}(M)$ with compact closure ($\Lambda \in \mathcal{B}_c(M)$), it clearly follows from (16) that

$$\Gamma_\Lambda = \bigsqcup_{n=0}^{\infty} \Gamma_\Lambda^{(n)}$$

the σ-algebra $\mathcal{B}(\Gamma_\Lambda)$ being defined by the disjoint union of the σ-algebras $\mathcal{B}\left(\Gamma_Y^{(n)}\right)$, $n \in \mathbb{N}_0$.

For each $\Lambda \in \mathcal{B}_c(M)$ there is a natural measurable mapping $p_\Lambda : \Gamma \to \Gamma_\Lambda$. Similarly, given any pair $\Lambda_1, \Lambda_2 \in \mathcal{B}_c(M)$ with $\Lambda_1 \subset \Lambda_2$ there is a natural mapping $p_{\Lambda_2,\Lambda_1} : \Gamma_{\Lambda_2} \to \Gamma_{\Lambda_1}$. They are defined, respectively, by

$$\begin{aligned} p_\Lambda : \Gamma &\longrightarrow \Gamma_\Lambda \\ \gamma &\longmapsto \gamma_\Lambda := \gamma \cap \Lambda \end{aligned} \qquad \begin{aligned} p_{\Lambda_2,\Lambda_1} : \Gamma_{\Lambda_2} &\longrightarrow \Gamma_{\Lambda_1} \\ \gamma &\longmapsto \gamma_{\Lambda_1} \end{aligned}$$

It can be shown that $(\Gamma, \mathcal{B}(\Gamma))$ coincides (up to an isomorphism) with the projective limit of the measurable spaces $(\Gamma_\Lambda, \mathcal{B}(\Gamma_\Lambda))$, $\Lambda \in \mathcal{B}_c(M)$, with respect to the projection p_Λ, i.e., $\mathcal{B}(\Gamma)$ is the smallest σ-algebra on Γ with respect to which all projections p_Λ, $\Lambda \in \mathcal{B}_c(M)$, are measurable.

Let now μ be a measure on the underlying measurable space $(M, \mathcal{B}(M))$ and consider for each $n \in \mathbb{N}$ the product measure $\mu^{\otimes n}$ on $(M^n, \mathcal{B}(M^n))$. Since $\mu^{\otimes n}(M^n \backslash \widetilde{M^n}) = 0$, one may consider for each $\Lambda \in \mathcal{B}_c(M)$ the restriction of $\mu^{\otimes n}$ to $(\widetilde{\Lambda^n}, \mathcal{B}(\widetilde{\Lambda^n}))$, which is a finite measure, and then the image measure $\mu_\Lambda^{(n)}$ on $(\Gamma_\Lambda^{(n)}, \mathcal{B}(\Gamma_\Lambda^{(n)}))$ under the mapping sym_Λ^n,

$$\mu_\Lambda^{(n)} := \mu^{\otimes n} \circ (\mathrm{sym}_\Lambda^n)^{-1}$$

For $n = 0$ we set $\mu_\Lambda^{(0)} := 1$. Now, one may define a probability measure $\pi_{\mu,\Lambda}$ on $(\Gamma_\Lambda, \mathscr{B}(\Gamma_\Lambda))$ by

$$\pi_{\mu,\Lambda} := \sum_{n=0}^{\infty} \frac{\exp(-\mu(\Lambda))}{n!} \mu_\Lambda^{(n)} \tag{17}$$

The family $\{\pi_{\mu,\Lambda} : \Lambda \in \mathscr{B}_c(M)\}$ of probability measures yields a probability measure on $(\Gamma, \mathscr{B}(\Gamma))$ with the $\pi_{\mu,\Lambda}$ as projections. This family is consistent, that is,

$$\pi_{\mu,\Lambda_1} = \pi_{\mu,\Lambda_2} \circ p_{\Lambda_2,\Lambda_1}^{-1}, \quad \forall \Lambda_1, \Lambda_2 \in \mathscr{B}_c(M), \Lambda_1 \subset \Lambda_2$$

and thus, by the version of Kolmogorov's theorem for the projective limit space $(\Gamma, \mathscr{B}(\Gamma))$, the family $\{\pi_{\mu,\Lambda} : \Lambda \in \mathscr{B}_c(M)\}$ determines uniquely a measure π_μ on $(\Gamma, \mathscr{B}(\Gamma))$ such that

$$\pi_{\mu,\Lambda} = \pi_\mu \circ p_\Lambda^{-1}, \quad \forall \Lambda \in \mathscr{B}_c(M)$$

The next step is to compute the characteristic functional of the measure π_μ. Given a $\varphi \in \mathscr{D}(M)$ we have $\operatorname{supp}\varphi \subset \Lambda$ for some $\Lambda \in \mathscr{B}_c(M)$, meaning that

$$\langle \gamma, \varphi \rangle = \langle p_\Lambda(\gamma), \varphi \rangle, \quad \forall \gamma \in \Gamma$$

Thus

$$\int_\Gamma e^{i\langle \gamma,\varphi\rangle} d\pi_\mu(\gamma) = \int_{\Gamma_\Lambda} e^{i\langle \gamma,\varphi\rangle} d\pi_{\mu,\Lambda}(\gamma)$$

and the definition (17) of the measure $\pi_{\mu,\Lambda}$ yields for the right-hand side of the equality

$$\sum_{n=0}^{\infty} \frac{\exp(-\mu(\Lambda))}{n!} \int_{\Lambda^n} e^{i(\varphi(x_1)+\cdots+\varphi(x_n))} d\mu^{\otimes n}(x) = \sum_{n=0}^{\infty} \frac{\exp(-\mu(\Lambda))}{n!} \left(\int_\Lambda e^{i\varphi(x)} d\mu(x) \right)^n$$

which corresponds to the Taylor expansion of the characteristic function (15) of the infinite-dimensional Poisson measure

$$\exp\left(\int_\Lambda (e^{i\varphi(x)} - 1)\, d\mu(x) \right)$$

This shows that the probability measure on $(\mathscr{D}'(M), \mathscr{C}_\sigma(\mathscr{D}'(M)))$ given by (15) is actually supported on generalized functions of the form $\sum_{x\in\gamma} \delta_x$, $\gamma \in \Gamma$. Thus, the infinite-dimensional Poisson measure π_μ can either be considered as a measure on $(\Gamma, \mathscr{B}(\Gamma))$ or on $(\mathscr{D}', \mathscr{C}_\sigma(\mathscr{D}'(M)))$. Notice that, in contrast to Γ, $\mathscr{D}'(M) \supset \Gamma$ is a linear space. Since $\pi_\mu(\Gamma) = 1$, the measure space $(\mathcal{D}'(M), \mathcal{C}_\sigma(\mathcal{D}'(M)), \pi_\mu)$ can, in this way, be regarded as a linear extension of the Poisson space $(\Gamma, \mathcal{B}(\Gamma), \pi_\mu)$.

Appendix 2. Complete Monotonicity of the Mittag-Leffler Function for Complex Arguments

A positive C^∞-function f is said to be completely monotone if for each $k \in \mathbb{N}_0$

$$(-1)^k f^{(k)}(t) \geq 0, \quad \forall t > 0$$

According to Bernstein's theorem (see e.g. [25, Chap. XIII.4 Theorem 1]), for functions f such that $f(0^+) = 1$ the complete monotonicity property is equivalent to the existence of a probability measure ν on $\mathbb{R}_0^+$ such that

$$f(t) = \int_0^\infty e^{-t\tau}\, d\nu(\tau) < \infty, \quad \forall t > 0$$

Pollard in [26] proved the complete monotonicity of $E_\alpha, 0 < \alpha < 1$, for non-positive real arguments showing that

$$E_\alpha(-t) = \int_0^\infty e^{-t\tau}\, d\nu_\alpha(\tau), \quad \forall t \geq 0 \tag{18}$$

for ν_α being the probability measure on $\mathbb{R}_0^+$

$$d\nu_\alpha(\tau) := \alpha^{-1}\tau^{-1-1/\alpha} f_\alpha(\tau^{-1/\alpha})\, d\tau \tag{19}$$

where f_α is the α-stable probability density given by

$$\int_0^\infty e^{-t\tau} f_\alpha(\tau)\, d\tau = e^{-t^\alpha}, \quad 0 < \alpha < 1$$

The complete monotonicity property and the integral representation (18) of E_α may be extended to complex arguments.

Lemma 1 *For any $z \in \mathbb{C}$ such that* $\mathrm{Re}(z) \geq 0$, *the following representation holds*

$$E_\alpha(-z) = \int_0^\infty e^{-z\tau}\, d\nu_\alpha(\tau), \quad 0 < \alpha \leq 1$$

Proof According to [26], for each $0 < \alpha < 1$ fixed, for all $t \geq 0$ one has

$$
\begin{aligned}
E_\alpha(-t) &= \int_0^\infty e^{-t\tau}\, d\nu_\alpha(\tau), \\
&= \sum_{n=0}^{\infty} \frac{(-t)^n}{n!} \int_0^\infty \tau^n\, d\nu_\alpha(\tau)
\end{aligned}
\tag{20}
$$

Comparing (20) with the Taylor expansion (7) of E_α, one concludes that the moments of the measure ν_α are given by

$$
m_n(\nu_\alpha) := \int_0^\infty \tau^n\, d\nu_\alpha(\tau) = \frac{n!}{\Gamma(\alpha n + 1)}, \quad n \in \mathbb{N}_0
$$

For complex values z let

$$
I(-z) := \int_0^\infty e^{-z\tau}\, d\nu_\alpha(\tau)
$$

which is finite provided $\mathrm{Re}(z) \geq 0$. For each $z \in \mathbb{C}$ such that $\mathrm{Re}(z) \geq 0$ one then obtains

$$
I(-z) = \sum_{n=0}^{\infty} \frac{(-z)^n}{n!} \left(\int_0^\infty \tau^n\, d\nu_\alpha(\tau) \right) = \sum_{n=0}^{\infty} \frac{(-z)^n}{n!} m_n(\nu_\alpha) = \sum_{n=0}^{\infty} \frac{(-z)^n}{\Gamma(\alpha n + 1)} = E_\alpha(-z)
$$

leading to the integral representation

$$
E_\alpha(-z) = \int_0^\infty e^{-z\tau} d\nu_\alpha(\tau)
$$

for all $z \in \mathbb{C}$ such that $\mathrm{Re}(z) \geq 0$. ■

References

1. Goldin, G.A.: Nonrelativistic current algebras as unitary representations of groups. J. Math. Phys. **12**, 462–487 (1971)
2. Goldin, G.A., Grodnik, J., Powers, R.T., Sharp, D.H.: Nonrelativistic current algebra in the N/V limit. J. Math. Phys. **15**, 88–100 (1974)
3. Goldin, G.A.: Lectures on diffeomorphism groups in quantum physics. In: Govaerts, J., Hounkonnou, M.N., Msezane, A.Z. (eds.) Contemporary Problems in Mathematical Physics, Proceedings of the Third International Workshop, pp. 3–93. World Scientific, Hackensack, NJ, USA (2004)
4. Goldin, G.A.: Local current algebras: early ideas and new quantum-theoretical directions. Bulg. J. Phys. **33**, 81–97 (2006)
5. Haubold, H.J., Mathai, A.M., Saxena, R.K.: Mittag-Leffler functions and their applications. J. Appl. Math. 2011, ID 298628 (2011)
6. Oliveira, M.J., Ouerdiane, H., Silva, J.L., Mendes, R.V.: The fractional Poisson measure in infinite dimensions. arXiv:1002.2124

7. Sixdeniers, J.M., Penson, K.A., Solomon, A.I.: Mittag-Leffler coherent states. J. Physics A: Math. Gen. **32**, 7543–7563 (1999)
8. Sirin, H., Ertik, H., Büyükkılıç, F., Demirhan, D.: Investigation of Bose-Einstein condensation based on fractality using fractional mathematics, J. Stat. Mech.: Theor. Exp. P10022 (2010)
9. Cipriano, F., Ouerdiane, H., Vilela, R.: Stochastic solution of a KPP-type nonlinear fractional differential equation. Fract. Calcul. Appl. Anal. **12**, 47–57 (2009)
10. Mainardi, F., Gorenflo, R., Scalas, E.: A fractional generalization of the Poisson process. Vietnam J. Math. **32**, 53–64 (2004)
11. Gelfand, I.M., Ya. Vilenkin, N.: Generalized Functions, vol. IV, Academic Press, New York, London (1968)
12. Albeverio, S., Kondratiev, Y., Röckner, M.: Analysis and geometry on configuration spaces. J. Funct. Anal. **154**, 444–500 (1998)
13. Albeverio, S., Kondratiev, Y., Röckner, M.: Analysis and geometry on configuration spaces: the Gibbsian case. J. Funct. Anal. **157**, 242–291 (1998)
14. Shimomura, H.: Poisson measures on the configuration space and unitary representations of the group of diffeomorphisms. J. Math. Kyoto Univ. **34**, 599–614 (1994)
15. Parthasarathy, K.R.: Probability Measures on Metric Spaces. Academic Press, New York (1967)
16. Kondratiev, Y., Kuna, T.: Harmonic analysis on configuration space I. General theory. Infin. Dimens. Anal. Quantum Probab. Relat. Top. **5**, 201–233 (2002)
17. Kondratiev, Y., Kuna, T., Oliveira, M.J.: Analytic aspects of Poissonian white noise analysis. Methods Funct. Anal. Topology **8**, 15–48 (2002)
18. Kondratiev, Y., Kuna, T., Oliveira, M.J.: On the relations between Poissonian white noise analysis and harmonic analysis on configuration spaces. J. Funct. Anal. **213**, 1–30 (2004)
19. T. Kuna; Studies in Configuration Space Analysis and Applications, PhD thesis, Bonner Mathematische Schriften Nr. 324, University of Bonn (1999)
20. Lenard, A.: Correlation functions and the uniqueness of the state in classical statistical mechanics. Comm. Math. Phys. **30**, 35–44 (1973)
21. Lenard, A.: States of classical statistical mechanical systems of infinitely many particles I. Arch. Rational Mech. Anal. **59**, 219–239 (1975)
22. Lenard, A.: States of classical statistical mechanical systems of infinitely many particles II. Arch. Rational Mech. Anal. **59**, 241–256 (1975)
23. Fichtner, K.-H., Winkler, G.: Generalized Brownian motion, point processes and stochastic calculus for random fields. Math. Nachr. **161**, 291–307 (1993)
24. M. J. Oliveira; Configuration Space Analysis and Poissonian White Noise Analysis, Ph. D. Thesis, Lisbon (2002). http://www.math.uni-bielefeld.de/igk/study-materials/corpo_2
25. Feller, W.: An introduction to probability theory and its applications, vol. John Wiley & Sons, New York, II, Second Edition (1971)
26. Pollard, H.: The complete monotonic character of the Mittag-Leffler function $E_\alpha(-x)$. Bull. Amer. Math. Soc. **54**, 1115–1116 (1948)

Dynamical Properties of a Cosmological Model with Diffusion

M.P. Ramos and A.J. Soares

Abstract The description of the dynamics of particles undergoing diffusion in general relativity has been an object of interest in the last years. Most recently a new cosmological model with diffusion has been studied in which the evolution of the particle system is described by a Fokker-Planck equation. This equation is then coupled to a modified system of Einstein equations, in order to satisfy the energy conservation condition. Continuing with this work, we study in the present paper a spatially homogeneous and isotropic spacetime model with diffusion velocity. We write the system of ordinary differential equations of this particular model and obtain the solutions for which the scale factor in the Robertson Walker metric is linear in time. We analyse the asymptotic behavior of the subclass of spatially flat solutions. The system representing the homogeneous and isotropic model with diffusion is rewritten using dynamical variables. For the subclass of spatially flat solutions we were able to determine all equilibrium points and analyse their local stability properties.

Keywords Cosmology · Diffusion · Robertson Walker metric · Dynamical systems

1 Introduction

A new model to describe the dynamics of particles undergoing diffusion in general relativity is given in [1]. In this model the evolution of the particle system is described by a Fokker-Planck equation without friction on the tangent bundle of the spacetime (M, g). In general relativity, the matter field is specified by the energy-momentum tensor $\mathbf{T}_{\mu\nu}$ and the geometry of the spacetime is given by the Einstein field equations,

M.P. Ramos
Department of Mathematics and Applications, University of Minho,
Guimarães, Portugal
e-mail: mpr@math.uminho.pt

A.J. Soares (✉)
Centre of Mathematics, University of Minho, Braga, Portugal
e-mail: ajsoares@math.uminho.pt

P. Gonçalves and A.J. Soares (eds.), *From Particle Systems to Partial Differential Equations II*, Springer Proceedings in Mathematics & Statistics 129,
DOI 10.1007/978-3-319-16637-7_12

$$R_{\mu\nu} - \frac{1}{2} g_{\mu\nu} R = \mathbf{T}_{\mu\nu}, \tag{1}$$

where $R_{\mu\nu}$ and R are the Ricci tensor and Ricci scalar respectively. In kinetic theory, the matter field is described by the one-particle distribution function f in the phase space and the energy momentum tensor is given by a suitable integral of f over the velocity (or momentum) variable. When considering the existence of diffusion, f is the solution of the Fokker-Planck equation. On the other hand, a matter source with diffusion cannot be the only contributor to the Einstein equations (1), because the kinetic energy of the particles is not preserved when undergoing diffusion. Consequently, the energy-momentum tensor will not satisfy the energy conservation condition $\nabla_\mu \mathbf{T}^{\mu\nu} = 0$ (∇_μ denoting the Levi Civita covariant derivative), which is a requirement of the Bianchi identity and Einstein equations (1). In order to overcome this difficulty, a term in the cosmological scalar field ϕ is added on the left hand side of (1) and the corresponding energy-momentum tensor $T_{\mu\nu}$ satisfies the modified Einstein's equations

$$R_{\mu\nu} - \frac{1}{2} g_{\mu\nu} R + \phi g_{\mu\nu} = T_{\mu\nu}, \tag{2}$$

The scalar field ϕ is not an ordinary matter field but rather a background field interacting with the fluid particles and therefore causing their diffusion. It is determined by the one particle distribution function f, via the equation $\nabla_\nu \phi = \nabla^\mu T_{\mu\nu}$. In doing so, the coupling of Fokker-Planck equation to the Einstein's equations respects the Bianchi identity $\nabla^\mu (R_{\mu\nu} - \frac{1}{2} g_{\mu\nu} R) = 0$ and the energy momentum tensor satisfies the energy conservation condition $\nabla_\mu T^{\mu\nu} = 0$.
It is shown in [2] that the evolution of the particle system described by a Fokker-Planck equation without friction leads to an energy-momentum tensor satisfying

$$T^{\mu\nu} = \rho u^\mu u^\nu + p(g^{\mu\nu} + u^\mu u^\nu), \quad J^\mu = n u^\mu, \tag{3}$$

$$\nabla_\mu T^{\mu\nu} = \sigma n u^\nu, \quad \nabla_\mu (n u^\mu) = 0. \tag{4}$$

The scalar functions ρ, p, n, σ represent respectively the rest-frame energy density, the pressure, the number density of the fluid and the diffusion constant, while the vector u^μ satisfying the condition $u^\mu u_\mu = -1$ is the four-velocity of the fluid. Equations in (3) are projected along the direction of u^μ and onto a hypersurface orthogonal to the direction of u^μ, giving

$$\nabla_\mu (\rho u^\mu) + p \nabla_\mu u^\mu = \sigma n, \tag{5}$$

$$(\rho + p) u^\mu \nabla_\mu u^\nu + u^\nu u^\mu \nabla_\mu p + \nabla^\nu p = 0, \tag{6}$$

$$\nabla_\mu (n u^\mu) = 0. \tag{7}$$

The Euler equation (6) is the same as in the diffusion-free case. The continuity equation (5) is affected by the presence of diffusion, due to the fact that the diffusion force $\sigma n u^\mu$ acts on the direction of the matter flow.

In the model studied in [2], the pressure and the energy density satisfy the following linear relation,

$$p = (\gamma - 1)\rho, \tag{8}$$

where $\frac{2}{3} \leq \gamma < 2$. In order to transform the system (5)–(6) into a complete system we introduce the following equation of state,

$$\rho = n^{\gamma} S, \tag{9}$$

where the entropy S satisfies the condition

$$u^{\mu} \nabla_{\mu} S = \sigma n^{1-\gamma}. \tag{10}$$

Equations (5), (6), (8), (9) and (10) constitute a complete system for describing the evolution of the matter field variables $(n,\ S,\ u^{\mu})$. These equations coupled to the modified Einstein equations (1) are the equations of the cosmological model with fluid matter undergoing velocity diffusion given in [2].
In this work we study the solutions to the cosmological model described above that represent a spatially homogeneous and isotropic spacetime. Continuing with the work done in [2], in Sect. 2 we write the ODE system that describes these solutions and we solve this system explicitly for the case where the scale factor $a(t)$ in the metric is linear in time, for all values of the spatial curvature and all values of γ. In Sect. 3 we study the asymptotic behavior of the subclass of spatially flat solutions of the model, we find conditions on the initial data for which singularities may or may not occur. A dynamical system formulation for the model is given in Sect. 4. In particular we write the dynamical system representing the spatially flat solutions of this model in this section. The dynamical system in question has two ordinary differential equations. We determine all the fixed points of this system and show that the interior point is associated to the self similar solution given in Sect. 2. We obtain the phase portrait of the two dimensional system and study the local stability of the fixed points on the boundary. Finally, in Sect. 5 we state the conclusions and make some closing remarks. For sake of completeness, we include an appendix with some tools from dynamical systems theory that are used in the analysis developed in Sect. 4 of the present paper.

2 Spatially Homogeneous and Isotropic Solutions

In this section we present the model equations and we solve them explicitly when the scale factor is linear in time. We consider the Robertson-Walker metric [3, 4],

$$ds^2 = -dt^2 + a(t)^2 \left[\frac{dr^2}{1 - kr^2} + r^2 d\Omega^2 \right], \tag{11}$$

where $k = 0$ corresponds to spatially flat solutions, $k = -1$ corresponds to a space with negative spatial curvature, and $k = 1$ corresponds to a space with positive spatial curvature. In the case $k = 0$, we may introduce a cartesian system of coordinates such that

$$ds^2 = -dt^2 + a(t)^2(dx^2 + dy^2 + dz^2), \quad \text{and} \quad a_0 := a(0) = 1.$$

For any $k = 0, \pm 1$, the equations for the scale factor $a(t)$, the entropy $S(t)$ of the fluid and the cosmological scalar field $\phi(t)$ are

$$\dot{a} = Ha, \tag{12a}$$

$$\dot{S} = \sigma n_0^{1-\gamma} \left(\frac{a_0}{a(t)}\right)^{3-3\gamma}, \tag{12b}$$

$$\dot{\phi} = -\sigma n_0 \left(\frac{a_0}{a(t)}\right)^3, \tag{12c}$$

$$\dot{H} = \frac{1}{3}\left[\phi - \left(\frac{3}{2}\gamma - 1\right)\rho\right] - H^2, \tag{12d}$$

where

$$H^2 = \frac{1}{3}(\rho + \phi) - \frac{k}{a(t)^2} \tag{12e}$$

is the Hubble function, and

$$\rho(t) = \left(\frac{n_0 a_0^3}{a(t)^3}\right)^\gamma S(t) \tag{12f}$$

is the rest energy density of the fluid. Moreover, γ is the parameter of the equation of state, $\sigma > 0$ is the diffusion constant, and $n_0 > 0$ is the initial (at time $t = 0$) particle density. The initial data set consists of (a_0, H_0, S_0, ϕ_0), where $a_0 = 1$ for $k = 0$, and H_0, S_0, ϕ_0 are *positive* numbers such that (12e) is satisfied at time $t = 0$, i.e.,

$$H_0^2 = \frac{1}{3}(n_0^\gamma S_0 + \phi_0) - \frac{k}{a_0^2}. \tag{13}$$

Using (12e) we may rewrite (12d) in the following two forms

$$\dot{H} = -\frac{\gamma}{2}\rho + \frac{k}{a^2}, \tag{14}$$

$$\dot{H} = \frac{\gamma}{2}\phi - \frac{3\gamma}{2}H^2 - \frac{3\gamma - 2}{2a^2}k. \tag{15}$$

Equations (12) constitute the model equations studied in the present paper. In general, such equations can not be solved explicitly. However, in the particular case that the scale factor $a(t)$ is linear in time, we find the following explicit solution

$$a(t) = a_0 + \alpha_k t, \tag{16a}$$

$$\phi(t) = \frac{\sigma n_0 a_0^3}{2\alpha_k} a(t)^{-2}, \tag{16b}$$

$$S(t) = \left(\frac{\sigma n_0^{1-\gamma} a_0^{3-3\gamma}}{\alpha_k (3\gamma - 2)} \right) a(t)^{3\gamma-2}, \tag{16c}$$

where α_k is the real solution of the polynomial equation

$$\alpha^3 + k\alpha - \frac{\gamma \sigma n_0 a_0^3}{2(3\gamma - 2)} = 0.$$

Note that $\alpha_k > 0$, for all $k = 0, \pm 1$. In particular, for $k = 0$ the solution (16) becomes

$$a(t) = 1 + \left(\frac{\gamma \sigma n_0}{2(3\gamma - 2)} \right)^{1/3} t, \tag{17a}$$

$$\phi(t) = \left(\sqrt{\frac{3\gamma - 2}{\gamma}} \frac{\sigma n_0}{2} \right)^{2/3} a(t)^{-2}, \tag{17b}$$

$$S(t) = \left(\frac{2\sigma^2 n_0^{2-3\gamma}}{\gamma (3\gamma - 2)^2} \right)^{1/3} a(t)^{3\gamma-2}. \tag{17c}$$

3 Asymptotic Behavior of Spatially Flat Solutions

In this section we consider the model equations (12) written in the previous section, and we analyse the asymptotic behavior of the subclass of spatially flat solutions, corresponding to the case $k = 0$. Since $a(0) = 1$, the scale factor is positive in some maximal interval $[0, T)$ and, by (14), H is strictly decreasing on $[0, T)$ (which implies that $a(t)$ cannot blow up in finite time). If $T = +\infty$, the solution is singularity free in the future; if $T < \infty$, i.e., $a(T) = 0$, the solution is singular at the time $t = T$. In this section we look for sufficient conditions on the initial data for which one of the two possibilities (existence or absence of a future singularity) may occur.

First, we observe that ϕ is strictly decreasing on $[0, T)$. If $\phi(t)$ is positive on $[0, T)$, then by (12e) H cannot vanish in this interval. Whence H is positive, i.e., $a(t)$ is increasing, on $[0, T)$ and therefore $a(T) > 0$. Therefore a singularity cannot form in the interval of time in which ϕ remains positive. In contrast, if ϕ vanishes at some time, then a singularity will form. In fact, by (15) we infer that if there exists $\bar{t}$ such that $\phi(\bar{t}) < 0$, then $\dot{H} \leq \gamma/2\phi(\bar{t}) - 3\gamma H^2/2$, whence there exists $\bar{t} < t_* < +\infty$ such that $H \to -\infty$ as $t \to t_*$ (which implies $\lim_{t\to t_*} a(t) = 0$).

Let $\bar{t}$ be the maximal time such that $\phi(t) > 0$ for $t \in [0, \bar{t})$. If $\bar{t} < +\infty$, then $\phi(\bar{t}) = 0$ must hold. By (15), $\dot{H} \geq -3\gamma H^2/2$, for all $t \in [0, \bar{t})$. Since $H > 0$ in this interval, we obtain

$$H(t) \geq \frac{H_0}{1 + \frac{3}{2}\gamma H_0 t}, \quad t \in [0, \bar{t}),$$

whence, for $t \in [0, \bar{t})$,

$$a(t) \geq (1 + \frac{3}{2}\gamma H_0 t)^{\frac{2}{3\gamma}}. \tag{18}$$

On the other hand, since $H(t) \leq H_0$, we have

$$a(t) \leq \exp(H_0 t). \tag{19}$$

By the previous discussion, in order to see whether or not a singularity is formed in the future, we may equivalently check whether ϕ vanishes or not in finite time.

Proposition 1 *For*

$$\phi_0 \geq \frac{2\sigma n_0}{3H_0(2-\gamma)}, \tag{20}$$

there is no future singularity. On the other hand, for

$$\phi_0 < \frac{\sigma n_0}{3H_0} \tag{21}$$

a singularity forms in finite time in the future.

Proof Considering (18) then by (12c) we obtain

$$\begin{aligned}
\phi(t) &= \phi_0 - \sigma n_0 \int_0^t \frac{ds}{a(s)^3} \geq \phi_0 - \sigma n_0 \int_0^t \frac{ds}{(1 + \frac{3}{2}\gamma H_0 s)^{\frac{2}{\gamma}}} \\
&= \phi_0 + \frac{2\sigma n_0}{3H_0(2-\gamma)} \left[(1 + \frac{3}{2}\gamma H_0 t)^{1-\frac{2}{\gamma}} - 1 \right] \\
&\geq \phi_0 + \frac{2\sigma n_0}{3H_0(2-\gamma)} \left[(1 + \frac{3}{4}\gamma H_0 \bar{t})^{1-\frac{2}{\gamma}} - 1 \right], \quad t \in \left(\frac{\bar{t}}{2}, \bar{t} \right).
\end{aligned}$$

It follows that when ϕ_0 satisfies (20), $\phi(t)$ is uniformly strictly positive on $[0, \bar{t})$ and therefore it cannot vanish at $t = \bar{t}$. Thus $\bar{t}$ cannot be finite, therefore no future singularity is formed. Suppose now that $\phi(t) \geq 0$ for all $t \geq 0$. Then, by (19), we find that

$$\phi(t) \leq \phi_0 - \sigma n_0 \int_0^t e^{-3H_0 s} ds = \phi_0 + \frac{\sigma n_0}{3H_0}(\exp(-3H_0 t) - 1).$$

If (21) holds, then $\phi(t) \to \phi_\infty < 0$ as $t \to +\infty$, which is a contradiction. Whence a singularity must form in finite time. □

Note that by using (13), we may rewrite (20) as

$$S_0 \geq \frac{4\sigma^2 n_0^{2-\gamma}}{3(2-\gamma)^2}\phi_0^{-2} - \frac{\phi_0}{n_0^\gamma} \tag{22}$$

which is always satisfied for

$$\phi_0 \geq \left(\frac{4\sigma^2 n_0^2}{3(2-\gamma)^2}\right)^{1/3}. \tag{23}$$

Thus we have the following result.

Corollary 1 *If the initial datum for the cosmological scalar field verifies (23), there is no singularity in the future.*

Note that the arguments used so far do not consider the equation (12b) for the entropy S. In order to improve our singularity analysis, it is crucial to use (12b). The qualitative dynamics of the cosmological model in the past depends on which of the following two mutually exclusive behaviours the solution verifies.

Case 1. The entropy $S(t)$ vanishes at some negative time t_0, while $a(t_0)$ is still positive.
Case 2. The factor $a(t)$ vanishes at some negative time t_0, while $S(t_0)$ is still positive.

Observe that in case 1, the solution is unphysical for $t < t_0$, even if the metric of spacetime remains smooth, because ρ becomes negative. This unphysical region of spacetime can be avoided by matching the metric at the time t_0 with the de Sitter solution $a_{DS}(t) = C\exp(\sqrt{\phi(t_0)/3}\,t)$, where the constant C is such that $a_{DS}(t_0) = a(t_0)$. The resulting cosmological model has no big-bang singularity and is vacuum up to the time t_0 (since $\rho = 0$ and $a(t) = a_{DS}(t)$ for $t \leq t_0$), at which time the vacuum energy $\phi(t_0)$ starts to be converted into matter energy ρ. On the other hand, in case 2, a big-bang singularity forms in the past.
We give sufficient conditions for the occurrence of each of the two possible scenarios described in cases 1 and 2. Let us introduce the times t_a and t_S defined by

$$t_a = \inf\{t < 0 : a(\tau) > 0, \text{ for all } t < \tau < 0\},$$
$$t_S = \inf\{t < 0 : S(\tau) > 0, \text{ for all } t < \tau < 0\},$$

with $t_a, t_S < 0$. Note that, from (18) and (19) we have

$$\left(1+\frac{3}{2}\gamma H_0 t\right)^{\frac{2}{3\gamma}} \leq a(t) \leq e^{H_0 t}, \quad \text{for all } t \in (\max\{t_a, t_S\}, 0) \tag{24}$$

In particular, the lower bound in (24) implies that

$$t_a > -\frac{2}{3\gamma H_0}. \tag{25}$$

In the following proposition, we state the qualitative dynamics of the cosmological model when the solution behaves as in case 1.

Proposition 2 *If $t_a \leq t_S$ (case 1 above) then one of the following conditions holds*

(1a) When $\gamma > 1$, we have $S_0 < \dfrac{\sigma n_0^{1-\gamma}}{3H_0(\gamma-1)}\left[1-\exp\left(2(1-\gamma)/\gamma\right)\right]$;

(1b) When $\gamma \leq 1$, we have $S_0 < \dfrac{2\sigma n_0^{1-\gamma}}{3(3\gamma-2)H_0}$.

Proof If we consider $t_a \leq t_S$, then $S(t_a) \leq 0$. For $\gamma > 1$, using the upper bound of (24) in (12b) we have

$$S(t) \geq S_0 - \frac{\sigma n_0^{1-\gamma}}{3H_0(\gamma-1)}\left(1-\exp\left(3H_0(\gamma-1)t\right)\right).$$

On the other hand, if we consider (25), this last inequality leads to

$$S(t_a) > S_0 - \frac{\sigma n_0^{1-\gamma}}{3H_0(\gamma-1)}\left(1-\exp\left(2(1-\gamma)/\gamma\right)\right),$$

which then gives (1a). The proof when $\gamma \leq 1$ is similar. In this case we use the lower bound of (24) in (12b) to get

$$S(t) \geq S_0 - \frac{\sigma n_0^{1-\gamma}}{3(3\gamma-2)H_0}\left(1-(1+3\gamma/2H_0 t)^{3-2/\gamma}\right),$$

so that, by (25), this last condition implies that

$$S(t_a) > S_0 - \frac{\sigma n_0^{1-\gamma}}{3(3\gamma-2)H_0}$$

which in turn gives (1b). □

4 Qualitative Dynamics of Solutions

In this section we introduce dynamical variables, that will be used to transform the system (12) into an autonomous dynamical system. Then, for $k = 0$, we develop a qualitative analysis of the resulting system in order to study the dynamics of spatially flat solutions of (12). The fixed points of the dynamical system represent solutions of (12) and the stability properties of the fixed points give qualitative information about the evolution of the solutions of the cosmological system (12).

4.1 The Dynamical System

For each $k = -1, 0, 1$ we define the variable by $D = \sqrt{\rho}$, that is

$$D = \sqrt{3H^2 - \phi + \frac{3k}{a^2(t)}}, \tag{26}$$

and introduce the dimensionless expansion-normalized variables

$$\chi_\rho = \arctan\left(\frac{\rho}{D^2}\right), \quad \chi_H = \arctan\left(\frac{H}{D}\right), \quad \chi_\psi = \arctan\left(\frac{\dot{\phi}}{D^3}\right), \quad \chi_a = \arctan\left(\frac{k}{D^2 a^2}\right). \tag{27}$$

The constraint equation (12e) implies that $\tan \chi_\rho = 1$, that is $\chi_\rho = \pi/4$. Let us also define a new time variable τ, defined in terms of the cosmological time t by

$$\tau = \frac{Dt}{\cos \chi_H \cos \chi_\psi \cos \chi_a}.$$

The new time derivative is therefore given by

$$\frac{d}{d\tau}(\cdot) = \frac{1}{D} \cos \chi_H \cos \chi_\psi \cos \chi_a \frac{d}{dt}(\cdot)$$

and we use the simple notation $(\cdot)' = \frac{d}{d\tau}(\cdot)$. The system (12), in terms of the new variables χ_H, χ_ψ and χ_a, transforms to

$$\chi_H' = \frac{1}{2\sqrt{2}} \cos \chi_H \left[\sin \chi_H \sin \chi_\psi \cos \chi_H \cos \chi_a + 2 \sin \chi_a \cos^2 \chi_H \cos \chi_\psi\,, \right.$$
$$\left. + \gamma \cos \chi_\psi \cos \chi_a (3 \sin^2 \chi_H - \cos^2 \chi_H)\right], \tag{28a}$$

$$\chi_\psi' = \frac{3}{2\sqrt{2}} \sin \chi_\psi \cos \chi_\psi \left[\sin \chi_\psi \cos \chi_H \cos \chi_a + (3\gamma - 2) \sin \chi_H \cos \chi_\psi \cos \chi_a \right], \tag{28b}$$

$$\chi_a' = \frac{\sqrt{2}}{2} \sin\chi_a \cos^2\chi_a \Big((3\gamma - 2)\sin\chi_H \cos\chi_\psi + \sin\chi_\psi \cos\chi_H\Big). \tag{28c}$$

Finally we introduce the state space $\mathscr{X}$, for the dynamical system (28), defined by

$$\mathscr{X} = \Big\{(\chi_H, \chi_\psi, \chi_a) \in \Big(-\frac{\pi}{2}, \frac{\pi}{2}\Big) \times \Big(-\frac{\pi}{2}, 0\Big) \times \Big(-\frac{\pi}{2}, \frac{\pi}{2}\Big)\Big\}. \tag{29}$$

The dynamical system (28) admits a smooth extension to the closure $\overline{\mathscr{X}}$ of the state space. From (28c) we see that the 2-dimensional plane $\chi_a = 0$ is an invariant plane. The flow induced on this plane describes the dynamics of spatially flat solutions ($k = 0$). In the next subsection, we study this flow in some detail.

4.2 Dynamics of Spatially Flat Solutions

The flow induced on the "roof" of $\mathscr{X}$ is described by the reduced dynamical system obtained by setting $\chi_a = 0$ in (28), that is

$$\chi_H' = \frac{1}{2\sqrt{2}} \cos\chi_H \Big(\sin\chi_H \cos\chi_H \sin\chi_\psi + \gamma\cos\chi_\psi (3\sin^2\chi_H - \cos^2\chi_H)\Big), \tag{30a}$$

$$\chi_\psi' = \frac{3}{2\sqrt{2}} \sin\chi_\psi \cos\chi_\psi \Big[\sin\chi_\psi \cos\chi_H + (3\gamma - 2)\sin\chi_H \cos\chi_\psi\Big]. \tag{30b}$$

The state space for this dynamical system is $\overline{\mathscr{X}_{\mathrm{up}}}$, where

$$\mathscr{X}_{\mathrm{up}} = \Big\{(\chi_H, \chi_\psi) \in \Big(-\frac{\pi}{2}, \frac{\pi}{2}\Big) \times \Big(-\frac{\pi}{2}, 0\Big)\Big\}.$$

The dynamical system (30) describes the flow of the spatially flat solutions of (12) which we will study in what follows. We shall say that an orbit Γ of the dynamical system (30) is typical if there exists a one parameter family of orbits having the same α- and ω-limit set of Γ. If no orbit other than Γ admits the same limit sets of Γ, we shall say that Γ is isolated. The qualitative behavior of the orbits of the dynamical system (30) is depicted in Fig. 1. Our next goal is to prove the principal features of this behavior as well as to analyse the physical interpretation of the flow depicted in Fig. 1 (in terms of solutions of the Einstein equations).

Fixed points. The dynamical system (30) possesses eight fixed points, seven of which are located on the boundary and one in the interior. They are represented in Fig. 1 and listed in Table 1. The interior fixed point U_γ is associated to the self-similar solution (17) which has been characterized in Sect. 2, with $a(t)$ being a linear function on time. Since the other fixed points are located at the boundary of the state space, they no longer correspond to exact solutions of (12), but to limiting states when one or more variables take an extreme value.

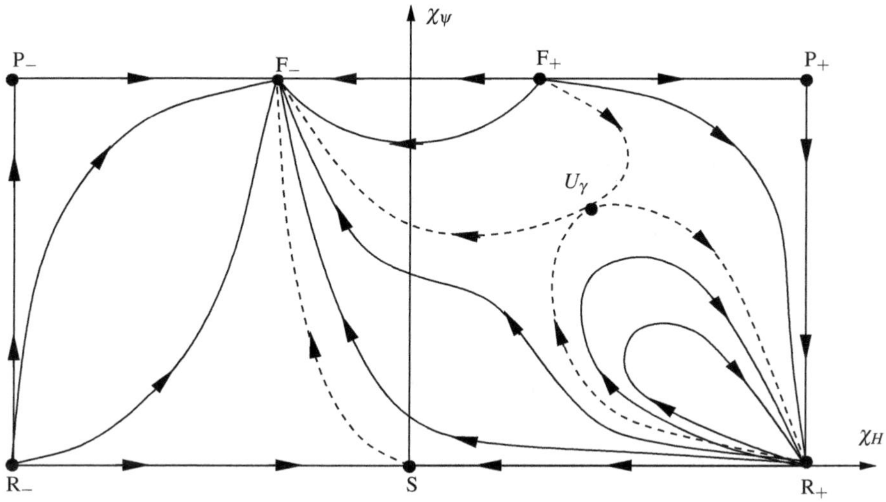

Fig. 1 Phase portrait of the dynamical system (30). *Full lines* represent typical orbits and *dashed lines* represent isolated orbits

Table 1 Fixed points of the dynamical system (30) in the state space $\overline{\mathscr{X}_{\text{up}}}$. U_γ is the unique interior fixed point

Fixed point	χ_H	χ_ψ
P_-	$-\frac{\pi}{2}$	0
F_-	$-\frac{\pi}{6}$	0
F_+	$\frac{\pi}{6}$	0
P_+	$\frac{\pi}{2}$	0
R_+	$\frac{\pi}{2}$	$-\frac{\pi}{2}$
S	0	$-\frac{\pi}{2}$
R_-	$-\frac{\pi}{2}$	$-\frac{\pi}{2}$
U_γ	$\arctan\sqrt{\frac{\gamma}{2}}$	$-\arctan\sqrt{\frac{\gamma}{2}}\,(3\gamma-2)$

We state the following conclusions about the fixed points U_γ, $F_\pm$ and $P_\pm$ of the dynamical system (30) and the corresponding solutions of the cosmological system (12).

- U_γ: At this fixed point we have $\chi_\psi = -\arctan\sqrt{\frac{\gamma}{2}}\,(3\gamma-2)$ and $\chi_H = \arctan\sqrt{\frac{\gamma}{2}}$. Hence, $H/D = \sqrt{\frac{\gamma}{2}}$ and $\dot{\phi}/D^3 = -\sqrt{\frac{\gamma}{2}}(3\gamma-2)$. Substituting these relations, together with (26), into equations (12a), (12b) and (12c), we obtain a differential system whose solution is given by (17).
- $F_\pm$: Since $H/D = \pm 1/\sqrt{3}$ at these fixed points, the cosmological scalar field ϕ is identically zero. Thus the orbits that converge to (resp. emanate from) the fixed point F_- (resp. F_+) identify the diffusion-free perfect fluid solutions with zero cosmological constant, i.e., the Friedmann-Lemaître solutions

$$a_{H_0}(t) = \left(1 + \frac{3}{2}\gamma H_0 t\right)^{\frac{2}{3\gamma}}. \tag{31}$$

In particular, the fixed point F_+ is associated to the one-parameter family of expanding solutions $\{a_{H_0}(t)\}_{H_0>0}$, while F_- is associated to the one-parameter family of contracting solutions $\{a_{H_0}(t)\}_{H_0<0}$.

- $P_\pm$: At these fixed points we have $\chi_\psi = 0$ and $\chi_H = \pm\frac{\pi}{2}$. Hence, $\phi = const = \Lambda$ and $H/D \to \pm\infty$ as the point $P_\pm$ is approached, which implies that $D \to 0$. Thus $H^2 \to \Lambda/3$ and we obtain that P_+ is associated to the expanding de Sitter vacuum solution

$$a_\Lambda^+(t) = \exp\left(\sqrt{\frac{\Lambda}{3}}t\right), \tag{32}$$

while P_- is associated to the contracting de Sitter vacuum solution

$$a_\Lambda^-(t) = \exp\left(-\sqrt{\frac{\Lambda}{3}}t\right). \tag{33}$$

The identification of the solutions corresponding to the fixed points $R_\pm$ is not as straightforward. Unfortunately we were unable to identify the orbits that emanate and converge to the fixed points $R_\pm$ and S with solutions of system (12). Posteriorly, this has led us to reformulate the dynamical variables in order to overcome this difficulty. However, at the time of the presentation of this work at the conference *Particle Systems and PDEs II*, the dynamical variables used to study the model under investigation here were in fact the ones we describe in this section. A new version of the dynamical treatment of the system (12) can be found in [5].

The flow on the boundary of $\mathscr{X}_{\text{up}}$. The analysis of flow induced on the one-dimensional boundary of $\mathscr{X}_{\text{up}}$ is straightforward. The left (right) side $\chi_H = -\pi/2$ ($\chi_H = \pi/2$) consists of an orbit starting at R_- (P_+) and ending at P_- (R_+); the bottom side $\chi_\psi = -\pi/2$ consists of two orbits, connecting the points $R_\pm$ to S; finally the top side $\chi_\psi = 0$ consists of three orbits, one connecting P_- to F_-, one connecting F_+ to F_- and one connecting F_+ to P_+. Next we discuss the local stability properties of the fixed points. This analysis is straightforward, except for the fixed points $R_\pm$, which are not hyperbolic.

The interior fixed point U_γ. Point U_γ is a saddle: the matrix of the linearized dynamical system around this fixed point has real eigenvalues of opposite sign, see Fig. 2. Hence there exist exactly two interior orbits that have U_γ as ω-limit point and exactly two orbits that have U_γ as α-limit point. This implies, in particular, that the solution (17) is unstable in the class of spatially flat Robertson-Walker solutions.

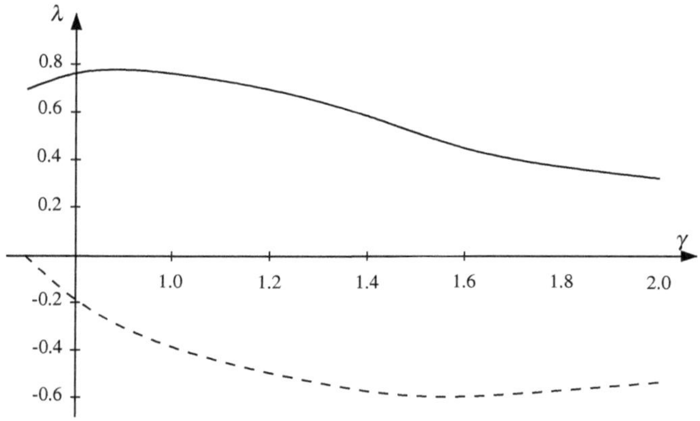

Fig. 2 Eigenvalues λ associated to the fixed point U_γ, versus different values of γ. *Full line* for the positive eigenvalue and *dashed line* for the negative eigenvalue

4.3 Stability Properties of Spatially Flat Solutions

Next we analyze the local stability properties of some of the fixed points on the boundary. To this purpose it is convenient to consider the dynamical system (30) in the extended state space $\mathscr{X}_\varepsilon$ defined by

$$\mathscr{X}_\varepsilon = \left\{(\chi_H, \chi_\psi)\colon -\frac{\pi}{2} - \varepsilon < \chi_H < \frac{\pi}{2} + \varepsilon,\ -\frac{\pi}{2} - \varepsilon < \chi_\psi < \varepsilon\right\}, \quad with\ \overline{\mathscr{X}} \subset \mathscr{X}_\varepsilon,$$

where $\varepsilon > 0$ is small. This extension allows us to perform the linearization procedure to study the local stability of the fixed points at the boundary of $\mathscr{X}$. A straightforward calculation shows that the fixed points $\mathrm{P}_\pm$ and S are hyperbolic saddles in the extended state space, while F_- and F_+ are respectively a hyperbolic sink and a hyperbolic source in the extended state space. Combining this information with the structure of the flow on the boundary we obtain the following result.

Lemma 1 *The fixed point* F_- *(resp.* F_+*) is the* ω*-limit (resp.* α*-limit) of a one parameter family of interior orbits, while* S *is the* α*-limit of exactly one interior orbit. There exists no interior orbit whose* α *or* ω*-limit set contain the points* $\mathrm{P}_\pm$.

The fixed points R_- and R_+ are not hyperbolic. In fact, the matrix of the linearized system around these two fixed points vanishes. In general this feature is the signal of a possible complicated behavior of the dynamical system near the fixed point. This is not so however for the fixed point R_-, which we will prove to be the source of a one parameter set of interior orbits. The latter statement is part of claim (iii) in the following theorem.

Theorem 1 *The following holds:*

(i) $\chi_H' < 0$ *for* $\chi_H = 0$, *and* $-\pi/2 < \chi_\psi \le 0$. *In particular interior orbits can cross the line* $\chi_H = 0$ *only from the right to the left. Moreover, the* ω*-limit of each interior orbit which intersects the region* $\chi_H < 0$ *is the fixed point* F_-*;*
(ii) *There exists exactly one orbit* $\mathrm{S} \to \mathrm{F}_-$ *whose* α*-limit is* S *and whose* ω*-limit is* F_-*;*
(iii) *If an orbit* Γ *intersects the region* $\mathcal{L}$ *on the left of* $\mathrm{S} \to \mathrm{F}_-$, *then the whole orbit* Γ *is contained in this region. Moreover, the* ω*-limit of* Γ *is* F_-, *while the* α*-limit is* R_-. *There exist a one parameter family of orbits having the same properties as* Γ*;*
(iv) *If an orbit* Γ *intersects the region* $\mathcal{R}$ *on the right of* $\mathrm{S} \to \mathrm{F}_-$, *then the whole orbit* Γ *is contained in this region. The* α*-limit of* Γ *is contained in the region* $\chi_H > 0$. *There exists a one parameter family of orbits having the same behavior as* Γ.

Proof We have $\chi'_{H|_{\chi_H=0}} = -\frac{\gamma}{2}\cos\chi_\psi$, by which the first part of the claim (i) follows. To prove the second part, we observe that $\chi'_\psi > 0$, for $(\chi_H, \chi_\psi) \in \left(-\frac{\pi}{2}, 0\right) \times \left(-\frac{\pi}{2}, 0\right)$. It follows that for any $0 < \varepsilon < \pi/2$, the set

$$S_\varepsilon = \left\{(\chi_H, \chi_\psi) \in \left[-\frac{\pi}{2}, 0\right] \times \left[-\frac{\pi}{2} + \varepsilon, 0\right]\right\}$$

is future invariant. Since any interior orbit that crosses the line $\chi_H = 0$ must intersect S_ε, for some ε, it follows by LaSalle invariance principle [6, Th. 4.11] that the ω-limit set of any such orbit must be contained in the set $\{x \in S_\varepsilon : \chi'_\psi = 0\}$, by which it follows immediately that it must coincide with the point F_-. Next we prove (ii). We have shown in Lemma 1 that there is exactly one interior orbit that has S as α-limit. To show that its ω-limit is the point F_-, we use that the eigenvector corresponding to the positive eigenvalue of the linearized system around S in the extended state space is given by

$$v = \left(-\frac{\gamma}{4}, 1\right).$$

Since $v_1 = -\frac{\gamma}{4} < 0$, the unstable manifold of S intersects the region $\chi_H < 0$ and hence, since this region is future invariant, the whole orbit starting from S must be contained in it. Thus by (i) its ω-limit set must coincide with F_-. As to (iii), we notice that the region to the left (as well as the region on the right) of the orbit $\mathrm{S} \to \mathrm{F}_-$ is an invariant set. Since $-\chi_\psi$ is strictly monotone decreasing on this region, the Monotonicity Principle [6, Th. 4.12] gives that the ω-limit set of orbits in $\mathcal{L}$ is contained in the set

$$\left\{x \in \overline{\mathcal{L}} \setminus \mathcal{L} : \lim_{y \to x} \chi_\psi \neq -\pi/2\right\},$$

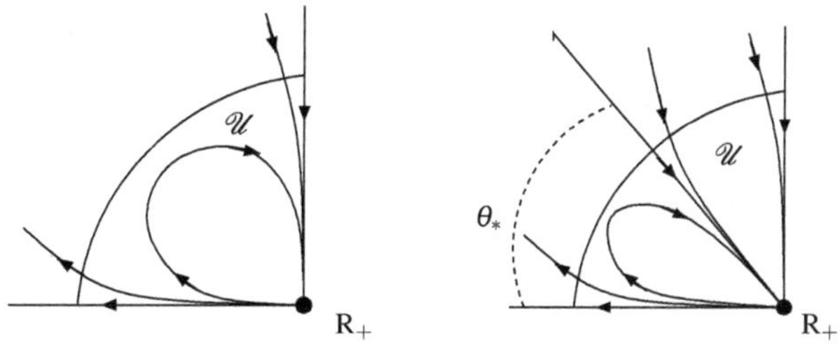

Fig. 3 Phase portrait of the dynamical system (30) in a corner neighborhood $\mathscr{U}$ of R_+ for $2/3 < \gamma \leq 1$ (*left*) and $1 < \gamma < 2$ (*right*). Each orbit is typical

while the α-limit set is contained in the set

$$\left\{x \in \overline{\mathcal{L}} \setminus \mathcal{L} : \lim_{y \to x} \chi_\psi \neq 0\right\}.$$

The claim (iii) follows immediately from the structure of the flow on the boundary of $\mathcal{R}$. Finally, (iv) follows by the fact that $\mathcal{R}$ is an invariant set and that χ_H is monotone decreasing on $\chi_H < 0$. □

In preparation for the global analysis of orbits that intersect the region $\chi_H > 0$, we describe the behavior of the flow near the point R_+. We define a *corner neighborhood* of R_+ to be the intersection of a neighborhood of R_+ in the extended state space with the interior of $\mathscr{X}_{up}$.

Proposition 3 *There exists a corner neighborhood $\mathscr{U}$ of the fixed point R_+ where the qualitative behavior of the flow is as depicted in Fig. 3. In particular, for each $x \in \mathscr{U}$, the orbit Γ_x passing through x verifies one (and only one) of the following statements:*

1. *the α- and ω-limit set of Γ_x consist of the point R_+;*
2. *the α-limit set of Γ_x consists of the point R_+, and $\mathscr{U}$ contains no ω-limit points of Γ_x;*
3. *the ω-limit set of Γ_x consists of the point R_+, and $\mathscr{U}$ contains no α-limit points of Γ_x.*

Proof Let us first shift the fixed point R_+ to the origin by introducing the new variables $\overline{\chi}_H = \chi_H - \pi/2$ and $\overline{\chi}_\psi = \chi_\psi + \pi/2$. The state space becomes

$$\overline{\mathscr{X}}_{up} = \left\{(\overline{\chi}_H, \overline{\chi}_\psi) \in (-\pi, 0) \times \left(0, \frac{\pi}{2}\right)\right\}$$

while the dynamical system becomes

$$\begin{aligned}\overline{\chi}'_H &= -\frac{1}{2\sqrt{2}} \sin\overline{\chi}_H \Big(\sin\overline{\chi}_H \cos\overline{\chi}_H \cos\overline{\chi}_\psi + \gamma \sin\overline{\chi}_\psi (3\cos^2\overline{\chi}_H - \sin^2\overline{\chi}_H)\Big) \\ &:= P(\overline{\chi}_H, \overline{\chi}_\psi),\end{aligned}$$

$$\overline{\chi}'_\psi = -\frac{3}{2\sqrt{2}} \sin\overline{\chi}_\psi \cos\overline{\chi}_\psi \Big[\sin\overline{\chi}_H \cos\overline{\chi}_\psi + (3\gamma - 2)\cos\overline{\chi}_H \sin\overline{\chi}_\psi \Big]$$
$$:= Q(\overline{\chi}_H, \overline{\chi}_\psi).$$

We study the previous dynamical system in a neighborhood of $(\overline{\chi}_H, \overline{\chi}_\psi) = (0,0)$. Let

$$P^{(2)}(\overline{\chi}_H, \overline{\chi}_\psi) := \frac{1}{2}\left[\frac{\partial^2 P}{\partial \overline{\chi}_H^2}(0,0)\overline{\chi}_H^2 + \frac{\partial^2 P}{\partial \overline{\chi}_\psi^2}(0,0)\overline{\chi}_\psi^2 + 2\frac{\partial^2 P}{\partial \overline{\chi}_H \partial \overline{\chi}_\psi}(0,0)\overline{\chi}_H\overline{\chi}_\psi \right]$$
$$= -\frac{1}{2\sqrt{2}}\overline{\chi}_H\Big(\overline{\chi}_H + 3\gamma\overline{\chi}_\psi\Big),$$
$$Q^{(2)}(\overline{\chi}_H, \overline{\chi}_\psi) := \frac{1}{2}\left[\frac{\partial^2 Q}{\partial \overline{\chi}_H^2}(0,0)\overline{\chi}_H^2 + \frac{\partial^2 Q}{\partial \overline{\chi}_\psi^2}(0,0)\overline{\chi}_\psi^2 + 2\frac{\partial^2 Q}{\partial \overline{\chi}_H \partial \overline{\chi}_\psi}(0,0)\overline{\chi}_H\overline{\chi}_\psi \right]$$
$$= -\frac{1}{\sqrt{2}}\overline{\chi}_H\overline{\chi}_\psi\Big(\overline{\chi}_H + 3(\gamma-1)\overline{\chi}_\psi\Big),$$
$$\mathcal{S}(\theta) := \Big[xQ^{(2)}(x,y) - yP^{(2)}(x,y)\Big]_{\substack{x=\cos\theta \\ y=\sin\theta}}$$
$$= -\frac{1}{\sqrt{2}}\sin\theta\cos\theta\Big[\cos\theta + 3\sin\theta(\gamma-1)\Big]$$

Since $\mathcal{S}(\theta)$ is not identically null, it follows by [7, Th. 2, pag. 140] that the solutions of $\mathcal{S}(\theta) = 0$ identify the directions along which an orbit may converge to or emanate from the fixed point $\mathrm{R}_+ = (0,0)$. For $\gamma = 1$ the only such directions are the axes $\overline{\chi}_\psi = 0$, $\overline{\chi}_H = 0$. For $\gamma \neq 1$ we have the additional direction

$$\theta_* = -\arctan\left(\frac{1}{3(\gamma-1)}\right).$$

Notice that for $\gamma < 1$, this direction does not intersect the state space $\overline{\mathcal{X}}_{\mathrm{up}}$, which is the reason for the different behavior depicted in Fig. 3. Let us assume $\gamma \neq 1$ (the case $\gamma = 1$ can be analyzed similarly). The direction θ_* and the axes $\overline{\chi}_\psi = 0$, $\overline{\chi}_H = 0$ divide any neighborhood of $(0,0)$ into six sectors, as shown in Fig. 4. By the direction of the flow on the separatrixes, the sectors *II*, *III*, *V* and *VI* are parabolic. To identify the character of the sectors *I* and *IV*, we use that, by Theorem [7, Th. 7, pag. 305], the number e of elliptic sectors and the number h of hyperbolic sectors in the neighborhood of a fixed point satisfy the relation (proved by Bendixson in [8])

$$e - h = 2(i-1), \tag{34}$$

where i is the (Morse) index of the fixed point. For the fixed point R_+ we have $i = 2$ (computed with Mathematica), by which it follows that $e = 2$ and $h = 0$. Hence the sectors I and IV are elliptic and the qualitative behavior of the orbits is depicted in Fig. 4. □

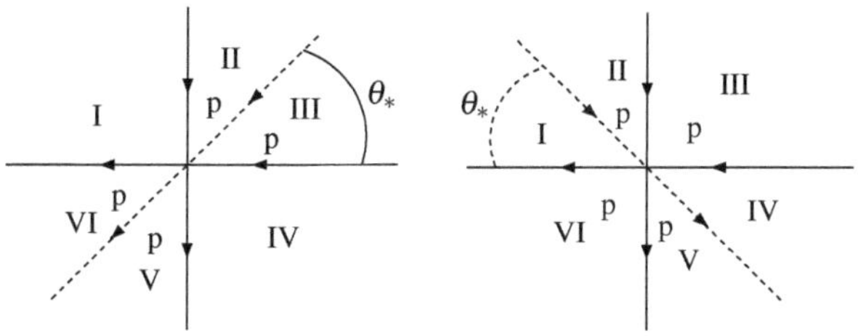

Fig. 4 The six sectors in a neighborhood of R_+ in the extended state space for $2/3 < \gamma < 1$ (*left*) and $1 < \gamma < 2$ (*right*). For $\gamma = 1$ there exist only four sectors, separated by the axes. The sectors labelled by "p" are parabolic, while the sectors *I* and *IV* may be both hyperbolic or both elliptic. Bendixson formula (34) entails that they are both elliptic

5 Summary

In this work we studied a model developed within the Einstein theory of relativity with a cosmological scalar field. This scalar field can be viewed as a background medium in which diffusion takes place. In particular we considered the Robertson-Walker spacetime, so that the model studied here is homogeneous and isotropic. The matter field variables are solutions of a non linear system of ordinary differential equations on the variable time. All solutions for which the scalar function of the metric is linear in time were obtained explicitly. In order to further understand the dynamical nature of other solutions the system was rewritten in terms of normalized dynamical variables. We obtained all fixed points of the dynamical system, one interior fixed point and seven fixed points on the boundary of the phase space. In particular the only interior equilibrium point is associated to the solution in which the scalar function of the metric is linear in time. The seven equilibrium points on the boundary of the state space correspond to limiting states when one or more variables take an extreme value, these limiting states being the Friedmann-Lemaître metrics and the de Sitter vacuum metrics.

Acknowledgments The paper is partially supported by the Research Centre of Mathematics of the University of Minho, through the National Funds from the "Fundação para a Ciência e a Tecnologia", Project PEstOE/MAT/UI0013/2014.

Appendix—Tools from Dynamical Systems

In this appendix we include some concepts and tools from the theory of dynamical systems that are used in the paper. The content of this appendix is mainly based on the books [6, 7].

We consider an autonomous dynamical system of the form

$$\dot{x} = f(x), \tag{35}$$

where $f : \mathbb{R}^n \longrightarrow \mathbb{R}^n$ is a $\mathscr{C}^1$ vector field and $x : \mathbb{R}_0^+ \longrightarrow \mathbb{R}^n$ is a function, $x = x(t)$. In the particular case of a planar dynamical system ($n = 2$) we introduce the following notation

$$x = (x, y)^{\mathrm{T}}, \quad f_1(x) = P(x, y), \quad f_2(x) = Q(x, y), \tag{36}$$

and rewrite Eq. (35) in the form

$$\dot{x} = P(x, y), \qquad \dot{y} = Q(x, y). \tag{37}$$

Theorem 2 *Consider the planar dynamical system (37) with $P(x, y)$ and $Q(x, y)$ being analytic functions of (x, y) in some open subset $E \subset \mathbb{R}^2$ containing the origin. Assume that the Taylor expansions of P and Q about the origin begin with mth-degree terms $P^{(m)}(x, y)$ and $Q^{(m)}(x, y)$ with $m \geq 1$. Then any orbit of the planar dynamical system (37) that approaches the origin as $t \to +\infty$ either spirals toward the origin as $t \to +\infty$ or it tends toward the origin in a definite direction $\theta = \theta_0$ as $t \to +\infty$. If the function $g(x, y) = xQ^{(m)}(x, y) - yP^{(m)}(x, y)$ is not identically null, then all directions of the approach $\theta = \theta_0$ satisfy the relation*

$$\cos\theta_0 Q^{(m)}(\cos\theta_0, \sin\theta_0) - \sin\theta_0 P^{(m)}(\cos\theta_0, \sin\theta_0) = 0. \tag{38}$$

Moreover, if one orbit of the system (37) spirals toward the origin as $t \to +\infty$ then all trajectories of (37) in a deleted neighborhood of the origin spiral toward the origin as $t \to +\infty$.

For details on Theorem 2, see Ref. [7], page 140, Th. 2.
Now, we recall the definition of a sector, with reference to the planar dynamical system (37), as well as the possible character of a sector. If the Taylor expansions of P and Q about the origin begin with mth-degree terms $P^{(m)}$ and $Q^{(m)}$ and if the function

$$g(\theta) = \cos\theta\, Q^{(m)}(\cos\theta, \sin\theta) - \sin\theta\, P^{(m)}(\cos\theta, \sin\theta)$$

is not identically null, from Theorem 2 it follows that there are at most $2(m + 1)$ directions, obtained as the solutions of the equation $g(\theta) = 0$, along which a orbit of the system (37) may approach the origin. Thus, the solution curves of the system (37) that approach the origin along these directions divide a neighborhood of the origin into a finite number of open regions called *sectors*. Three types of sectors can occur, namely either a *hyperbolic*, *parabolic* or an *elliptic* sector when it is topologically equivalent to the sector represented in Fig. 5, picture (a), (b) and (c), respectively, where the directions of the flow need not to be preserved. Moreover, the trajectories that lie on the boundary of a hyperbolic sector are called *separatrixes*.
Another important concept that is used in this paper is the Morse index of a fixed point of the dynamical system (37). We begin with the definition of the index of a Jordan curve. Let $f = (P, Q)^{\mathrm{T}}$ be a $\mathscr{C}^1$ vector field on an open subset $E \subset \mathbb{R}^2$ and

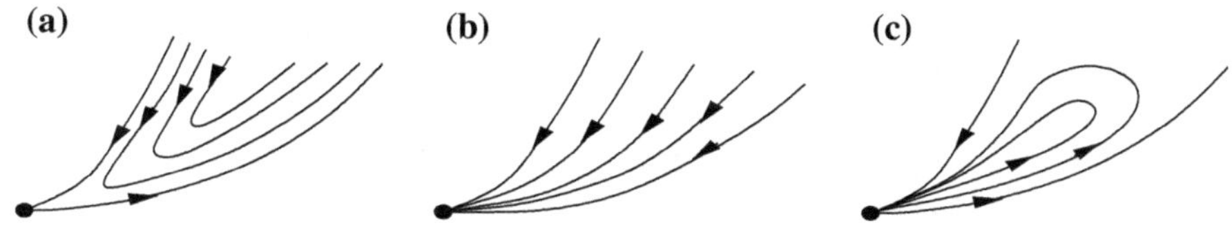

Fig. 5 Sector in the neighborhood of the origin of hyperbolic type (**a**), parabolic type (**b**) and elliptic type (**c**)

let $\mathscr{C}$ be a Jordan curve contained in E, such that the system (37) has no fixed point on $\mathscr{C}$. The *index of* $\mathscr{C}$ *relative to* f is the integer $i(\mathscr{C})$ computed as

$$i(\mathscr{C}) = \frac{1}{2\pi} \oint_{\mathscr{C}} \frac{P dQ - Q dP}{P^2 + Q^2}.$$

Let x_0 be an isolated fixed point of system (37) and assume that the Jordan curve $\mathscr{C}$ contains x_0 and no other fixed points of (37) on its interior. The *Morse index of* x_0 *with respect to* f is defined by

$$i(x_0) = i(\mathscr{C}).$$

The following result is very convenient for the evaluation of the Morse index of a fixed point. It is stated for a fixed point at the origin but it is also valid for an arbitrary fixed point x_0.

Theorem 3 *Consider the planar dynamical system (37) with $P(x, y)$ and $Q(x, y)$ being analytic functions of (x, y) in some open subset $E \subset \mathbb{R}^2$ containing the origin. If the origin is an isolated fixed point of the system (37) then the Morse index of the origin, say i, satisfies the relation*

$$i = 1 + \frac{1}{2}(e - h), \tag{39}$$

where e and h indicate the number of elliptic and hyperbolic sectors, respectively, in a neighborhood of the origin.

For details on Theorem 3, see Ref. [7], page 305, Th. 7.
As a consequence of Theorem 3, it follows that the number h of hyperbolic sectors and the number e of elliptic sectors have the same parity.
Further concepts and properties that are used in the analysis developed in Sect. 4 are those related to the α- and ω- limit sets of and orbit. We come back to the dynamical system (35) and assume that, for each $x_0 \in \mathbb{R}^n$, the system has a unique global solution $x \in \mathscr{C}^1(\mathbb{R})$ such that $x(0) = x_0$. We say that an equilibrium point x^* is an ω-limit point of the solution $x(t)$ if there exists a sequence $t_n \to +\infty$ such that $\lim_{n \to +\infty} x(t_n) = x^*$. The set of all ω-limit points of the solution $x(t)$ is called its ω-limit set. Analogously, by considering a sequence $t_n \to -\infty$, such that

$\lim_{n \to -\infty} x(t_n) = x^*$, we define the concepts of an α-limit point and the α-limit set of a solution $x(t)$. Since solutions with the same orbit have equal ω- and α-limit sets, we will refer to ω- and α-limit sets of an orbit γ, and we will denote them by $\omega(\gamma)$ and $\alpha(\gamma)$, respectively. The following two theorems state important results on the limit sets $\omega(\gamma)$ and $\alpha(\gamma)$, that are used in Sect. 4.

Theorem 4 (LaSalle Invariance Principle) *Let $S \subset \mathbb{R}^n$ be a compact and positively invariant subset of the dynamical system (35), and $Z : S \longrightarrow \mathbb{R}$ a $\mathscr{C}^1$ monotone function along the flow of the dynamical system. Let γ be an orbit in S. Then*

$$\omega(\gamma) \subseteq \Big\{ x \in S : \; Z'(x) = 0 \Big\},$$

where $Z' = \Delta Z \cdot f$.

For details on Theorem 4, see Ref. [6], page 103, Th. 4.11.

Theorem 5 (Monotonicity Principle) *Let $S \subset \mathbb{R}^n$ be an invariant subset of the dynamical system (35) and $Z : S \longrightarrow \mathbb{R}$ a $\mathscr{C}^1$ strictly monotonically decreasing function along the flow of the dynamical system. Let a and b be defined by $a = \inf\{Z(x) : \; x \in S\}$ and $b = \sup\{Z(x) : \; x \in S\}$. Let γ be an orbit in S. Then*

$$\alpha(\gamma) \subseteq \Big\{ s \in \partial S : \; \lim_{x \to s} Z(x) \neq a \Big\}, \quad \omega(\gamma) \subseteq \Big\{ s \in \partial S : \; \lim_{x \to s} Z(x) \neq b \Big\},$$

For details on Theorem 5, see Ref. [6], page 103, Th. 4.12. Finally, the following theorem states a crucial result about the limit sets $\omega(\gamma)$ and $\alpha(\gamma)$ in the particular case of a dynamical system in $\mathbb{R}^2$.

Theorem 6 (Generalized Poincaré-Bendixson) *Consider the dynamical system (35) on $\mathbb{R}^2$, and suppose that the dynamical system has only a finite number of equilibrium points. Then, for any orbit γ of the dynamical system, each one of the limit sets $\omega(\gamma)$ and $\alpha(\gamma)$ can only be one of the following: an equilibrium point; a periodic orbit the union of equilibrium points and heteroclinic cycles.*

For details on Theorem 6, see Ref. [7], page 101, Th. 4.10, or Ref. [6], page 245, Th. 2.

References

1. Calogero, S.: A kinetic theory of diffusion in general relativity with cosmological scalar field. J. Cosmol. Astropart. Phys. **11**(JCAP11), 1–16 (2011)
2. Calogero, S.: Cosmological models with fluid matter undergoing velocity diffusion. J. Geom. Phys. **62**(11), 2208–2213 (2012)
3. Stephani, H., Kramer, D., MacCallum, M., Hoenselaers, C., Herlt, E.: Exact Solutions of Einstein's Field Equations. Cambridge University Press, Cambridge (2003)

4. Wald, R.M.: General Relativity. University of Chicago Press, Chicago (1984)
5. Alho, A., Calogero, S., Ramos, M.P.M., Soares A.J.: Dynamics of Robertson-Walker spacetimes with diffusion. Annals Phys **354**, 475–488 (2015)
6. Wainwright, J., Ellis, G.F.R.: Dynamical Systems in Cosmology. Cambridge University Press, Cambridge (1997)
7. Perko, L.: Differential Equations and Dynamical Systems, 3rd edn. Springer, New York (2000)
8. Bendixson, I.: Sur les courbes définies par des équations différentielles. Acta Mathematica **24**, 1–88 (1901)

The Structure of Shock Waves in Dissipative Hyperbolic Models

Srboljub Simić

Abstract The study is devoted to the shock structure problem in hyperbolic systems of balance laws, where the dissipation is taken into account through relaxation. These models typically arise by extending the set of state variables, as well as governing equations. The existence of physically admissible solution is related to stability properties of equilibrium states and their transcritical bifurcation. The main examples will be the hyperbolic model of isothermal viscoelasticity, 13 moments equations for monatomic gases and the binary mixture of non-reacting ideal gases. The problems which arise in models with mixed type of dissipation, i.e. both relaxation and diffusion, are tackled as well.

Keywords Shock structure · Hyperbolic systems · Stability

1 Introduction

Shock waves are regarded as localized abrupt changes of state variables which move through space. Mathematical models in continuum physics are often given the form of hyperbolic systems of conservation laws [5]. In this case shock waves are represented as moving singular surfaces on which jumps of state variables occur. When the model comprises certain type of dissipation, the jumps are regularized—transformed into continuous solutions of field equations with narrow regions where state variables have steep gradients.

Dissipative mechanisms can be modeled in different ways. The most common one is through diffusive type relations which lead to parabolic systems of balance laws [13]. The term parabolic is not used here in a strict mathematical sense, but as an indication that the system predicts infinite speed of propagation of disturbances, in contrast to finite speeds of propagation in hyperbolic systems. Such parabolic systems can be regarded as perturbations of hyperbolic systems of conservation

S. Simić (✉)
Department of Mechanics, Faculty of Technical Sciences, University of Novi Sad,
Trg Dositeja Obradovića 6, 21000 Novi Sad, Serbia
e-mail: ssimic@uns.ac.rs

P. Gonçalves and A.J. Soares (eds.), *From Particle Systems to Partial Differential Equations II*, Springer Proceedings in Mathematics & Statistics 129,
DOI 10.1007/978-3-319-16637-7_13

laws, which can be recovered in the so-called vanishing viscosity limit [7] (the name borrowed from the physically best known example). They also admit existence of continuous traveling wave solutions which asymptotically connect two states of the system corresponding to the jumps of state variables in hyperbolic systems [8]. In this case it is said that the shock wave is endowed with a structure, i.e. we have a shock structure instead of a shock wave.

Another way of including dissipation into the model is through relaxation mechanism [14]. Namely, the system of field equations can be extended by the evolution equations which describe the trend of the fluxes towards the equilibrium values [18]. This introduces several new features in the model: (a) the state space becomes extended—the fluxes are treated as state variables; (b) it retains the hyperbolicity, at least in a certain region of the extended state space; (c) the characteristic speeds (speeds of propagation of the disturbances) of the extended system may not coincide with the ones of the basic hyperbolic system. It is also common to regard the characteristic time in which relaxation towards equilibrium occur—so-called relaxation time—as small compared to the time needed for significant changes of the state variables. In such a way extended hyperbolic systems of balance laws appear as singularly perturbed systems and one can recover the underlying hyperbolic system of conservation laws in the limit of infinitely small relaxation time. These models gained less popularity because they lie at the border of continuum approach—they tend to capture non-equilibrium phenomena in a more precise way, and thus bridge the gap between macroscopic and mesoscopic level of description. However, the main feature of these systems, which is of interest for the present study, is that they also admit the existence of the shock structure [27].

The aim of this review is to show recent results related to the existence of the shock structure in dissipative hyperbolic models. The main feature that will be analyzed is the bifurcation of equilibrium states and their stability properties. In Sect. 2 examples of particular dissipative hyperbolic systems will be analyzed: hyperbolic model of isothermal viscoelasticity, 13 moments model of viscous and heat-conducting monatomic gas and binary multi-temperature mixture of ideal gases. It will be shown that exchange of stability properties of equilibrium states coincide with the bifurcation of solution of Rankine-Hugoniot equations of the underlying equilibrium (non-dissipative) subsystem. Also, it will be shown that the existence of the shock structure for weak shocks can be proved by solving the bifurcation equation. Motivated by the aforementioned examples, the mathematical framework for stability and bifurcation analysis of the shock structure in hyperbolic systems will be delineated in Sect. 3. Moreover, extension to mixed hyperbolic-parabolic models will also be discussed.

2 Shock Structure in Hyperbolic Systems of Balance Laws

The main assumption in the present study of the shock structure will be that shock waves, i.e. singular surfaces (shock fronts) moving through space, are plane waves moving in direction orthogonal to the shock front at constant speed. As a consequence, the analysis can be confined to wave propagation in one space dimension, say x, and the traveling waves can be represented as functions of a single variable $\xi = x - st$, where s is the constant speed of the shock wave. This will reduce the original system of governing equations to a system of ordinary differential equations (ODE's) which will be studied by the methods of dynamical systems theory.

2.1 Isothermal Viscoelasticity

The simplest example to be studied will be the model of isothermal viscoelasticity, proposed by Suliciu [25]. Starting point, i.e. non-dissipative hyperbolic system of conservation laws, is the celebrated $p-$system [7] which models elastic response of the medium

$$\begin{aligned} \partial_t u^1 - \partial_x u^2 &= 0, \\ \partial_t u^2 - \partial_x p(u^1) &= 0. \end{aligned} \tag{1}$$

In (1) u^1 and u^2 are state variables and ∂_t and ∂_x denote partial derivatives with respect to time and space variable, respectively. Characteristic speeds of the system are ($\mathbf{u} = (u^1, u^2)^T$ stands for the vector of state variables)

$$\lambda_1(\mathbf{u}) = -\sqrt{p'(u^1)}, \quad \lambda_2(\mathbf{u}) = \sqrt{p'(u^1)}. \tag{2}$$

Certain assumptions have to be made about the function $p(u^1)$ in order to assure hyperbolicity of the system, $p'(u^1) \neq 0$, and genuine nonlinearity of the characteristic speeds, $p''(u^1) > 0$ (or $p''(u^1) < 0$). Rankine-Hugoniot equations, which govern the jump of state variables, for (1) read

$$\begin{aligned} s\left(u_0^1 - u_1^1\right) + \left(u_0^2 - u_1^2\right) &= 0, \\ s\left(u_0^2 - u_1^2\right) + \left(p(u_0^1) - p(u_1^1)\right) &= 0, \end{aligned} \tag{3}$$

where s is the speed of shock, and subscripts 0 and 1 denote the states of the system in front and behind the shock, respectively. Equation (3) have trivial solution, $u_0^1 = u_1^1$, $u_0^2 = u_1^2$, for any s. Nontrivial solution bifurcates from the trivial one in the neighborhood of the critical value of the shock speed, i.e. the one which coincides with the characteristic speed: $s = \lambda_1(u_0^1)$, or $s = \lambda_2(u_0^1)$. However, nontrivial solutions of Rankine-Hugoniot equations are not physically admissible for any s. In this study we shall use the Lax condition to select the admissible solution. For the shock wave moving in positive x direction it reads

$$\sqrt{p'(u_0^1)} < s < \sqrt{p'(u_1^1)}. \tag{4}$$

Note finally that the simplicity of the model allows reduction of the Rankine-Hugoniot equations (3) to a single equation of the form

$$s^2 \left(u_0^1 - u_1^1\right) - \left(p(u_0^1) - p(u_1^1)\right) = 0. \tag{5}$$

To comprise dissipation this model can be extended by the equation which takes into account viscous effects. By introducing non-equilibrium variable v, which plays the role of stress, Suliciu's model is written in the following form [25]

$$\begin{aligned} \partial_t u^1 - \partial_x u^2 &= 0, \\ \partial_t u^2 - \partial_x v &= 0, \\ \partial_t v - \mu \partial_x u^2 &= -\frac{1}{\tau}\left(v - p(u^1)\right). \end{aligned} \tag{6}$$

where τ is a small parameter and $\mu > 0$. When $\tau \to 0$, $v \to p(u^1)$, so that $v_E = p(u^1)$ determines the *equilibrium manifold*; on equilibrium manifold the system (6) is reduced to (1), thus called the *equilibrium subsystem*. Important feature of extended model (6) is that its characteristic speeds differ from the ones of equilibrium subsystem ($\mathbf{U} = (u^1, u^2, v)^T$ being the extended vector of state variables)

$$\Lambda_1(\mathbf{U}) = -\sqrt{\mu}, \quad \Lambda_2(\mathbf{U}) = 0, \quad \Lambda_3(\mathbf{U}) = \sqrt{\mu}, \tag{7}$$

but the entropy dissipation inequality requires that the *subcharacteristic condition* [3, 4] has to be satisfied on equilibrium manifold $\mathbf{U}_E = (u^1, u^2, p(u^1))^T$

$$\Lambda_1(\mathbf{U}_E) \leq \lambda_1(\mathbf{u}) < \lambda_2(\mathbf{u}) \leq \Lambda_3(\mathbf{U}_E). \tag{8}$$

Analysis of the shock structure starts with the assumption that solution has the form of a traveling wave, $\mathbf{U} = \mathbf{U}(\xi)$, where $\xi = (x - st)/\tau$ is the scaled independent variable. By denoting $\dot{\mathbf{U}} = d\mathbf{U}/d\xi$, (6) is reduced to the following set consisted of two algebraic equations and a single ordinary differential equation (ODE), named *shock structure equations*

$$\begin{aligned} s(u^1 - u_1^1) + (u^2 - u_1^2) &= 0, \\ s(u^2 - u_1^2) + (v - p(u_1^1)) &= 0, \\ s\dot{v} + \mu \dot{u}^2 &= v - p(u^1). \end{aligned} \tag{9}$$

Algebraic equations $(9)_{1,2}$ emerged from $(6)_{1,2}$ through integration, using the downstream equilibrium data $\mathbf{U}_1 = \mathbf{U}(\infty) = (u_1^1, u_1^2, p(u_1^1))^T$. Taking $(9)_{1,2}$ to express u^2 and v in terms of u^1, the shock structure problem can be reformulated in terms of a single ODE

$$\dot{u}^1 = \theta(u^1, s) = \frac{s^2 \left(u^1 - u_1^1\right) - p(u^1) + p(u_1^1)}{s\left(s^2 - \mu\right)}, \tag{10}$$

in which the shock speed s is a parameter. This equation reveals two important features of the shock structure problem. First, stationary points of (10) are located on the equilibrium manifold, thus recovering the Rankine-Hugoniot equations (5) (or (3)) of the equilibrium subsystem. Second, the shock structure equations become singular whenever the shock speed meets the characteristic speed (7) of the extended system (6).

The shock structure is a continuous solution which asymptotically connects equilibrium states of the system at infinity

$$\mathbf{U}_0 = \mathbf{U}(-\infty) = (u_0^1, u_0^2, p(u_0^1))^T, \quad \mathbf{U}_1 = \mathbf{U}(\infty) = (u_1^1, u_1^2, p(u_1^1))^T. \tag{11}$$

To that end it is important to determine the stability properties of the stationary points. Starting from (10), the linearized variational equation can be derived in the neighborhood of downstream equilibrium

$$\dot{y} = \Theta(u_1^1, s)y, \quad \Theta(u_1^1, s) = \frac{\partial \theta}{\partial u^1}(u_1^1, s) = \frac{s^2 - p'(u_1^1)}{s\left(s^2 - \mu\right)}, \tag{12}$$

where $y = u^1 - u_1^1$. Stability of the stationary point is determined by the sign of $\Theta(u_1^1, s)$, and it depends on the shock speed s. It is obvious that $\Theta(u_1^1, s^*) = 0$ for

$$s^* = \pm\sqrt{p'(u_1^1)}. \tag{13}$$

Moreover, $\Theta(u_1^1, s)$ changes the sign in the neighborhood of the *critical shock speed* s^*

$$\frac{\partial \Theta}{\partial s}(u_1^1, s^*) = -\frac{2}{\mu - p'(u_1^1)} \neq 0, \tag{14}$$

thus indicating that downstream equilibrium changes its stability. Namely, the following conclusion can be drawn

- u_1^1 is unstable for $s < s^* = \lambda_2(u_1^1)$,
- u_1^1 is stable for $s > s^* = \lambda_2(u_1^1)$,

which can be related to Lax condition (4).

The existence of the shock structure can be proved by further expansion of the variational equation. Retaining the lowest order nonlinear terms, in this case the second order ones, the following equation is obtained

$$\dot{y} \approx \frac{1}{2(\mu - p'(u_1^1))}\left(-2\varepsilon y + \frac{p''(u_1^1)}{\sqrt{p'(u_1^1)}}y^2\right), \tag{15}$$

where $\varepsilon = s - s^*$. Equation (15) itself can be regarded as *bifurcation equation* and it describes the transcritical bifurcation of equilibria. It also yields continuous solution, at least for small ε, which asymptotically connects upstream equilibrium u_0^1 with downstream one u_1^1, provided the critical value of the shock speed corresponds to the genuinely nonlinear characteristic speed of the equilibrium subsystem. This implies the existence of the shock structure for weak shocks.

The analysis of Suliciu's model of isothermal viscoelasticity reveals the following three important properties [21]:

1. equilibrium states correspond to the solutions of Rankine-Hugoniot equations on equilibrium manifold (i.e. of the equilibrium subsystem);
2. they change stability properties when shock speed crosses the characteristic speed of the equilibrium subsystem;
3. stationary points obey transcritical bifurcation pattern.

Our aim is to show that these properties hold also in more involved dissipative systems, and to delineate the general mathematical structure which lies beneath these features.

2.2 13 Moments Model

The 13 moments model was proposed by Grad [10] within the kinetic theory of monatomic gases. Later it became a part of an approach known as *extended thermodynamics* [18], a continuum theory which bridges the gap between macro and meso scale. One of the important features of the model is its hyperbolicity, at least in certain region of the extended state space. Nucleus of the model is consisted of conservation laws of mass, momentum and energy. In one-dimensional case the equations are

$$\begin{aligned} \partial_t \rho + \partial_x(\rho v) &= 0, \\ \partial_t(\rho v) + \partial_x(\rho v^2 - \sigma + p) &= 0, \\ \partial_t \left(\frac{1}{2}\rho v^2 + \rho e\right) + \partial_x \left(\left(\frac{1}{2}\rho v^2 + \rho e\right) v - \sigma v + p v + q\right) &= 0, \end{aligned} \tag{16}$$

where ρ, v, e, p, σ and q are mass density, velocity, internal energy density, pressure, stress and heat flux of the gas, respectively. They are adjoined with thermal and caloric equations of state

$$p = p(\rho, T) = \rho \frac{k_B}{m} T, \quad e = e(\rho, T) = \frac{3}{2} \frac{k_B}{m} T, \tag{17}$$

where T is temperature of the gas, k_B the Boltzmann constant and m the atomic mass. If the stress and the heat flux can be neglected, $\sigma = 0$, $q = 0$, then (16) reduce to classical Euler's gas dynamics equations, and the vector of state variables is $\mathbf{u} = (\rho,\ v,\ T)^T$. The characteristic speeds in the Euler's model are

$$\lambda_1(\mathbf{u}) = v - c_s, \quad \lambda_2(\mathbf{u}) = v, \quad \lambda_3(\mathbf{u}) = v + c_s, \qquad c_s = \left(\frac{5}{3}\frac{k_B}{m}T\right)^{1/2}, \tag{18}$$

c_s being the local speed of sound. Admissibility of the shock waves, moving with speed s, is provided by the Lax condition

$$\lambda_3(\mathbf{u}_0) < s < \lambda_3(\mathbf{u}_1), \tag{19}$$

where $\mathbf{u}_0$ and $\mathbf{u}_1$ are the states in front and behind the shock, respectively, which are determined as solutions of Rankine-Hugoniot equations.

Classical way of introducing dissipation into Euler's equations (16) is to propose the constitutive relations of Navier-Stokes-Fourier type for the stress and heat flux

$$\sigma = \frac{4}{3}\mu \partial_x v, \quad q = -\kappa \partial_x T, \tag{20}$$

where μ and κ are viscosity and heat-conductivity, respectively. Although widely accepted, this model has two major drawbacks: it predicts infinite speeds of propagation of disturbances and it is valid only in the neighborhood of local equilibrium state. Another way to introduce dissipation is to extend the system of governing equations by the balance laws for the stress and the heat flux. In one space dimension they read

$$\begin{aligned} \partial_t\left(\rho v^2 + p - \sigma\right) + \partial_x\left(\rho v^3 + 3pv - 3\sigma v + \frac{6}{5}q\right) &= \frac{1}{\tau_\sigma}\sigma, \\ \partial_t\left(\frac{1}{2}\rho v^3 + \frac{5}{2}pv - \sigma v + q\right) & \\ + \partial_x\left(\frac{1}{2}\rho v^4 + 4pv^2 - \frac{5}{2}\sigma v^2 + \frac{16}{5}qv - \frac{7}{2}\frac{p}{\rho}\sigma + \frac{5}{2}\frac{p^2}{\rho}\right) &= -\frac{1}{\tau_q}\left(q - \frac{3}{2}\sigma v\right). \end{aligned} \tag{21}$$

In (21) τ_σ and τ_q are small parameters—relaxation times. In the course of derivation of this model there appears the closure problem, which can be solved by the Grad's method [10], application of the entropy inequality (extended thermodynamics [18]) and by the use of maximum entropy principle [6]. This problem will not be discussed in this study. The extended set of state variables now becomes $\mathbf{U} = (\rho,\ v,\ T,\ \sigma,\ q)^T$. The equilibrium manifold is determined by $\mathbf{U}_\mathrm{E} = (\rho,\ v,\ T,\ 0,\ 0)^T$ and the equilibrium subsystem becomes the system of Euler's equations (16). Characteristic speeds of the 13 moments model (16)–(21) on equilibrium manifold are

$$\begin{aligned} &\Lambda_1(\mathbf{U}_\mathrm{E}) = v - 1.6503c_s, \quad \Lambda_2(\mathbf{U}_\mathrm{E}) = v - 0.6297c_s, \quad \Lambda_3(\mathbf{U}_\mathrm{E}) = v, \\ &\Lambda_4(\mathbf{U}_\mathrm{E}) = v + 0.6297c_s, \quad \Lambda_5(\mathbf{U}_\mathrm{E}) = v + 1.6503c_s. \end{aligned} \tag{22}$$

They fulfill the subcharacteristic condition

$$\Lambda_1(\mathbf{U}_E) < \lambda_1(\mathbf{u}) < \lambda_3(\mathbf{u}) < \Lambda_5(\mathbf{U}_E). \tag{23}$$

The stability and bifurcation analysis of the 13 moments model is given in detail in [22]. We shall give here only the results which are the most important for the present study.

Assuming the traveling wave solution and exploiting the conservation laws (16), the problem can be reduced to a set of two ODE's of the first order

$$\frac{du}{d\xi} = F(u, T, M_0), \quad \frac{dT}{d\xi} = G(u, T, M_0), \tag{24}$$

where u and T are dimensionless velocity and temperature, respectively, ξ is dimensionless space variable and M_0 is Mach number in upstream equilibrium. The Mach number is regarded as a parameter in the model and $M_0 = 1$ corresponds to the highest characteristic speed of the equilibrium subsystem. Explicit form of (24) can be found in [22]. Equilibrium states, that are at the same time the stationary points of (24), are given by

$$u_0 = M_0, \quad T_0 = 1, \quad u_1 = \frac{3 + M_0^2}{4M_0}, \quad T_1 = \frac{1}{16}\left(14 - \frac{3}{M_0^2} + 5M_0^2\right), \tag{25}$$

subscripts 0 and 1 corresponding to upstream and downstream equilibrium, respectively. They coincide with solutions of the Rankine-Hugoniot equations of the equilibrium subsystem. It was shown by Gilbarg and Paolucci [8] in the case of Navier-Stokes-Fourier model that the shock structure can be constructed as a heteroclinic orbit which asymptotically connects stationary points. The result was confirmed by Grad [11] and Weiss [27] for 13 moments model.

In [22] it was shown by means of linear stability analysis that stationary points change their stability properties while M_0 is changed. Namely, there is one eigenvalue at each stationary point which changes the sign for the critical value of Mach number $M_0 = 1$, while the other one remains negative. Consequently, upstream equilibrium, which is a stable node for $M_0 < 1$, becomes a saddle for $M_0 > 1$, and vice versa for downstream equilibrium. Since $M_0 > 1$ corresponds to physically admissible solutions of the Rankine-Hugoniot equations, heteroclinic orbit asymptotically connects upstream saddle to downstream stable node. Moreover, two stability inequalities emerge

$$0 < \lambda_{01}(M_0) \quad \Leftrightarrow \quad \lambda_{11}(M_0) < 0, \tag{26}$$

which are equivalent to the Lax condition (19) (Fig. 1).

Bifurcation analysis in the the spirit of the previous subsection can be performed in this case also. It is based upon center manifold reduction procedure and leads to

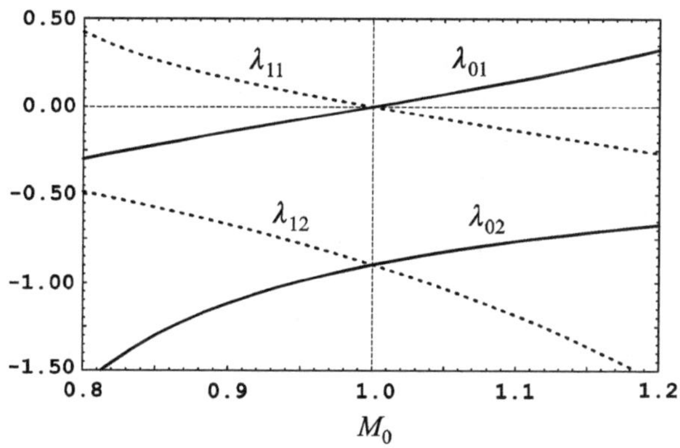

Fig. 1 Eigenvalues of the linearized variational equations of 13 moments model in upstream (λ_{01}, λ_{02}) and downstream (λ_{11}, λ_{12}) equilibrium

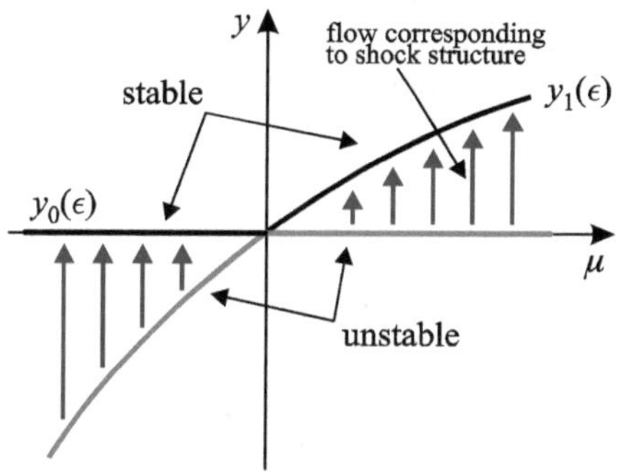

Fig. 2 Bifurcation diagram for the 13 moments model: y is lumped state variable and $\varepsilon = M_0 - 1$

the following bifurcation equation [22]

$$\frac{dy}{d\xi} \approx \frac{10}{7}\varepsilon y - \frac{20}{7}y^2, \tag{27}$$

where y is lumped state variable obtained in the course of normal form reduction and $\varepsilon = M_0 - 1$. Like in the case of isothermal viscoelasticity, we obtain again a transcritical bifurcation pattern (Fig. 2). It was shown in [22] that the same stability results, and the very same bifurcation equation (27), are obtained also in the case of higher order hyperbolic dissipative models of gas dynamics, like 14 and 21 moments models. This indicates the general pattern which has to be sought in the analysis of weak shocks and corresponding shock structures in hyperbolic systems.

2.3 Binary Multi-temperature Mixture of Ideal Gases

Multi-temperature (MT) model of mixtures of Euler fluids is developed within the framework of extended thermodynamics. It is deeply rooted in the principles of rational thermodynamics, proposed by Truesdell [26], which roughly state that behavior of each constituent is described by the same balance laws as for a single fluid, as long as we take into account the influence of other constituents upon it. On the other hand, it is a continuation of the mixture theory developed by Müller [17] which removed the paradox of infinite speeds present in classical thermodynamics of irreversible processes. At the same time it is in agreement with macroscopic equations derived from kinetic theory of gases [2].

Structure of the MT model of mixtures was studied by Ruggeri and Simić [19], and some of its properties were analyzed in [9, 20, 23, 24]. Recently, the shock structure problem was studied by Madjarević and Simić [16] for comparison with available experimental data. Thorough derivation of governing equations for a binary mixture may be found in [19] and in the paper by Madjarević [15] in the present volume. Therefore, our starting point will be the one-dimensional shock structure equations for a binary mixture of non-reacting Euler fluids with respect to a single independent variable $\xi = x - st$, s being the shock speed

$$
\begin{aligned}
&\frac{d}{d\xi}(\rho u) = 0, \\
&\frac{d}{d\xi}\left(\rho u^2 + p + \frac{J^2}{\rho c(1-c)}\right) = 0, \\
&\frac{d}{d\xi}\left\{\left(\frac{1}{2}\rho u^2 + \rho\varepsilon + p\right)u + \left(\frac{uJ}{\rho c(1-c)} + \frac{1}{\beta}\right)J\right\} = 0, \\
&\frac{d}{d\xi}(\rho c u + J) = 0, \\
&\frac{d}{d\xi}\left\{\rho c u^2 + \frac{J^2}{\rho c} + 2uJ + p_1\right\} = \hat{m}_1, \\
&\frac{d}{d\xi}\left\{\left(\frac{1}{2}\rho c\left(u + \frac{J}{\rho c}\right)^2 + \rho c\varepsilon_1 + p_1\right)\left(u + \frac{J}{\rho c}\right)\right\} = \hat{m}_1 u + \hat{e}_1.
\end{aligned}
\tag{28}
$$

In (28) ρ is the mixture mass density, $u = v - s$ is relative mixture velocity with respect to the shock front, $c = \rho_1/\rho$ is the mass concentration of the first constituent, $J = \rho_1 u_1 (= -\rho_2 u_2)$ is the diffusion flux of the first constituent expressed in terms of diffusion velocity $u_1 = v_1 - v$ $(u_2 = v_2 - v)$. Also, p and ε are pressure and internal energy density of the mixture, p_1 and ε_1 are the same partial quantities for the for the first constituent, while β determines the so-called thermal inertia

$$
\beta = \frac{1}{g_1 - g_2}, \quad g_\alpha = \varepsilon_\alpha + \frac{p_\alpha}{\rho_\alpha} + \frac{u_\alpha^2}{2}, \quad (\alpha = 1, 2). \tag{29}
$$

Mutual interaction between the constituents is described by the source terms $\hat{m}_1$ and $\hat{e}_1$ (for detailed analysis of the structure of source terms see Madjarević [15] in this volume).

The system of governing equations is consisted of two parts: equations $(28)_{1-3}$ are the conservation laws of mass, momentum and energy for the mixture, while $(28)_{4-6}$ are balance laws of mass, momentum and energy for the first constituent. Thus, the system fits into the concept of extended thermodynamics if one regard density ρ, (relative) velocity u and (average) temperature T as basic state variables, and take the concentration c, diffusion flux J and diffusion temperature $\Theta = T_2 - T_1$ as their extension. Therefore, the extended state vector reads $\mathbf{U} = (\rho, u, T, c, J, \Theta)^T$. Source terms determine the equilibrium manifold as $J = 0$, $\Theta = 0$, and $c = \text{const.}$ along the trajectory. In such a way the system of conservation laws for the mixture $(28)_{1-3}$ becomes the equilibrium subsystem.

The shock structure problem consists in finding the heteroclinic orbit in the extended phase space which connects two equilibrium states

$$\mathbf{U}(-\infty) = \mathbf{U}_0, \quad \mathbf{U}(\infty) = \mathbf{U}_1, \quad \dot{\mathbf{U}}(\pm\infty) = \mathbf{0}. \tag{30}$$

By the analysis of (28) one easily determines that equilibrium states in dimensionless form are

$$\mathbf{U}_0 = \begin{bmatrix} 1 \\ M_0 \\ 1 \\ c_0 \\ 0 \\ 0 \end{bmatrix}, \quad \mathbf{U}_1 = \begin{bmatrix} \frac{4M_0^2}{3+M_0^2} \\ \frac{3+M_0^2}{4M_0} \\ \frac{1}{16}\left(14 - \frac{3}{M_0^2} + 5M_0^2\right) \\ c_0 \\ 0 \\ 0 \end{bmatrix}, \tag{31}$$

where M_0 is the Mach number in upstream equilibrium calculated with respect to equilibrium sound speed of the mixture. Note that ρ, u and T satisfy Rankine-Hugoniot equations for equilibrium subsystem $(28)_{1-3}$, $c = c_0$ and $J = 0$, $\Theta = 0$ in both equilibrium states.

The shock structure equations (28) have a property which distinguishes them from the same kind of equations in isothermal elasticity (9) and in 13 moments model (24). Namely, although (28) consists of four conservation laws and two balance laws, the system cannot be reduced to a minimal set of ODE's, i.e. two ODE's, since the singularity appears at equilibrium states (stationary points) and in the interior of the domain. It is a consequence of the physical mechanism present in the model: lighter gas is heated first within the shock structure, but heavier one becomes hotter in the rear part of the profile. Thus, diffusion temperature changes the sign within the profile and there is a point $-\infty < \xi^* < \infty$ where $\Theta(\xi^*) = 0$.

Among several possibilities for solution of the problem we decided to analyze the whole system (28) and determine the character of stationary points in the whole extended phase space. General form of the shock structure equations is

$$\frac{d}{d\xi}\mathbf{F}(\mathbf{U}) = \mathbf{Q}(\mathbf{U}) \quad \Leftrightarrow \quad \mathbf{A}(\mathbf{U})\frac{d\mathbf{U}}{d\xi} = \mathbf{Q}(\mathbf{U}), \tag{32}$$

where $\mathbf{A}(\mathbf{U}) = D\mathbf{F}(\mathbf{U})$ is Jacobian matrix. By linearizing (32) in the neighborhood of $\mathbf{U}_0$ one obtains

$$\mathbf{A}(\mathbf{U}_0)\frac{d\Delta\mathbf{U}}{d\xi} = \mathbf{P}(\mathbf{U}_0)\Delta\mathbf{U}, \tag{33}$$

where $\Delta\mathbf{U} = \mathbf{U} - \mathbf{U}_0$ and $\mathbf{P}(\mathbf{U}_0) = D\mathbf{Q}(\mathbf{U}_0)$. Eigenvalue problem and characteristic equation corresponding to (33) then read

$$\lambda\mathbf{A}(\mathbf{U}_0)\mathbf{R} = \mathbf{P}(\mathbf{U}_0)\mathbf{R}; \quad \det(\mathbf{P}(\mathbf{U}_0) - \lambda\mathbf{A}(\mathbf{U}_0)) = 0, \tag{34}$$

$\mathbf{R}$ being the right eigenvector. The same can be done at the equilibrium state $\mathbf{U}_1$. Since there are four conservation laws in the system, there will be four eigenvalues $\lambda_i \equiv 0$ at both equilibrium states. Consequently, stationary points are not hyperbolic in the sense of dynamical system theory [12]. Among remaining two eigenvalues there is one which changes the sign in the neighborhood of the critical value of Mach number $M_0 = 1$, while the last one is negative in the same neighborhood. In Fig. 3 the graphs of the eigenvalues are shown for the case of the mixture of Helium and Xenon. They confirm the observations given above.

Although the behavior of the eigenvalues indicate possible transcritical bifurcation of equilibra, i.e. there exist a single eigenvalue which changes the sign when the shock speed crosses the value which coincides with the highest characteristic speed of the equilibrium subsystem, we shall not pursue with this analysis. Namely, in the first two cases (isothermal viscoelasticity and 13 moments model) stationary points were hyperbolic and the hyperbolicity was lost only for certain value of the parameter (Mach number M_0), which formally permits application of the center manifold reduction. In the MT model stationary points are not hyperbolic and the center manifold reduction is not straightforward, but this does not prevent numerical

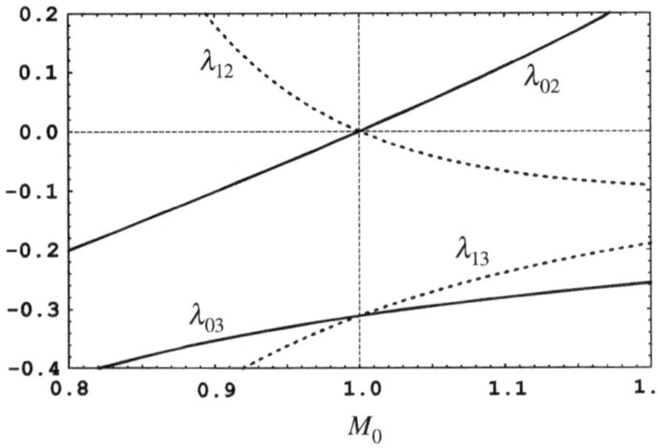

Fig. 3 Eigenvalues of the linearized variational equations of MT model for binary mixture of Helium and Xenon in upstream (λ_{02}, λ_{03}) and downstream (λ_{12}, λ_{13}) equilibrium

construction of the shock structure [15, 24]. However, one has to keep in mind that hyperbolicity of stationary points in first two cases was achieved by reduction of the system of shock structure equations—analysis of the original system also yields non-hyperbolic stationary points in the extended phase space. This fact indicates that there exist a common mathematical structure underlying the existence of the shock structure.

3 Stability Aspects of the Shock Structure Problem

Motivated by the common properties of the stationary points in dissipative hyperbolic models, we shall try to extract in this section their common features related to the shock structure problem. The attention will be restricted to the properties of the eigenvalues in hyperbolic and mixed hyperbolic-parabolic models.

3.1 Stability of Equilibrium States in Hyperbolic Models

Consider the hyperbolic system of conservation laws in one space dimension

$$\partial_t \mathbf{u} + \partial_x \mathbf{F}(\mathbf{u}) = \mathbf{0}, \tag{35}$$

where $\mathbf{u} \in \mathbf{R}^n$ is the vector of state variables and $\mathbf{F}(\mathbf{u}) \in \mathbf{R}^n$ is the vector of fluxes. To (35) one may adjoin the eigenvalue problem

$$(-\lambda \mathbf{I} + D\mathbf{F}(\mathbf{u}))\mathbf{r} = \mathbf{0}, \tag{36}$$

where $\mathbf{I}$ is identity matrix, $D\mathbf{F}(\mathbf{u})$) is Jacobian matrix of $\mathbf{F}$ and $\mathbf{r} \in \mathbf{R}^n$. If all the eigenvalues $\lambda_i(\mathbf{u})$ are real, the system (35) is said to be hyperbolic in $t-$direction and the eigenvalues are called characteristic speeds.

Systems of conservation laws admit weak solutions with jumps, where the jumps of state variables are determined as solutions of Rankine-Hugoniot equations

$$[\mathbf{F}(\mathbf{u})] = s\,[\mathbf{u}]\,, \tag{37}$$

s being the shock speed and $[(\,)] = (\,)_1 - (\,)_0$, subscripts indicating the states in front (0) and behind (1) the shock. Nontrivial solutions of Rankine-Hugoniot equations, which are of interest for our analysis, bifurcate from the characteristic speeds, i.e. they exist in the neighborhood of $s = \lambda_i(\mathbf{u}_0)$. However, not all of them are admissible and one must take into account appropriate selection rules. For our purposes it will be sufficient to use the Lax condition (where we implicitly assumed that all the eigenvalues are distinct)

$$\lambda_1(\mathbf{u}_0) < \ldots < \lambda_i(\mathbf{u}_0) < s < \lambda_{i+1}(\mathbf{u}_0) < \ldots < \lambda_n(\mathbf{u}_0),$$
$$\lambda_1(\mathbf{u}_1) < \ldots < \lambda_{i-1}(\mathbf{u}_1) < s < \lambda_i(\mathbf{u}_1) < \ldots < \lambda_n(\mathbf{u}_1),$$

which can be combined into a single inequality for the so-called $i-$shock

$$\lambda_i(\mathbf{u}_0) \ < \ s \ < \ \lambda_i(\mathbf{u}_1). \tag{38}$$

Common way of introducing dissipation in continuum physics is to endow conservation laws (35) with diffusive term. In mathematical literature it is usually called (vanishing) viscosity approach and the model has the following structure [7]

$$\partial_t \mathbf{u} + \partial_x \mathbf{F}(\mathbf{u}) = \varepsilon \partial_x (\mathbf{B}(\mathbf{u}) \partial_x \mathbf{u}), \tag{39}$$

where $\mathbf{B}(\mathbf{u})$ is the viscosity matrix and $\varepsilon > 0$ a small parameter. Shock waves in these systems are regularized, i.e. endowed with a continuous structure. When the traveling wave solution is assumed, the shock structure can be represented as a heteroclinic orbit in the phase space. Existence of the shock structure is naturally related to the Lax admissibility condition (38), but the existence of heteroclinic orbits is not a prerogative admissible equilibrium states—they exist even for non-admissible solutions of Rankine-Hugoniot equations of (37) (see [22] for more details in the context of Navier-Stokes-Fourier model).

The hyperbolicity of the model can be preserved in the dissipative case also if we extend the set of state variables and derive the extended set of governing equations in the form of the system of balance laws

$$\partial_t \mathbf{U} + \partial_x \hat{\mathbf{F}}(\mathbf{U}) = \frac{1}{\tau} \mathbf{Q}(\mathbf{U}), \tag{40}$$

with the following structure

$$\mathbf{U} = \begin{pmatrix} \mathbf{u} \\ \mathbf{v} \end{pmatrix}, \quad \begin{matrix} \mathbf{u} \in \mathbf{R}^n, \\ \mathbf{v} \in \mathbf{R}^m, \end{matrix} \quad \hat{\mathbf{F}}(\mathbf{U}) = \begin{pmatrix} \mathbf{f}(\mathbf{u}, \mathbf{v}) \\ \mathbf{g}(\mathbf{u}, \mathbf{v}) \end{pmatrix}, \quad \mathbf{Q}(\mathbf{U}) = \begin{pmatrix} \mathbf{0} \\ \mathbf{q}(\mathbf{u}, \mathbf{v}) \end{pmatrix}. \tag{41}$$

In (41) $\mathbf{U} \in \mathbf{R}^N$, $N = n + m$, is the extended set of state variables, $\hat{\mathbf{F}}(\mathbf{U})$ is the corresponding vector of fluxes, $\mathbf{Q}(\mathbf{U})$ is the vector of source terms and $\tau > 0$ a small parameter—relaxation time. It is important assumption that there exist an *equilibrium manifold* in the extended phase space determined by relation

$$\mathbf{q}(\mathbf{u}, \mathbf{v}) = \mathbf{0}. \tag{42}$$

We shall also assume that (42) can be solved for $\mathbf{v}$, so that equilibrium manifold is explicitly determined

$$\mathbf{q}(\mathbf{u}, \mathbf{v}) = \mathbf{0} \quad \Rightarrow \quad \mathbf{v}_{\mathrm{E}} = \mathbf{h}(\mathbf{u}) \ \text{as} \ \tau \to 0. \tag{43}$$

On equilibrium manifold system (40) is reduced to an *equilibrium subsystem*

$$\partial_t \mathbf{u} + \partial_x \mathbf{f}(\mathbf{u}, \mathbf{h}(\mathbf{u})) = \mathbf{0}, \quad \mathbf{f}(\mathbf{u}, \mathbf{h}(\mathbf{u})) = \mathbf{F}(\mathbf{u}), \tag{44}$$

which coincides with the system of conservation laws (35). To the differential part of (40) one may adjoin the eigenvalue problem

$$(-\Lambda \mathbf{I} + D\hat{\mathbf{F}}(\mathbf{U}))\mathbf{R} = \mathbf{0}, \tag{45}$$

whose eigenvalues $\Lambda_j(\mathbf{U})$, if they are real, ought to satisfy the subcharacteristic condition on equilibrium manifold

$$\min_{1 \leq j \leq N} \Lambda_j(\mathbf{u}, \mathbf{h}(\mathbf{u})) \leq \lambda_i(\mathbf{u}) \leq \max_{1 \leq j \leq N} \Lambda_j(\mathbf{u}, \mathbf{h}(\mathbf{u})). \tag{46}$$

Shock structure is assumed to be a traveling wave solution of (40) which asymptotically connects equilibrium states. To that end we assume the solution in the form

$$\mathbf{U} = \mathbf{U}(\xi), \quad \xi = \frac{x - st}{\tau}, \tag{47}$$

and reduce the system of balance laws (40) to a system of ODE's—the shock structure equations

$$\frac{d}{d\xi}\left[-s\mathbf{U} + \hat{\mathbf{F}}(\mathbf{U})\right] = \mathbf{Q}(\mathbf{U}). \tag{48}$$

They are equipped with the boundary conditions

$$\mathbf{U}(-\infty) = \mathbf{U}_0, \quad \mathbf{U}(\infty) = \mathbf{U}_1, \quad \dot{\mathbf{U}}(\pm\infty) = \mathbf{0}, \tag{49}$$

for which it is assumed

$$\mathbf{v}(-\infty) = \mathbf{v}_{E0} = \mathbf{h}(\mathbf{u}_0), \quad \mathbf{v}(\infty) = \mathbf{v}_{E1} = \mathbf{h}(\mathbf{u}_1), \tag{50}$$

i.e. the solution approach the equilibrium manifolds at infinity. By integrating the conservative part of the system (48), one obtains $-s\mathbf{u} + \mathbf{f}(\mathbf{u}, \mathbf{v}) = \text{const}$. Using the boundary data (50) the Rankine-Hugoniot equations for equilibrium subsystem (44) are recovered

$$\mathbf{f}(\mathbf{u}_1, \mathbf{h}(\mathbf{u}_1)) - s\mathbf{u}_1 = \mathbf{f}(\mathbf{u}_0, \mathbf{h}(\mathbf{u}_0)) - s\mathbf{u}_0. \tag{51}$$

Therefore, we may conclude that shock structure connects equilibrium states of the system (40), i.e. stationary points of (48), which correspond to the solution of the Rankine-Hugoniot equations of the equilibrium subsystem (44). In view of that fact, critical value of the shock speed s ought to be the one from which nontrivial solution of Rankine-Hugoniot equations bifurcate, i.e. the one which coincides with the characteristic speed of the equilibrium subsystem.

Linear stability analysis of equilibrium states (stationary points), performed by the spectral analysis of linearized variational equations derived from the shock structure equations (48), yields that there are exactly n eigenvalues (corresponding to the conservative part of the system) which are identical zeros

$$\lambda^{(i_k)}(\mathbf{U}_0) = \lambda^{(i_k)}(\mathbf{U}_1) \equiv 0, \quad k = 1, \ldots, n. \tag{52}$$

Remaining m eigenvalues (corresponding to the non-equilibrium part of the system) obey the following relations

$$\begin{aligned} \lambda^{(j_1)}(\mathbf{U}_0) < \lambda^{(j_2)}(\mathbf{U}_0) < \ldots < \lambda^{(j_{m-1})}(\mathbf{U}_0) < 0 < \lambda^{(j_m)}(\mathbf{U}_0), \\ \lambda^{(j_1)}(\mathbf{U}_1) < \lambda^{(j_2)}(\mathbf{U}_1) < \ldots < \lambda^{(j_{m-1})}(\mathbf{U}_1) < \lambda^{(j_m)}(\mathbf{U}_1) < 0, \end{aligned} \tag{53}$$

i.e. the highest eigenvalue has different signs at upstream and downstream equilibria. Moreover, it is the only eigenvalue which changes the sign and it occurs when the shock speed crosses the highest characteristic speed of the equilibrium subsystem

$$\lambda^{(j_m)}(\mathbf{U}_0, \lambda_n(\mathbf{u}_0)) = 0, \quad \frac{d\lambda^{(j_m)}}{ds}(\mathbf{U}_0, \lambda_n(\mathbf{u}_0)) \neq 0. \tag{54}$$

This result, drawn from several convincing particular cases, which indicates the occurrence of transcritical bifurcation pattern, is rather a conjecture than a genuine theorem. The proof of this statement, which provides the general pattern recognized in previous section, as well as proper bifurcation analysis, is still an open problem for hyperbolic systems of balance laws. In the case of viscous dissipative systems one may refer to a recent result of Achleitner and Szmolyan [1].

The promising aspect of the inequalities (53) is that stationary points $\mathbf{U}_0$ and $\mathbf{U}_1$ behave like saddle and stable node, respectively, albeit in generalized sense since they are non-hyperbolic. This permits the numerical solution of the shock structure problem (48)–(49) as an initial value problem [27].

3.2 Stability of Equilibrium States in Mixed Hyperbolic-Parabolic Models

The final part of this review is actually related to the question: is there a generalization of the stability and bifurcation analysis to the systems with both relaxation and diffusive dissipative mechanisms? It is motivated by the analysis of MT model in viscous approximation [15]. We shall restrict here to a simple remark tempting to indicate the general structure.

By the system of balance laws with relaxation and diffusion we shall assume the following system of equations

$$\begin{aligned} \partial_t \mathbf{u} + \partial_x \mathbf{f}(\mathbf{u}, \mathbf{v}, \mathbf{w}) &= \mathbf{0}, \\ \partial_t \mathbf{v} + \partial_x \mathbf{g}(\mathbf{u}, \mathbf{v}, \mathbf{w}) &= \frac{1}{\tau}\mathbf{q}(\mathbf{u}, \mathbf{v}), \\ \mathbf{B}(\mathbf{u})\partial_x \mathbf{u} &= \frac{1}{\varepsilon}\mathbf{w}, \end{aligned} \tag{55}$$

for $\mathbf{w} \in \mathbf{R}^n$ and $\mathbf{u}$ and $\mathbf{v}$ as in previous subsection. The part $(55)_3$ recovers the diffusive mechanism in which $\mathbf{B}(\mathbf{u})$ is viscosity matrix. The equilibrium manifold is determined by two relations

$$\mathbf{q}(\mathbf{u}, \mathbf{v}), \quad \mathbf{w} = \mathbf{0}. \tag{56}$$

Omitting the complete analysis, we shall jump to the main result related to the linear stability analysis of equilibrium states based upon shock structure equations. There are still n eigenvalues which are identical zeros, like in (52). There is also one eigenvalue with different signs in upstream and downstream equilibria, but it is not anymore the highest one

$$\begin{aligned} \lambda^{(j_1)}(\mathbf{U}_0) < \lambda^{(j_2)}(\mathbf{U}_0) < \ldots < \lambda^{(j_{k-1})}(\mathbf{U}_0) < 0 < \lambda^{(j_k)}(\mathbf{U}_0) < \cdots < \lambda^{(j_{m+n})}(\mathbf{U}_0), \\ \lambda^{(j_1)}(\mathbf{U}_1) < \lambda^{(j_2)}(\mathbf{U}_1) < \ldots < \lambda^{(j_{k-1})}(\mathbf{U}_1) < \lambda^{(j_k)}(\mathbf{U}_1) < 0 < \ldots < \lambda^{(j_{m+n})}(\mathbf{U}_1). \end{aligned}$$

However, it retains the property of changing the sign (54) when the shock speed crosses the critical value corresponding to the highest characteristic speed of the equilibrium subsystem. The fact that both stationary points are generalized saddles in the extended phase space has important consequences on numerical solution of the shock structure problem. Heteroclinic orbit cannot be constructed as solution of the initial value problem, but one must solve it as a boundary value problem using some appropriate discretization scheme, such as finite differences, and taking proper initial guess.

4 Conclusion

This study was devoted to the analysis of the shock structure in dissipative hyperbolic models. The aim was to show that the existence of physically admissible shock structures is related to stability properties of equilibrium states. To that end three particular hyperbolic systems were studied: isothermal viscoelasticity, 13 moments model and multi-temperature model of binary mixture of Euler fluids. In each of these cases it was shown that stationary points change their stability properties when the shock speed crosses the critical value which corresponds to a characteristic speed of the equilibrium (non-dissipative) subsystem. In first two examples it was also shown that equilibrium states obey the transcritical bifurcation pattern. These examples motivated a formulation of the general result, which remains to be proved for hyperbolic dissipative systems, but has appropriate counterpart in parabolic models. Along with this open problem, there has to be determined under which conditions equilibrium states obey the transcritical bifurcation in hyperbolic and mixed hyperbolic-parabolic models.

Acknowledgments This work was supported by the Ministry of Education, Science and Technological Development, Republic of Serbia, through the project "Mechanics of nonlinear and dissipative systems—contemporary models, analysis and applications", Project No. ON174016.

References

1. Achleitner, F., Szmolyan, P.: Saddlenode bifurcation of viscous profiles. Physica D **241**, 1703–1717 (2012)
2. Bisi, M., Martalò, G., Spiga, G.: Multi-temperature euler hydrodynamics for a reacting gas from a kinetic approach to rarefied mixtures with resonant collisions. EPL (Europhys. Lett.) **95**, 55002 (2011)
3. Boillat, G., Ruggeri, T.: Hyperbolic principal subsystems: entropy convexity and subcharacteristic conditions. Arch. Ration. Mech. Anal. **137**, 305–320 (1997)
4. Chen, G.-Q., Levermore, C.D., Liu, T.-P.: Hyperbolic conservation laws with stiff relaxation terms and entropy. Commun. Pure Appl. Math. **XLVII**, 787–830 (1994)
5. Dafermos, C.M.: Hyperbolic Conservation Laws in Continuum Physics. Springer, Berlin (2010)
6. Dreyer, W.: Maximisation of the entropy in non-equilibrium. J. Phys. A: Math. Gen. **20**, 6505–6517 (1987)
7. Evans, L.C.: Partial Differential Equations. American Mathematical Society, Providence (2010)
8. Gilbarg, D., Paolucci, D.: The structure of shock waves in the continuum theory of fluids. J. Ration. Mech. Anal. **2**, 617–642 (1953)
9. Gouin, H., Ruggeri, T.: Identification of an average temperature and a dynamical pressure in a multitemperature mixture of fluids. Phys. Rev. E **78**, 016303 (2009)
10. Grad, H.: On the kinetic theory of rarefied gases. Commun. Pure Appl. Math. **2**, 331–407 (1949)
11. Grad, H.: The profile of a steady plane shock wave. Commun. Pure Appl. Math. **5**, 257–300 (1952)
12. Guckenheimer, J., Holmes, P.: Nonlinear Oscillations, Dynamical Systems and Bifurcations of Vector Fields. Springer, New York (1986)
13. Liu, I.-S.: Continuum Mechanics. Springer, Berlin (2002)
14. Liu, T.-P.: Hyperbolic conservation laws with relaxation. Commun. Math. Phys. **108**, 153–175 (1987)
15. Madjarević, D.: Shock structure and temperature overshoot in macroscopic multi-temperature model of binary mixtures. Present volume
16. Madjarević, D., Simić, S.: Shock structure in helium-argon mixture—a comparison of hyperbolic multi-temperature model with experiment. EPL (Europhys. Lett.) **102**, 44002 (2013)
17. Müller, I.: A thermodynamic theory of mixtures of fluids. Arch. Rational Mech. Anal. **28**, 1–39 (1968)
18. Müller, I., Ruggeri, T.: Rational Extended Thermodynamics. Springer, New York (1998)
19. Ruggeri, T., Simić, S.: On the hyperbolic system of a mixture of eulerian fluids: a comparison between single-and multi-temperature models. Math. Methods Appl. Sci. **30**, 827–849 (2007)

20. Ruggeri, T., Simić, S.: Average temperature and maxwellian iteration in multi-temperature mixtures of fluids. Phys. Rev. E **80**, 026317 (2009)
21. Simić, S.S.: A note on shock profiles in dissipative hyperbolic and parabolic models. Publication de l'Institut Mathématique, Nouvelle Série **84**(98), 97–107 (2008)
22. Simić, S.S.: Shock structure in continuum models of gas dynamics: stability and bifurcation analysis. Nonlinearity **22**, 1337–1366 (2009)
23. Simić, S.: Shock structure in the mixture of gases: stability and bifurcation of equilibria. In: Juan, C., Moreno, P. (eds.) Thermodynamics—Kinetics of Dynamic Systems, pp. 179–204, InTech, Rijeka (2011)
24. Simić, S.: Shock structure problem in multi-temperature gaseous mixtures. Note Mat. **32**, 207–225 (2012)
25. Suliciu, I.: On modelling phase transitions by means of rate-type constitutive equations. Shock wave structure. Internat. J. Engng. Sci. **28**, 829–841 (1990)
26. Truesdell, C.: Rational Thermodyn. McGraw-Hill, New York (1969)
27. Weiss, W.: Continuous shock structure in extended thermodynamics. Phys. Rev. E **52**, R5760–R5763 (1995)

Diffusion Coefficient for the Disordered Harmonic Chain Perturbed by an Energy Conserving Noise

Marielle Simon

Abstract We investigate the macroscopic behavior of the disordered harmonic chain of oscillators, through energy diffusion. The hamiltonian dynamics of the disordered system is perturbed by a degenerate conservative noise. After rescaling space and time diffusively, energy fluctuations in equilibrium evolve according to a linear heat equation (Simon, Equilibrium fluctuations for the disordered harmonic chain perturbed by an energy conserving noise, 2013). Here we concentrate on the diffusion coefficient, given by the non-gradient Varadhan's approach, and equivalently defined through the Green-Kubo formula. We compare the two approaches and investigate the convergence of the diffusion coefficient in a vanishing noise limit.

Keywords Disordered harmonic chain · Diffusion coefficient · Green-Kubo

1 Introduction

This work is based on [1], and addresses diffusion problems for harmonic chains of oscillators with random defects. The purely deterministic disordered harmonic chain of N oscillators was introduced in [2] and since then has attracted a lot of interest. After the first analyses of [2, 3], Ajanki and Huveneers [4] study this disordered chain when coupled at the boundaries to Langevin heat baths, with respective temperatures T_R and T_L. They prove an anomalous heat transport in the following sense: if J_N denotes the total energy current across the chain, then

$$\mathbb{E}\left[\int J_N \, \mathrm{d}\mu_{ss}^N\right] \sim (T_R - T_L) N^{-3/2}$$

M. Simon (✉)
Departamento de Matemática, PUC, Rua Marquês de São Vicente 225, Rio de Janeiro 22453-900, Brazil
e-mail: marielle.simon@mat.puc-rio.br

P. Gonçalves and A.J. Soares (eds.), *From Particle Systems to Partial Differential Equations II*, Springer Proceedings in Mathematics & Statistics 129,
DOI 10.1007/978-3-319-16637-7_14

in the limit $N \to \infty$, where $\mathbb{E}$ states for the expectation w.r.t. the random environment and μ_{ss}^N is the non-equilibrium stationary state for the dynamical system.

Here we study the diffusive behavior for the disordered harmonic chain, but perturbed by an energy conserving noise. Thanks to the stochastic perturbation, the conductivity of the one-dimensional chain should become finite and positive. Besides, some homogenization effect occurs and the conductivity does not depend on the statistics of the disorder in the thermodynamic limit.

The disorder effect has already been investigated for lattice gas dynamics, for example in [5–8]. These papers share one feature: the models are non-gradient due to the presence of the environment. Non-gradient systems are usually solved by establishing a microscopic Fourier law up to a small fluctuating term, following the method initially developed by Varadhan in [9], and then generalized to non-reversible dynamics [10]. The study of disordered chains of oscillators perturbed by a conservative noise has appeared more recently [11–13]. In these papers, the thermal conductivity is studied via the Green-Kubo formula. Here, the diffusion coefficient is furthermore defined through fluctuating hydrodynamics.

In [14], we have obtained the diffusive scaling limit for a homogeneous chain of coupled harmonic oscillators perturbed by a noise, which randomly flips the sign of the velocities, so that the energy is conserved but not the momentum. Our first motivation was to investigate the same chain of harmonic oscillators, still perturbed by the velocity-flip noise, but now provided with i.i.d. random masses. In [14], a system of non-linear homogeneous hydrodynamic equations has been derived thanks to the relative entropy method, and one of the major ingredient for the proof was an exact *fluctuation-dissipation equation* (see for example [15]).

The disorder assumption makes all previous computations pointless: in particular, the fluctuation-dissipation equations are not directly solvable any more. To overcome this difficulty, one replaces these exact equations by approximations: more precisely, there exists a sequence of local functions for which an approximate fluctuation-dissipation decomposition holds, in the sense that the difference has a small space-time variance with respect to the dynamics in equilibrium. The main ingredients of the usual non-reversible non-gradient method are: a *spectral gap* for the symmetric part of the dynamics, and a *sector condition* for the total generator.

Our model has special features that enforce the Varadhan's method to be considered with new perspectives. In particular, the symmetric part of the generator is poorly ergodic, and does not have a spectral gap when restricted to microcanonical manifolds. Moreover, due to the degeneracy of the noise, the asymmetric part of the generator is not controlled by the symmetric part (in technical terms, the sector condition does not hold), with the only velocity-flip noise. Besides, remark that the energy current depends on the disorder, and has to be approximated by a fluctuation-dissipation equation which takes into account the fluctuations of the disorder itself.

Because of the high degeneracy of the velocity-flip noise, we add a second stochastic perturbation, that exchanges velocities (divided by the square root of mass) and positions at random independent Poissonian times, so that a *weak sector condition* can be proved (see [1, Proposition 5.7]). However, the spectral gap estimate and the usual sector condition still do not hold when adding the exchange noise, meaning that the stochastic perturbation remains very degenerate. Due to the harmonicity of the chain, the generator of the dynamics preserves the degree of polynomials, and even a degenerate noise is sufficient to apply Varadhan's method. The sector condition and the non-gradient decomposition are only needed for a specific class of functions. Contrary to the standard approach, we do not need to prove any general result concerning the so-called *closed forms* (we refer to [16, 17] for the general theory). This is a clear advantage of our model: since some difficult technical parts are in some sense simplified, the usual approach to non-gradient problems becomes slightly neater.

We show in addition that the diffusion coefficient can be equivalently defined by the Green-Kubo formula. This space-time variance of the current at equilibrium is only formal in the sense that a double limit (in space and time) has to be taken. As in [11], we prove here that the limit is well-defined, and that the homogenization effect occurs for the Green-Kubo formula: for almost every realization of the disorder, the thermal conductivity exists, is independent of the disorder, is positive and finite.

Finally, let us introduce $\gamma > 0$ the intensity of the flip noise, and $\lambda > 0$ the intensity of the exchange noise. We denote the diffusion coefficient by $D(\lambda, \gamma)$ when obtained through the variational formula in the Varadhan's method, and by $\overline{D}(\lambda, \gamma)$ when defined through the Green-Kubo formula. We prove in [1] that the two conductivities are equal: $D(\lambda, \gamma) = \overline{D}(\lambda, \gamma)$, when the two intensities λ, γ are positive. Furthermore, the Green-Kubo formula remains well-defined when $\lambda = 0$, namely: $\overline{D}(0, \gamma)$ exists, is finite and positive. Finally, $D(\lambda, \gamma)$ tends to $\overline{D}(0, \gamma)$ as λ vanishes. The existence question for $D(0, \gamma)$, when defining through hydrodynamics (or even fluctuating hydrodynamics) remains open. We start in the following section by introducing the model together with notations and definitions.

2 The Harmonic Chain Perturbed by Stochastic Jump Noises

We introduce the harmonic hamiltonian system described by the sequence $\{p_x, r_x\}$, where p_x stands for the momentum of the oscillator at site x, and r_x represents the distance between oscillator x and oscillator $x + 1$. Each atom $x \in \mathbb{Z}$ has a mass M_x, the velocity of atom x is given by p_x/M_x. We assume the disorder $\mathbf{M} := \{M_x\}_{x\in\mathbb{Z}}$ to be a collection of real i.i.d. positive random variables that are bounded from above and below by positive constants. The equations of motions are given by

$$\frac{\mathrm{d}p_x}{\mathrm{d}t} = r_x - r_{x-1}, \qquad \frac{\mathrm{d}r_x}{\mathrm{d}t} = \frac{p_{x+1}}{M_{x+1}} - \frac{p_x}{M_x}, \qquad x \in \mathbb{Z}.$$

The dynamics conserves the total energy

$$\mathscr{E} := \sum_{x\in\mathbb{Z}} \left\{ \frac{p_x^2}{2M_x} + \frac{r_x^2}{2} \right\}.$$

To overcome the lack of ergodicity of deterministic chains, we add a stochastic perturbation to this new dynamics, so that the diffusion coefficient can be defined through Varadhan's approach (Theorem 3.3). The noise can be easily described: at independently distributed random Poissonian times, the quantity $p_x/\sqrt{M_x}$ and the interdistance r_x are exchanged, or the momentum p_x is flipped into $-p_x$.

Even if Theorem 3.3 could be proved *mutatis mutandis* for this harmonic chain, for pedagogical reasons we now focus on a simplified model (as in [18]), which has exactly the same features and involves less painful computations. From now on, we study the dynamics on the new configurations $\{\eta_x\}_{x\in\mathbb{Z}}$ written as

$$m_x \mathrm{d}\eta_x = (\eta_{x+1} - \eta_{x-1})\mathrm{d}t, \tag{1}$$

where $\mathbf{m} := \{m_x\}_{x\in\mathbb{Z}}$ is the new disorder with the same characteristics as before. It is convenient to change the variable η_x into $\omega_x := \sqrt{m_x}\eta_x$, and the total energy reads

$$\mathscr{E} = \sum_{x\in\mathbb{Z}} \omega_x^2.$$

Let us now introduce the corresponding stochastic energy conserving dynamics: the evolution is described by (1) between random exponential times, and at each ring one of the following interactions can happen:

a. Exchange noise—two nearest neighbour variables ω_x and ω_{x+1} are exchanged;
b. Flip noise—the variable ω_x at site x is flipped into $-\omega_x$.

We now describe the dynamics on the finite torus $\mathbb{T}_N := \{0, ..., N\}$, meaning that boundary conditions are periodic. The configuration $\{\omega_x\}_{x\in\mathbb{T}_N}$ evolves according to a dynamics which can be divided into two parts, a deterministic one and a stochastic one. The space of configurations of our system is given by $\Omega_N = \mathbb{R}^N$. The product and translation invariant measure that describes the disorder $\mathbf{m}$ on the space $\Omega_{\mathscr{D}} = [C^{-1}, C]^{\mathbb{Z}}$ is denoted by $\mathbb{P}$ and its expectation is denoted by $\mathbb{E}$. For a fixed disorder field $\mathbf{m} = \{m_x\}_{x\in\mathbb{Z}}$, the dynamics can be entirely defined by the generator of the Markov process $\{\omega_x(t)\ ;\ x \in \mathbb{T}_N\}_{t\geqslant 0}$, that is

$$\mathscr{L}_N^{\mathbf{m}} = \mathscr{A}_N^{\mathbf{m}} + \gamma \mathscr{S}_N^{\text{flip}} + \lambda \mathscr{S}_N^{\text{exch}} \tag{2}$$

where,

$$\mathscr{A}_N^{\mathbf{m}} = \sum_{x \in \mathbb{T}_N} \left\{ \left(\frac{\omega_{x+1}}{\sqrt{m_x m_{x+1}}} - \frac{\omega_{x-1}}{\sqrt{m_{x-1} m_x}} \right) \frac{\partial}{\partial \omega_x} \right\},$$

and, for all functions $f : \Omega_{\mathscr{D}} \times \Omega_N \to \mathbb{R}$,

$$\mathscr{S}_N^{\text{flip}} f(\mathbf{m}, \omega) = \sum_{x \in \mathbb{T}_N} f(\mathbf{m}, \omega^x) - f(\mathbf{m}, \omega),$$

$$\mathscr{S}_N^{\text{exch}} f(\mathbf{m}, \omega) = \sum_{x \in \mathbb{T}_N} f(\mathbf{m}, \omega^{x,x+1}) - f(\mathbf{m}, \omega).$$

Here, the configuration ω^x is the configuration obtained from ω by flipping the momentum of particle x: $(\omega^x)_z = \omega_z$ if $z \neq x$ and $\omega_x^x = -\omega_x$. The configuration $\omega^{x,x+1}$ is obtained from ω by exchanging the momenta of particles x and $x+1$: $(\omega^{x,x+1})_z = \omega_z$ if $z \neq x, x+1$ while $\omega_x^{x,x+1} = \omega_{x+1}$, and $\omega_{x+1}^{x,x+1} = \omega_x$. We denote the total generator of the noise by $\mathscr{S}_N := \gamma \mathscr{S}_N^{\text{flip}} + \lambda \mathscr{S}_N^{\text{exch}}$, where $\gamma, \lambda > 0$ are two positive parameters which regulate the respective strengths of noises.

One quantity is conserved: the total energy $\sum \omega_x^2$. The following translation invariant product Gibbs measures μ_β^N on Ω_N are invariant for the process:

$$\mathrm{d}\mu_\beta^N(\omega) := \prod_{x \in \mathbb{T}_N} \sqrt{\frac{2\pi}{\beta}} \exp\left(-\frac{\beta}{2}\omega_x^2\right) \mathrm{d}\omega_x.$$

In the following, the expectation of f with respect to μ_β^N is denoted by $\langle f \rangle_\beta$. The index β stands for the inverse temperature: $\langle \omega_0^2 \rangle_\beta = 1/\beta$. From the definition, our model is not reversible with respect to the measure μ_β^N. Precisely, $\mathscr{A}_N^{\mathbf{m}}$ is an antisymmetric operator in $\mathbf{L}^2(\mu_\beta^N)$, whereas $\mathscr{S}_N$ is symmetric.

We denote by Ω the space of configurations in the infinite line, that is $\Omega := \mathbb{R}^{\mathbb{Z}}$, and by μ_β the product Gibbs measure on $\mathbb{R}^{\mathbb{Z}}$. Hereafter, for every $\beta > 0$, we denote by $\mathbb{P}_\beta^\star$ the probability measure on $\Omega_{\mathscr{D}} \times \Omega$ defined by $\mathbb{P}_\beta^\star := \mathbb{P} \otimes \mu_\beta$. We notice that $\mathbb{P}_\beta^\star$ is translation invariant and we write $\mathbb{E}_\beta^\star$ for the corresponding expectation.

Since the dynamics conserves the total energy, there exist instantaneous currents of energy $j_{x,x+1}$ such that $\mathscr{L}_N^{\mathbf{m}}(\omega_x^2) = j_{x,x+1}(\mathbf{m}, \omega) - j_{x-1,x}(\mathbf{m}, \omega)$. The quantity $j_{x,x+1}$ is the amount of energy flowing between the particles x and $x+1$, and is equal to

$$j_{x,x+1}(\mathbf{m}, \omega) = \frac{2\omega_x \omega_{x+1}}{\sqrt{m_x m_{x+1}}} + \lambda(\omega_{x+1}^2 - \omega_x^2).$$

We write $j_{x,x+1} = j_{x,x+1}^A + j_{x,x+1}^S$ where $j_{x,x+1}^A$ (resp. $j_{x,x+1}^S$) is the current associated to the antisymmetric (resp. symmetric) part of the generator:

$$j^A_{x,x+1}(\mathbf{m},\omega) = \frac{2\omega_x\omega_{x+1}}{\sqrt{m_x m_{x+1}}}, \qquad j^S_{x,x+1}(\mathbf{m},\omega) = \lambda(\omega_{x+1}^2 - \omega_x^2).$$

Unfortunately the current cannot be directly written as the gradient of a local function, neither by an exact fluctuation-dissipation equation involving local functions (except if masses are equal). We also define the *static compressibility* that is equal to $\chi_\beta := \langle\omega_0^4\rangle_\beta - \langle\omega_0^2\rangle_\beta^2 = 2\beta^{-2}$.

2.1 Cylinder Functions and Dirichlet Form

For every $x \in \mathbb{Z}$ and f a measurable function on $\Omega_{\mathcal{D}} \times \Omega$, we consider the translated function $\tau_x f$, which is the function on $\Omega_{\mathcal{D}} \times \Omega$ defined by: $\tau_x f(\mathbf{m},\omega) := f(\tau_x\mathbf{m}, \tau_x\omega)$, where $\tau_x\mathbf{m}$ and $\tau_x\omega$ are the disorder and particle configurations translated by $x \in \mathbb{Z}$, respectively: $(\tau_x\mathbf{m})_z := m_{x+z}$, and $(\tau_x\omega)_z = \omega_{x+z}$. For a fixed positive integer ℓ, we define $\Lambda_\ell := \{-\ell, ..., \ell\}$. If the box is centered at site $x \in \mathbb{Z}$, we denote it by $\Lambda_\ell(x) := \{-\ell+x, ..., \ell+x\}$. If f is a measurable function on $\Omega_{\mathcal{D}} \times \Omega$, the *support* of f, denoted by Λ_f, is the smallest subset of $\mathbb{Z}$ such that $f(\mathbf{m},\omega)$ only depends on $\{m_x, \omega_x \; ; \; x \in \Lambda_f\}$ and f is called a *cylinder (or local) function* if Λ_f is finite. In that case, we denote by s_f the smallest positive integer s such that Λ_s contains the support of f and then $\Lambda_f = \Lambda_{s_f}$. For every cylinder function $f : \Omega_{\mathcal{D}} \times \Omega \to \mathbb{R}$, consider the formal sum

$$\Gamma_f := \sum_{x\in\mathbb{Z}} \tau_x f$$

which does not make sense but for which

$$(\nabla_x f)(\mathbf{m},\omega) := f(\mathbf{m},\omega^x) - f(\mathbf{m},\omega),$$
$$(\nabla_{x,x+1} f)(\mathbf{m},\omega) := f(\mathbf{m},\omega^{x,x+1}) - f(\mathbf{m},\omega)$$

are well-defined.

Definition 2.1 We denote by $\mathcal{C}$ the set of measurable cylinder functions φ on $\Omega_{\mathcal{D}} \times \Omega$, such that

1. for all $\omega \in \Omega$, the random variable $\mathbf{m} \mapsto \varphi(\mathbf{m},\omega)$ is continuous on $\Omega_{\mathcal{D}}$;
2. for all $\mathbf{m} \in \Omega_{\mathcal{D}}$, the function $\omega \mapsto \varphi(\mathbf{m},\omega)$ belongs to $\mathbf{L}^2(\mu_\beta)$ and has null average with respect to μ_β.

Definition 2.2 We introduce the set of *quadratic* cylinder functions on $\Omega_{\mathcal{D}} \times \Omega$, denoted by $\mathcal{Q} \subset \mathcal{C}$, and defined as follows: $f \in \mathcal{Q}$ if there exists a sequence $\{\psi_{i,j}(\mathbf{m})\}_{i,j\in\mathbb{Z}}$ of real cylinder measurable functions on $\Omega_{\mathcal{D}}$ such that

1. for all $i,j \in \mathbb{Z}$, $\omega \in \Omega$, the random variable $\mathbf{m} \mapsto \psi_{i,j}(\mathbf{m},\omega)$ is continuous;
2. $\psi_{i,j}$ vanishes for all but a finite number of pairs (i,j),

3. f is written as

$$f(\mathbf{m}, \omega) = \sum_{i \in \mathbb{Z}} \psi_{i,i}(\mathbf{m})(\omega_{i+1}^2 - \omega_i^2) + \sum_{\substack{i,j \in \mathbb{Z} \\ i \neq j}} \psi_{i,j}(\mathbf{m})\omega_i\omega_j, \tag{3}$$

In other words, quadratic functions are homogeneous polynomials of degree two in the variable ω, that have null average with respect to μ_β for every $\mathbf{m} \in \Omega_{\mathscr{D}}$. An other definition through *Hermite polynomials* is given in [1]. We are now ready to define two sets of functions that will play further a crucial role.

Definition 2.3 Let $\mathscr{C}_0$ be the set of cylinder functions φ on $\Omega_{\mathscr{D}} \times \Omega$ such that there exists a finite subset Λ of $\mathbb{Z}$, and cylinder, measurable functions $\{F_x, G_x\}_{x \in \Lambda}$ defined on $\Omega_{\mathscr{D}} \times \Omega$, that verify

$$\varphi = \sum_{x \in \Lambda} \Big\{ \nabla_x(F_x) + \nabla_{x,x+1}(G_x) \Big\},$$

and such that, for all $x \in \Lambda$,

1. for all $\omega \in \Omega$, $\mathbf{m} \mapsto F_x(\mathbf{m}, \omega)$ and $\mathbf{m} \mapsto G_x(\mathbf{m}, \omega)$ are continuous on $\Omega_{\mathscr{D}}$;
2. for all $\mathbf{m} \in \Omega_{\mathscr{D}}$, $\omega \mapsto F_x(\mathbf{m}, \omega)$ and $\omega \mapsto G_x(\mathbf{m}, \omega)$ belong to $\mathbf{L}^2(\mu_\beta)$.

Let $\mathscr{Q}_0 \subset \mathscr{C}_0$ be the set of such functions φ, with the additional assumption that the cylinder functions F_x, G_x are homogeneous polynomials of degree two in ω.

Before giving a few properties of these two spaces, let us now consider operators $\mathscr{L}^{\mathbf{m}}$, $\mathscr{A}^{\mathbf{m}}$ and $\mathscr{S}$ acting on functions $f \in \mathscr{C}$ in the same way as (2), except that the sums now run on the whole line $\mathbb{Z}$. For a finite subset Λ_ℓ of $\mathbb{Z}$ defined as above, we denote by $\mathscr{L}^{\mathbf{m}}_{\Lambda_\ell}$, resp. $\mathscr{S}_{\Lambda_\ell}$, the restriction of the generator $\mathscr{L}^{\mathbf{m}}$, resp. $\mathscr{S}$, to the box Λ_ℓ, assuming periodic boundary conditions.

Definition 2.4 Let $\mathscr{C}_0$ (respectively $\mathscr{Q}_0$) be the set of cylinder (respectively quadratic cylinder) functions φ on $\Omega_{\mathscr{D}} \times \Omega$ such that there exists a finite subset $\Lambda \Subset \mathbb{Z}$, and cylinder functions $\{F_x, G_x\}_{x \in \Lambda}$ satisfying

$$\varphi = \sum_{x \in \Lambda} \nabla_x(F_x) + \nabla_{x,x+1}(G_x).$$

If φ belongs to $\mathscr{Q}_0$, we assume the cylinder functions F_x, G_x to be quadratic.

Finally we introduce the *Dirichlet form* associated to the generator: for any $x \in \mathbb{Z}$ and $f, g \in \mathscr{C}$, let us define $\mathscr{D}_\ell(\mu_\beta; f) := \big\langle (-\mathscr{L}^{\mathbf{m}}_{\Lambda_\ell}) f, f \big\rangle_\beta = \big\langle (-\mathscr{S}_{\Lambda_\ell}) f, f \big\rangle_\beta$.

2.2 Semi-inner Products and Diffusion Coefficient

For cylinder functions $g, h \in \mathscr{C}$, let us introduce

$$\langle\langle g, h\rangle\rangle_{\beta,\star} := \sum_{x\in\mathbb{Z}} \mathbb{E}^\star_\beta[g\,\tau_x h] \quad\text{and}\quad \langle\langle g\rangle\rangle_{\beta,\star\star} := \sum_{x\in\mathbb{Z}} x\,\mathbb{E}^\star_\beta[g\,\omega_x^2] \tag{4}$$

which are well-defined because g and h belong to $\mathscr{C}$ and therefore all but a finite number of terms vanish. Notice that $\langle\langle\cdot,\cdot\rangle\rangle_{\beta,\star}$ is a semi inner product.

Definition 2.5 We define the *diffusion coefficient* $D(\beta)$ for $\beta > 0$ as equal to

$$\lambda + \frac{1}{\chi_\beta}\inf_{f\in\mathscr{Q}}\sup_{g\in\mathscr{Q}}\Big\{\langle\langle f, -\mathscr{S}f\rangle\rangle_{\beta,\star} + 2\langle\langle j^A_{0,1} - \mathscr{A}^{\mathbf{m}}f, g\rangle\rangle_{\beta,\star} - \langle\langle g, -\mathscr{S}g\rangle\rangle_{\beta,\star}\Big\}.$$

The first term in the sum is only due to the exchange noise, whereas the second one comes from the hamiltonian part of the dynamics. Formally, this formula reads

$$D(\beta) = \lambda + \frac{1}{\chi_\beta}\langle\langle j^A_{0,1}, (-\mathscr{L}^{\mathbf{m}})^{-1} j^A_{0,1}\rangle\rangle_{\beta,\star}, \tag{5}$$

but the last term is ill-defined because $j^A_{0,1}$ is not in the range of $\mathscr{L}^{\mathbf{m}}$. More rigorously, we should define $\langle\langle j^A_{0,1}, (-\mathscr{L}^{\mathbf{m}})^{-1} j^A_{0,1}\rangle\rangle_{\beta,\star}$ as

$$\limsup_{z\to 0}\langle\langle j^A_{0,1}, (z - \mathscr{L}^{\mathbf{m}})^{-1} j^A_{0,1}\rangle\rangle_{\beta,\star}.$$

The scalar product above is now well-defined, and the problem is reduced to prove convergence as $z \to 0$. From Hille-Yosida Theorem (see [19, Proposition 2.1] for instance) (5) is equal to the infinite volume Green-Kubo formula:

$$\overline{D}(\beta) = \lambda + \frac{1}{\chi_\beta}\lim_{\substack{z\to 0\\ z>0}}\mathbb{E}\left[\int_0^{+\infty} e^{-zt}\left\langle\sum_{x\in\mathbb{Z}} j^A_{x,x+1}(t), j^A_{0,1}(0)\right\rangle_\beta \mathrm{d}t\right]. \tag{6}$$

In Sect. 4, we prove that (6) converges, inspired by [11]. Assuming the convergence in the Green-Kubo formula, one can easily see that $\overline{D}(\beta)$ does not depend on β. We denote

$$L(z) := \frac{1}{\chi_\beta}\int_0^{+\infty} e^{-zt}\,\langle\langle j^A_{0,1}(t), j^A_{0,1}(0)\rangle\rangle_{\beta,\star}\mathrm{d}t.$$

The function L is smooth on $(0,+\infty)$. The Hilbert space generated by the set of local functions and the inner product $\langle\langle\cdot,\cdot\rangle\rangle_{\beta,\star}$ is denoted by $\mathbf{L}^2_\star$. We define $h_z := h_z(\mathbf{m},\omega;\beta)$ as the solution of the resolvent equation (in $\mathbf{L}^2_\star$) $(z - \mathscr{L}^{\mathbf{m}})h_z = j^A_{0,1}$. Then,

$$L(z) = \frac{1}{\chi_\beta}\langle\langle h_z, j^A_{0,1}\rangle\rangle_{\beta,\star} = \frac{\beta^2}{2}\langle\langle h_z, j^A_{0,1}\rangle\rangle_{\beta,\star}. \tag{7}$$

Observe that if ω is distributed according to μ_β then $\beta^{1/2}\omega$ is distributed according to μ_1. Since $h_z(\mathbf{m}, \omega; 1) = h_z(\mathbf{m}, \omega; \beta)$ and $j^A_{x,x+1}$ is a homogeneous function of degree two in ω, it follows that the diffusion coefficient does not depend on β.

From now on, we assume $\beta = 1$. This assumption is justified since we are going to deal only with quadratic functions (as defined before). For instance, when one result is stated for the scalar product $\langle\langle\cdot\rangle\rangle_{1,\star}$, the same argument in the proof can be rewritten for any $\beta > 0$, after multiplying the process $\{\omega_x(t)\}$ by $\beta^{-1/2}$.

3 Non-gradient Varadhan Approach

In this section we are going to identify the diffusion coefficient D given in Definition 2.5. Roughly speaking, D is the asymptotic component of the energy current $j_{x,x+1}$ in the direction of the gradient $\omega^2_{x+1} - \omega^2_x$, and makes the expression below vanish:

$$\inf_{f\in\mathcal{Q}} \limsup_{N\to\infty} \limsup_{t\to\infty} \frac{1}{tN}\mathbb{E}^\star_1\Big[\Big(\int_0^t \sum_{x\in\mathbb{T}_N} \big[j_{x,x+1} - D(\omega^2_{x+1} - \omega^2_x) - \mathcal{L}^{\mathbf{m}}(\tau_x f)\big]\mathrm{d}s\Big)^2\Big].$$

3.1 *An Insight Through Additive Functionals of Markov Processes*

Consider a continuous time Markov process $\{Y_s\}_{s\geqslant 0}$ on a complete and separable metric space E, with an invariant measure π. We denote by $\langle\cdot\rangle_\pi$ the inner product in $\mathbf{L}^2(\pi)$ and by $\mathcal{L}$ the infinitesimal generator of the process. The adjoint of $\mathcal{L}$ in $\mathbf{L}^2(\pi)$ is denoted by $\mathcal{L}^\star$. Fix a function $V : E \to \mathbb{R}$ in $\mathbf{L}^2(\pi)$ such that $\langle V\rangle_\pi = 0$. Theorem 2.7 in [20] gives conditions which guarantee a CLT for

$$\frac{1}{\sqrt{t}}\int_0^t V(Y_s)\mathrm{d}s$$

and shows that the limiting variance equals

$$\sigma^2(V, \pi) = 2 \lim_{z\to 0^+} \big\langle V, (z - \mathcal{L})^{-1} V\big\rangle_\pi .$$

Let the generator $\mathcal{L}$ be decomposed as $\mathcal{L} = \mathcal{S} + \mathcal{A}$, where $\mathcal{S} = (\mathcal{L} + \mathcal{L}^\star)/2$ and $\mathcal{A} = (\mathcal{L} - \mathcal{L}^\star)/2$ denote, respectively, the symmetric and antisymmetric parts of $\mathcal{L}$. Let $\mathcal{H}_1$ be the completion of $\mathbf{L}^2(\pi)$ with respect to the semi-norm $\|\cdot\|_1$:

$$\|f\|_1^2 := \big\langle f, (-\mathcal{L})f\big\rangle_\pi = \big\langle f, (-\mathcal{S})f\big\rangle_\pi .$$

Let $\mathscr{H}_{-1}$ be the dual space of $\mathscr{H}_1$ w.r.t. $\mathbf{L}^2(\pi)$, in other words, the Hilbert space generated by suitably regular functions and the norm $\|\cdot\|_{-1}$ defined by

$$\|f\|_{-1}^2 := \sup_g \left\{ 2\langle f, g\rangle_\pi - \|g\|_1^2 \right\},$$

where the supremum is carried over all local functions g. Formally, $\|f\|_{-1}$ can also be thought as $\langle f, (-\mathscr{S})^{-1} f\rangle_\pi$. The following result is a rigorous estimate of the time variance in terms of the $\mathscr{H}_{-1}$ norm, which is proved in [20, Lemma 2.4].

Lemma 3.1 *Given $T > 0$ and a mean zero function V in $\mathbf{L}^2(\pi) \cap \mathscr{H}_{-1}$,*

$$\mathbb{E}_\pi \left[\sup_{0\leqslant t\leqslant T} \left(\int_0^t V(s)\mathrm{d}s \right)^2 \right] \leqslant 24T\|V\|_{-1}^2. \tag{8}$$

Then, we should take V proportional to

$$\sum_{x\in\mathbb{T}_N} \left[j_{x,x+1} - D(\omega_{x+1}^2 - \omega_x^2) - \mathscr{L}^{\mathbf{m}}(\tau_x f) \right]$$

and then take the limit as $N \to \infty$. In the right-hand side of (8) we obtain a variance that depends on N, and the main task is to show that this variance converges. Precisely, we can prove that the limit of the variance results in a semi-norm, which is denoted by $|||\cdot|||_1$ and defined in (9). The final step consists in minimizing this semi-norm on a well-chosen subspace, through orthogonal projections in Hilbert spaces. The hard point is that $|||\cdot|||_1$ only depends on the symmetric part of the generator $\mathscr{S}$, and the latter is really degenerate (it does not have a spectral gap).

In [1], we prove that the variance $\langle f, (-\mathscr{S})^{-1} f\rangle_1$ is well defined for every function f in $\mathscr{Q}_0$. In Sect. 3.2, we relate the previous limiting variance (taking the limit as N goes to infinity) to the suitable semi-norm. Finally, Sect. 3.3 focuses on the diffusion coefficient and its different expressions.

3.2 Limiting Variance and Semi-norm

We return to the case $\beta = 1$. We look for a variational formula for the variance

$$(2\ell)^{-1}\mathbb{E}\left\langle \left(-\mathscr{S}_{\Lambda_\ell}\right)^{-1} \sum_{|x|\leqslant \ell_\varphi} \tau_x\varphi, \sum_{|x|\leqslant \ell_\varphi} \tau_x\varphi \right\rangle_1$$

where $\varphi \in \mathscr{Q}_0$ and $\ell_\varphi = \ell - s_\varphi - 1$. We first introduce a semi-norm on $\mathscr{Q}_0$. For any cylinder function φ in $\mathscr{Q}_0$, let us define

$$|||\varphi|||_1^2 = 2 \sup_{g\in\mathscr{Q}} \left\{ \langle\langle \varphi, g\rangle\rangle_{1,\star} + \frac{\langle\langle\varphi\rangle\rangle_{1,\star\star}^2}{\lambda\chi_1} - \frac{\lambda}{4}\mathbb{E}_1^\star\left[\left(\nabla_{0,1}\Gamma_g\right)^2\right] - \frac{\gamma}{4}\mathbb{E}_1^\star\left[\left(\nabla_0\Gamma_g\right)^2\right]\right\}$$

$$= \sup_{\substack{g\in\mathscr{Q}\\ a\in\mathbb{R}}} \left\{2\langle\langle\varphi, g\rangle\rangle_{1,\star} + 2a\langle\langle\varphi\rangle\rangle_{1,\star\star} - \mathbb{E}\left[\mathscr{D}_0\left(\mu_1; a\omega_0^2 + \Gamma_g\right)\right]\right\}. \tag{9}$$

This formula can be formally restated as

$$|||\varphi|||_1^2 = \langle\langle\varphi, (-\mathscr{S})^{-1}\varphi\rangle\rangle_{1,\star} + \frac{2}{\lambda\,\chi_1}\langle\langle\varphi\rangle\rangle_{1,\star\star}^2. \tag{10}$$

Since φ belongs to $\mathscr{Q}_0$, one can prove that the first term in the right-hand side of (10) is well-defined (Proposition 4.4 in [1]). We are now in position to state the key result of the non-gradient Varadhan approach.

Theorem 3.2 *Consider a quadratic cylinder function $\varphi \in \mathscr{Q}_0$. Then*

$$\lim_{\ell\to\infty} (2\ell)^{-1}\mathbb{E}\Big\langle\Big(- \mathscr{S}_{\Lambda_\ell}\Big)^{-1} \sum_{|x|\leqslant\ell_\varphi} \tau_x\varphi, \sum_{|x|\leqslant\ell_\varphi} \tau_x\varphi\Big\rangle_1 = |||\varphi|||_1^2.$$

3.3 Hilbert Space and Projections

We can easily define from $|||\cdot|||_1$ a semi-inner product on $\mathscr{C}_0$ through polarization. Denote by $\mathscr{N}$ the kernel of the semi-norm $|||\cdot|||_1$ on $\mathscr{C}_0$. Then, the completion of $\mathscr{Q}_0|_{\mathscr{N}}$ denoted by $\mathscr{H}_1$ is a Hilbert space. Let us explain how the well-known Varadhan's approach is modified. Usually, the Hilbert space on which orthogonal projections are performed is the completion of $\mathscr{C}_0|_{\mathscr{N}}$, in other words it involves all local functions. Then, the standard procedure aims at proving that each element of that Hilbert space can be approximated by a sequence of functions in the range of the generator plus an additional term which is proportional to the current. The crucial steps for obtaining this decomposition consist in: first, controlling the antisymmetric part of the generator by the symmetric one for every cylinder function, and second, proving a strong result on germs of closed forms. These two key points are not valid in our model, but they can be proved when restricted to quadratic functions. It turns out that these weak versions are sufficient, since we are looking for a fluctuation-dissipation approximation that involves quadratic functions only.

In [1], we show that $\mathscr{H}_1$ is the completion of $\mathscr{S}\mathscr{Q}|_{\mathscr{N}} + \{j_{0,1}^S\}$. In other words, all elements of $\mathscr{H}_1$ can be approximated by $aj_{0,1}^S + \mathscr{S}g$ for some $a \in \mathbb{R}$ and $g \in \mathscr{Q}$. This is not irrelevant since the symmetric part of the generator preserves the degree of polynomial functions. The sum of the two subspaces $\{j_{0,1}^S\}$ and $\overline{\mathscr{S}\mathscr{Q}|_{\mathscr{N}}}$ is orthogonal. Nevertheless, this decomposition is not satisfactory, because we want the fluctuating term to be on the form $\mathcal{L}^{\mathbf{m}}(f)$, and not $\mathcal{S}(f)$. In order to make this replacement, we need to prove the weak sector condition, that gives a control of $|||\mathcal{A}^{\mathbf{m}} g|||_1$ by $|||\mathcal{S} g|||_1$, when g is a quadratic function. The only trouble is that this new decomposition is not orthogonal any more, so that we can not express the diffusion coefficient as a variational formula. This problem is solved by clever projections into a suitable Hilbert space. The main theorem given in [1] is:

Theorem 3.3 *For every $g \in \mathcal{Q}_0$, there exists a unique constant $a \in \mathbb{R}$, such that*

$$g + a(\omega_1^2 - \omega_0^2) \in \overline{\mathcal{L}^{\mathbf{m}} \mathcal{Q}} \quad in\ \mathcal{H}_1.$$

In particular, there exists a unique number D, and a sequence $\{f_k\} \in \mathcal{Q}$ such that

$$|||j_{0,1} - D(\omega_1^2 - \omega_0^2) - \mathcal{L}^{\mathbf{m}}(f_k)|||_1 \xrightarrow[k\to\infty]{} 0.$$

Finally, one can prove more formulas for coefficient D defined in Theorem 3.3, and relate it to Definition 2.5, by following the argument given by instance in [21].

4 Convergence of Green-Kubo Formula

Remind that the *Green-Kubo formula* predicted by linear response theory is

$$\overline{\kappa}(z) := \lambda + \frac{1}{2}\langle\langle j_{0,1}^A, (z - \mathcal{L}^{\mathbf{m}})^{-1} j_{0,1}^A\rangle\rangle_{1,\star}. \tag{11}$$

Hereafter, we extend the inner-product $\langle\langle\cdot\rangle\rangle_{1,\star}$ (originally defined on $\mathcal{C}$) to the Hilbert space generated by the set of square integrable functions and denoted by $\mathbf{L}_\star^2$.

4.1 Existence of the Green-Kubo Formula

In this paragraph we prove the existence and finiteness of the Green-Kubo formula. The argument is based on the paper [11], where the author generalizes [22, 23].

Theorem 4.1 *The z-vanishing limit $\overline{D} := \lim \overline{\kappa}(z)$ exists, is finite and positive.*

Proof Recall (7). We have to prove that $\langle\langle h_z, j_{0,1}^A\rangle\rangle_{1,\star}$ converges as z vanishes, and that the limit is finite and non-negative. Then, from (11) it will follow that $\overline{D} \geqslant \lambda > 0$ and $\overline{D}$ is positive. We denote by $\|\cdot\|_1$ the semi-norm corresponding to the symmetric part of the generator due to the flip noise

$$\|f\|_1^2 = \langle\langle f, (-\gamma\mathcal{S}^{\text{flip}})f\rangle\rangle_{1,\star}$$

and $\mathscr{H}_\star$ is the Hilbert space obtained by the completion of $\mathbf{L}^2_\star$ w.r.t. that semi-norm. We multiply the resolvent equation by h_z and integrate with respect to $\langle\langle\cdot\rangle\rangle_{1,\star}$:

$$z\langle\langle h_z, h_z\rangle\rangle_{1,\star} + \|h_z\|_1^2 + \langle\langle h_z, (-\lambda\mathscr{S}^{\text{exch}})h_z\rangle\rangle_{1,\star} = \langle\langle h_z, j^A_{0,1}\rangle\rangle_{1,\star}.$$

Let us notice that $(-\gamma\mathscr{S}^{\text{flip}})(j^A_{0,1}) = 2\gamma j^A_{0,1}$. As a consequence, the Cauchy-Schwarz inequality for the scalar product $\langle\langle\cdot, (-\gamma\mathscr{S}^{\text{flip}})\cdot\rangle\rangle_{1,\star}$ on the right-hand side gives $\|h_z\|_1^2 \leqslant C$ for some positive constant C. Since $\{h_z\}_z$ is bounded in $\mathscr{H}_\star$, we can extract a weakly converging subsequence in $\mathscr{H}_\star$. We continue to denote this subsequence by $\{h_z\}_z$ and we denote by h_0 the limit.

Now we are going to show that the convergence is stronger (see (4) in Lemma 4.2 below) and that the limit is independent of the subsequence. Since the generator $\mathscr{L}^{\mathbf{m}}$ conserves the degree of homogeneous polynomial functions, we know that the solution of the resolvent equation is expected to be on the form

$$h_z(\omega) = \sum_{x,y\in\mathbb{Z}^2} \varphi_z(x,y)\omega_x\omega_y,$$

where $\varphi_z : \mathbb{Z}^2 \to \mathbb{R}$ is a square-summable symmetric function. Let $h_z = h_z^= + h_z^{\neq}$ be the decomposition of h_z according to the two subspaces $\mathscr{Q}^=$ and $\mathscr{Q}^{\neq}$, where $\mathscr{Q}^=$ is generated by $\{\omega_x^2, x \in \mathbb{Z}\}$ and $\mathscr{Q}^{\neq}$ is generated by $\{\omega_x\omega_y, x \neq y\}$. The main point in the following argument is that all gradient terms vanish in $\mathbf{L}^2_\star$.

First, one can see how the spaces $\mathscr{Q}^=$ and $\mathscr{Q}^{\neq}$ are mapped by the generators:

$$\begin{aligned}
\mathscr{A}^{\mathbf{m}} &: \mathscr{Q}^= \to \mathscr{Q}^{\neq} & \mathscr{A}^{\mathbf{m}} &: \mathscr{Q}^{\neq} \to \mathscr{Q} \\
\mathscr{S}^{\text{flip}} &: \mathscr{Q}^= \to \{0\} & \mathscr{S}^{\text{flip}} &: \mathscr{Q}^{\neq} \to \mathscr{Q}^{\neq} \\
\mathscr{S}^{\text{exch}} &: \mathscr{Q}^= \to \mathscr{Q}^= & \mathscr{S}^{\text{exch}} &: \mathscr{Q}^{\neq} \to \mathscr{Q}^{\neq}
\end{aligned}$$

Moreover, if $f \in \mathscr{Q}^=$, then $\mathscr{A}^{\mathbf{m}}(f)$ is a gradient in $\mathscr{Q}^{\neq}$, and $\mathscr{S}^{\text{exch}}(f)$ is a gradient in $\mathscr{Q}^=$. With all these considerations, the resolvent equation rewrites in $\mathbf{L}^2_\star$ as

$$\begin{cases} zh_z^= - \lambda\mathscr{S}^{\text{exch}}(h_z^=) = 0 \\ zh_z^{\neq} - \lambda\mathscr{S}^{\text{exch}}(h_z^{\neq}) - \gamma\mathscr{S}^{\text{flip}}(h_z^{\neq}) - \mathscr{A}^{\mathbf{m}}(h_z^{\neq}) = j^A_{0,1}. \end{cases}$$

The first equation means that $h_z^= = 0$ in $\mathbf{L}^2_\star$ and therefore the solution h_z of the resolvent equation is an element of $\mathscr{Q}^{\neq}$. As a consequence, we can write $(-\gamma\mathscr{S}^{\text{flip}})(h_z) = 2\gamma h_z$, and this remark is one of the key points in the following argument.

Lemma 4.2 *All the properties below are satisfied:*

1. $\lim_{z\to 0} z\langle\langle h_z, h_z\rangle\rangle_{1,\star} = 0$
2. $\{h_z\}$ *weakly converges as* z *goes to 0 towards* h_0 *in* $\mathbf{L}^2_\star$
3. $\langle\langle j^A_{0,1}, h_0\rangle\rangle_{1,\star} = \langle\langle h_0, (-\mathscr{S})h_0\rangle\rangle_{1,\star}$
4. $\langle\langle (h_z - h_0), (-\mathscr{S})(h_z - h_0)\rangle\rangle_{1,\star}$ *vanishes as* z *goes to 0*
5. *the weak limit of* $\{h_z\}$ *does not depend on the subsequence.*

We briefly prove the five points: (1) and (2) come from the fact that $(-\gamma\mathscr{S}^{\text{flip}})(h_z) = 2\gamma h_z$. To get (3), we multiply the resolvent equation by $h_{z'}$ and integrate:

$$z\langle\langle h_{z'}, h_z\rangle\rangle_{1,\star} + \langle\langle h_{z'}, (-\mathscr{S})h_z\rangle\rangle_{1,\star} + \langle\langle h_{z'}, (-\mathscr{A}^{\mathbf{m}})h_z\rangle\rangle_{1,\star} = \langle\langle h_{z'}, j^A_{0,1}\rangle\rangle_{1,\star}.$$

We first take the limit as $z' \to 0$ and then as $z \to 0$, and we use (1) and (2) to obtain (3). In the same way, multiplying the resolvent equation by h_z gives

$$z\langle\langle h_z, h_z\rangle\rangle_{1,\star} + \langle\langle h_z, (-\mathscr{S})h_z\rangle\rangle_{1,\star} = \langle\langle h_z, j^A_{0,1}\rangle\rangle_{1,\star}.$$

The first term of the left-hand side vanishes as z goes to 0, and the right-hand side converges to $\langle\langle h_0, (-\mathscr{S})h_0\rangle\rangle_{1,\star}$. This implies (4), that is

$$\langle\langle (h_z - h_0), (-\mathscr{S})(h_z - h_0)\rangle\rangle_{1,\star} \xrightarrow[z\to 0]{} 0.$$

The uniqueness of the limit follows by a standard argument with same ideas as before. We have proved the first part: the limit exists. To obtain its finiteness, we are going to give an upper bound, using the following variational formula:

$$\langle\langle j^A_{0,1}, (z - \mathscr{L}^{\mathbf{m}})^{-1} j^A_{0,1}\rangle\rangle_{1,\star} = \sup_f \left\{ 2\langle\langle f, j^A_{0,1}\rangle\rangle_{1,\star} - \|f\|^2_{1,z} - \|\mathscr{A}^{\mathbf{m}} f\|^2_{-1,z} \right\},$$

where the supremum is carried over local functions and the two norms $\|\cdot\|_{\pm 1,z}$ are

$$\|f\|^2_{\pm 1,z} = \langle\langle f, (z - \mathscr{S})^{\pm 1} f\rangle\rangle_{1,\star}.$$

For the upper bound, we neglect the term coming from the antisymmetric part $\mathscr{A}^{\mathbf{m}} f$:

$$\langle\langle j^A_{0,1}, (z - \mathscr{L}^{\mathbf{m}})^{-1} j^A_{0,1}\rangle\rangle_{1,\star} \leqslant \langle\langle j^A_{0,1}, (z - \mathscr{S})^{-1} j^A_{0,1}\rangle\rangle_{1,\star}.$$

In the right-hand side we can also neglect the part coming from the exchange symmetric part $\mathscr{S}^{\text{exch}}$, and remind that $\mathscr{S}^{\text{flip}}(j^A_{0,1}) = -2j^A_{0,1}$. This gives an explicit finite upper bound. Then, we have from Lemma 4.2, Property (3) that

$$\lim_{z\to 0} \langle\langle j^A_{0,1}, (z - \mathscr{L}^{\mathbf{m}})^{-1} j^A_{0,1}\rangle\rangle_{1,\star} = \langle\langle j^A_{0,1}, h_0\rangle\rangle_{1,\star} = \langle\langle h_0, (-\mathscr{S})h_0\rangle\rangle_{1,\star} \geqslant 0.$$

4.2 Equivalence of the Definitions and Vanishing Noise Limit

Finally, we can rigorously prove [1] the equality between the variational formula for the diffusion coefficient and the Green-Kubo formula, precisely:

Theorem 4.3 *For every $\lambda > 0$ and $\gamma > 0$,*

$$\overline{D} := \lambda + \frac{1}{2} \lim_{\substack{z\to 0 \\ z>0}} \langle\langle j^A_{0,1}, (z - \mathscr{L}^{\mathbf{m}})^{-1} j^A_{0,1} \rangle\rangle_{1,\star},$$

coincides with the diffusion coefficient D defined in Theorem 3.3.

Moreover, it can be easily shown that the homogenized Green-Kubo formula also converges if the strength λ of the exchange noise vanishes. First, we turn (11) into a new definition that highlights the dependence on $\lambda > 0$. For that purpose we introduce new notations: we define $\mathscr{S}_0 := \gamma \mathscr{S}^{\text{flip}}$, $\mathscr{S}_\lambda := \mathscr{S}_0 + \lambda \mathscr{S}^{\text{exch}}$, and then

$$\begin{cases} \mathscr{L}^{\mathbf{m}}_0 := \mathscr{A}^{\mathbf{m}} + \mathscr{S}_0 \\ \mathscr{L}^{\mathbf{m}}_\lambda := \mathscr{A}^{\mathbf{m}} + \mathscr{S}_\lambda = \mathscr{L}^{\mathbf{m}}_0 + \lambda \mathscr{S}^{\text{exch}} \end{cases} \text{and } J_0(\mathbf{m})(\omega) := \frac{\omega_0 \omega_1}{\sqrt{m_0 m_1}} = j^A_{0,1}(\mathbf{m}, \omega).$$

Let us introduce the homogenized Green-Kubo formula for both noises:

$$\overline{\kappa}(\lambda, z) := \langle\langle J_0(\mathbf{m}), (z - \mathscr{L}^{\mathbf{m}}_\lambda)^{-1} J_0(\mathbf{m}) \rangle\rangle_{1,\star} \tag{12}$$

and the homogenized Green-Kubo formula for flip noise only:

$$\overline{\kappa}_0(z) := \langle\langle J_0(\mathbf{m}), (z - \mathscr{L}^{\mathbf{m}}_0)^{-1} J_0(\mathbf{m}) \rangle\rangle_{1,\star}. \tag{13}$$

According to the previous paragraph, we already know that the Green-Kubo formulas (12) and (13) converge as z goes to 0. Then, the following diffusion coefficients are well defined, for all $\lambda > 0$,

$$\overline{D}(\lambda) := \lambda + \lim_{z\to 0} \overline{\kappa}(\lambda, z), \qquad \overline{D}_0 := \lim_{z\to 0} \overline{\kappa}_0(z).$$

The main result of this subsection is stated in the following theorem, proved in [1].

Theorem 4.4 *The function $\lambda \mapsto \overline{D}(\lambda)$ is continuous at 0.*

Let us remark that the theorem above does not imply the existence of the hydrodynamics diffusion coefficient $D(0, \gamma)$. This question remains open.

Acknowledgments This problem was suggested by Cédric Bernardin, and I am grateful to him for helpful and valuable remarks. I warmly thank Makiko Sasada and Stefano Olla for their interest and constructive discussions on this work.

References

1. Simon, M.: Equilibrium fluctuations for the disordered harmonic chain perturbed by an energy conserving noise, eprint arXiv:1402.3617 (2014)
2. Casher, A., Lebowitz, J.L.: Heat flow in regular and disordered harmonic chains. J. Math. Phys. **12**(8), 1701–1711 (1971)
3. Dhar, A.: Heat conduction in the disordered harmonic chain revisited. Phys. Rev. Lett. **86**(26), 5882–5885 (2001)
4. Ajanki, O., Huveneers, F.: Rigorous scaling law for the heat current in disordered harmonic chain. Commun. Math. Phys. **301**(3), 841–883 (2011)
5. Faggionato, A., Martinelli, F.: Hydrodynamic limit of a disordered lattice gas. Prob. Theory Relat. Fields **127**(4), 535–608 (2003)
6. Jara, M., Landim, C.: Quenched non-equilibrium central limit theorem for a tagged particle in the exclusion process with bond disorder. Ann. Inst. Henri Poincaré Probab. Stat. **44**(2), 341–361 (2008)
7. Mourragui, M., Orlandi, E.: Lattice gas model in random medium and open boundaries: hydrodynamic and relaxation to the steady state. J. Stat. Phys. **136**(4), 685–714 (2009)
8. Quastel, J.: Bulk diffusion in a system with site disorder. Ann. Probab. **34**(5), 1990–2036 (2006)
9. Varadhan, S.R.S.: Nonlinear diffusion limit for a system with nearest neighbor interactions. II, Asymptotic problems in probability theory: stochastic models and diffusions on fractals. Pitman Res. Notes Math. Ser. **283**, 75–128 (1993)
10. Komoriya, K.: Hydrodynamic limit for asymmetric mean zero exclusion processes with speed change. Ann. Inst. H. Poincaré Probab. Stat. **34**(6), 767–797 (1998)
11. Bernardin, C.: Thermal conductivity for a noisy disordered harmonic chain. J. Stat. Phys. **133**(3), 417–433 (2008)
12. Bernardin, C., Huveneers, F.: Small perturbation of a disordered harmonic chain by a noise and an anharmonic potential. Probab. Theory Relat. Fields **157**(1–2), 301–331 (2013)
13. Dhar, A., Kannan, V., Lebowitz, J.L.: Heat conduction in disordered harmonic lattices with energy conserving noise. Phys. Rev. E **83**, 021108 (2011)
14. Simon, M.: Hydrodynamic limit for the velocity-flip model. Stoch. Processes Appl. **123**, 3623–3662 (2013)
15. Landim, C., Yau, H.T.: Fluctuation-dissipation equation of asymmetric simple exclusion processes. Probab. Theory Relat. Fields **108**, 321–356 (1997)
16. Kipnis, L., Landim, C.: Scaling limits of interacting particle systems. Grundlehren der Mathematischen Wissenschaften [Fundamental Principles of Mathematical Sciences]. Springer, Berlin (1999)
17. Sasada, M.: Hydrodynamic limit for exclusion processes with velocity. Markov Process. Relat. Fields **17**(3), 391–428 (2011)
18. Bernardin, C., Stoltz, G.: Anomalous diffusion for a class of systems with two conserved quantities. Nonlinearity **25**, 1099–1133 (2012)
19. Ethier, S.N., Kurtz, T.G.: Markov processes. Wiley Series in Probability and Mathematical Statistics. Wiley, New York (1986)
20. Komorowski, T., Landim, C., Olla, S.: Fluctuations in Markov processes. Grundlehren der Mathematischen Wissenschaften [Fundamental Principles of Mathematical Sciences]. Springer (2012)
21. Olla, S., Sasada, M.: Macroscopic energy diffusion for a chain of anharmonic oscillators, eprint arXiv:1109.5297v3 (2013)
22. Benabou, G.: Homogenization of Ornstein-Uhlenbeck process in random environment. Commun. Math. Phys. **266**(3), 699–714 (2006)
23. Kipnis, C., Varadhan, S.R.S.: Central limit theorem for additive functionals of reversible Markov processes and applications to simple exclusions. Commun. Math. Phys. **104**, 1–19 (1986)

Conditioned Stochastic Particle Systems and Integrable Quantum Spin Systems

Gunter M. Schütz

Abstract We consider from a microscopic perspective large deviation properties of several stochastic interacting particle systems, using their mapping to integrable quantum spin systems. A brief review of recent work is given and several new results are presented: (i) For the general disordered symmetric exclusion process (SEP) on some finite lattice conditioned on no jumps into some absorbing sublattice and with initial Bernoulli product measure with density ρ we prove that the probability $S_{\rho(t)}$ of no absorption event up to microscopic time t can be expressed in terms of the generating function for the particle number of a SEP with particle injection and empty initial lattice. Specifically, for the symmetric simple exclusion process on $\mathbb{Z}$ conditioned on no jumps into the origin we obtain the explicit first and second order expansion in ρ of $S_{\rho(t)}$ and also to first order in ρ the optimal microscopic density profile under this conditioning. (ii) For the disordered ASEP on the finite torus conditioned on a very large current we show that the effective dynamics that optimally realizes this rare event does not depend on the disorder, except for the time scale. (iii) For annihilating and coalescing random walkers we obtain the generating function of the number of annihilated particles up to time t, which turns out to exhibit some universal features.

Keywords Exclusion processes · Large deviations · Conditioned dynamics · Exact results · Integrable models

1 Introduction

It is long known that the generator of some paradigmatic stochastic interacting particle systems such as the asymmetric simple exclusion process (ASEP) [46] on a one-dimensional lattice Λ of L sites can be mapped to the Hamiltonian H of an

G.M. Schütz (✉)
Institute of Complex Systems II, Forschungszentrum Jülich GmbH,
52425 Jülich, Germany
e-mail: g.schuetz@fz-juelich.de

P. Gonçalves and A.J. Soares (eds.), *From Particle Systems to Partial Differential Equations II*, Springer Proceedings in Mathematics & Statistics 129,
DOI 10.1007/978-3-319-16637-7_15

integrable quantum many-body system [1, 65]. The non-diagonal matrix elements $h_{\eta'\eta} \leq 0$ of H are the negative transition rates for a transition from configuration η to η' of the particle system and each diagonal element $h_{\eta\eta} \geq 0$ is the sum of transition rates out of a configuration η, i.e., the negative sum of the non-diagonal elements $h_{\eta'\eta}$ of column η. The interpretation as a quantum many body system becomes evident by a suitable choice of tensor basis [47, 65]. Consider a particle system with local state space S_i for each lattice site i and full state space $S = S_1 \times \cdots \times S_L$. For definiteness we assume finite cardinality $0 < |S_i| < \infty \; \forall i$, even though much of what is discussed below can be generalized to infinite state spaces. To each local configuration $\eta(i) \in S_i$ with $i \in \Lambda$ of the particle system one assigns a canonical basis vector $|\eta(i))$ of the vector space $\mathbb{C}^{|S_i|}$ and an associated dual basis vector $(\eta(i)|$ with inner product $(\eta(i)|\eta'(i)) = \delta_{\eta(i),\eta'(i)}$. The particle configuration $\eta = \{\eta(1), \eta(2), \ldots, \eta(L)\} \in S$ on the lattice is then represented by the tensor vectors $|\,\eta\,\rangle = |\eta(1)) \otimes |\eta(2)) \otimes \cdots \otimes |\eta(L))$ and their duals by $\langle\,\eta\,|$. Thus a probability measure $P(\eta)$ is given by a probability vector $|\,P\,\rangle$ whose components are $P(\eta) = \langle\,\eta\,|\,P\,\rangle$. The semi-group property of the Markov process is reflected in $|\,P(t)\,\rangle = \mathrm{e}^{-Ht}|\,P(0)\,\rangle$ for the time evolution of an initial measure $P_0(\eta) = \langle\,\eta\,|\,P_0\,\rangle$. The expectation $\langle f(t)\,\rangle$ of a random variable $f(\eta_t)$ is given by the inner product $\langle f(t)\,\rangle = \langle\,s\,|\hat{f}|\,P(t)\,\rangle$ where $\langle\,s\,| = \sum_{\eta\in S}\langle\,\eta\,|$ is called the summation vector and $\hat{f} = \sum_{\eta\in S} f(\eta)|\,\eta\,\rangle\langle\,\eta\,|$ is a diagonal matrix (using the quantum mechanical shorthand $|\cdot\rangle\langle\cdot| \equiv |\cdot\rangle \otimes \langle\cdot|)$ with $f(\eta)$ on the diagonal. Because of probability conservation $\langle\,s\,|$ is a left eigenvector of H with eigenvalue 0. Under this mapping the generator of the ASEP is represented by the quantum Hamiltonian of a spin-1/2 Heisenberg ferromagnet with an Dzyaloshinskii–Moriya interaction with imaginary amplitude [28, 40], see next section.

Using this mapping one can employ rigorous methods borrowed from condensed matter theory to obtain information about the particle system. However, due to the requirement of probability conservation and positivity of the classical transition rates this mapping is restricted to quantum Hamiltonians with two special properties:

(i) All non-diagonal matrix elements of H must satisfy $h_{\eta'\eta} \in (-\infty, 0]$.
(ii) The lowest eigenvalue of H is equal to zero for all lattice sizes L.

One can turn the mapping around and ask which integrable quantum Hamiltonian, or, in fact, which general finite-dimensional matrix can be mapped to the generator of a Markov chain.

The answer follows from the two properties (i) and (ii) by using similarity transformations and the fact that adding a multiple of a unit matrix to H is a "harmless" operation in the time evolution operator e^{-Ht}: Imagine an irreducible square matrix A which has—after a suitable similarity transformation—the property $a_{\eta'\eta} \leq 0$ for all its off-diagonal elements $\eta \neq \eta'$. Then we can subtract a multiple λ of the unit matrix $\mathbb{1}$ such that $M = \mathrm{e}^{-(A-\lambda\mathbb{1})t}$ has only strictly positive matrix elements. It follows from the Perron-Frobenius theorem that the left largest eigenvector $\langle\,0\,|$ of M (and hence the left eigenvector to the lowest eigenvalue a_0 of A), has strictly positive components $\Delta(\eta)$. Next we define the diagonal matrix Δ with these components on the diagonal. Since all components are strictly positive, its inverse exists. Consider now $H := \Delta A \Delta^{-1} - a_0$. Since Δ is diagonal and $\Delta(\eta) > 0$ for all η, we have that $h_{\eta'\eta} = a_{\eta'\eta}\Delta(\eta')/\Delta(\eta) \leq 0$. Moreover, the lowest eigenvalue of H is 0 due to the subtraction of $a_0\mathbb{1}$ from $\Delta A \Delta^{-1}$. Therefore H is the transition matrix of some Markov chain. This implies that an integrable quantum many-body system that satisfies property (i) (perhaps after some similarity transformation), but *not* property (ii), can still be mapped to a Markov chain which can then be analyzed using the mathematical machinery of quantum integrability.

There is a second reason to consider quantum Hamiltonians with property (i), but not (ii). This comes from the fact that *conditioning* a process on some event may naturally lead to such a generator. In order to see this consider first conditioning a process on the event that a nonempty subset $X \subset S$ has never been reached until some positive time t, when starting from an initial distribution with support in $A = S\backslash X$. We denote configurations in X by ξ and configurations in A by α. This "event conditioning" can be studied by making the set X absorbing and frozen, i.e., all transition rates $w_{\alpha\xi}$ from X to A and transition rates $w_{\xi'\xi}$ within X are defined to be zero. The generator H of such a process has a block tridiagonal structure $0_{|X|} \oplus G + H_{XA}$ with a zero-block $0_{|X|}$ of dimension $|X|$ (corresponding to the absence of transitions within X), a block G of dimension $|A|$ containing all off-diagonal matrix elements $h_{\eta'\eta}$ for transitions within sector A and all diagonal elements for transitions out of A, and the matrix H_{XA} which has as only non-zero elements $h_{\xi\alpha}$, which are the negative transition rates from A to X. We also define a truncated basis where the basis vectors $|\,\eta^A\,\rangle$ ($\langle\,\eta^A\,|$) are obtained from $|\,\eta\,\rangle$ ($\langle\,\eta\,|$) (spanning the vector space $\mathbb{C}^{|S|}$) by deleting the 0-components corresponding to configurations in X. The truncated canonical basis vectors span $\mathbb{C}^{|A|}$. The truncated probability vector is denoted by $|\,P_0^A\,\rangle$ and the truncated summation vector by $\langle\,s^A\,| = \sum_{\eta\in A}\langle\,\eta^A\,|$.

The block tridiagonal structure of H has as a consequence that starting from an initial measure P_0 with support in A the conditional probability of being in a state $\alpha \in A$ at time t, given that the process has never been in any configuration $\xi \in X$ until time t, is given by

$$\tilde{P}(\eta, t) = \frac{1}{S_{P_0}(t)}\langle\,\eta^A\,|\mathrm{e}^{-Gt}|\,P_0^A\,\rangle \tag{1}$$

where $S_{P_0}(t) = \langle\,s^A\,|\mathrm{e}^{-Gt}|\,P_0^A\,\rangle$ is the survival probability of never having left A. By construction, G satisfies property (i), but not (ii). The matrix $\hat{H} = \Delta G \Delta^{-1} - g_0$ where g_0 is the lowest eigenvalue of G and Δ the associated left eigenvector is then Doob's h-transform [58]. If G is the quantum Hamiltonian of some integrable model then also $\hat{H}$ is integrable.

A different type of conditioning that has received much attention in recent years arises from considering joint processes (η_t, Q_t) where η_t is an interacting particle system evolving as described above and Q_t is a separate random variable that has no independent dynamics, but is incremented by some value $c_{\eta'\eta}$ whenever a transition from η to η' occurs. Thus η_t evolves autonomously, i.e., independently of Q_t, while Q_t contains information about the history of η_s up to time t. Interesting questions arise from conditioning the process at time t on reaching some fixed value $Q_t = Q$. Important applications include large deviations for the current or the activity in exclusion processes [22, 44] (where the increments are integers) or the derivation of fluctuation theorems [30, 32, 43] (where the increments are real-valued). In this paper we restrict ourselves to processes with integer increments so that $Q \in \mathbb{Z}$.

In order to see the link to quantum Hamiltonians we go to a fluctuating ensemble where instead of the joint probability $P(\eta, Q, t) = \text{Prob}\,[\,\eta_t = \eta, Q_t = Q\,]$ one studies the generating function $Y(\eta, s, t) = \sum_Q e^{sQ} P(\eta, Q, t)$ with generalized "chemical potential" $s \in \mathbb{R}$. We refer to this conditioning at time t as *grand canonical conditioning*. One finds (see below) that

$$Y(\eta, s, t) = \frac{1}{Y_s(t)} \langle\, \eta\, | e^{-H(s)t} |\, P_0\, \rangle \tag{2}$$

with normalization factor $Y_s(t) = \langle\, s\, | e^{-H(s)t} |\, P_0\, \rangle$ where the weighted generator $H(s)$ is obtained from H by multiplying the off-diagonal elements $h_{\eta'\eta}$ by a factor $e^{sc_{\eta'\eta}}$ where $c_{\eta'\eta}$ is the amount by which the counting variable Q has changed in the transition $\eta \to \eta'$. The diagonal elements of H remain unchanged. Obviously, also $H(s)$ satisfies property (i), but not (ii). Notice that $H(0) = H$.

In many cases of interest the integrability of the original Hamiltonian is not affected by this modification. The matrix $H_{eff}(s) = \Delta(s) H(s) \Delta^{-1}(s) - \varepsilon_0(s)$, where $\varepsilon_0(s)$ is the lowest eigenvector of $H(s)$ and $\Delta(s)$ the associated left eigenvector, is a generalized Doob's h-transform [7, 20, 36]. It defines a new process that we shall call *effective* process (or effective dynamics). For processes with finite state space it can be interpreted as that process which makes the untypical large deviation of Q parametrized by s typical [31]. Evidently, if $H(s)$ is the quantum Hamiltonian of some integrable model then also the effective process is integrable.

We also define the conditioned history of some random variable f which is the expectation $\langle f(t)\, \rangle^c$ of f at time t of the process conditioned at time $T \geq t \geq 0$. For the event conditioning this quantity is given by

$$\langle f(t)\, \rangle^c = \langle\, s_A\, | e^{-G(T-t)} \hat{f} e^{-Gt} |\, P_0\, \rangle / S(t) \tag{3}$$

and for grand canonical conditioning by a similar expression with G replaced by $H(s)$, summation vector $\langle\, s_A \,|$ replaced by the summation vector $\langle\, s\, |$ over the full state space S of the particle system and $S(t)$ replaced by $Y_s(t)$.

For both types of conditioning we present some new results which are motivated by recent developments in the field. For selfcontainedness and for reference we discuss in Sect. 2 the grandcanonical conditioning in more detail and briefly review some results that we obtained earlier. Then we present our new results in Sects. 3.1–3.3 with a brief introduction into the history of the model that we treat.

2 Conditioning on Current and Activity

Following the seminal papers [9, 24], the large deviation theory for the ASEP has been developed in considerable detail, resulting in a powerful macroscopic fluctuation theory (MFT) that is capable of describing large deviations of the current and density profiles of quite general lattice gases on a macroscopic level, see [10] for a recent review. Here the notion of "macroscopic" refers to the scale where the lattice constant a is taken to zero and the lattice site k is taken to infinity with finite rescaled macroscopic space variable $x = ka$. The microscopic time is taken to infinity under rescaling by a factor $1/a^z$ where $z = 1$ for Eulerian scaling and $z = 2$ for diffusive scaling. The *typical* macroscopic density profile of the ASEP, i.e., without conditioning on some large deviation, is obtained from the usual hydrodynamic limit [57].

However, large deviations can also be studied directly on a microscopic level. To illustrate this we focus on large deviations of the integrated hopping activity in the asymmetric simple exclusion process (ASEP) on the torus $\mathbb{T}_L := \mathbb{Z}/L\mathbb{Z}$. Particles jump according to the following rules:

1. Each particle attempts to jump independently of the other particles after an exponentially distributed random time with parameter $1/(w_+ + w_-)$ with probability $w_+/(w_+ + w_-)$ to the next site on the right (clockwise) and $w_-/(w_+ + w_-)$ to the next site on the left (counterclockwise)
2. The hopping attempt is rejected if the site to which the particle tries to move is occupied.

We introduce the hopping asymmetry $q^2 := w_+/w_-$ and time scale $w_0 = \sqrt{w_+ w_-}$ and assume without loss of generality a hopping bias to the right, i.e., $q > 1$.

The local state space of the ASEP is $S_i = \{0, 1\}$ for all $i \in \mathbb{T}_L$, corresponding to a vacant or occupied site i with occupation variable $n_i = 0$ or $n_i = 1$. The full state space therefore is therefore $S = \{0, 1\}^L$ with microscopic configurations $\mathbf{n} = \{n_1, \ldots, n_L\}$. Following the quantum Hamiltonian formalism outlined above we introduce the usual Pauli matrices

$$\sigma^x = \begin{pmatrix} 0 & 1 \\ 1 & 0 \end{pmatrix}, \quad \sigma^y = \begin{pmatrix} 0 & -i \\ i & 0 \end{pmatrix}, \quad \sigma^z = \begin{pmatrix} 1 & 0 \\ 0 & -1 \end{pmatrix}, \tag{4}$$

and the two-dimensional unit matrix $\mathbb{1}$. From these we construct operators a_k acting locally on site k of the lattice in terms by the tensor product

$$a_k := \mathbb{1} \otimes \ldots \otimes a \otimes \ldots \otimes \mathbb{1} \tag{5}$$

with the arbitrary 2×2 matrix a at position k in the product. This allows us to define local particle creation and annihilation operators $\sigma_k^{\pm} = (\sigma_k^x \pm i\sigma_k^y)$ and projectors $\hat{n}_k = (\mathbf{1} - \sigma_k^z)/2$ on particles and $\hat{v}_k = \mathbf{1} - \hat{n}_k$ on vacancies on site k where $\mathbf{1} = \mathbb{1}^{\otimes L}$ is the unit matrix of dimension 2^L. Here the local basis vector is chosen such that $|0) = (1, 0)^T$ and $|1) = (0, 1)^T$ where the superscript T denotes transposition.

Correspondingly $(0| = (1, 0)$ and $(1| = (0, 1)$ for the dual basis. A state $\mathbf{n} \in S$ is then represented by $|\,\mathbf{n}\,\rangle = |n_1) \otimes \ldots \otimes |n_L)$ and the generator is given by [28, 65]

$$H = -\sum_{k=1}^{L} w_+(\sigma_k^+\sigma_{k+1}^- - \hat{n}_k\hat{v}_{k+1}) + w_-(\sigma_k^-\sigma_{k+1}^+ - \hat{v}_k\hat{n}_{k+1}) \tag{6}$$

with the identification of site $L+1$ with site 1. The total particle number N is conserved, which is expressed in the commutation relation $[H, \hat{N}] = 0$ for the number operator $\hat{N} = \sum_{k=1}^{L} \hat{n}_k$. It is well-known that (6) defines an integrable model related to the six-vertex model [28, 40].

The unique stationary distribution of the process with N particles is the uniform measure with $P^*(\mathbf{n}) = 1/\binom{L}{N}$. We refer to this invariant measure as canonical. From these measures one can construct the grandcanonical Bernoulli product measure with $P^*(\mathbf{n}) = \rho^{N(\mathbf{n})}(1-\rho)^{L-N(\mathbf{n})}$ where $N(\mathbf{n}) = \sum_{k=1}^{L} n_k$. The grandcanonical measure has the vector representation $|\,\rho\,\rangle := ((1-\rho, \rho)^T)^{\otimes L}$, reflecting its product structure. In the thermodynamic limit $L \to \infty$ with $\rho = N/L$ fixed the canonical measure and its grand canonical counterpart become equivalent. The stationary current in the grandcanonical measure is given by $j^* = (w_+ - w_-)\rho(1-\rho)$.

Now we consider the time-integrated number of jumps $J_k^+(t)$ from site k to $k+1$ up to time t, and analogously the time-integrated number of jumps $J_k^-(t)$ from site $k+1$ to k up to time t. The time-integrated local current is then given by $J_k(t) = J_k^+(t) - J_k^-(t)$ and $A_k(t) = J_k^+(t) + J_k^-(t)$ is the time-integrated local activity and $j_k(t) := J_k(t)/t$, $a_k(t) := A_k(t)/t$ the respective time averages. We also define the space-integrated quantities $J^{\pm}(t) = \sum_{k=1}^{L} J_k^{\pm}(t)$ and analogously $J(t)$ and $A(t)$ as well as their space-time averages $j^{\pm}(t) = J^{\pm}(t)/(Lt)$ and analogously $j(t)$ and $a(t)$.

In order to review the general philosophy of the large deviation approach to these quantities we first recall that for the processes that we have in mind (such as the ASEP) the probability to observe for a long time interval t an untypical mean $j \neq j^*$ is exponentially small in L and t [22]. This is expressed in the asymptotic large deviation property $\text{Prob}\,[\,J(t) = J\,] \propto \exp(-f(j)Lt)$ where $f(j)$ is the rate function and one takes first the limit $t \to \infty$ and then $L \to \infty$. As a second step it is then natural to introduce a generalized fugacity $y = \mathrm{e}^s$ with generalized chemical potential s and to study the generating function $Y_s^c(t) := \sum_{J\in\mathbb{Z}} y^J \text{Prob}\,[\,J(t) = J\,]$. In general, this quantity depends on the initial distribution which will be indicated below when appropriate. The cumulant function $g(s) = \lim_{L\to\infty} \lim_{t\to\infty} \ln Y_s^c(t)/(Lt)$ is the Legendre transform of the rate function, i.e., $g^c(s) = \max_j[js - f(j)]$. The intensive variable s is thus conjugate to the mean current density j. Similar large deviation properties hold for the activity $a(t)$ with generating function $Y_s^a(t)$ and the hopping activities $j^{\pm}(t)$ with generating functions $Y_s^{\pm}(t)$ and their local counterparts $Y_s^{\cdot}(k, t)$.

Following [22, 30] one has for an initial distribution P_0 of particles

$$Y_s^{\cdot}(k, t) = \langle\, s\,|\mathrm{e}^{-H^{\cdot}(k,s)t}|\,P_0\,\rangle \tag{7}$$

where $H^{\cdot}(k, s) = H + F^{\cdot}(k, s)$ with

$$F^{+}(k, s) = (1 - e^{s}) w_{+} \sigma_k^{+} \sigma_{k+1}^{-}, \quad F^{-}(k, s) = (1 - e^{s}) w_{-} \sigma_k^{-} \sigma_{k+1}^{+} \tag{8}$$

and $F^{c}(k, s) = F^{+}(k, s) - F^{-}(k, -s)$ and $F^{a}(k, s) = F^{+}(k, s) + F^{-}(k, s)$. These are examples for the general construction (2), detailed in [30]. For the ASEP and other integrable models $g^{\cdot}(s)$ and other large deviation functions, e.g. associated with particle creation and annihilation events, have been computed with the Bethe ansatz [2, 19, 22, 23, 68] or free fermion methods [54].

The MFT provides also information about the optimal macroscopic density profile $q(x, t)$ to realize at time t a prolonged untypical behaviour of the integrated current at time $T \geq t$, i.e., the density profile that one is most likely to observe at time t in the interval $[0, T]$. One expects the optimal macroscopic density profile to be the large scale limit of the conditional history (3) of the local particle number. To our knowledge no microscopic computation of this large deviation property has been performed yet.

Using MFT for the weakly asymmetric exclusion process (WASEP), where $w_{+} - w_{-} = \nu/L$ is small, Bodineau and Derrida [12] have considered the time range $t = cT$ with $c \in (0, 1)$. In the limit $T \to \infty$ and $L \to \infty$ they found a dynamical phase transition at some critical $s_c < 0$ (i.e., for conditioning on untypically low currents $j < j^{*}$), where the optimal macroscopic density profile $q(x, t)$ changes from a constant $q(x, t) = \rho$ to a travelling wave of the form $q(x, t) = \rho(x - vt)$. For strictly positive s (i.e., for any current j conditioned on $j > j^{*}$) the macroscopic density profile is time-independent and flat and hence equal to the typical profile $\rho(x, t) = \rho$. More recently, this phenomenon was described in more detail and also studied numerically with Monte-Carlo simulations [25, 34]. A dynamical phase transition was also identified for conditioned non-conservative dynamics [13].

The macroscopic approach is extremely powerful, but has the drawback that no information is available on the expected non-trivial space-time realization of the ASEP conditioned on currents larger than the typical value j^{*}. Also, no information can be obtained about fluctuations of the traveling wave for $s < 0$. This gap was partially filled in a series of papers [7, 52, 53, 66, 68], exploiting the integrability and symmetries of the weighted generator $H(s)$.

Significantly, this microscopic approach shows that for conditioning on high currents or high activities (more precisely, in the limit $s \to \infty$ on the time scale set by $w_0 = e^{-s} q^{-1}$) there is a different dynamical phase transition that does not appear in the density profile, but in the fluctuations. The dynamical exponent for density fluctuations changes from $z = 3/2$ for typical behaviour to $z = 1$ for atypical behaviour for times $t = cT$ with $c \in (0, 1)$ and $T \to \infty$. The associated effective microscopic dynamics, obtained from the generalized Doob's transform [7, 20, 36], have nearest neighbour jumps, but the jump rate depends on the whole configuration of the particles in the same way as in a process introduced by Spohn [69]. The stationary distribution of the effective process with N particles located on sites $k_1, \ldots, k_N \in \mathbb{T}_L$ has the determinantal form

$$P_L(\mathbf{n}) = \frac{2^{N(N-1)}}{L^N} \prod_{1 \le i < j \le N} \sin^2\left(\pi \frac{k_i - k_j}{L}\right) \tag{9}$$

with long-range anticorrelations (correlations with negative amplitude)

$$C(k) := \langle\, n(0)n(k)\,\rangle - \rho^2 = -\frac{\sin^2(k\pi\rho)}{k^2\pi^2} \tag{10}$$

in the thermodynamic limit $L \to \infty$.

For a current lower than typical we considered in [7] the grand canonically conditioned ASEP at $t = T$ finite for a special value of the conjugate variable s and a special family of initial measures that are inhomogeneous product measures mimicking a microscopically sharp shock and some site k and a soft antishock extending over the whole lattice $\mathbb{T}_L$. It turns out that under the conditioned dynamics this measure evolves into a time-dependent convex combination of such measures with microscopic shock positions at sites l where the weights for each such measure are the transition probabilities of a biased simple random walk starting from site k at $t = 0$ and ending at site l at time $t = T$ and explicitly known random walk transition rates [7]. This is an exact microscopic result reminiscent of the typical shock dynamics proved in [6] for special shock densities. Thus under these conditions one has on macroscopic level a travelling wave with shock moving around the ring with finite speed, reminiscent of what is predicted by MFT at times cT with $c \neq 1$ for general shock densities. Indeed, for generic shock densities one may still expect a microscopically stable shocks under conditioned dynamics, as was argued to be the case for typical microscopic behaviour in [59] by using numerical evidence and the heuristic domain wall approximation [41].

In addition to this macroscopic information one finds that the shock position performs diffusive fluctuations around its mean position with an explicitly computed shock diffusion coefficient. Moreover, the shock remains microscopically sharp at all times and there are no correlations seen from the shock position. The techniques used to derive these results are an adaptation of our earlier work [6] on the microscopic structure of shocks under unconditioned dynamics. This approach, which uses the q-deformed non-abelian symmetries of the integrable weighted generator, has a probabilistic interpretation as self-duality [15, 27, 35, 64].

3 New Applications of Conditioning

After this brief review of microscopic large deviation results we consider now some models that have a long history of study, but where there is renewed interest from the viewpoint of large deviations.

3.1 SEP Conditioned on Not Entering a Sublattice

Recently Meerson et al. [49] used macroscopic fluctuation theory (MFT) to study the absorption into a static target in a gas of diffusing particles with exclusion, modelled by the symmetric simple exclusion process (SSEP) [45]. The target is considered to be an absorber, i.e., when a particles hits the target it disappears from the dynamics. A quantity Meerson and collaborators are interested in is the probability $S(t)$ that no gas particle has hit the target until time t. This conditioning on an event is of the first type outlined in the introduction and is equivalent to the survival probability of the target. This problem has a long history of study, for non-interacting particles see e.g. [11, 17, 70] and more recently in the context of the search for an immobile target by a swarm of diffusive searchers [51] and in the study of the complete statistics of particle absorption by the absorber [50]. In the absence of exclusion the survival probability for an initial density ρ of particles is of the form $\tilde{S}_\rho(t) \propto \exp(-a\rho\sqrt{t})$ with some constant a.

For one dimension and in the presence of exclusion it was proofed rigorously [18] that the survival probability is bounded from above by the survival probability $\tilde{S}_\rho(t)$ without exclusion. Indeed, for an initial distribution that is a Bernoulli product measure with density ρ on all sites not covered by the target, MFT predicts the asymptotic behaviour [49]

$$S_\rho(t) \propto \mathrm{e}^{-f(\rho)\sqrt{t}} \tag{11}$$

where $f(\rho)$ plays the role of a large deviation function. By placing the right edge of the target at site 0, this problem is equivalent to conditioning the SSEP on the event that no particle has reached site 0 until time t. For this case Meerson et al. computed from MFT explicitly the first two terms in the Taylor expansion of (11) around $\rho = 0$, with the result

$$f(\rho) = \frac{2}{\sqrt{\pi}}\left[\rho + (\sqrt{2}-1)\rho^2 + \ldots\right] \tag{12}$$

where the dots represent terms of higher order in ρ. The first order term, corresponding to non-interacting particles, has been known for a long time, see [49] and references therein.

Here we take a microscopic approach and first generalize to the disordered symmetric exclusion process on any finite lattice Ξ, i.e., a lattice where hopping is allowed with strictly positive rate between fixed, but arbitrary pairs $\langle i\,, j\rangle$ of sites with $i, j \in \Xi$ and such that the hopping rates may depend in the link $\langle i\,, j\rangle$.[1] We show for arbitrary targets that the conditioned process is equivalent to an unconditioned SEP with

[1] To clarify our terminology: The symmetric simple exclusion process (SSEP) has nearest neighbour jumps only on a regular lattice such as $\mathbb{Z}^d$, whereas the general symmetric exclusion process (SEP) can have jumps between any pair of nodes on any graph. This terminology comes from physical intuition and is not mathematically precise, since the set of links on which jumps have non-zero rate may just as well be used to define an underlying graph on which one then would again have

particle injection. Then we focus on the case of Meerson et al. For the SSEP in one dimension the equivalent process was studied earlier [60] using a mathematically rigorous reformulation of the Bethe ansatz. This reformulation involves expressing the Bethe-wave function in terms of series expansions with simple convergence properties. This approach allows us to obtain the fluctuations of the conditioned process in a fully microscopic approach and to prove the (not fully rigorous) prediction (12) of MFT.

According to the construction outlined above, in one dimension (where $\Xi = \mathbb{Z}$) the SSEP conditioned on no particle reaching the origin is given by the truncated generator

$$G_{\mathbb{Z}^+} = -w \sum_{k=1}^{\infty} \left[\sigma_k^+ \sigma_{k+1}^- + \sigma_k^- \sigma_{k+1}^+ - \hat{n}_k \hat{v}_{k+1} - \hat{v}_k \hat{n}_{k+1} \right] + w\hat{n}_1 =: G_0 + w\hat{n}_1 \tag{13}$$

For general finite lattices Ξ and general targets, covering an arbitrary (but non-empty) sublattice $\Omega \subset \Xi$, and disordered symmetric hopping with exclusion between any pair of sites we by denote by $\Lambda \subset \Xi$ the sublattice of sites not covered by the target and between which hopping has non-zero rate. Hopping across a bond $\langle i\,, j \rangle$ with $i, j \in \Lambda$ occurs symmetrically with bond-dependent rate $w_{i,j} = w_{j,i}$. Furthermore we denote by $\mathcal{E} \subset \Lambda$ the exit sites which allow for hopping into the target in the unconditioned process. (For nearest neighbour hopping e.g. on a cubic lattice these sites form the external boundary of the target.)

Then one has as similar truncated generator G_Λ with projectors $\hat{n}_k$ on the set of sites $k \in \mathcal{E}$ from which jumps into the target can occur, viz.,

$$G_\Lambda = -\sum_{\langle i,j \rangle} w_{i,j} \left(\sigma_i^+ \sigma_j^- + \sigma_i^- \sigma_j^+ - \hat{n}_i \hat{v}_j - \hat{v}_i \hat{n}_j \right) + \sum_{i \in \mathcal{E}} d_i \hat{n}_i =: G_\Lambda^0 + G_\Lambda^e \tag{14}$$

where $d_i = \sum_{j \in \Omega} w_{i,j}$ is the sum of jump rates between i and the target region to which a particle in the unconditioned process can jump from site i. Consider now an initial distribution $|\, P_0 \,\rangle$ restricted to configurations with support on the sites in Λ outside the target, i.e., to all configurations with $n_l = 0$ for $l \in \Omega$. The survival probability $S_{P_0}(t)$ is then given by

$$S_{P_0}(t) = \langle\, s\, | \mathrm{e}^{-G_\Lambda t} |\, P_0 \,\rangle \tag{15}$$

where the summation vector $\langle\, s \,|$ is also restricted to configurations with support on the sites in Λ outside the target.

Our main result is the following.

(Footnote 1 continued)
only nearest neighbour jumps so that the process could also be called SSEP. However, we will only consider the SSEP on $\mathbb{Z}$ with bond-independent rates. This removes any ambiguity of notation. Moreover, we shall always speak of sites and lattices rather than of nodes and graphs.

Theorem 3.1 *Let Ξ be a finite lattice with M sites, Ω be the sublattice with K sites covered by fixed absorbers and $\Lambda = \Xi \setminus \Omega$ with $L = M - K$ sites. Then:*
(a) For initial product measure with density ρ one has

$$S_\rho(t) = \langle\, (1-\rho)^N \,\rangle_0 \tag{16}$$

where N is the total particle number under the evolution of an unconditioned SEP with injection of particles at the sites $i \in \mathcal{E}$ with rate $d_i = \sum_{j\in\Omega} w_{i,j}$, defined by the generator

$$H_\Lambda = -\sum_{\langle i,j\rangle} w_{i,j}\left(\sigma_i^+\sigma_j^- + \sigma_i^-\sigma_j^+ - \hat{n}_i\hat{v}_j - \hat{v}_i\hat{n}_j\right) - \sum_{i\in\mathcal{E}} d_i(\sigma_i^- - \hat{v}_i) =: H_\Lambda^0 + H_\Lambda^e \tag{17}$$

and $\langle\cdot\rangle_0$ represents expectation for the initial distribution representing the empty lattice.

(b) For initially N particles placed with probability 1 on sites $k_1, \ldots, k_N \in \Lambda$ one has

$$S_{k_1,\ldots,k_N}(t) = \langle\, v_{k_1} \ldots v_{k_N} \,\rangle_0 \tag{18}$$

for the process with open boundaries defined by (17) and initially empty lattice.

Proof Observe that $\langle\, s \,| = \langle\, 0 \,| U$ where $U = (\hat{v} + \sigma^x)^{\otimes L}$ and that $G_\Lambda^0 = H_\Lambda^0$ is symmetric under the action of the generators $S^{x,y,z} = \sum_{k\in\Lambda}\sigma_k^{x,y,z}$ of the Lie-group $SU(2)$ and therefore commutes with U [62]. Moreover, by explicit computation $U\hat{n}_iU^{-1} = \hat{v}_i - \sigma_i^+$. These relations are valid for any lattice. Therefore $\tilde{G}_\Lambda := UG_\Lambda U^{-1} = G_\Lambda^0 - \sum_{i\in\mathcal{E}} d_i(\sigma_i^+ - \hat{v}_i)$ which yields for an initial product measure $S_\rho(t) = \langle\, 0 \,| \mathrm{e}^{-\tilde{G}_\Lambda t} U |\, \rho \,\rangle$. Next we observe that $U|\, \rho \,\rangle = (1-\rho)^{\hat{N}}|\, s \,\rangle$ and $H_\Lambda = \tilde{G}_\Lambda^T$ where the superscript T denotes transposition. Transposition of (15) then proves (16). On the other hand, for fixed initial condition with N particles on sites $k_1, \ldots, k_N$ one has $U|\, k_1, \ldots, k_N \,\rangle = \hat{v}_{k_1} \cdots \hat{v}_{k_N}|\, s \,\rangle$ since $(\hat{v}+\sigma^x)|0) = |s)$ and $(\hat{v}+\sigma^x)|1) = |0) = \hat{v}|s)$. Transposition of (15) then yields (18). □

In order to avoid irrelevant technicalities regarding arbitrary infinite lattices we have restricted ourselves to finite lattices Ξ. However, all the steps involved in the previous proof remain valid for the SSEP on $\Xi = \mathbb{Z}^d$ by adapting duality arguments along the lines discussed by Liggett [45] to the case of open boundaries. In particular, for the SSEP on $\Xi = \mathbb{Z}$ with constant jump rates w and right target edge at site 0 we have $\Lambda = \mathbb{Z}^+$. The process (17) derived from $G_{\mathbb{Z}^+}$ corresponds to injection of particles at site 1 with rate w with generator

$$H = -w\sum_{k=1}^{\infty}\left[\sigma_k^+\sigma_{k+1}^- + \sigma_k^-\sigma_{k+1}^+ - \hat{n}_k\hat{v}_{k+1} - \hat{v}_k\hat{n}_{k+1}\right] - w(\sigma_1^- - \hat{v}_1) \tag{19}$$

and initial density $\rho_0 = 0$. This is the special case $\rho^* = 1$, $\rho_0 = 0$ of the SSEP with open boundaries where particles are injected at site 1 with rate $w\rho^*$ and extracted with rate $w(1-\rho^*)$ and the lattice has an initial product distribution with density ρ_0, see [60] with the (arbitrary) choice of time scale $w = 1$.

Corollary 3.2 *From (16) it follows that the lowest order terms in the Taylor expansion of* $\ln S_\rho(t)$ *in* ρ *are given by*

$$\ln S_\rho(t) = -\langle N \rangle_0 \rho + \frac{1}{2}\left(\langle N(N-1) \rangle_0 - \langle N \rangle_0^2\right) \rho^2 + \ldots \tag{20}$$

Using rigorous techniques, viz., the explicit solution of a system of ordinary linear differential equations by the rigorous reformulation of the (in general non-rigorous) Bethe ansatz which was mentioned above, these expansion coefficients have been computed in [60] for arbitrary boundary density ρ^* and arbitrary initial density ρ_0. They agree for the relevant case $\rho^* = 1$, $\rho_0 = 0$ with the MFT prediction (12) which is thus proved.

Next we discuss the conditioned time evolution of the local density in the general SEP.

Proposition 3.3 *For the SEP (14) conditioned on no absorption in the target up to time* $T \geq t \geq 0$ *let*

$$\phi_{\rho,T}(k,t) = \frac{\langle s \,| \mathrm{e}^{-G_\Lambda (T-t)} \hat{n}_k \mathrm{e}^{-G_\Lambda t} |\, \rho \rangle}{S_\rho(T)} \tag{21}$$

be the expected intermediate local particle number at time $t \in [0, T]$ *with initial Bernoulli product measure with density* ρ.

(a) This function has for all $\rho \in [0, 1]$ *and for all* $t \in [0, T]$ *the time reversal symmetry*

$$\phi_{\rho,T}(k,t) = \phi_{\rho,T}(k, T-t) \tag{22}$$

(b) and the property

$$\phi_{\rho,T}(k,t) = \frac{\langle s \,|(1-\rho)^{\hat{N}} \mathrm{e}^{-H_\Lambda t} (\hat{v}_k - \sigma_k^-) \mathrm{e}^{-H_\Lambda (T-t)} |\, 0 \rangle}{S_\rho(T)}. \tag{23}$$

Proof (a) The first statement (22) follows from a reversibility argument: One uses particle number conservation $G_\Lambda = y^{\hat{N}} G_\Lambda y^{-\hat{N}} = y^{\hat{V}} G_\Lambda y^{-\hat{V}}$, $\forall y \in \mathbb{C}$, and the time reversal symmetry $G_\Lambda = G_\Lambda^T$ of the conditioned evolution and notes that $|\, \rho \rangle = (1-\rho)^{\hat{V}} \rho^{\hat{N}} |\, s \rangle$. (b) The second equality (23) is obtained using the same transformations as in the proof of (3.1). □

For $\Lambda = Z^+$ (23) can be evaluated to first order in ρ using symmetries and elementary computations. We define

$$q_T(k,t) := -\rho \frac{\mathrm{d}}{\mathrm{d}\rho}|_{\rho=0}\phi_{\rho,T}(k,t) \tag{24}$$

and obtain

$$q_T(k,t) = \rho \sum_{l=1}^{\infty} \langle\, s\, | \hat{n}_l \mathrm{e}^{-Ht}(\hat{v}_k - \sigma_k^-)\mathrm{e}^{-H(T-t)} |\, 0\,\rangle \tag{25}$$

with H given by (19). This brings us into a position to state:

Proposition 3.4 *Let $q_T(k,t)$ be the conditional expected particle number at site k at time t to first order in the initial density ρ of the SSEP on $\mathbb{Z}$, conditioned on the event that no particle has crossed the origin up to time $T \geq t \geq 0$, with initial product measure with density ρ on $\mathbb{Z}^+$. One has*

$$q_T(k,t) = \rho f_k(2wt) f_k(2w(T-t)) \tag{26}$$

where

$$f_k(u) = \mathrm{e}^{-u} \sum_{n=-k+1}^{k} I_n(u) \tag{27}$$

with the modified Bessel function $I_n(\cdot)$.

Proof In order to prove this result one first observes that $\langle\, s\, | \hat{n}_m \mathrm{e}^{-Ht}(\hat{v}_k - \sigma_k^-) = -\langle\, s\, | \hat{v}_m \mathrm{e}^{-Ht}(\hat{v}_k - \sigma_k^-)$. Next one shows by direct computation that the vector $\langle\, m, t\, | := \langle\, s\, | \hat{v}_m \mathrm{e}^{-Ht}$ satisfies the lattice diffusion equation $\mathrm{d}/(\mathrm{d}t)\langle\, m, t\, | = w(\langle\, m+1, t\, | + \langle\, m-1, t\, | - 2\langle\, m, t\, |)$ with absorbing boundary condition $\langle\, 0, t\, | = 0$ $\forall t$. Hence $\langle\, m, t\, | = \mathrm{e}^{-2wt} \sum_{n=-m+1}^{m} I_n(2wt) \langle\, n, 0\, |$ which yields

$$q_T(k,t) = \rho\, \mathrm{e}^{-2wt} \sum_{n=-m+1}^{m} I_n(2wt) \langle\, s\, | \hat{v}_n (\hat{v}_k - \sigma_k^-) \mathrm{e}^{-H(T-t)} |\, 0\,\rangle . \tag{28}$$

Now observe that $\langle\, s\, | \hat{v}_n (\hat{v}_k - \sigma_k^-) = \langle\, s\, | \hat{v}_n \delta_{n,k}$ and therefore $q_T(k,t) = \rho f_k(2wt)$ $\langle\, s\, | \hat{v}_k \mathrm{e}^{-H(T-t)} |\, 0\,\rangle$. Now one repeats the previous computation for the time evolution of $\langle\, s\, | \hat{v}_k$ (with t replaced by $T - t$) and uses $\langle\, s\, | \hat{v}_n |\, 0\,\rangle = 1$. This yields (26). □

At large scales where u is large and $k = O(\sqrt{u})$ the function $f_k(u)$ converges to the error function. We rescale time $t = \tau T$, with $0 \leq \tau \leq 1$, and space $k = x\sqrt{wT}$. Then (26) becomes

$$\lim_{T\to\infty} q_T(k,t) = \rho\, \mathrm{erf}\left(\frac{x}{2\sqrt{\tau}}\right)\mathrm{erf}\left(\frac{x}{2\sqrt{1-\tau}}\right). \tag{29}$$

A similar result was obtained from MFT for the macroscopic optimal density profile of non-interacting random walks [49] conditioned on no absorption at the origin.

3.2 ASEP with Defect Bond and Site Disorder Conditioned on Maximal Current

The TASEP on $\mathbb{T}_L$ with a defect bond is the TASEP essentially as described above, with the difference that hopping attempts across bond $\langle L, 1\rangle$ happen with rate r and $0 < r < 1$. This process has a long history of study which started with the numerical observation of a defect-induced phase transition first in a closely related model of interface growth [75] and then with a numerical study by Janowsky and Lebowitz [37] who also reported a defect-induced phase transition from a phase with homogeneous density (on coarse-grained scale) to a phase-separated stationary state with two domains of different densities. A major issue that is still open from a mathematical perspective is whether at half-filling the phase transition occurs at some finite defect strength $r_c < 1$ or for arbitrarily small defects, i.e., at $r_c = 1$ [38]. In a cellular automaton version of the TASEP with discrete-time update [61] this question could be answered rigorously. In that model the phase transition sets in for any defect strength. Tang and Lyuksyutov [73] predicted this behaviour also for the original TASEP, using renormalization group arguments for the closely related directed polymer problem. For subsequent work on this problem, both in mathematical probability theory and statistical physics, see [5, 29, 55, 67, 71] and references therein.

Here we follow the line of investigation started in [52] and consider the ASEP with a defect conditioned on a large current, parametrized by the conjugate variable s. Hopping attempts across bond $\langle L, 1\rangle$ happen with rate $rw_\pm$ and $0 < r < 1$. It is convenient to define the defect strength r in terms of the positive variable $u := -\ln(r)$. Following the construction outlined above one has to study in the maximal current limit $s \to \infty$ the rescaled weighted generator

$$H^*(u) = -\sum_{k=1}^{L-1} \sigma_k^+\sigma_{k+1}^- - \mathrm{e}^{-u}\sigma_L^+\sigma_{L+1}^- \tag{30}$$

Solving for the conditioned stationary distribution and the conditioned dynamics looks at first sight like a difficult spatially inhomogeneous many-body problem. However, the inhomogeneity turns out to have a trivial effect. In order to see this we prove

Proposition 3.5 *The weighted generator (30) with bulk hopping rate 1 and defect hopping rate* $r = e^{-u}$ *is related to a homogeneous generator with hopping rate* $c(u) = r^{1/L} = e^{-u/L}$ *by the similarity transformation*

$$V_u H^*(u) V_u^{-1} = c(u) H^*(0). \quad (31)$$

with $V_u = \exp\left(-\frac{u}{L}\sum_{k=1}^{L} k\hat{n}_k\right)$.

Proof This is proved by straightforward computation, using the factorized form $V_u = \prod_{k=1}^{L} V_u(k)$ with $V_u(k) = e^{-\frac{u}{L}k\hat{n}_k}$ and the transformation property $e^{-\alpha\hat{n}}\sigma^{\pm}e^{\alpha\hat{n}} = e^{\pm\alpha}\sigma^{\pm}$ of the Pauli matrices. □

An obvious corollary is that all eigenvalues $\varepsilon_i(u)$ of the weighted generator (30) are given by

$$\varepsilon_i(u) = c(u)\varepsilon_i(0) \quad (32)$$

in terms of the eigenvalues of the homogeneous generator. For the left and right eigenvectors to these eigenvalues one has $\langle\, \varepsilon_i(u)\,| = \langle\, \varepsilon_i(0)\,| V_u$ and $|\, \varepsilon_i(u)\,\rangle = V_u^{-1}|\, \varepsilon_i(0)\,\rangle$. The eigenvalues and eigenvectors for $u = 0$ are computed in [52] using the integrability of the model by free fermion techniques.

Another immediate consequence is stated in the following theorem:

Theorem 3.6 *The effective dynamics for conditioning on maximal current in the disordered ASEP, given by the generalized Doob-transform* $H^*_{\mathit{eff}}(u) := \Delta_u H^*(u)\Delta_u^{-1} - \varepsilon_0(u)$ *of (30), has the property*

$$H^*_{\mathit{eff}}(u) = c(u) H^*_{\mathit{eff}}(0). \quad (33)$$

This follows from $\Delta_u = \Delta V_u$ with $\Delta = \Delta(0)$ given in [52]. The interpretation of this at first sight surprising fact is the following. The conditioning on maximal current selects those (untypical) realizations of the process where hopping attempts occur instantly, irrespective of the actual rate (which determines the typical mean frequency between hopping attempts). Only the time scale of the effective process (rescaled by e^s) is affected by thc defect. As an obvious but noteworthy corollary we find that the stationary distribution of the transformed process (33) does not depend on u and is given by (10).

This result can be extended to the fully disordered ASEP [74], which also has a long history of study lasting until today, see [14, 72] and references therein. We denote the set of site-dependent hopping rates $r_k = e^{-u_k}$ to the right by $\mathbf{u}$ and obtain under conditioning on maximal current

$$H^*(\mathbf{u}) = -\sum_{k=1}^{L} e^{-u_k}\sigma_k^+\sigma_{k+1}^-. \quad (34)$$

with the property

$$V_{\mathbf{u}} H^*(\mathbf{u}) V_{\mathbf{u}}^{-1} = c(\mathbf{u}) H^*(\mathbf{0}). \tag{35}$$

where

$$V_{\mathbf{u}} = \exp\left(-\sum_{k=1}^{L} \alpha_k \hat{n}_k \right), \quad c(\mathbf{u}) = \exp\left(-\frac{1}{L} \sum_{k=1}^{L} u_k \right) = \left(\prod_{k=1}^{L} r_k \right)^{\frac{1}{L}} \tag{36}$$

and $\alpha_k = -\sum_{i=1}^{k-1} (u_k - c(\mathbf{u}))$. Theorem (3.6) applies also to this case with the obvious substitution $u \to \mathbf{u}$. Hence hopping disorder is irrelevant in the ASEP under maximal current conditions.

3.3 Annihilating and Coalescing Random Walks

We consider non-conservative dynamics where random walks annihilate instantly when they meet, also called vicious random walkers [26]. There is a vast body of literature, both in statistical physics and probability theory, on this process and the related process of coalescing random walks where only one particle annihilates upon meeting.

Many different techniques (see e.g. [4, 21, 39, 42, 63] and references therein) allow for a detailed analysis of this problem. It has been known for a long time that on $\mathbb{Z}$ the average particle density approaches 0 proportional to $1/\sqrt{t}$, irrespective of the initial density (after some finite crossover time) [16]. In the spirit of [54] we study here for vicious walkers on $\mathbb{T}_L$ the annihilation activity $A^{ann}(t)$, i.e., the number of particles annihilated until time t. Thus we are dealing with the weighted generator

$$H^a(s) = H_0 + H_1^a(s) \tag{37}$$

with

$$H_0 = -\sum_{k=1}^{L} \left(w_+ (\sigma_k^+ \sigma_{k+1}^- - \hat{n}_k) + w_- (\sigma_k^- \sigma_{k+1}^+ - \hat{n}_{k+1}) \right) \tag{38}$$

$$H_1^a(s) = -(w_+ + w_-) \mathrm{e}^{2s} \sum_{k=1}^{L} \sigma_k^+ \sigma_{k+1}^+ \tag{39}$$

For definiteness we consider an even number of sites $L = 2K$. As initial distribution we take a product measure with density ρ.

We consider the generating function

$$Y_s(t) = \langle \, \mathrm{e}^{sA^{ann}(t)} \, \rangle = \langle \, s \, | \mathrm{e}^{-H^a(s)t} | \, \rho \, \rangle, \quad Y_s^* = \lim_{t\to\infty} Y_s(t) \tag{40}$$

and define a generalized large deviation function

$$g(s) := \lim_{t\to\infty} t^\alpha \lim_{L\to\infty} \frac{1}{L}(\ln Y_s^* - \ln Y_s(t)) =: \lim_{t\to\infty} t^\alpha (h_s^* - h_s(t)). \tag{41}$$

Observe the order of limits in this expression. Taking the limit $t \to \infty$ first would result trivially in $g(s) = 0$ since in a finite lattice the system reaches the trivial steady state with no particles in finite time. The total number of particles annihilated until time t is extensive in system size. This motivates the factor $1/L$ in (41). On the other hand, one expects the annihilation activity per site to decay in time. Thus one expects $\alpha > 0$ and some stationary constant Y_s^*. In fact, here we show that $\alpha = 1/2$ and we also present the full functions $g(s)$ and h_s^*.

Theorem 3.7 *For annihilating random walks as defined in (37) with product initial distribution with strictly positive density $\rho > 0$ the constant h_s^* defined in (41) takes the value*

$$h_s^* = 1 - \rho(1 - \mathrm{e}^s). \tag{42}$$

The generalized large deviation function $g(s)$ has a non-trivial limit for $\alpha = 1/2$ and does not depend on the hopping asymmetry $w_+ - w_-$ nor on the initial density ρ. Its explicit form is given by

$$g(s) = \frac{1}{4\sqrt{\pi(w_+ + w_-)t}} \sum_{m=1}^{\infty} \frac{(1 - \mathrm{e}^{-2s})^m}{m^{3/2}}. \tag{43}$$

Proof By definition $A^{ann}(t)$ is the number of particles annihilated until time t, i.e., $A^{ann}(t) = N(0) - N(t)$. Hence $Y_s(t) = \langle \, \mathrm{e}^{sA^{ann}(t)} \, \rangle = \langle \, \mathrm{e}^{s(N(0)-N(t))} \, \rangle$. On the other hand,

$$\langle \, \mathrm{e}^{s(N(0)-N(t))} \, \rangle = \langle \, s \, | \mathrm{e}^{-s\hat{N}} \mathrm{e}^{-Ht} \mathrm{e}^{s\hat{N}} | \, \rho \, \rangle. \tag{44}$$

Next we observe $\mathrm{e}^{s\hat{N}} | \, \rho \, \rangle = a_L(s) | \, \tilde{\rho}(s) \, \rangle$ with

$$\tilde{\rho}(s) := \mathrm{e}^s \rho / [1 - \rho(1 - \mathrm{e}^s)], \quad a_L(s) := [1 - \rho(1 - \mathrm{e}^s)]^L. \tag{45}$$

Below we shall drop the argument s of $\tilde{\rho}$. This yields

$$Y_s(t) = (h_s^*)^L \langle \, \mathrm{e}^{-sN(t)} \, \rangle_{\tilde{\rho}} \tag{46}$$

with the function h_s^* (42).

It remains to consider $g(s) = -\lim_{t\to\infty} t^\alpha \lim_{L\to\infty} 1/L \ln \langle \mathrm{e}^{-sN(t)} \rangle_{\tilde{\rho}}$. Since the particle number parity $(-1)^N$ is conserved we decompose the Bernoulli product measure with density $\tilde{\rho}$ into its even and odd components

$$P^+(\mathbf{n}) = \frac{1}{Z^+}\tilde{\rho}^{N(\mathbf{n})}(1-\tilde{\rho})^{L-N(\mathbf{n})} \sum_{M=0}^{K} \delta_{N(\mathbf{n}),2M} \tag{47}$$

$$P^-(\mathbf{n}) = \frac{1}{Z^-}\tilde{\rho}^{N(\mathbf{n})}(1-\tilde{\rho})^{L-N(\mathbf{n})} \sum_{M=0}^{K-1} \delta_{N(\mathbf{n}),2M+1} \tag{48}$$

where $N(\mathbf{n}) = \sum_{k=1}^{L} n(k)$ is the particle number in configuration $\mathbf{n}$ and

$$Z^\pm = \frac{1}{2}\left(1 \pm (1-2\tilde{\rho})^L\right). \tag{49}$$

The probability vectors corresponding to these distributions are denoted by $|\,\tilde{\rho}^\pm\,\rangle$. We also define the projected summation vectors

$$\langle\, s^\pm \,| = \sum_{\{\mathbf{n}:(-1)^{N(\mathbf{n})}=\pm 1\}} \langle\, \mathbf{n} \,|. \tag{50}$$

This decomposition gives

$$\langle\, \mathrm{e}^{-sN(t)} \,\rangle_{\tilde{\rho}} = Z^+ \Phi_s^+(t) + Z^- \Phi_s^-(t) \tag{51}$$

with $\Phi_s^\pm(t) = \langle\, \mathrm{e}^{-sN(t)} \,\rangle_{\tilde{\rho}^\pm}$.

The computation of $\Phi_s^+(t)$ was performed in [65], using the mapping of annihilating random walks to a system of free fermions [48], and then employing a further mapping to $K = L/2$ non-interacting spins [63]. This allows for a tensor representation of the even initial condition on $(\mathbb{C}^2)^{\otimes K}$

$$|\,\tilde{\rho}^+\,\rangle = \frac{1}{Z^+}\prod_{p=1}^{K}\left[(1-\tilde{\rho})^2 + \tilde{\rho}^2 \cot\frac{(2p-1)\pi}{2L}\sigma_p^-\right]|\,0\,\rangle, \tag{52}$$

of the even summation vector

$$\langle\, s^+ \,| = \langle\, 0\,| \prod_{p=1}^{K}\left[(1 + \cot\frac{(2p-1)\pi}{2L}\sigma_p^+\right], \tag{53}$$

and the even generator

$$H^+ = -(w_+ + w_-)\sum_{p=1}^{K}\left[\sin\frac{(2p-1)\pi}{L}\sigma_p^+ - \left(1 - \cos\frac{(2p-1)\pi}{L}\right)\hat{n}_p\right]. \tag{54}$$

Expressing the parameters D and α in [65] in terms of the present choice of parameters leads to the identification $\alpha = is$ and $D = (w_+ + w_-)/2$. This yields

$$\Phi_s^+(t) = \prod_{p=1}^{K} \left[1 - \gamma_p^+ (1 - e^{-2s}) e^{-2\varepsilon_p^+ t}\right] \tag{55}$$

with

$$\gamma_p^+ = \frac{\tilde{\rho}^2}{\tilde{\rho}^2 + (1-\tilde{\rho})^2 \tan^2 \frac{(2p-1)\pi}{2L}}, \quad \varepsilon_p^+ = 2D\left(1 - \cos\frac{(2p-1)\pi}{L}\right). \tag{56}$$

For the odd sector, which was not studied previously, we follow the approach of [63] to obtain

$$|\,\tilde{\rho}^-\,\rangle = \frac{L\tilde{\rho}(1-\tilde{\rho})}{Z^-} \prod_{p=1}^{K-1} \left[(1-\tilde{\rho})^2 + \tilde{\rho}^2 \cot\frac{p\pi}{L} \sigma_p^-\right] \sigma_0^- \,|\,0\,\rangle \tag{57}$$

and the odd summation vector

$$\langle\, s^-\,| = \langle\, 0\,| \sigma_0^+ \prod_{p=1}^{K-1} \left[(1 + \cot\frac{p\pi}{L} \sigma_p^+\right]. \tag{58}$$

The required normalization $\langle\, s^- \,|\, \tilde{\rho}^- \,\rangle = 1$ follows from applying the trigonometric product identities 6.1.1.7. and 6.1.1.8. in [56]. The generator on the odd subspace is similar to even case, but with the replacement $p \to p + 1/2$. This yields

$$\Phi_s^-(t) = e^{-s} \prod_{p=1}^{K} \left[1 - \gamma_p^- (1 - e^{-2s}) e^{-2\varepsilon_p^- t}\right] \tag{59}$$

with γ_p^-, ε_p^- similar to the corresponding even quantities (56), but with the replacement $p \to p + 1/2$. The exponential prefactor comes from the term involving σ_0^+.

In order to take the thermodynamic limit one expresses the product as the exponential of the sum of logarithms. Since all terms are well-behaved in the domain of finite s and sufficiently large t one can expand the logarithm in a power series. Then using Euler-MacLaurin one can replace the sums by integrals and realizes that the even and odd product converge to the same quantity. Taking the thermodynamic limit thus yields $g(s)$ as given in (43). □

Finally we consider a mixed process where particles coalesce with probability p or annihilate with probability $1 - p$ when they meet on the same site. The weighted generator for the integrated number of annihilated particles is given by $H^{ac}(s) = H_0 + H_1^{ac}(s)$ with annihilation part

$$H_1^{ac}(s) = -\sum_{k=1}^{L}\Big[(w_+ + w_-)(1-p)\mathrm{e}^{2s}\sigma_k^+\sigma_{k+1}^+ + pw_+\mathrm{e}^s\sigma_k^+\hat{n}_{k+1} + pw_-\mathrm{e}^s\hat{n}_k^+\sigma_{k+1}^+\Big]. \tag{60}$$

In the limit $p = 1$ one has coalescing particles [3, 8]. We define

$$Y_s^{ac}(t) := \mathrm{e}^{sA^{ac}(t)} = \langle\, s\,|\mathrm{e}^{-H^{ac}(s)t}|\,\rho\,\rangle \tag{61}$$

and in complete analogy to (41)

$$g^{ac}(s) := \lim_{t\to\infty} t^\alpha \lim_{L\to\infty}\frac{1}{L}(\ln Y_s^* - \ln Y_s^{ac}(t)) =: \lim_{t\to\infty} t^\alpha(h_s^* - h_s^{ac}(t)). \tag{62}$$

Here trivially the same function Y_s^* appears since the initial measure is the same and Y_s^* is just the asymptotic generating function for the annihilation of all particles.

It was shown in [33] that $H^{ac}(0) = BH^a(0)B^{-1}$ with $B = \mathrm{e}^{-\theta(S^+ - \hat{N})}$ and $\mathrm{e}^{-\theta} = 1 - p/2$. These transformations lead to $g^{ac}(s) = g(s')$ with some $s'(s,p)$. The lack of dependence of the generalized large deviation function $g(\cdot)$ on the hopping asymmetry is trivial in the sense that is well-known that for annihilating random walks the bias can be absorbed in a Galilei-transformation at large space-time scales [63]. However, there is a non-trivial universality property in the sense that $g(\cdot)$ does not depend on the initial density and on the branching ratio. The branching ratio enters only through the non-universal argument $s'(s,p)$. We conjecture that the same generalized large deviation function g will be found in other models such as the SSEP with pair annihilation with arbitrary rate, which is also an integrable model [1, 65].

Acknowledgments The author thanks B. Meerson, P.L. Krapivsky, V. Popkov and Andreas Schadschneider for inspiring discussions and the Galileo Galilei Institute for Theoretical Physics for hospitality. Partial support by DFG and the INFN during the completion of this work is gratefully acknowledged.

References

1. Alcaraz, F.C., Rittenberg, V.: Reaction-diffusion processes as physical realizations of Hecke Algebras. Phys. Lett. B **314**(3), 377–380 (1993)
2. Appert, C., Derrida, B., Lecomte, V., Van Wijland, F.: Universal cumulants of the current in diffusive systems on a ring. Phys. Rev. E **78**, 021122 (2008)
3. Arratia, R.: Limiting point processes for rescalings of coalescing and annihilating random walks on $\mathbb{Z}^d$. Ann. Probab. **9**(6), 909–936 (1981)

4. Ayyer, A., Mallick, K.: Exact results for an asymmetric annihilation process with open boundaries. J. Phys. A: Math. Theor. **43**(4), 045003 (2010)
5. Bahadoran, C.: Blockage hydrodynamics of one-dimensional driven conservative systems. Ann. Probab. **32**(1B), 805–854 (2004)
6. Belitsky, V., Schütz, G.M.: Diffusion and coalescence of shocks in the partially asymmetric exclusion process. Electron. J. Prob. 7, Paper No. 11, 1–21 (2002)
7. Belitsky, V., Schütz, G.M.: Microscopic structure of shocks and antishocks in the ASEP conditioned on low current. J. Stat. Phys. **152**, 93–111 (2013)
8. Ben Avraham, D.: Complete exact solution of diffusion-limited coalescence, $A + A - A$. Phys. Rev. Lett. **81**, 4756–4759 (1998)
9. Bertini, L., De Sole, A., Gabrielli, D., Jona Lasinio, G., Landim, C.: Macroscopic fluctuation theory for stationary non-equilibrium states. J. Stat. Phys. **107**, 635–675 (2002)
10. Bertini, L., De Sole, A., Gabrielli, D., Jona Lasinio, G., Landim, C.: Macroscopic fluctuation theory. arXiv:1404.6466v1 [cond-mat.stat-mech](2014)
11. Blumen, A., Klafter, J., Zumofen, G.: Target annihilation by random walkers. Phys. Rev. B **30**(9), 5379–5382 (1984)
12. Bodineau, T., Derrida, B.: Distribution of current in non-equilibrium diffusive systems and phase transitions. Phys. Rev. E **72**, 066110 (2005)
13. Bodineau, T., Lagouge, M.: Current large deviations in a driven dissipative model. J. Stat. Phys. **139**(2), 201–219 (2010)
14. Bodineau, T., Bahadoran, C.: Properties and conjectures for the flux of TASEP with site disorder. Braz. J. Prob. Stat. (2014) (To Appear)
15. Borodin, A., Corwin, I., Sasamoto, T.: From duality to determinants for Q-TASEP and ASEP. Ann. Probab. **42**(6), 2314–2382 (2014)
16. Bramson M., Griffeath, D.: Clustering and dispersion rates for some interacting particle systems on $\mathbb{Z}$. Ann. Probab. **8**(2), 183–213 (1980)
17. Burlatsky, S.F., Ovchinnikov, A.A.: Effect of reactant-fluctuation density on the kinetics of recombination, multiplication, and trapping processes. Sov. Phys. JETP **65**, 908–917 (1987)
18. Burlatsky, S.F., Moreau, M., Oshanin, G., Blumen, A.: Comment on "pair and triple correlations in diffusion-limited $A + B \to B$ reactions". Phys. Rev. Lett. **75**, 585 (1995)
19. Castro-Alvaredo, O., Chen, Y., Doyon, B., Marianne Hoogeveen, M.: Thermodynamic Bethe ansatz for non-equilibrium steady states: exact energy current and fluctuations in integrable QFT. J. Stat. Mech. P03011 (2014)
20. Chetrite, R., Touchette, H.: Nonequilibrium Markov processes conditioned on large deviations. Ann. Henri Poincaré 1–51 (2014). doi:10.1007/s00023-014-0375-8
21. Colomo, F., Isergin, A.G., Korepin, V.E., Tognetti, V.: Temperature correlation functions in the XX0 Heisenberg chain. Theor. Math. Phys. **94**(1), 11–38 (1993)
22. Derrida, B., Lebowitz, J.L.: Exact large deviation function in the asymmetric exclusion process. Phys. Rev. Lett. **80**(2), 209–213 (1998)
23. Derrida, B., Appert, C.: Universal large deviation function of the Kardar-Parisi-Zhang equation in one dimension. J. Stat. Phys. **94**, 1–30 (1999)
24. Derrida, B., Lebowitz, J.L., Speer, E.R.: Large deviation of the density profile in the steady state of the open symmetric simple exclusion process. J. Stat. Phys. **107**, 599–634 (2002)
25. Espigares, C.P., Garrido, P.L., Hurtado, P.I.: Dynamical phase transition for current statistics in a simple driven diffusive system. Phys. Rev. E **87**, 032115 (2013)
26. Fisher, M.E.: Walks, walls, wetting, and melting. J. Stat. Phys. **34**, 667–729 (1984)
27. Giardinà, C., Kurchan, J., Redig, F., Vafayi, K.: Duality and hidden symmetries in interacting particle systems. J. Stat. Phys. **135**, 25–55 (2009)
28. Gwa, L.H., Spohn, H.: Bethe solution for the dynamical-scaling exponent of the noisy Burgers equation. Phys. Rev. A **46**(2), 844–854 (1992)
29. Ha, M., Timonen, J., den Nijs, M.: Queuing transitions in the asymmetric simple exclusion process. Phys. Rev. E **68**, 056122 (2003)
30. Harris, R.J., Schütz, G.M.: Fluctuation theorems for stochastic dynamics. J. Stat. Mech. P07020 (2007)

31. Harris, R.J., Popkov, V., Schütz, G.M.: Dynamics of instantaneous condensation in the ZRP conditioned on an atypical current. Entropy **15**, 5065–5083 (2013)
32. Harris, R.J., Schütz, G.M.: Fluctuation theorems for stochastic dynamics interacting particle systems. Markov Process Relat. Fields. **20**, 3–44 (2014)
33. Henkel, M., Orlandini, E., Schütz, G.M.: Equivalences between stochastic systems. J. Phys. A: Math. Gen. **28**, 6335–6344 (1995)
34. Hurtado, P.I., Garrido, P.L.: Spontaneous symmetry breaking at the fluctuating level. Phys. Rev. Lett. **107**, 180601 (2011)
35. Imamura, T., Sasamoto, T.: Current moments of 1D ASEP by duality. J. Stat. Phys. **142**(5), 919–930 (2011)
36. Jack, R.L., Sollich, P.: Large deviations and ensembles of trajectories in stochastic models. Prog. Theor. Phys. Supp. **184**, 304–317 (2010)
37. Janowsky, S.A., Lebowitz, J.L.: Finite-size effects and shock fluctuations in the asymmetric simple-exclusion process. Phys. Rev. A **45**(2), 618–625 (1992)
38. Janowsky, S.A., Lebowitz, J.L.: Exact results for the asymmetric simple exclusion process with a blockage. J. Stat. Phys. **77**, 35–51 (1994)
39. Katori, M., Tanemura, H.: Complex Brownian motion representation of the Dyson model. Electron. Commun. Probab. **18**(4), 1–16 (2013)
40. Kim, D.: Bethe ansatz solution for crossover scaling functions of the asymmetric XXZ chain and the KPZ-type growth model. Phys. Rev. E **52**, 3512–3524 (1995)
41. Kolomeisky, A.B., Schütz, G.M., Kolomeisky, E.B., Straley, J.P.: Phase diagram of one-dimensional driven lattice gases with open boundaries. J. Phys. A: Math. Gen. **31**(33), 6911–6919 (1998)
42. Krapivsky, P.L., Redner, S., Ben-Naim, E.: A kinetic view of statistical physics. Cambridge University Press, Cambridge (2010)
43. Lebowitz, J.L., Spohn, H.: A Gallavotti-Cohen-Type symmetry in the large deviation functional for stochastic dynamics. J. Stat. Phys. **95**, 333–365 (1999)
44. Lecomte, V., Garrahan, J.P., Van Wijland, F.: Inactive dynamical phase of a symmetric exclusion process on a ring. J. Phys. A: Math. Theor. **45**, 175001 (2012)
45. Liggett, T.M.: Interacting Particle Systems. Springer, Berlin (1985)
46. Liggett, T.M.: Stochastic Interacting Systems: Contact Voter and Exclusion Processes. Springer, Berlin (1999)
47. Lloyd, P., Sudbury, A., Donnelly, P.: Quantum operators in classical probability theory: I. "Quantum spin" techniques and the exclusion model of diffusion. Stochast. Processes Appl. **61**(2), 205–221 (1996)
48. Lushnikov, A.A.: Binary reaction $1 + 1 \to 0$ in one dimension. Phys. Lett. A **120**(3), 135–137 (1987)
49. Meerson, B., Vilenkin, A., Krapivsky, P.L.: Survival of a static target in a gas of diffusing particles with exclusion. Phys. Rev. E **90**, 022120 (2014)
50. Meerson, B., Redner, S.: Large fluctuations in diffusion-controlled absorption. J. Stat. Mech. P08008 (2014)
51. Mejía-Monasterio, C., Oshanin, G., Schehr, G.: First passages for a search by a swarm of independent random searchers. J. Stat. Mech. P06022 (2011)
52. Popkov, V., Simon, D., Schütz, G.M.: ASEP on a ring conditioned on enhanced flux. J. Stat. Mech. P10007 (2010)
53. Popkov, V., Schütz, G.M.: Transition probabilities and dynamic structure factor in the ASEP conditioned on strong flux. J. Stat. Phys. **142**(3), 627–639 (2011)
54. Popkov, V., Schütz, G.M.: Large deviation functions in a system of diffusing particles with creation and annihilation. Phys. Rev. E **84**, 021131 (2011)
55. Popkov, V., Schadschneider, A., Schmidt, J.: TASEP with open boundary conditions and slow bond. Preprint (2014)
56. Prudnikov, A.P., Brychkov, Y.A., Marichev, O.I.: Integrals and Series, vol. 1. Gordon and Breach, New York (1986)

57. Rezakhanlou, F.: Hydrodynamic limit for attractive particle systems on $\mathbb{Z}^d$. Commun. Math. Phys. **140**, 417–448 (1991)
58. Rogers, L.C.G., Williams, D.: Diffusions, markov processes, and martingales. In: Foundations, vol. 1. Wiley, New york (1994)
59. Santen, L., Appert, C.: The asymmetric exclusion process revisited: fluctuations and dynamics in the domain wall picture. J. Stat. Phys. **106**(1–2), 187–199 (2002)
60. Santos, J.E., Schütz, G.M.: Exact time-dependent correlation functions for the symmetric exclusion process with open boundary. Phys. Rev. E **64**, 036107 (2001)
61. Schütz, G.: Generalized Bethe ansatz solution of a one-dimensional asymmetric exclusion process on a ring with blockage. J. Stat. Phys. **71**(3)-(4), 471–505 (1993)
62. Schütz, G., Sandow, S.: Non-abelian symmetries of stochastic processes: derivation of correlation functions for random vertex models and disordered interacting many-particle systems. Phys. Rev. E **49**, 2726–2744 (1994)
63. Schütz, G.M.: Diffusion-annihilation in the presence of a driving field. J. Phys. A: Math. Gen. **28**(12), 3405–3415 (1995)
64. Schütz, G.M.: Duality relations for the asymmetric exclusion process. J. Stat. Phys. **86**(5/6), 1265–1287 (1997)
65. Schütz, G.M.: Exactly solvable models for many-body systems far from equilibrium. In: Domb, C., Lebowitz, J. (eds.) Phase Transitions and Critical Phenomena, vol. 19, pp. 1–251. Academic Press, London (2001)
66. Schütz, G.M.: The Space-time structure of extreme current and activity events in the ASEP. In: Proceedings of the "International School and Workshop on Nonlinear Mathematical Physics and Natural Hazards". Springer (To Appear)
67. Seppäläinen, T.: Hydrodynamic profiles for the totally asymmetric exclusion process with a slow bond. J. Stat. Phys. **102**, 69–96 (2001)
68. Simon, D.: Construction of a coordinate Bethe ansatz for the asymmetric simple exclusion process with open boundaries. J. Stat. Mech. P07017 (2009)
69. Spohn, H.: Bosonization, vicinal surfaces, and hydrodynamic fluctuation theory. Phys. Rev. E **60**, 6411–6420 (1999)
70. Szabo, A., Zwanzig, R., Agmon, N.: Diffusion-controlled reactions with mobile traps. Phys. Rev. Lett. **61**, 2496–2499 (1988)
71. Szavits-Nossan, J., Uzelac, K.: Impurity-induced shocks in the asymmetric exclusion process with long-range hopping. J. Stat. Mech. P12019 (2009)
72. Szavits-Nossan, J.: Disordered exclusion process revisited: some exact results in the low-current regime. J. Phys. A: Math. Theor. **46**, 315001 (2013)
73. Tang, L.-H., Lyuksyutov, I.F.: Directed polymer localization in a disordered medium. Phys. Rev. Lett. **71**, 2745–2748 (1993)
74. Tripathy, G., Barma, M.: Driven lattice gases with quenched disorder: exact results and different macroscopic regimes. Phys. Rev. E **58**(2), 1911–1926 (1998)
75. Wolf, D.E., Tang, L.-H.: Inhomogeneous growth processes. Pyhs. Rev. Lett. **64**, 1591–1594 (1990)

MIX
Papier aus verantwortungsvollen Quellen
Paper from responsible sources
FSC® C105338

If you have any concerns about our products,
you can contact us on
ProductSafety@springernature.com

In case Publisher is established outside the EU,
the EU authorized representative is:
Springer Nature Customer Service Center GmbH
Europaplatz 3, 69115 Heidelberg, Germany

Printed by Libri Plureos GmbH
in Hamburg, Germany